VOLUME SEVEN HUNDRED AND TWELVE

METHODS IN ENZYMOLOGY

CRISPR-Cas Systems for RNA and Genome Editing - Part A

METHODS IN ENZYMOLOGY

VOLUME SEVEN HUNDRED AND TWELVE

METHODS IN ENZYMOLOGY

CRISPR-Cas Systems for RNA and Genome Editing - Part A

Edited by

AILONG KE
Department of Molecular Biology and Genetics
Cornell University
Ithaca, NY, United States

WEIXIN TANG
Department of Chemistry
University of Chicago
Chicago, IL, United States

ELSEVIER

ACADEMIC PRESS
An imprint of Elsevier

Academic Press is an imprint of Elsevier
50 Hampshire Street, 5th Floor, Cambridge, MA 02139, United States
525 B Street, Suite 1650, San Diego, CA 92101, United States
125 London Wall, London, EC2Y 5AS, United Kingdom

First edition 2025

ISBN: 978-0-443-31752-1
ISSN: 0076-6879

For information on all Academic Press publications
visit our website at https://www.elsevier.com/books-and-journals

Publisher: Zoe Kruze
Editorial Project Manager: Saloni Vohra
Production Project Manager: James Selvam
Cover Designer: Bakyalakshmi S

Typeset by MPS Limited, India

Contents

Contributors *xix*

1. Measuring double-strand break repair events in mammalian cells with multi-target CRISPR **1**

Alberto Marin-Gonzalez, Adam T. Rybczynski, Roger S. Zou, and Taekjip Ha

1. Introduction 2
   1.1 DSB repair in mammalian cells 2
   1.2 Current methods to induce DSBs in mammalian cells and their limitations 3
   1.3 Overview of multi-target CRISPR 4
2. Protocol 5
   2.1 Overview of the procedure and other comments 5
   2.2 Materials and reagents 5
   2.3 Large-scale delivery of multi-target CRISPR into HEK293T (~12 M cells) 7
   2.4 Small-scale delivery of multi-target CRISPR into HEK293T cells (~0.5 M cells) 9
   2.5 Light-activation of multi-target CRISPR 11
3. Variations to the protocol 12
   3.1 Validating multi-target CRISPR in different cell types 12
   3.2 Designing and validating new mgRNAs 14
4. Guidelines for sequencing and sequencing data analysis 16
   4.1 Sequencing recommendations 16
   4.2 Sequencing data analysis 17
5. Conclusions 20

Acknowledgements 20

Appendix A. Supporting information 20

References 20

2. General and robust sample preparation strategies for cryo-EM studies of CRISPR-Cas9 and Cas12 enzymes **23**

Satoshi N. Omura and Osamu Nureki

1. Introduction 24
2. Cas protein expression and purification 26

2.1 *E. coli* transformation 26
2.2 Culture growth and induction 27
2.3 Protein purification 27
3. Guide RNA preparation 29
3.1 Template preparation for in vitro transcription of gRNA 29
3.2 In vitro transcription 30
3.3 Gel purification 30
4. Complex reconstruction and purification 31
4.1 Target DNA design 31
4.2 Complex reconstruction 31
4.3 Complex purification 32
4.4 Sample concentration 33
5. Grid preparation for cryo-EM 33
5.1 Use of detergent 34
5.2 Use of amylamine 35
6. Summary 35
Acknowledgments 36
Author contributions 36
Declaration of interests 36
References 36

3. Visualizing the conformational landscape of CRISPR-Cas9 through kinetics-informed structural studies **41**

Grace N. Hibshman and David W. Taylor

1. Introduction 42
2. Kinetic characterization 44
2.1 Purification of Cas9 44
2.2 gRNA preparation 44
2.3 DNA preparation 44
2.4 Reconstitution of active Cas9-gRNA complex 46
2.5 Cleavage assay 46
2.6 Stopped-flow measurements of DNA unwinding 46
3. Cryo-EM structural studies 47
3.1 Sample Preparation 47
3.2 Cryo-EM Data Collection and Processing Pipeline 48
4. Alternative Approaches 51
5. Conclusion 52
Acknowledgements 52
References 52

4. Biochemical reconstitution of a type I-B CRISPR-associated transposon **55**

Shukun Wang and Leifu Chang

1. Introduction 56
2. Plasmid preparation 58
   2.1 pCascade 58
   2.2 TnsAB 63
   2.3 TnsC 63
   2.4 TniQ 63
   2.5 pDonor and pTarget 63
3. Purification of Cascade 64
   3.1 Transformation 64
   3.2 Inoculation 65
   3.3 Induction 65
   3.4 Cell collection 66
   3.5 Immobilized metal affinity chromatography using Ni-NTA beads 66
   3.6 Ion exchange chromatography 67
   3.7 Size exclusion chromatography 68
4. Purification of TnsAB and TnsC 68
   4.1 Transformation, inoculation, induction 69
   4.2 Cell collection and affinity chromatography using Streptactin beads 69
   4.3 Size exclusion chromatography 70
5. Purification of TniQ-TnsC complex 70
   5.1 Transformation 71
   5.2 Inoculation 71
   5.3 Induction 71
   5.4 Cell collection and Strep-tag affinity chromatography 71
   5.5 Size exclusion chromatography 72
6. *In vitro* transposition assay 72
   6.1 Qualitative detection by PCR 73
   6.2 Quantitative detection by qPCR 74
   6.3 Qualitative detection by transformation 76
7. Conclusion 77

Acknowledgments 77

Author contributions 77

References 77

5. Determining the biochemical function of type IV CRISPR ribonucleoprotein complexes and accessory proteins **81**

Andrew A. Williams, Olivine Redman, Hannah Domgaard, Matthew J. Armbrust, and Ryan N. Jackson

1. Introduction 82
2. Plasmid clearance assay to assess in vivo type IV-A immunity 85
   2.1 Plasmid clearance assay materials and equipment 87
   2.2 Plasmid clearance assay protocol day 1 – transform the type IV-A system 87
   2.3 Plasmid clearance assay protocol day 2 – make a small liquid starter culture 89
   2.4 Plasmid clearance assay protocol day 3 – make cells competent 89
   2.5 Plasmid clearance assay protocol day 3 or 4 – perform plasmid curing assay 90
3. Expressing and purifying the type IV-A Csf ribonucleoprotein complex 91
   3.1 Csf complex expression materials and equipment 91
   3.2 Csf complex expression day 1 – bacterial transformation 91
   3.3 Csf complex expression day 2 – small culture growth 92
   3.4 Csf complex expression day 3 – large culture growth 94
   3.5 Csf complex expression day 4 – harvesting 94
   3.6 Csf complex purification materials and equipment 94
   3.7 Csf complex purification – lysis and strep-affinity chromatography 95
   3.8 Csf complex purification – concentrate and size-exclusion chromatography 96
4. Expressing and purifying the CasDinG helicase 96
   4.1 CasDinG expression materials and equipment 97
   4.2 CasDinG expression days 1 and 2 – bacterial transformation and liquid culture 97
   4.3 CasDinG expression day 3 – large culture growth 98
   4.4 CasDinG expression day 3 – harvesting cells 98
   4.5 CasDinG purification materials and equipment 98
   4.6 CasDinG purification – lysis and strep-affinity chromatography 99
   4.7 CasDinG purification – desalting and heparin chromatography 100
   4.8 CasDinG purification – concentration and size-exclusion chromatography 100
5. Purification of type IV-A Csf complex bound to DNA targets and the CasDinG helicase 101
   5.1 Materials for SEC of Csf complexes bound to target DNA and CasDinG 101
   5.2 SEC of Csf complex bound to target DNA - preparation 104

5.3 SEC of Csf complex bound to target DNA – protocol 104
5.4 SEC of CasDinG recruited to R-loop-bound Csf complexes - preparation 105
5.5 SEC of CasDinG recruited to R-loop-bound Csf complexes – protocol 105
6. Determining Csf complex or CasDinG affinity to nucleic acid targets with fluorescence anisotropy 106
6.1 Materials and equipment 107
6.2 Set up and perform fluorescence anisotropy 107
6.3 Fluorescence anisotropy troubleshooting 110
7. CasDinG helicase assay 110
7.1 Materials 111
7.2 Duplex substrate annealing 111
7.3 Prepare controls 112
7.4 Time course protocol 112
8. Conclusion 114
References 115

6. Analysis of metal-dependent DNA nicking activities by Cas endonucleases 117

Giang T. Nguyen, Akshara Raju, and Dipali G. Sashital

1. Introduction 118
2. Cleavage assays using negatively supercoiled plasmid DNA substrates 120
2.1 Plasmid substrate preparation 121
2.2 Initiate cleavage by mixing Cas RNP and DNA in the presence of Mg^{2+} 122
2.3 Initiate cleavage through the addition of Mg^{2+} ion to pre-bound RNP-DNA complex 129
2.4 DNA binding time-course analyzed by determining fraction cleaved 131
3. Cleavage assays using linear DNA substrates 134
3.1 Linear substrate preparation 135
3.2 Procedure 137
3.3 Cleavage assays and separation of nicked products for linear substrates 137
4. Summary and conclusion 139
Acknowledgments 139
References 139

7. Purification and in vivo, cell-free, and in vitro characterization of CRISPR-Cas12a2 **143**

Friso T. Schut, Thomson Hallmark, Oleg Dmytrenko, Ryan N. Jackson, and Chase L. Beisel

1. Introduction 144
2. gRNA design for Cas12a2 145
3. Cas12a2 activity in *E. coli* 147
   3.1 Plasmid interference assay 147
   3.2 Growth-based assays 151
4. Cas12a2 activity in cell-free transcription-translation (TXTL) 155
   4.1 Design of DNA expression constructs for TXTL reactions 156
   4.2 DNA preparation for TXTL reactions 157
   4.3 TXTL pre-expression of SuCas12a2 with a gRNA 158
   4.4 TXTL collateral cleavage assay 164
5. Expression and purification of Cas12a2 165
   5.1 Cas12a2 expression 165
   5.2 Cas12a2 purification 168
6. Cas12a2 *in vitro* cleavage assays 172
   6.1 Cas12a2 cleavage visualization 173
   6.2 Sequence-specific target RNA design 174
   6.3 Sequence non-specific, collateral substrate design 176

References 180

8. Experimental strategy for characterization of novel TnpB orthologs **183**

Gytis Druteika, Tautvydas Karvelis, and Virginijus Šikšnys

1. Introduction 184
2. Methods 186
   2.1 Bioinformatic curation of TnpB and guide RNA 186
   2.2 Construction of randomized TAM (7 N) plasmid library 186
   2.3 TnpB and reRNA expression and complex assembly 187
   2.4 Plasmid library cleavage 188
   2.5 Capture of cleavage products *via* adapter ligation 188
   2.6 Illumina library generation and deep sequencing 190
   2.7 Cleavage position and TAM identification 191
3. Concluding remarks 193

Acknowledgments 193

References 193

9. Hydrolytic endonucleolytic ribozyme (HYER): Systematic identification, characterization and potential application in nucleic acid manipulation **197**

Zi-Xian Liu and Jun-Jie Gogo Liu

1. Introduction 198
2. Identification of GII introns 200
3. The structural studies and the mechanisms of GII intron splicing and propagation 201
4. GII intron-derived tools for DNA manipulation 203
5. Protocol 204
   5.1 Materials 204
   5.2 Bioinformatics pipeline 207
   5.3 *In vitro* DNA cleavage activity characterization 208
   5.4 *In vivo* activity characterization 212
   5.5 Structure determination of HYERs 217
   5.6 Engineering and optimization of HYER1 218
6. Perspectives on the application of HYERs 219

Acknowledgments 220

References 220

10. Methods for Cas13a expression and purification for use in CRISPR diagnostics **225**

Her Xiang Chai, Rebecca S. Bamert, and Gavin J. Knott

1. Introduction 226
2. Before you begin 228
   2.1 Preparation of purification buffers 228
   2.2 Preparing working solutions for RNA detection assay 230
3. Key materials and equipment tables 230
   3.1 Protein purification 230
   3.2 Nucleic acid detection buffer components 232
   3.3 Oligonucleotides 232
   3.4 Materials and equipment 232
   3.5 Stock solution 233
4. Expression of LbuCas13a 234
   4.1 TB media preparation 234
   4.2 Transformation 234
   4.3 Overnight culture 234

4.4 Expression (Day 1) 234
4.5 Expression (Day 2) 235
5. Purification of LbuCas13a 235
5.1 Purification (Day 1) 235
5.2 Purification (Day 2) 236
6. Benchmarking LbuCas13a in an RNA detection assay 239
6.1 Preparation of LbuCas13a RNP (Ribonucleic acid protein) complex 239
6.2 Preparing target RNA and substrate reporter mix 239
6.3 Conducting assay 240
7. Expected outcomes from purification 240
8. Troubleshooting and optimisation 241
8.1 Overgrown cell pellet after overnight expression 241
8.2 Degradation of protein during purification 242
8.3 Incomplete cleavage of MBP tag 242
8.4 RNase contamination 242
8.5 Poor limit of detection 242
9. Concluding remarks 243
References 243

11. Using CRISPR for viral nucleic acid detection 245

Maaike S. Grimm and Cameron Myhrvold

1. Introduction 246
2. Existing nucleic acid detection methods 247
3. Class 2 CRISPR-Cas systems 249
3.1 Cas9 249
3.2 Cas12 249
3.3 Cas13 251
4. First-generation detection methods 253
5. Single-step detection methods 254
6. Amplification-free methods 257
7. Multiplexed detection methods 259
8. How to perform CRISPR-based viral detection 261
8.1 Comparison of CRISPR-Cas viral nucleic acid detection methods 263
9. Challenges and future improvements 263
10. Conclusion 268
References 269

12. Preparation of high-purity RNPs of CRISPR-based DNA base editors **277**

Mitchell J. McAndrew, Madeleine B. King, and Audrone Lapinaite

1. Introduction 278
2. Materials 281
   2.1 Expression of recombinant DNA base editors 281
   2.2 Purification of ABE and CBE enzymes 282
   2.3 In vitro transcription (IVT) of sgRNA 284
   2.4 In vitro DNA deamination assay 286
3. Methods 287
   3.1 Transformation of ABE or CBE plasmid (20 h) 287
   3.2 Expression of recombinant DNA base editors (48 h) 288
   3.3 Purification of ABE and CBE enzymes (14.5 h) 289
   3.4 *In vitro* transcription (IVT) of sgRNA (30 h) 296
   3.5 *In vitro* DNA deamination assay (7-8 h) 301
4. Summary and conclusion 304

Acknowledgement 305

Funding 305

Data availability 305

5. Buffer Appendix 305
   5.1 *Purification buffers* 305

References 313

13. Bacterial directed evolution of CRISPR base editors **317**

Reilly Q. Mach and Shannon M. Miller

1. Introduction and background 318
   1.1 Introduction to genome editing proteins 318
   1.2 Previous engineering efforts 320
   1.3 Benefits and limitations of base editing compared to other genome editing methods 321
   1.4 Benefits and limitations of plate-based selection compared to other directed evolution methods 321
   1.5 Experimental design 322
   1.6 Reagents 322
   1.7 Equipment 325
   1.8 Software 326
   1.9 Primers 327
   1.10 Plasmids 327

2. Optimization of selection strategy 327
2.1 Identifying compatible antibiotic resistance gene mutations for base editing 327
2.2 Designing base editor constructs 329
2.3 Mock selection to validate system and determine starting points 330
2.4 Tuning stringency of the selection 331
3. Methods to generate diversity 332
3.1 Site saturation mutagenesis 332
3.2 Error-prone PCR 334
3.3 DNA shuffling 336
4. Performing plate-based selection 338
4.1 Transforming into the selection strain and performing selection 339
4.2 Determining library size and coverage 340
5. Analyzing outcomes 341
5.1 Genotyping of individual variants 341
5.2 Determining success of a selection 342
5.3 Repeating selection iteratively and sub-cloning 343
5.4 Human cell base editing 344
5.5 Determining the end of evolution 344
6. Conclusions 345
Appendix A. Supplementary material 345
References 345

14. Genome editing with programmable base editors in human cells **351**

Nicola R.B. Osgood, Natalie M. Zawalick, Courtney B. Sawyer, Quinn T. Cowan, Sifeng Gu, S.J. Mawson, Brodie L. Ranzau, Lehan Li, Melissa Gymrek, Alon Goren, and Alexis C. Komor

1. Key resources table 352
2. Introduction 355
3. Selecting a base editor 357
3.1 Method 1 – selecting a base editor 361
4. Designing guide RNAs (gRNAs) 362
4.1 Method 2a – designing a gRNA: PAM-first approach 363
4.2 Examples 364
4.3 Method 2b – designing a gRNA: edit-first approach 365
4.4 Examples 366
4.5 Method 3 – anneal and phosphorylate oligonucleotides 369
4.6 Method 4 - golden gate cloning 370
4.7 Method 5 – transformation, inoculation, and sequencing 371

5. Transfecting mammalian cells and harvesting DNA 373
 5.1 Method 6a – forward transfection of adherent mammalian cells in 48-well plate 374
 5.2 Method 6b – reverse transfection of adherent mammalian cells in 48-well plate 377
 5.3 Method 6c – electroporation of suspension mammalian cells in a 12-well plate 379
 5.4 Method 6d – piggyBac™ BE transposition and gRNA transduction of human iPSCs in a 6-well plate 380
 5.5 Method 7 – lysing adherent mammalian cells following transfection in a 48-well plate 384
6. Next-generation sequencing to quantify base editing efficiency 385
 6.1 Method 8 – next generation sequencing sample preparation 388
7. Enriching for cells with high BE expression or activity 397
 7.1 Method 9a – FACS of adherent mammalian cells for base editing enrichment 398
 7.2 Method 9b - Namocell® sorting for cell line generation with adherent mammalian cells using base editing 401
Acknowledgments 403
Competing interests 403
Appendix A. Supporting information 403
References 403

15. Prime editing in bacteria with BacPE **405**

Hongyuan Zhang and Quanjiang Ji

1. Introduction 405
2. Before you begin: design pegRNA for genome editing in *E. coli* 406
 2.1 Materials and equipment 409
 2.2 Step-by-step method details: perform prime editing in *E. coli* 410
3. Concluding remarks 416
4. Limitations 417
5. Declaration of AI and AI-assisted technologies in the writing process 417
References 417

16. A quick guide to evaluating prime editing efficiency in mammalian cells **419**

Chengfang Liu, Sifan Cheng, Junjie Zhu, Lina Zhou, and Jia Chen

1. Introduction 420
2. Mechanism of prime editor 421

3. Construction of plasmids expressing pegRNA and nicking-sgRNA 424
  3.1 Reagents and materials 424
  3.2 Oligonucleotide sequences 424
  3.3 Protocol 425
  3.4 Notes 428
4. Transfection 429
  4.1 Methods of cell transfection 429
  4.2 Transfection based on liposomes 429
  4.3 Notes 431
5. Examination of editing efficiency 431
  5.1 Methods for assessing editing efficiency 431
  5.2 Examining of editing efficiency based on NGS 431
  5.3 Result 434
6. Conclusion 434
References 435

17. Using Prime Editing Guide Generator (PEGG) for high-throughput generation of prime editing sensor libraries **437**

Samuel I. Gould and Francisco J. Sánchez-Rivera

1. Introduction 438
2. Materials 440
3. Methods 441
  3.1 Installation 441
  3.2 Mutation input formatting 441
  3.3 Reference genome loading 443
  3.4 Running PEGG to generate pegRNAs 443
  3.5 Sensor design options 445
  3.6 Exhaustive description of PEGG output 445
  3.7 Generating non- and safe-targeting pegRNAs with the library module 447
  3.8 Oligonucleotide design 448
  3.9 Visualization of pegRNAs 449
4. Notes 449
Acknowledgements 450
Competing interests 450
References 450

18. Cloning and validating systems for high throughput molecular recording **453**

Anqi Zhao and Michelle M. Chan

1. Introduction 454
2. Before you begin 455
    2.1 Target site design considerations 455
    2.2 Lineage tracing vector design 457
3. Materials and equipment 457
4. Time course experiment to assess the cut site performance 459
    4.1 Plasmid construction 460
    4.2 Cell line engineering 463
    4.3 Time course experiment 464
5. Target site - triple guide plasmid generation 466
    5.1 Step 1: Triple guide cassette assembly 466
    5.2 Step 2: Target site assembly 468
    5.3 Step 3: lineage tracing vector plasmid library construction 468
6. Lineage tracing cell line construction and validation 469
7. An example of conducting in vitro lineage tracing in gastruloids 470
8. Limitations 471

References 472

19. Delivery of genome editors with engineered virus-like particles **475**

Christopher Lu, Yuanhang Li, Jacob Ryan Cummings, and Samagya Banskota

1. Introduction 476
    1.1 Therapeutic gene editing technologies 476
    1.2 Delivery of genome editors and comparisons of various delivery methods 480
    1.3 VLPs as a delivery platform 481
2. Materials and methods 490
    2.1 Key resources – materials and equipment 490
    2.2 eVLP production and purification 492
    2.3 Cellular assays for eVLP validation 499
    2.4 eVLP characterization 502
3. Optimization and troubleshooting 507
    3.1 Transfection 507
    3.2 Characterization assays 508
4. Summary 509

Acknowledgements 510

Declaration of interests 510
References 510

20. CRISPRoff epigenome editing for programmable gene silencing in human cell lines and primary T cells **517**

Rithu K. Pattali, Izaiah J. Ornelas, Carolyn D. Nguyen, Da Xu, Nikita S. Divekar, and James K. Nuñez

1. Introduction 518
2. Key considerations prior to use of crisproff 520
2.1 Comparing epigenetic editing for gene silencing to gene knockout strategies 520
2.2 Delivery method of CRISPRoff 521
2.3 Gene targets and cell types 521
2.4 Assessing efficiency and specificity of CRISPRoff 522
3. Selection of single guide RNA and cloning 522
3.1 Design and selection of sgRNAs for CRISPRoff 522
3.2 Cloning sgRNA plasmid 523
4. Targeted gene silencing in Hek 293T cells using crisproff plasmid 523
4.1 Required materials 524
4.2 Delivery of CRISPRoff by plasmid transfection 525
5. Crisproff mRNA *in vitro* synthesis 529
5.1 Required materials 530
5.2 *In vitro* synthesis of CRISPRoff mRNA 531
6. Considerations and troubleshooting 534
7. Targeted gene silencing in Hek 293T cells using Crisproff mRNA 535
7.1 Required materials 535
7.2 CRISPRoff mRNA nucleofection in HEK 293T cells 536
8. Troubleshooting and tips 538
9. Targeted gene silencing in Htert Rpe-1 cells using Crisproff mRNA 539
9.1 Required materials 539
9.2 CRISPRoff mRNA transfection in hTERT RPE-1 cells 540
10. Troubleshooting and tips 541
11. Targeted gene silencing in primary human donor derived T cells using Crisproff mRNA 542
11.1 Required materials 542
11.2 CRISPRoff mRNA and sgRNAs co-nucleofection in primary human T cells 544
12. Troubleshooting and tips 546
13. Conclusion 546
Acknowledgments 548
References 548

Contributors

Matthew J. Armbrust
Department of Chemistry and Biochemistry, Utah State University, Logan, UT, United States

Rebecca S. Bamert
Department of Biochemistry and Molecular Biology, Monash Biomedicine Discovery Institute, Monash University, Clayton, VIC, Australia

Samagya Banskota
Department of Biomedical Engineering; Biological Design Center, Boston University, Boston, MA, United States

Chase L. Beisel
Helmholtz Institute for RNA-based Infection Research (HIRI), Helmholtz Centre for Infection Research (HZI); Medical Faculty, University of Wurzburg, Wurzburg, Germany

Her Xiang Chai
Department of Biochemistry and Molecular Biology, Monash Biomedicine Discovery Institute, Monash University, Clayton, VIC, Australia

Michelle M. Chan
Lewis-Sigler Institute for Integrative Genomics; Department of Molecular Biology, Princeton University, Princeton, NJ, United States

Leifu Chang
Department of Biological Sciences; Purdue Institute for Cancer Research, Purdue University, West Lafayette, IN, United States

Jia Chen
Gene Editing Center, School of Life Science and Technology, ShanghaiTech University; Shanghai Clinical Research and Trial Center, Shanghai, P.R. China

Sifan Cheng
Gene Editing Center, School of Life Science and Technology, ShanghaiTech University, Shanghai, P.R. China

Quinn T. Cowan
Department of Chemistry and Biochemistry, University of California San Diego, La Jolla, CA, United States

Jacob Ryan Cummings
Department of Biomedical Engineering; Biological Design Center, Boston University, Boston, MA, United States

Nikita S. Divekar
Department of Molecular and Cell Biology, University of California, Berkeley, Berkeley, CA, United States

Oleg Dmytrenko
Helmholtz Institute for RNA-based Infection Research (HIRI), Helmholtz Centre for Infection Research (HZI), Würzburg, Germany

Hannah Domgaard
Department of Chemistry and Biochemistry, Utah State University, Logan, UT, United States

Gytis Druteika
Institute of Biotechnology, Life Sciences Center, Vilnius University, Vilnius, Lithuania

Alon Goren
Department of Medicine, University of California San Diego, La Jolla, CA, United States

Samuel I. Gould
Department of Biology; David H. Koch Institute for Integrative Cancer Research, Massachusetts Institute of Technology, Cambridge, MA, United States

Maaike S. Grimm
Department of Molecular Biology, Princeton University, Princeton, NJ, United States

Sifeng Gu
Department of Chemistry and Biochemistry, University of California San Diego, La Jolla, CA, United States

Melissa Gymrek
Department of Medicine; Department of Computer Science and Engineering, University of California San Diego, La Jolla, CA, United States

Taekjip Ha
Program in Cellular and Molecular Medicine, Boston Children's Hospital; Department of Pediatrics, Harvard Medical School; Howard Hughes Medical Institute, Boston, MA, United States

Thomson Hallmark
Department of Chemistry and Biochemistry, Utah State University, Logan, UT, United States

Grace N. Hibshman
Interdisciplinary Life Sciences Graduate Programs; Department of Molecular Biosciences, University of Texas at Austin, Austin, TX, United States

Ryan N. Jackson
Department of Chemistry and Biochemistry, Utah State University, Logan, UT, United States

Quanjiang Ji
School of Physical Science and Technology; Gene Editing Center, School of Life Science and Technology, ShanghaiTech University; Shanghai Clinical Research and Trial Center, Shanghai, P.R. China

Tautvydas Karvelis
Institute of Biotechnology, Life Sciences Center, Vilnius University, Vilnius, Lithuania

Madeleine B. King
School of Molecular Sciences, Arizona State University, Tempe, AZ; Gavin Herbert Eye Institute - Center for Translational Vision Research, University of California Irvine, School of Medicine, Irvine, CA, United States

Gavin J. Knott
Department of Biochemistry and Molecular Biology, Monash Biomedicine Discovery Institute, Monash University, Clayton, VIC, Australia

Alexis C. Komor
Department of Chemistry and Biochemistry; Moores UCSD Cancer Center; Sanford Stem Cell Institute, University of California San Diego, La Jolla, CA, United States

Audrone Lapinaite
Gavin Herbert Eye Institute - Center for Translational Vision Research; Department of Ophthalmology, University of California Irvine, School of Medicine; Department of Biomedical Engineering, University of California Irvine, Irvine, CA, United States

Lehan Li
Department of Chemistry and Biochemistry, University of California San Diego, La Jolla, CA, United States

Yuanhang Li
Biological Design Center; Department of Mechanical Engineering, Boston University, Boston, MA, United States

Chengfang Liu
Gene Editing Center, School of Life Science and Technology, ShanghaiTech University, Shanghai, P.R. China

Jun-Jie Gogo Liu
Beijing Frontier Research Center for Biological Structure, Center of Synthetic and Systems Biology, State Key Laboratory of Membrane Biology, Tsinghua-Peking Center for Life Sciences, School of Life Sciences, Tsinghua University, Beijing, P.R. China

Zi-Xian Liu
Beijing Frontier Research Center for Biological Structure, Center of Synthetic and Systems Biology, State Key Laboratory of Membrane Biology, Tsinghua-Peking Center for Life Sciences, School of Life Sciences, Tsinghua University, Beijing, P.R. China

Christopher Lu
Department of Biomedical Engineering; Biological Design Center, Boston University, Boston, MA, United States

Reilly Q. Mach
The Scripps Research Institute

Alberto Marin-Gonzalez
Program in Cellular and Molecular Medicine, Boston Children's Hospital; Department of Pediatrics, Harvard Medical School; Howard Hughes Medical Institute, Boston, MA, United States

S.J. Mawson
Department of Chemistry and Biochemistry, University of California San Diego, La Jolla, CA, United States

Mitchell J. McAndrew
School of Molecular Sciences, Arizona State University, Tempe, AZ; Gavin Herbert Eye Institute - Center for Translational Vision Research, University of California Irvine, School of Medicine, Irvine, CA, United States

Shannon M. Miller
The Scripps Research Institute

Cameron Myhrvold
Department of Molecular Biology; Department of Chemical and Biological Engineering; Omenn-Darling Bioengineering Institute; Department of Chemistry, Princeton University, Princeton, NJ, United States

Carolyn D. Nguyen
Department of Molecular and Cell Biology, University of California, Berkeley, Berkeley, CA, United States

Giang T. Nguyen
Department of Hematology, St. Jude Children's Research Hospital, Memphis, TN, United States

Osamu Nureki
Department of Biological Sciences, Graduate School of Science, The University of Tokyo, Bunkyo-ku, Tokyo, Japan

James K. Nuñez
Department of Molecular and Cell Biology, University of California, Berkeley, Berkeley; Chan Zuckerberg Biohub San Francisco, San Francisco, CA, United States

Satoshi N. Omura
Department of Biological Sciences, Graduate School of Science, The University of Tokyo, Bunkyo-ku, Tokyo, Japan

Izaiah J. Ornelas
Department of Molecular and Cell Biology, University of California, Berkeley, Berkeley, CA, United States

Nicola R.B. Osgood
Department of Chemistry and Biochemistry, University of California San Diego, La Jolla, CA, United States

Rithu K. Pattali
Department of Molecular and Cell Biology, University of California, Berkeley, Berkeley, CA, United States

Akshara Raju
Roy J. Carver Department of Biochemistry, Biophysics and Molecular Biology, Iowa State University, Ames, IA, United States

Brodie L. Ranzau
Department of Chemistry and Biochemistry, University of California San Diego, La Jolla, CA, United States

Olivine Redman
Department of Chemistry and Biochemistry, Utah State University, Logan, UT, United States

Adam T. Rybczynski
Program in Cellular and Molecular Medicine, Boston Children's Hospital; Department of Pediatrics, Harvard Medical School, Boston, MA, United States

Dipali G. Sashital
Roy J. Carver Department of Biochemistry, Biophysics and Molecular Biology, Iowa State University, Ames, IA, United States

Courtney B. Sawyer
Department of Medicine, University of California San Diego, La Jolla, CA, United States

Friso T. Schut
Helmholtz Institute for RNA-based Infection Research (HIRI), Helmholtz Centre for Infection Research (HZI), Würzburg, Germany

Francisco J. Sánchez-Rivera
Department of Biology, David H. Koch Institute for Integrative Cancer Research, Massachusetts Institute of Technology, Cambridge, MA, United States

David W. Taylor
Interdisciplinary Life Sciences Graduate Programs; Department of Molecular Biosciences, University of Texas at Austin, Austin, TX, United States

Shukun Wang
Department of Biological Sciences, Purdue University, West Lafayette, IN, United States

Andrew A. Williams
Department of Chemistry and Biochemistry, Utah State University, Logan, UT, United States

Da Xu
Department of Molecular and Cell Biology, University of California, Berkeley, Berkeley, CA, United States

Natalie M. Zawalick
Department of Chemistry and Biochemistry, University of California San Diego, La Jolla, CA, United States

Hongyuan Zhang
School of Physical Science and Technology, ShanghaiTech University, Shanghai, P.R. China

Anqi Zhao
Lewis-Sigler Institute for Integrative Genomics; Department of Molecular Biology, Princeton University, Princeton, NJ, United States

Lina Zhou
Gene Editing Center, School of Life Science and Technology, ShanghaiTech University, Shanghai, P.R. China

Junjie Zhu
Gene Editing Center, School of Life Science and Technology, ShanghaiTech University, Shanghai, P.R. China

Roger S. Zou
Department of Medicine, Massachusetts General Hospital, Boston, MA, United States

Virginijus Šikšnys
Institute of Biotechnology, Life Sciences Center, Vilnius University, Vilnius, Lithuania

CHAPTER ONE

Measuring double-strand break repair events in mammalian cells with multi-target CRISPR

Alberto Marin-Gonzalez[a,b,c], Adam T. Rybczynski[a,b], Roger S. Zou[d], and Taekjip Ha[a,b,c,*]

[a]Program in Cellular and Molecular Medicine, Boston Children's Hospital, Boston, MA, United States
[b]Department of Pediatrics, Harvard Medical School, Boston, MA, United States
[c]Howard Hughes Medical Institute, Boston, MA, United States
[d]Department of Medicine, Massachusetts General Hospital, Boston, MA, United States
*Corresponding author. e-mail address: taekjip.ha@childrens.harvard.edu

Contents

1. Introduction 2
   1.1 DSB repair in mammalian cells 2
   1.2 Current methods to induce DSBs in mammalian cells and their limitations 3
   1.3 Overview of multi-target CRISPR 4
2. Protocol 5
   2.1 Overview of the procedure and other comments 5
   2.2 Materials and reagents 5
   2.3 Large-scale delivery of multi-target CRISPR into HEK293T (~12 M cells) 7
   2.4 Small-scale delivery of multi-target CRISPR into HEK293T cells (~0.5 M cells) 9
   2.5 Light-activation of multi-target CRISPR 11
3. Variations to the protocol 12
   3.1 Validating multi-target CRISPR in different cell types 12
   3.2 Designing and validating new mgRNAs 14
4. Guidelines for sequencing and sequencing data analysis 16
   4.1 Sequencing recommendations 16
   4.2 Sequencing data analysis 17
5. Conclusions 20
Acknowledgements 20
Appendix A. Supporting information 20
References 20

Abstract

A mechanistic understanding of the different pathways involved in the repair of DSBs is a timely, yet challenging task. CRISPR-Cas9 is a powerful tool to induce DNA double-strand breaks (DSB) at defined genomic locations to study the ensuing repair response, but Cas9 studies are typically limited by i) low-throughput induction of DSB,

Methods in Enzymology, Volume 712
ISSN 0076-6879, https://doi.org/10.1016/bs.mie.2025.01.011

by targeting only one or a few genomic sites, or ii) the use of genetically integrated reporter systems, which do not always reflect endogenous phenotypes. To address these limitations, we developed multi-target CRISPR, a Cas9-based tool to controllably induce DSBs in high-throughput at endogenous sites, by leveraging repetitive genomic regions. In this Chapter, we describe how to design and execute a multi-target CRISPR experiment. We also detail how to analyze next-generation sequencing data for characterization of DSB repair events at multiple cut sites. We envision that multi-target CRISPR will become a valuable tool for the study of mammalian DSB repair mechanisms.

1. Introduction

1.1 DSB repair in mammalian cells

DNA double-strand breaks (DSBs) are one of the most toxic forms of DNA lesions. When misrepaired, DSBs can result in mutations, genomic translocations, and epigenetic changes that can compromise cellular function (Ciccia & Elledge, 2010; Schumacher, Pothof, Vijg, & Hoeijmakers, 2021; Scully, Panday, Elango, & Willis, 2019). Consequently, mammalian cells have evolved sophisticated pathways to repair DSB and preserve genome integrity. The two main pathways are (classical) non-homologous end-joining (NHEJ) and homologous recombination (HR) (Scully et al., 2019). NHEJ is fast and error prone and consists of a ligation of the two broken DNA ends with little end processing. HR is slower but more accurate than NHEJ, and it operates by copying an undamaged sequence template into the lesion. NHEJ is operational during the entire cell cycle, whereas HR is only active in the S/G2 phases, when a sister chromatid is present. Two additional pathways are microhomology-mediated end joining and single-strand annealing, both of which are highly mutagenic and are prone to cause deletions by bridging together homologous sequences from both sides of the break (Bhargava, Onyango, & Stark, 2016; Ramsden, Carvajal-Garcia, & Gupta, 2022).

A comprehensive characterization of the mechanisms behind mammalian DSB repair holds great promise for biomedical applications. For example, manipulation of DSB repair pathways can open new avenues for improved genome editing technologies. In this respect, inhibition of key NHEJ factors, such as 53BP1, DNA-PKcs or DNA ligase IV, have been employed to increase the efficiency of HR-mediated Cas9 gene editing in cultured mammalian cells (Yeh, Richardson, & Corn, 2019). Cancer treatments can also benefit from an in-depth understanding of

DSB repair mechanisms. For instance, cancer cells with deficiency of the HR factors BRCA1 or BRCA2 are vulnerable to inhibition of PARP, and this lethality is commonly exploited in chemotherapy treatments (Dias, Moser, Ganesan, & Jonkers, 2021). Discovery of new forms of synthetic lethality in DSB repair is an active area of research with promising applications towards the development of new drugs (Setton et al., 2021).

1.2 Current methods to induce DSBs in mammalian cells and their limitations

Despite decades of work, several major aspects of mammalian DSB repair are only beginning to be understood. This is well illustrated by recent works that have identified new interactions between DSB repair factors (Hussmann et al., 2021), uncovered unexpected roles of chromatin modifications in DSB repair (Yilmaz et al., 2021), or captured complex 3D genome structures formed during the repair process (Arnould et al., 2021; Ochs et al., 2019). Many of these findings on DSB repair processes have been catalyzed by the use of CRISPR-Cas9 as a unique tool to induce DSBs with increased efficacy and/or precision. Cas9 has the advantage of inducing clean DSB at defined genomic locations, in contrast to methods based on ionizing radiation or chemical mutagens, which induce mixed DNA lesions in a delocalized manner. Type II endonucleases can also induce DSB at defined DNA sites, but their target sites are constrained by the specific enzyme used, whereas Cas9 target sites can be easily programmed via rational guide RNA (gRNA) design (Vítor, Huertas, Legube, & de Almeida, 2020). Moreover, Cas9 has generally higher cleavage efficiency than other endonucleases and, in combination with caged gRNAs that enable light-induced activation, Cas9 can induce DSBs with high temporal control (within seconds) (Liu et al., 2020).

Despite these strengths, studies of repair using Cas9-induced DSBs are often limited to one or few target sites, which fail to capture the heterogeneity of DSB repair responses present in different chromatin contexts and preclude the extraction of general findings, raising the question of whether certain observations are due to the particular choice of cut site. Reporter systems have been employed to address this issue, but findings on exogenously integrated loci are not always applicable to endogenous sites (Allen et al., 2019; Shen et al., 2018). With these ideas in mind, we developed multi-target CRISPR, a tool to induce DSBs in high-throughput at endogenous sites (Zou et al., 2022).

1.3 Overview of multi-target CRISPR

We reasoned that directing Cas9 to transposable elements in a controlled manner would allow us to multiplex Cas9 activity at endogenous sites (Fig. 1A). To that end, we developed a computational pipeline that searches for 20 bp-long sequences in the human genome that have up to three mismatches from a consensus Alu SINE sequence, a transposable element that is highly abundant in the human genome (Deininger, 2011). Using the hg38 genome assembly as reference, we identified ~40,000 such multi-target gRNAs (mgRNA), each of them containing between 2 and over 1000 predicted on-target sites (Fig. 1B). An analogous search on the *mm10* genome yielded ~8500 mgRNAs for *Mus musculus*. Using instead a consensus sequence of the B4 transposable element as input yielded ~1800 mgRNAs for *Mus musculus*. Importantly, the target locations of mgRNAs often span diverse chromatin contexts, thus allowing us to study the interplay between different chromatin modifications and Cas9 break repair (Zou et al., 2022).

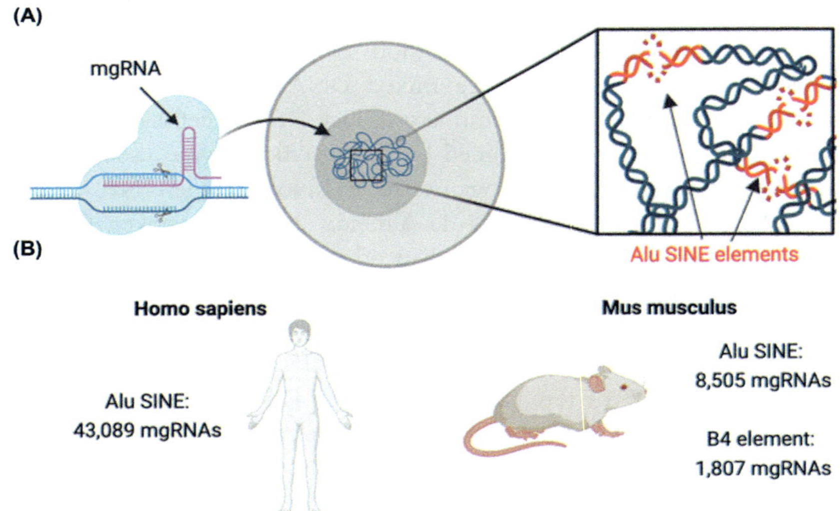

Fig. 1 Scheme of multi-target CRISPR. (A) By targeting transposable elements, such as Alu SINE, multi-target CRISPR induces hundreds of double-strand breaks at endogenous sites. (B) We applied our computational pipeline to identify mgRNAs in *Homo sapiens* and *Mus musculus* using different transposable element consensus sequences as input.

We have used mgRNAs to measure DSB repair processes in multiple cell lines with high-throughput and temporal resolution (Deshpande et al., 2023; Zou et al., 2022). Here, we describe how to implement multi-target CRISPR for DSB repair studies. First, we present a comprehensive protocol to deliver multi-target into HEK293T cells and prepare the cells for ChIP or IF assays, followed by some instructions on the use of light-activated caged gRNAs for multi-target CRISPR. Then, we present some general aspects to consider before doing a multi-target CRISPR experiment, regarding the choice of mgRNA sequence and the validation of multi-target CRISPR delivery and activity in different cell lines. We finish with some recommendations and guidelines on how to sequence and analyze ChIP-Seq and other NGS experiments in the context of multi-target CRISPR.

2. Protocol

2.1 Overview of the procedure and other comments

In this section, we present some general procedures for implementing multi-target CRISPR. We describe a protocol for a large-scale delivery of multi-target CRISPR into 1.2×10^7 HEK293T cells and a protocol for small-scale delivery into 5×10^5 cells. The former is the preferred delivery method when combining multi-target CRISPR with assays such as ChIP-Seq or Hi-C, which require large cell numbers; the latter is generally preferred for immunofluorescence (IF) or cell viability assays which typically require 10^6 cells or less (Fig. 2).

We have successfully employed the large-scale delivery protocol for cell numbers between 10^6 and 1.5×10^7 and the small-scale delivery protocol for cell numbers between 2×10^5 and 10^6. The only aspect to take into consideration when modifying the protocol for different cell numbers is that cells should be plated at a confluency of between ~50–80 % after the electroporation. To apply the protocol to different cell types, please see Section 3.1.

2.2 Materials and reagents

2.2.1 Equipment

1. Lonza 4D-Nucleofector® Core Unit (Lonza, AAF-1003B) and 4D-Nucleofector® X Unit (Lonza, AAF-1003X).
2. Thermocycler.
3. UV flashlight uvBeast New V3 365 nm Mini (6x).
4. Benchtop centrifuge.
5. Cell scrapers.

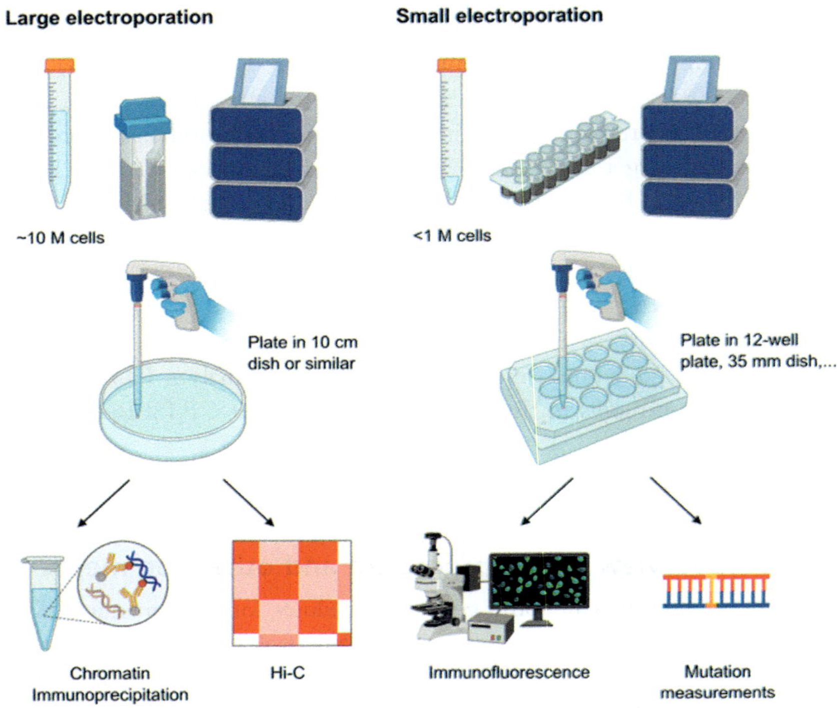

Fig. 2 Scheme of large and small electroporations and downstream assays. Large electroporations (left) fit ~10 M cells and are used in combination with assays such as ChIP or Hi-C that require large (several millions) cell numbers. Small electroporations (right) are performed using a maximum of 1 M cells and they can be used for IF or mutation measurements.

2.2.2 Reagents

1. IDT Alt-R® CRISPR-Cas9 tracrRNA (IDT #1072532), Alt-R® Cas9 Electroporation Enhancer (IDT #1075915) and nuclease-free duplex buffer (IDT #11-05-01-03).
2. IDT Alt-R™ CRISPR-Cas9 crRNA. Protospacer sequence `CCTGTAGTCCCAGCTACTGG` for AluGG.
3. IDT Alt-R™ S.p. Cas9 Nuclease V3, 100 µg (IDT #1081058).
4. Lonza SF Cell Line 4D-Nucleofector™ X Kit L (Lonza #V4XC-2024) and/or SF Cell Line 4D-Nucleofector™ X Kit S (Lonza # V4XC-2032). These include SF solution, supplement and nucleofection cuvettes.
5. HEK293T cells (ATCC CRL-3216).

6. Dulbecco's Modified Eagle's Medium, DMEM (Corning #10-013-CV).
7. Heat-inactivated fetal bovine serum, FBS (Corning #35-011-CV).
8. Collagen I, rat tail (Gibco # A1048301).
9. Phosphate buffered saline, PBS 1x (Corning #21-040-CV).
10. Pierce™ 16 % Formaldehyde (w/v), Methanol-free (ThermoScientific #28908).
11. Dialysis buffer: 20 mM HEPES pH 7.5, 500 mM KCl, 20 % glycerol.

2.3 Large-scale delivery of multi-target CRISPR into HEK293T (~12 M cells)

In this section, we describe a protocol to nucleofect multi-target CRISPR into ~1.2×10^7 HEK293T cells and plate them in a 10 cm plate in preparation for ChIP.

1. Reconstitute IDT tracrRNA, crRNA, and electroporation enhancer using IDT duplex buffer to a concentration of 100 μM.
2. Add 2 μL tracrRNA and 2 μL crRNA in a 0.5 mL PCR tube. Mix well. Incubate at 95 C for 3 min in order to denature the RNAs. Let the mixture cool at room temperature for 5 min. Then add 3 μL IDT Cas9 at 10 mg/mL to tracr:crRNA, add 8 μL dialysis buffer (20 mM HEPES pH 7.5, 500 mM KCl, 20 % glycerol).
3. Mix well and incubate for at least 15 min at room temperature in order for the RNP to assemble. Once formed, the RNP is stable at room temperature for several hours.
4. Prepare nucleofection solution: add 73.8 μL SF solution and 16.2 μL supplement to a clean PCR tube. Mix well. Keep at room temperature.
5. Prepare collagen plates. Dilute 60 μL collagen in 6 mL PBS for a 1:100 dilution. Mix well and add 3 mL diluted collagen into each of two 10-cm plates. Incubate at 37 C for > 5 min. When ready for plating, aspirate collagen from plates.
6. Harvest ~12 M HEK293T cells via trypsinization and collect them in a 15 mL tube. Cells should be at ~80-90 % confluency and > 95 % viability at the time of harvest for optimal results.
7. As no-electroporation control, harvest ~8 M HEK293T cells via trypsinization. Transfer the cells to a separate 15 mL tube.
   a. Note: nucleofection causes cell death. To account for this, and to aim for an equal number of cells in both nucleofection and no-nucleofection control conditions, we typically harvest fewer cells for control than for nucleofection. We found that a 2:3 ratio of control to nucleofected cells typically results in similar cell numbers after the treatment.

8. Spin down all cells at 200 g for 3 min at room temperature.
9. While cells spin, transfer 10 mL of complete DMEM into a clean 15 mL tube. This will be used to pick up cells after nucleofection.
10. Discard supernatant from spun cells and wash cells with 3 mL PBS. Repeat spin.
11. Discard supernatant from spun cells and plate the no-electroporation control cells in one of the collagen pre-coated plates using 10 mL complete DMEM.
12. Resuspend the electroporation cells in 90 μL nucleofection solution. Mix gently until achieving a homogenous suspension and transfer 110 μL into the PCR tube containing the RNP. Add 2 μL electroporation enhancer.
13. Mix by gently pipetting up and down five times. Load 127 μL cells-RNP mixture into a large, 130 μL nucleofection cuvette.
14. Place the cuvette in the nucleofection machine and run the program CA-189.
15. To pick up the electroporated cells, transfer 100 μL complete DMEM from 15 mL tube (see point 9 above) into the nucleofection cuvette. Mix by gently pipetting up and down a couple of times. Transfer all the contents from the nucleofection cuvette to the 15 mL tube containing 10 mL complete DMEM.
16. Repeat previous step.
17. Plate 10 mL electroporated cells in the collagen pre-coated 10 cm plate.
18. Allow at least three hours for cells to adhere to plate before harvesting for ChIP.

2.3.1 Harvest and fix for ChIP

1. If cells are to be used for ChIP, remove media and gently wash the plate with 3 mL warm PBS.
2. Remove PBS and add 5 mL DMEM only (with no FBS). Scrape cells and transfer to a clean 15 mL tube.
3. Add another 5 mL DMEM (no FBS) to the plate and repeat the scrape. Transfer to 15 mL tube for a final of 10 mL of cells at ~1 M/mL.
4. To fix cells, add 721 μL of 16% formaldehyde for final of 1% formaldehyde. Mix for 10 min at room temperature with rotation.
5. To quench the crosslinking reaction, add 750 μL 2 M glycine in PBS for a final of 130 mM glycine. Mix for 4 min at 4 C with rotation.
6. Spin down the cells at 1200 g for 3 min at 4 C. Discard supernatant and add 10 mL ice-cold PBS.

7. Repeat previous step for a total of two PBS washes.
8. Spin down (same spin) and discard all PBS. Use P200 to get rid of all the remaining PBS.
9. Pellets are now ready for ChIP. Alternatively, flash-freeze the pellet in liquid nitrogen and store at −80 C. Pellets are stable at −80 C for over a year.
10. We aim for a minimum of 4×10^6 cells per ChIP reaction. Thus, at the time of ChIP, a single pellet can be split into several ChIP reactions depending on the number of cells used for electroporation. In this case, where we nucleofected $\sim 1.2 \times 10^7$ cells, pellets can be used for two ChIP reactions.

For a detailed ChIP protocol, we refer the reader to (Zou & Ha, 2021). We follow that protocol, except for the library preparation part. To prepare ChIP-Seq libraries, we use the NEB® Ultra™ II DNA Library Prep Kit (NEB E7645) following the provider's guidelines but using half volumes. We amplify the adapter-ligated libraries using the NEB® dual indexed oligos kit (either NEB 6440 or NEB 7600), with half volumes (25 μL PCR volume) and run the PCR to 13 cycles (with the exception of γH2AX where we use 10cycles).

2.4 Small-scale delivery of multi-target CRISPR into HEK293T cells (~0.5 M cells)

Here we explain how to deliver multi-target CRISPR into 5×10^5 HEK293T cells and plate them into two imaging dishes after nucleofection in preparation for IF.

1. Resuspend IDT tracrRNA, crRNA and electroporation enhancer to concentration of 100 μM using duplex hybridization buffer.
2. Mix 1 μL tracrRNA and 1 μL crRNA in a 0.5 mL PCR tube. Mix well by pipetting. Incubate at 95 C for 3 min. Let the mixture cool at room temperature for 5 min
3. Add 1.5 μL IDT Cas9 at 10 mg/mL to tracr:crRNA and 1.5 μL dialysis buffer. Mix well and incubate for at least 15 min at room temperature in order for the RNP to assemble.
4. To prepare nucleofection solution, add 16.4 μL SF solution and 3.6 μL supplement to a PCR tube. Mix well and keep at room temperature.
5. Prepare collagen plates. Dilute 40 μL collagen in 4 mL PBS for a 1:100 dilution. Mix well and add 200 μL diluted collagen into each of four 35 mm imaging dishes. Incubate at 37 C for > 5 min
6. Harvest $\sim 5 \times 10^5$ HEK293T cells via trypsinization and transfer into in a 15 mL tube. Cells should be at ~80–90 % confluency and 95 % viability at the time of harvest for optimal results.

7. As no-electroporation control, harvest ~3.3×10^5 HEK293T cells. Transfer them to a separate 15 mL tube.
8. We typically use a 2:3 ratio of control to nucleofected cells (see also Section 2.3).
9. Centrifuge cells at 200 g for 3 min at room temperature.
10. Transfer 400 μL complete DMEM into a clean 15 mL tube.
11. When the spin finishes, discard supernatant and wash cells with 1 mL PBS. Repeat spin.
12. During this spin, aspirate the collagen from the pre-coated plates in preparation for plating.
13. Discard supernatant and resuspend the no-electroporation control cells in 400 μL complete DMEM. Plate 200 μL cells into each of two collagen pre-coated 35 mm imaging dishes.
14. Resuspend the electroporation cells in 20 μL nucleofection solution. Mix gently until achieving a homogenous suspension and transfer 21.5 μL into the PCR tube containing the RNP. Add 1 μL electroporation enhancer.
15. Mix by gently pipetting up and down five times. Load 21.5 μL cells-RNP mixture into a small, 20 μL nucleofection cuvette.
16. Place the cuvette in the nucleofection machine and run the program CA-189.
17. To pick up the electroporated cells, transfer 20 μL complete DMEM from 15 mL tube into the nucleofection cuvette. Mix by gently pipetting up and down a couple of times. Transfer all contents from the nucleofection cuvette to the 15 mL tube containing 400 μL of complete DMEM.
18. Repeat previous step.
19. Plate 200 μL electroporated cells into each of the two collagen pre-coated imaging dishes.

2.4.1 Fix for immunofluorescence

1. Dilute 16 % formaldehyde 1:4 in PBS to make 4 % formaldehyde.
2. Remove media from dishes and add 2 mL pre-warmed 4 % formaldehyde in PBS. Incubate at room temperature for 15 min
3. Remove formaldehyde and wash the cells by gently adding 2 mL warm PBS. Incubate for 5 min
4. Repeat previous wash twice, for a total of three washes.
5. Store the dishes containing cells in PBS at 4 C. Fixed cells are stable at 4 C for a couple of weeks. Alternatively, proceed with IF protocol.

2.5 Light-activation of multi-target CRISPR

Caged gRNAs allow for very-fast light-induced CRISPR activation (Liu et al., 2020). Multi-target CRISPR can be combined with very-fast CRISPR by introducing caged groups in the PAM-distal side of the gRNA (see also Section 3.2. for caged gRNA design and validation). For caged gRNA delivery, we recommend protecting the caged gRNA and the assembled RNP from light by staying away from natural light and by covering the RNP with foil during the assembly incubation (step 3 in Sections 2.3 and 2.4). During the electroporation, we typically turn off the light of the cell culture hood to prevent light exposure, and after the electroporation, we cover the plate of electroporated cells with foil to protect them from light (Fig. 3A). Also, when using caged gRNA we use a no-light instead of a no-electroporation control, that is, cells that were electroporated but not subjected to light treatment, to ensure there is no leakage of the caged groups.

1. Deliver multi-target CRISPR following the guidelines from Sections 2.3 or 2.4 using a caged gRNA instead of a regular crRNA.
2. Incubate for 3–12 h to allow Cas9 binding.
   a. Note: shorter or longer times can negatively affect the results.
3. In a cell culture hood, take the lid off the plate and shine light for ~1 min. Keep the UV lamp(s) at a distance of ~3 cm (1″) from the top of the plate. For 6-wells and smaller dishes we use a single UV lamp (Fig. 3B). For a 10 cm plate, we use six UV lamps held together by a custom-designed 3D-printed holder (Fig. 3C).
   a. Note: UV radiation is hazardous. When performing this step, protect your eyes and skin with the proper equipment.
4. Place the plates back in the incubator until ready for harvest.

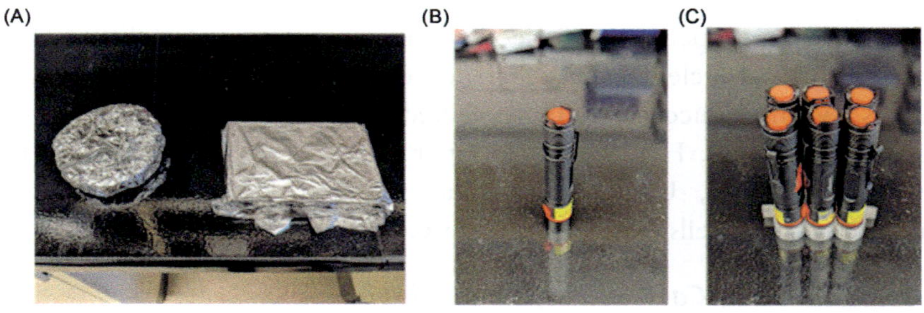

Fig. 3 Materials for light-activated multi-target CRISPR. (A) Tissue culture dishes are covered with foil after electroporating a caged gRNA in order to prevent unwanted UV-induced activation. (B) A single UV lamp is used to activate caged gRNAs in small plates such as 6-wells or smaller. (C) A set of six UV lamps is used to activate caged gRNAs in 10 cm dishes.

3. Variations to the protocol

In this section, we detail how to generalize the above protocol for different cell lines and how to pick and validate new mgRNAs.

3.1 Validating multi-target CRISPR in different cell types

Before employing multi-target CRISPR in a new cell line, we ensure that both nucleofection and Cas9 activity are efficient.

3.1.1 Nucleofection efficiency

We recommend delivering multi-target CRISPR via nucleofection of pre-assembled Cas9 - tracrRNA:crRNA RNP. In our hands, this method yields the highest cleavage efficiency and most reproducible results. Thus, in the protocols section below, we describe how to deliver multi-target CRISPR via nucleofection method. However, depending on the particular application, alternative methods such as plasmid or RNP transfections are also possible, and might even be a preferred choice for cells that are difficult to nucleofect.

High nucleofection efficiency is key for a successful multi-target CRISPR experiment. For every cell line used, we advise to test different nucleofection protocols by nucleofecting the GFP reporter plasmid provided by Lonza in the electroporation kit and measuring the electroporation efficiency by looking at presence or absence of fluorescence signal in cells under a widefield microscope (or similar), or via flow cytometry (for a more quantitative measurement) (Fig. 4A). We aim for a nucleofection efficiency of at least 75 % before proceeding with the actual experiment. Because nucleofections are stressful to the cells, we also measure cell viability after nucleofection of the GFP plasmid – or after a mock nucleofection with no cargo – and make sure that viability is at least ~75 %. Thus, we aim to optimize both nucleofection efficiency and viability (see Fig. 4B and C). In our experience, nucleofection of multi-target CRISPR in cell lines with low nucleofection efficiency or viability can yield suboptimal results. We have successfully nucleofected multi-target CRISPR into four different cell lines: HEK293T (SF kit, program CA-189), K562 (SF kit, program FF-120), U-2 OS (SE kit, program DN-100) and mouse embryonic stem cells (P3 kit, program CG-104).

3.1.2 Validating Cas9 activity in cellulo

Having found an efficient nucleofection protocol, we then recommend testing Cas9 cleavage efficiency. This can be done via two ways: by looking at cell proliferation and by measuring MRE11 recruitment to select cut sites.

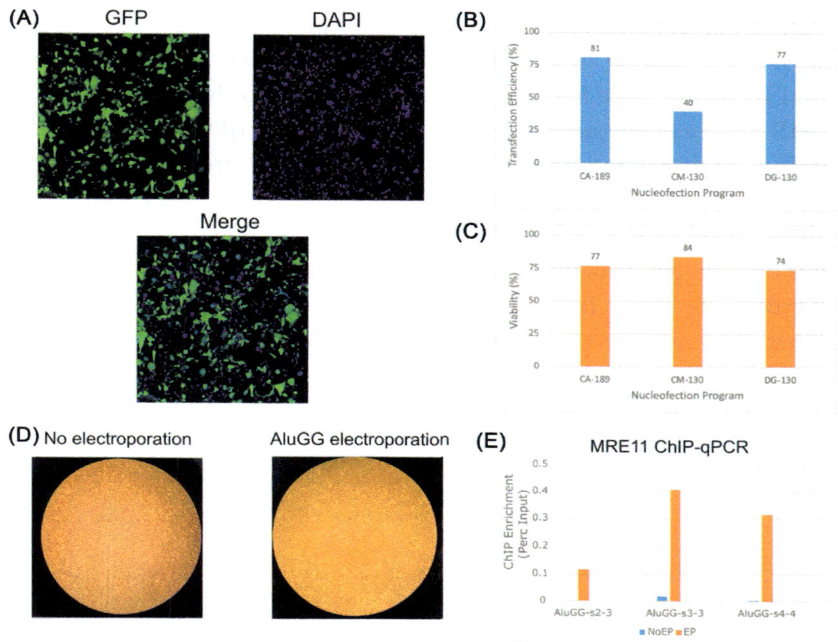

Fig. 4 Validation of nucleofections and multi-target CRISPR activity in HEK293T. (A) Microscope images after electroporation of HEK293T with GFP reporter plasmid. Images were collected in a widefield microscope, 24 h after electroporation. Quantification of transfection efficiency (B) and viability (C) after electroporation of HEK293T cells using different nucleofection programs (x-axis). Transfection efficiency was quantified from fluorescence images such as the ones shown in (A). Viability was measured via trypan blue staining. (D) ~200 k HEK293T cells were electroporated with AluGG multi-target CRISPR and then plated in a 6-well plate (right); ~200 k non-electroporated cells were plated in a separate 6-well dish as control (left). Cells were imaged under a 10x optical microscope 48 h after plating. (E) HEK293T cells were electroporated with AluGG multi-target CRISPR and were collected 3 h after electroporation, fixed with formaldehyde, and then quenched with glycine. ChIP was performed using an MRE11 antibody (novus NB100–142) following previously described protocols and putting aside 1% of chromatin prior to immunoprecipitation. qPCR was done using the NEB® qPCR master mix and the primers shown in Table 1. qPCR data was quantified using the percentage input method. A no-electroporation sample was used as a control to check for absence of MRE11 in undamaged conditions.

For mgRNAs with high enough number of targets (~50 or more), efficient Cas9 activity should cause cell cycle arrest (Fig. 4D). Tracking cell proliferation after multi-target CRISPR nucleofection is a straightforward approach to qualitatively validate Cas9 cleavage. As control, we use either a no-nucleofection sample, a mock nucleofection or a Cas9 complexed with a control, non-targeting gRNA.

Checking MRE11 recruitment via ChIP-qPCR is a more quantitative approach for testing Cas9 cleavage efficiency. 3 h after multi-target CRISPR delivery, we harvest cells, perform ChIP for MRE11, and check for MRE11 enrichment at three or four sites using qPCR. Note that for mgRNAs with a high number of cut sites (~50 or more), it is challenging to measure indel formation. This is because indels typically require >10 h to form (Brinkman et al., 2018; Liu et al., 2020), and at such long time points one can get substantial cell death as a consequence of a large number of DSBs. Therefore, for validating mgRNA activity with such a large number of targets, we recommend measuring MRE11 recruitment, which happens at shorter time-scales, rather than indel formation (Fig. 4E). For MRE11 ChIP-qPCR after AluGG mgRNA delivery, we typically use the qPCR primers shown in the Table 1.

We quantify using the percent input method, with the formula:

$$Perc\ Input = 100\ x\ 2^{(cq_{QC}+Log\,(0.01,2)-cq_{ChIP})}$$

where cq_{QC} is the quantification cycle of the input sample (taken as 1 % of chromatin prior to immunoprecipitation) and cq_{ChIP} is the quantification cycle of the ChIP sample.

3.2 Designing and validating new mgRNAs

Selecting an appropriate mgRNA is an important aspect of a multi-target CRISPR experiment. If possible, we recommend using one of the mgRNAs that we have already validated in either human or mouse cells (Table 2). These mgRNAs have around 100 target sites, a number that in our experience is large enough to enable high-throughput DSB studies without causing cell death in the short-term (~4 h). Also, these mgRNAs showed good cleavage efficiency and their target sites were well mapped by NGS as determined from MRE11 ChIP-Seq ((Zou et al., 2022) and data not shown).

If a different mgRNA from the validated mgRNA sequences above is required, the sequence will need to be properly tested. As initial characterizations, we recommend determining the effect of mgRNA activity on cell viability and measuring MRE11 recruitment to select on-target sites via ChIP-qPCR, as explained in Section 3.1.2. Enrichments on the order of ~0.2 % of input and higher are indicative that the mgRNA is working efficiently. In that case, one can proceed to prepare MRE11 ChIP-Seq libraries and sequence. To further validate the mgRNA, one should analyze the sequencing data to ensure the reads map well to the reference

Table 1 qPCR primers used to measure MR11 recruitment at AluGG mgRNA sites in Fig. 4E.

Primer pair label	Forward primer	Reverse primer
AluGG-s2-3	TGGTGTACACTCCTTCCAGA	TTTTCACATGCTTGCTTCACA
AluGG-s3-3	CAGTGGAGGAATTGGCCCA	CACTGTACCTGGCCCTTCC
AluGG-s4-4	TCTCTACTTCCCTATATGGCCAC	GTGCATGGCCACCTTTAGT

Table 2 Validated mgRNAs. * a caged version has been validated for AluGG: CC[NPOM-dT]G[NPOM-dT]AG[NPOM-dT]CCCAGCUACUGG.

mgRNA label	Protospacer sequence	Organism	# on-target sites
AluGG*	CCTGTAGTCCCAGCTACTGG	Human	126
AluCT	CCTGTAGTCCCAGCTACTCT	Human	145
AluTA	CCTGTAGTCCCAGCTACTTA	Human	117
mAlu106	CGCCTGTAATCCCAGCACTC	Mouse	106

genome. Because they are designed to target repetitive regions, some mgRNA might cut at regions that are too degenerate for repair events to be measured via NGS. Using 2×50 bp sequencing length and the AluGG mgRNA, we typically get ~90% or more of aligned reads with MAPQ > 25 for MRE11 ChIP-Seq. If the number of reads with high quality alignment is low, consider increasing read length to 2×75 bp, and if the problem persists, it might be preferred to discard the mgRNA altogether. If the MRE11 ChIP-Seq is successfully aligned to the reference genome, call MRE11 peaks using macs2 software (Zhang et al., 2008) and compute the average MRE11 peak around the best sites (see Section 4.2. for how to get average profiles when using mgRNAs). You should get profiles similar to the ones we reported in (Zou et al., 2022).

3.2.1 Designing and validating a caged mgRNA

To design a caged mgRNA for time-course measurements, we follow the same design principles as for mono-target (non-multi-target) caged gRNA explained in (Liu et al., 2020; Zou & Ha, 2021). In short, identify a gRNA that has 2 or 3 uracil nucleotides in the PAM-distal region (12 bp or more away from the PAM, counting from PAM). Avoid gRNA sequences where the uracil nucleotides are adjacent to each other. The candidate caged gRNA will contain NPOM-dT groups in the place of the PAM-distal uracils.

We validate this design of our caged gRNA in three steps. First, we generate the regular, non-caged gRNA and make sure it works efficiently (following the protocol in Section 3.2). Second, we generate a gRNA that mimics the caged gRNA, by placing adenines (rA) or guanines (rG) in place of the PAM-distal uracils in order to simulate the clash imposed by the NPOM groups. These mismatches introduced in the gRNA should prevent Cas9 cleavage. If such cleavage is prevented, we then order the caged gRNA containing the NPOM-dT groups in the place of the PAM-distal uracils. We then electroporate the Cas9-caged gRNA RNP as explained in Section 2.5. and test light-induced cleavage via MRE11 ChIP-qPCR. When doing this validation step, make sure to include a no-electroporation (no RNP delivery) as well as a no-light control (RNP delivery, but not UV light). If the caged gRNA is functioning properly, these two controls should yield no MRE11 signal, thus confirming that there is no leakage; that is, Cas9 has no activity without UV light.

4. Guidelines for sequencing and sequencing data analysis

4.1 Sequencing recommendations

A key feature of mgRNAs is that they can be combined with methods based on NGS, such as ChIP-Seq or ATAC-Seq, to measure DSB repair events. This is because Alu SINE elements are relatively short (280 bp), and the regions flanking the Alu SINE elements have high sequence diversity. So, we find that when using paired-end NGS, at least one of the reads typically falls in the region of high sequence diversity and can be unambiguously mapped to the reference genome. To ensure good quality mapping in NGS experiments, we strongly recommend doing paired-end sequencing with read length of at least 50 bp. However, for some mgRNAs, this length might need to be increased (see Section 3.2.1). We also recommend a minimum depth of 15 M paired-end reads per ChIP-Seq sample for narrow profiles (such as MRE11 or Cas9) and ~25 M paired-end reads for broad profiles (such as 53BP1 or γH2AX).

4.2 Sequencing data analysis

When combined with NGS assays such as ChIP-Seq for DNA repair proteins, multi-target CRISPR can yield rich data sets that inform on repair processes in high-throughput. Starting from fastq files, we align paired end reads to the reference GRCh38 genome using bowtie2 (Langmead & Salzberg, 2012), and then use samtools (Li et al., 2009) to filter by quality remove singletons and convert to bam file, remove PCR duplicates, sort and index. This can be done using the following commands:

```
Ref="[path to reference genome]"
File="[input fastq files, of the form: File_R1fastq.gz/
File_R2.fastq.gz]"
Gen="[label for reference genome, e.g. 'hg38']"

bowtie2 -p 4 -q --local -X 1000 -x $Ref -1 "$File"_R1.fastq.
gz -2 "$File"_R2.fastq.gz -S "$File"_"$Gen".sam
samtools view -h -S -b -F 0 × 08 -q 25 "$File"_"$Gen".sam >
"$File"_"$Gen"_unsorted.bam.
samtools fixmate -m "$File"_"$Gen"_unsorted.bam
"$File"_"$Gen"_fixmate.bam
samtools sort "$File"_"$Gen"_fixmate.bam > "$File"_"$Gen"_
sorted.bam
samtools markdup -r "$File"_"$Gen"_sorted.bam "$File"_
"$Gen"_final.bam
samtools index "$File"_"$Gen"_final.bam
samtools flagstat "$File"_"$Gen"_final.bam > "$File"_"
$Gen"_flagstats.txt
```

The resulting bam files can be converted into bigwig files using the deeptools software.

```
bamCompare -b1 [electroporation bam file] -b2 [no-electro-
poration bam file] --operation subtract -p max/2 -o [name for
output bigwig file]
```

We also use macs2 to call MRE11 peaks from the bam files, in order to obtain the best cleaved Cas9 sites.

```
macs2 callpeak -t [electroporation MRE11 bam file] -c [no-
electroporation MRE11 bam file] -outdir [path to output] --
name [name of output] -f BAMPE -g hs
```

As an initial analysis we typically compute average ChIP-Seq profiles for repair proteins, in order to check for enrichment of the protein of interest and measure the width of its spread. We then calculate average profiles around sites with the best cleavage efficiency (as determined, e.g. by MRE11 enrichment), or around all on-target sites. This can be done in an easy manner with deeptools, the input requirements being the bigWig files described above and a bed file with the regions to be analyzed. To obtain the bed file for the best cleaved sites, we extract the cut sites from the macs2 xls output file, then sort by fold enrichment, then select the best 100 sites according to fold enrichment and then output the list in bed format. This can be done with the command:

```
inpfile="[path to macs2 output xls file, to be used as input]"
outfile="[path to output file with sorted MRE11 sites]"
numpeak=[number of peaks, e.g. 100]
grep -vE "^#" $inpfile | awk 'BEGIN {OFS="\t"} NR> 2 {print
$1, $5, $5+1, $10, $7}' | sort -g -r -k5 | head -$numpeak>
$outfile
```

Alternatively, one might want to analyze the regions around on-target sites. If no MRE11 ChIP-seq profile is available for the cell line and mgRNA of interest, analyzing on-target sites might be the prefered option. In this article, the bed files containing the genomic locations of all on-target sites for the mgRNAs that are shown in Table 2 are placed in the section Appendix A. Supporting information. Alternatively, if other mgRNA is to be used, one can extract a bed file containing all on-target sites using the align mgRNA csv files generated by our original code (Zou et al., 2022) as input. The bed files with on-target sites can be obtained by running the command:

```
prot_seq="[protospacer sequence with PAM, e.g. CCTGTAGTC
CCAGCTACTGGNGG]"
align_file="[path to align file, e.g., align_files/psearch_
hg38_align.csv]"
out_file="[name of output file, e.g. AluGG_on-tgt.bed]"
grep "$prot_seq" $align_file | awk 'BEGIN {FS=","; OFS=
"\t"} {print $5, $6, $6+1}' > $out_file
```

We now have all the files required to plot the enrichment profiles. This can be done using the deeptools commands computeMatrix and plotProfile:

```
computeMatrix reference-point -S [path to bigWig] -R [path
to bed file] -a [window before cut] -b [window after cut]
-bs [binsize] -samplesLabel "[sample label]" -o [path to
matrix (output)]
plotProfile -m [path to matrix] -refPointLabel "Cas9 Cut
Site" -z "[Legend label]" -y "ChIP Enrichment" --color
[color] -o [path to output file]
```

We ran the code above on our previously published ChIP-Seq data on HEK293T for the MRE11, γH2AX, 53BP1 repair factors for non-electroporated samples and samples collected 3 h after AluGG electroporation. The results are shown in Fig. 5.

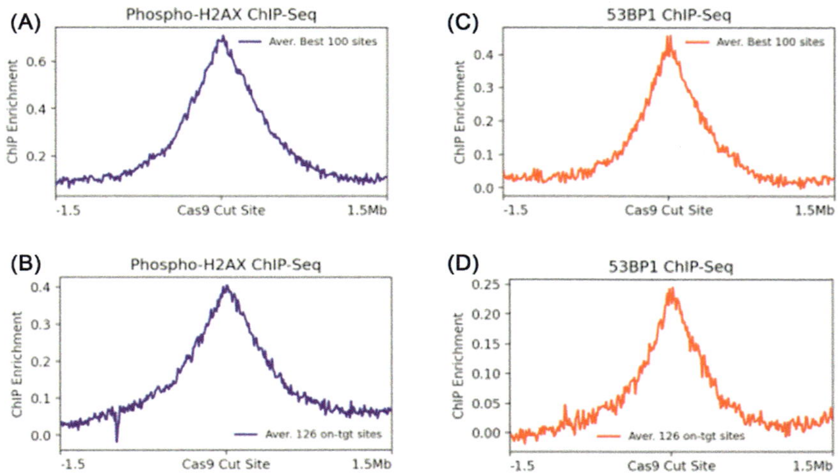

Fig. 5 Sequencing data analysis. γH2AX (A, B) and 53BP1 (C, D) ChIP-Seq profiles at 3 h after AluGG mgRNA delivery. Profiles were computed by averaging around the best 100cut sites (according to MRE11 enrichment) (A, C), or around all predicted Cas9 on-target sites (B, D) and substracting the corresponding averaged signal from a no-electroporation sample. Profiles were computed and plotted using deeptools (Ramírez, Dündar, Diehl, Grüning, & Manke, 2014). MRE11, γH2AX and 53BP1 ChIP-Seq datasets were obtained from (Zou et al., 2022) (accession codes SRR19744268, SRR19744267 and SRR19744256 for MRE11; SRR19744263 and SRR19744262 for γH2AX; and SRR19744202 and SRR19744266 for 53BP1).

5. Conclusions

We developed multi-target CRISPR as a tool to multiplex Cas9 activity at endogenous sites (Zou et al., 2022). In this chapter, we have described a protocol for multi-target CRISPR delivery and activation in HEK293T cells. We also explained general guidelines for multi-target CRISPR experiments, including how to apply this tool to different cell lines, how to select the proper mgRNA, and how to perform sequencing data analysis. The present protocol can be combined with different assays such as ChIP-Seq, IF, or sequencing of Cas9-induced mutations to interrogate different features related to Cas9 break repair. In summary, multi-target CRISPR allows us to introduce hundreds of DSB in mammalian cells in a synchronized manner. We expect that these features, together with its ease of use and availability, will make multi-target CRISPR a method of choice for DSB repair studies.

Acknowledgements

This work was supported by grants from the National Institutes of Health (R35 GM 122569 and U01 DK 127432 to T.H) and the National Science Foundation (EFMA 193303 to T.H.). A.M.-G. is a Howard Hughes Medical Institute (HHMI) Awardee of the Life Sciences Research Foundation. T.H. is an investigator of the HHMI.

Appendix A. Supporting information

Supplementary data associated with this article can be found in the online version at https://doi.org/10.1016/bs.mie.2025.01.011.

References

Allen, F., Crepaldi, L., Alsinet, C., Strong, A. J., Kleshchevnikov, V., De Angeli, P., ... Parts, L. (2019). Predicting the mutations generated by repair of Cas9-induced double-strand breaks. *Nature Biotechnology, 37*(1), 64–72. https://doi.org/10.1038/nbt.4317.

Arnould, C., Rocher, V., Finoux, A.-L., Clouaire, T., Li, K., Zhou, F., ... Legube, G. (2021). Loop extrusion as a mechanism for formation of DNA damage repair foci. *Nature, 590*(7847), 660–665. https://doi.org/10.1038/s41586-021-03193-z.

Bhargava, R., Onyango, D. O., & Stark, J. M. (2016). Regulation of single-strand annealing and its role in genome maintenance. *Trends in Genetics, 32*(9), 566–575. https://doi.org/10.1016/j.tig.2016.06.007.

Brinkman, E. K., Chen, T., de Haas, M., Holland, H. A., Akhtar, W., & van Steensel, B. (2018). Kinetics and fidelity of the repair of Cas9-induced double-strand DNA breaks. *Molecular Cell, 70*(5), 801–813.e806. https://doi.org/10.1016/j.molcel.2018.04.016.

Ciccia, A., & Elledge, S. J. (2010). The DNA damage response: Making it safe to play with knives. *Molecular Cell, 40*(2), 179–204. https://doi.org/10.1016/j.molcel.2010.09.019.

Deininger, P. (2011). Alu elements: Know the SINEs. *Genome Biology, 12*(12), 236. https://doi.org/10.1186/gb-2011-12-12-236.

Deshpande, R. A., Marin-Gonzalez, A., Barnes, H. K., Woolley, P. R., Ha, T., & Paull, T. T. (2023). Genome-wide analysis of DNA-PK-bound MRN cleavage products supports a sequential model of DSB repair pathway choice. *Nature Communications, 14*(1), 5759. https://doi.org/10.1038/s41467-023-41544-8.

Dias, M. P., Moser, S. C., Ganesan, S., & Jonkers, J. (2021). Understanding and overcoming resistance to PARP inhibitors in cancer therapy. *Nature Reviews Clinical Oncology, 18*(12), 773–791. https://doi.org/10.1038/s41571-021-00532-x.

Hussmann, J. A., Ling, J., Ravisankar, P., Yan, J., Cirincione, A., Xu, A., ... Adamson, B. (2021). Mapping the genetic landscape of DNA double-strand break repair. *Cell, 184*(22), 5653–5669.e5625. https://doi.org/10.1016/j.cell.2021.10.002.

Langmead, B., & Salzberg, S. L. (2012). Fast gapped-read alignment with Bowtie 2. *Nature Methods, 9*(4), 357–359. https://doi.org/10.1038/nmeth.1923.

Li, H., Handsaker, B., Wysoker, A., Fennell, T., Ruan, J., Homer, N., ... Genome Project Data Processing, S. (2009). The sequence alignment/map format and SAMtools. *Bioinformatics (Oxford, England), 25*(16), 2078–2079. https://doi.org/10.1093/bioinformatics/btp352.

Liu, Y., Zou, R. S., He, S., Nihongaki, Y., Li, X., Razavi, S., ... Ha, T. (2020). Very fast CRISPR on demand. *Science (New York, N. Y.), 368*(6496), 1265–1269. https://doi.org/10.1126/science.aay8204.

Ochs, F., Karemore, G., Miron, E., Brown, J., Sedlackova, H., Rask, M.-B., ... Lukas, C. (2019). Stabilization of chromatin topology safeguards genome integrity. *Nature, 574*(7779), 571–574. https://doi.org/10.1038/s41586-019-1659-4.

Ramírez, F., Dündar, F., Diehl, S., Grüning, B. A., & Manke, T. (2014). deepTools: A flexible platform for exploring deep-sequencing data. *Nucleic Acids Research, 42*(W1), W187–W191. https://doi.org/10.1093/nar/gku365.

Ramsden, D. A., Carvajal-Garcia, J., & Gupta, G. P. (2022). Mechanism, cellular functions and cancer roles of polymerase-theta-mediated DNA end joining. *Nature Reviews. Molecular Cell Biology, 23*(2), 125–140. https://doi.org/10.1038/s41580-021-00405-2.

Schumacher, B., Pothof, J., Vijg, J., & Hoeijmakers, J. H. J. (2021). The central role of DNA damage in the ageing process. *Nature, 592*(7856), 695–703. https://doi.org/10.1038/s41586-021-03307-7.

Scully, R., Panday, A., Elango, R., & Willis, N. A. (2019). DNA double-strand break repair-pathway choice in somatic mammalian cells. *Nature Reviews. Molecular Cell Biology, 20*(11), 698–714. https://doi.org/10.1038/s41580-019-0152-0.

Setton, J., Zinda, M., Riaz, N., Durocher, D., Zimmermann, M., Koehler, M., ... Powell, S. N. (2021). Synthetic lethality in cancer therapeutics: The next generation. *Cancer Discovery, 11*(7), 1626–1635. https://doi.org/10.1158/2159-8290.CD-20-1503.

Shen, M. W., Arbab, M., Hsu, J. Y., Worstell, D., Culbertson, S. J., Krabbe, O., ... Sherwood, R. I. (2018). Predictable and precise template-free CRISPR editing of pathogenic variants. *Nature, 563*(7733), 646–651. https://doi.org/10.1038/s41586-018-0686-x.

Vítor, A. C., Huertas, P., Legube, G., & de Almeida, S. F. (2020). Studying DNA double-strand break repair: An ever-growing toolbox [review]. *Frontiers in Molecular Biosciences,* 7(24), 1–16. https://doi.org/10.3389/fmolb.2020.00024.

Yeh, C. D., Richardson, C. D., & Corn, J. E. (2019). Advances in genome editing through control of DNA repair pathways. *Nature Cell Biology, 21*(12), 1468–1478. https://doi.org/10.1038/s41556-019-0425-z.

Yilmaz, D., Furst, A., Meaburn, K., Lezaja, A., Wen, Y., Altmeyer, M., ... Soutoglou, E. (2021). Activation of homologous recombination in G1 preserves centromeric integrity. *Nature, 600*(7890), 748–753. https://doi.org/10.1038/s41586-021-04200-z.

Zhang, Y., Liu, T., Meyer, C. A., Eeckhoute, J., Johnson, D. S., Bernstein, B. E., ... Liu, X. S. (2008). Model-based analysis of ChIP-Seq (MACS). *Genome Biology, 9*(9), R137. https://doi.org/10.1186/gb-2008-9-9-r137.

Zou, R. S., & Ha, T. (2021). Chapter Eleven – Light activation and deactivation of Cas9 for DNA repair studies. In B. F. Eichman (Vol. Ed.), *Methods in enzymology: Vol. 661*, (pp. 219–249). Academic Press. ⟨https://doi.org/10.1016/bs.mie.2021.08.007⟩.

Zou, R. S., Marin-Gonzalez, A., Liu, Y., Liu, H. B., Shen, L., Dveirin, R. K., ... Ha, T. (2022). Massively parallel genomic perturbations with multi-target CRISPR interrogates Cas9 activity and DNA repair at endogenous sites. *Nature Cell Biology, 24*(9), 1433–1444. https://doi.org/10.1038/s41556-022-00975-z.

CHAPTER TWO

General and robust sample preparation strategies for cryo-EM studies of CRISPR-Cas9 and Cas12 enzymes

Satoshi N. Omura and Osamu Nureki[*,1]
Department of Biological Sciences, Graduate School of Science, The University of Tokyo, Bunkyo-ku, Tokyo, Japan
*Corresponding author. e-mail address: nureki@bs.s.u-tokyo.ac.jp

Contents

1. Introduction 24
2. Cas protein expression and purification 26
   2.1 *E. coli* transformation 26
   2.2 Culture growth and induction 27
   2.3 Protein purification 27
3. Guide RNA preparation 29
   3.1 Template preparation for in vitro transcription of gRNA 29
   3.2 In vitro transcription 30
   3.3 Gel purification 30
4. Complex reconstruction and purification 31
   4.1 Target DNA design 31
   4.2 Complex reconstruction 31
   4.3 Complex purification 32
   4.4 Sample concentration 33
5. Grid preparation for cryo-EM 33
   5.1 Use of detergent 34
   5.2 Use of amylamine 35
6. Summary 35
Acknowledgments 36
Author contributions 36
Declaration of interests 36
References 36

[1] **Lead Contact**
Further information and requests for resources and reagents should be directed to and will be fulfilled by the Lead Contact.

Methods in Enzymology, Volume 712
ISSN 0076-6879, https://doi.org/10.1016/bs.mie.2025.01.052

Abstract

Cas9 and Cas12 are RNA-guided DNA endonucleases derived from prokaryotic CRISPR-Cas adaptive immune systems that have been repurposed as versatile genome-engineering tools. Computational mining of genomes and metagenomes has expanded the diversity of Cas9 and Cas12 enzymes that can be used to develop versatile, orthogonal molecular toolboxes. Structural information is pivotal to uncovering the precise molecular mechanisms of newly discovered Cas enzymes and providing a foundation for their application in genome editing. In this chapter, we describe detailed protocols for the preparation of Cas9 and Cas12 enzymes for cryo-electron microscopy. These methods will enable fast and robust structural determination of newly discovered Cas9 and Cas12 enzymes, which will enhance the understanding of diverse CRISPR-Cas effectors and provide a molecular framework for expanding CRISPR-based genome-editing technologies.

1. Introduction

CRISPR-Cas systems provide prokaryotes with adaptive immunity by inducing specific cleavage of foreign nucleic acids. These systems are currently divided into two classes (classes 1 and 2) and six types (types I–VI) depending on their ribonucleoprotein effector complex (Hille et al. 2018; Koonin, Gootenberg, & Abudayyeh, 2023). In class 1 systems, invading nucleic acids are degraded by an effector complex consisting of multiple Cas proteins and a CRISPR RNA (crRNA). In class 2 systems, comprising types II, V, and VI, the effector complex consists of a single multidomain Cas protein and a guide RNA (gRNA), which can be either a crRNA or a crRNA:trans-activating crRNA (tracrRNA) duplex. The type II Cas9 from *Streptococcus pyogenes* (SpCas9) associates with a cognate gRNA and cleaves double-stranded DNA (dsDNA) at sites that are complementary to a 20-nucleotide guide segment in the gRNA and flanked by an NGG (where N is any nucleotide) protospacer adjacent motif (PAM) (Gasiunas, Barrangou, Horvath, & Siksnys, 2012; Jinek et al. 2012). By contrast, Cas12a from *Acidaminococcus sp.* (AsCas12a), a member of the diverse type V effectors, binds to the gRNA and cleaves dsDNA at sites flanked by PAMs with the nucleotide sequence TTTV (where V is A, G, or C) (Zetsche et al. 2015). Cas9 contains two nuclease domains, HNH and RuvC, which cleave the target strand (TS) and non-target strand (NTS) in dsDNA targets, respectively, whereas Cas12a cleaves both the TS and the NTS with a single RuvC nuclease domain (Gasiunas et al. 2012; Jinek et al. 2012; Swarts and Jinek, 2019; Swarts, van der Oost, & Jinek, 2017).

Because of their simplicity of programming, versatility, and robust activity in mammalian cells, SpCas9 and AsCas12a are widely utilized as programmable genome-editing tools (Cong et al. 2013; Zetsche et al. 2015).

The application of Cas9 and Cas12a as programmable genome-editing tools has revolutionized many fields of biology, leading to advancements spanning from fundamental research to new gene therapy treatments (Doudna, 2020; Zhang, 2019). The various Cas enzymes are still limited in their applicability, however, because of several functional features. One notable limitation is the requirement for specific PAM sequences to enable binding to target DNA, which constrains the sequences that are accessible for editing. Additionally, both Cas9 and Cas12a exhibit some tolerance of mismatches between the gRNA and the target DNA, which can lead to unintended genomic alterations such as large deletions or insertions (Fu et al. 2013; Hsu et al. 2013; Kuscu, Arslan, Singh, Thorpe, & Adli, 2014; Pattanayak et al. 2013; Pacesa et al. 2022; Wu et al. 2014). Furthermore, neither SpCas9 (1368 amino acids) nor AsCas12a (1307 amino acids) can be efficiently packaged into a single adeno-associated virus (AAV) vector because of the large size of their genes, which hampers their clinical application for in vivo gene therapy (Wang, Zhang, & Gao, 2020).

Structural studies have illuminated many aspects of the molecular mechanisms of Cas9 and Cas12, providing a molecular framework to address the limitations of these enzymes. In 2014, the crystal structures of Cas9 were reported, providing detailed insights into the Cas9 gRNA-dependent DNA targeting and cleavage mechanism (Anders, Niewoehner, Duerst, & Jinek, 2014; Jinek et al. 2014; Nishimasu et al. 2014). By leveraging this structural information, we and others successfully engineered SpCas9 variants with altered PAM specificities, such as SpCas9-NG and SpG/SpRY, significantly broadening their target scope (Nishimasu et al. 2018; Walton, Christie, Whittaker, & Kleinstiver, 2020). Moreover, structure-guided rational engineering to mitigate off-target effects has yielded high-fidelity SpCas9 variants, based on the idea that the tolerance to mismatches in substrate DNA can be reduced by eliminating non-specific contacts between the Cas9 protein and the bound DNA target (Kleinstiver et al. 2016; Slaymaker et al. 2016; Chen et al. 2017; Vakulskas et al. 2018). In addition, naturally occurring small Cas enzymes show great promise as miniature genome-editing tools that can be packaged into a single AAV vector (Aliaga Goltsman et al. 2022; Bigelyte et al. 2021; Burstein et al. 2017; Edraki et al. 2019; Harrington et al. 2018; Kim et al. 2017; Ran et al. 2015; Wu et al. 2021; Chen et al. 2023). Recently, by combining deep mutational scanning and structure-guided design, we engineered a minimal Cas12f effector (422 amino acids) with enhanced in vivo activity, providing a promising genome-editing platform that might be applicable for gene therapy (Hino et al. 2023).

Ever-expanding genomic and metagenomic databases are uncovering a growing number and diversity of novel Cas9 and Cas12 enzymes, further broadening the landscape of potential alternatives for genome editing. The structural and biochemical characterization of naturally occurring CRISPR-Cas enzymes and their engineered variants remains pivotal for advancing next-generation gene-editing technologies. Rapid and reliable methodologies for the structural determination of Cas proteins are therefore critical. Recent advances in cryo-electron microscopy (cryo-EM) have enabled rapid structural determination without the need for protein crystallization. In this chapter, we provide detailed protocols for Cas9 and Cas12 heterologous expression and purification, gRNA preparation by in vitro transcription, effector complex reconstruction, and cryo-EM grid optimization.

2. Cas protein expression and purification

Cryo-EM single-particle analysis requires target proteins to be produced with high yield and high purity. To achieve this, we typically express Cas proteins using the pE-SUMO vector (Addgene #80749) in the *E. coli* strain Rosetta 2 DE3 (Fig. 1A). The expressed construct contains an N-terminal His_6-tag and a SUMO protease site, followed by a Cas protein. Inclusion of the SUMO protease site is optional; however, it enhances the expression of the downstream Cas protein. Expressed proteins are purified by affinity chromatography and ion exchange chromatography. The protocol is generally based on previously published procedures with minor modifications (Hino et al. 2023; Kurihara et al. 2022; Nakagawa et al. 2022; Omura et al. 2023; Takeda et al. 2021).

2.1 *E. coli* transformation

Gently mix 50 ng plasmid DNA with 10 μL freshly thawed, chemically competent Rosetta 2 DE3 cells and incubate on ice for 5 min. Induce heat shock by incubating the cells at 42 °C for 40 s, and immediately return the cells to ice for an additional 3 min. Add 100 μL Luria broth (LB) to the cells and incubate at 37 °C for 1 h. After incubation, spread 100 μL of the culture onto LB agar containing 30 μg/mL kanamycin and incubate overnight at 37 °C.

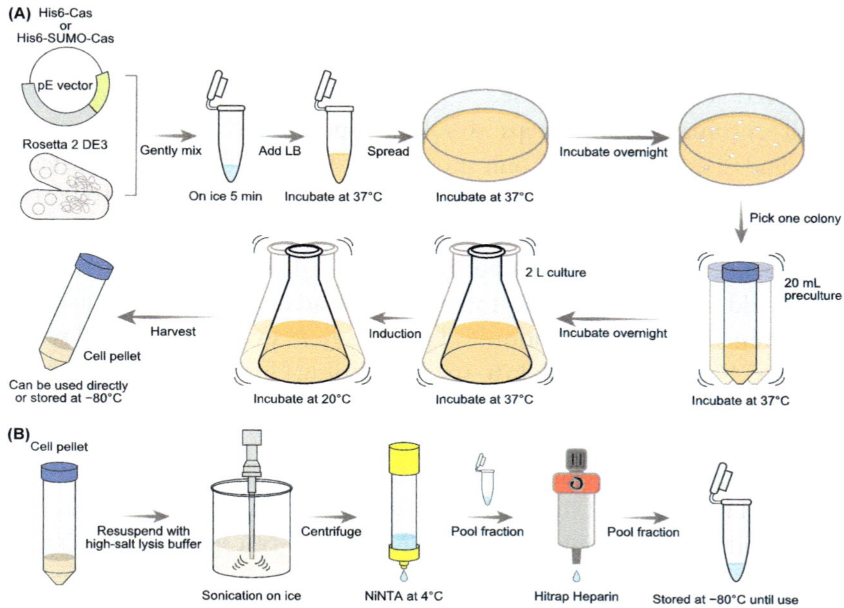

Fig. 1 Overview of Cas protein expression and purification. (A) Overview of recombinant Cas protein expression. (B) Overview of Cas protein purification.

2.2 Culture growth and induction

Pick a single colony from the LB agar to inoculate 20 mL LB containing 30 μg/mL kanamycin and 30 μg/mL chloramphenicol. Incubate the preculture at 37 °C in a shaking incubator (200 rpm) overnight. Transfer 20 mL of the preculture to 2 L prewarmed LB supplemented with 30 μg/mL kanamycin in a 5 L flask. Incubate the culture at 37 °C in a shaking incubator (120 rpm) until the optical density at 600 nm (OD600) reaches 0.6–0.8. Place the culture on ice for 15 min. Induce protein expression by adding 0.1 mM isopropyl β-D-thiogalactopyranoside (IPTG) and incubating at 20 °C for 20 h with shaking.

2.3 Protein purification

Harvest the cells by transferring a 1 L portion of the culture into a separate bottle and centrifuging at 5000 × *g* for 10 min on a fixed-angle rotor. Carefully decant the supernatant and resuspend the cell pellet in 20–40 mL ice-chilled lysis buffer (20 mM HEPES-NaOH, pH 7.6, 20 mM imidazole, and 1 M NaCl). Note that we typically use a lysis buffer with high salt concentration (1 M NaCl) to remove endogenous nucleic acids that bind non-specifically to the Cas proteins. The resuspended cell pellets can either be used directly for further purification or flash frozen in liquid nitrogen and stored at −80 °C until use.

All chromatographic steps should be performed at 4 °C. Lyse the resuspended cell pellets by sonication while keeping the sample on ice to avoid thermal denaturation. Clarify the lysate by centrifuging at 10,000 × *g* for 30 min at 4 °C in 50 mL Falcon tubes, then decant the supernatant. Equilibrate 3 mL Ni-NTA Superflow resin (QIAGEN) with 30 mL lysis buffer, add the cleared lysate to the equilibrated resin, and incubate for 30–60 min with gentle agitation to keep the resin evenly suspended within the lysate. Settle the resin by low-speed centrifugation at 1500 × *g* for 3 min and then carefully aspirate the supernatant. Load the resin into an Econo-Column (Bio-Rad) for protein elution.

After washing the column with 30 mL wash buffer (20 mM HEPES-NaOH, pH 7.6, 20 mM imidazole, and 300 mM NaCl), perform gravity flow elution with 12 mL elution buffer (20 mM HEPES-NaOH, pH 7.6, 300 mM imidazole, and 300 mM NaCl), collecting the eluate into five different fractions: E1 (1.5 mL), E2 (1.5 mL), E3 (3 mL), E4 (3 mL), and E5 (3 mL). Measure the absorbance of each fraction at 280 nm (A_{280}), using the elution buffer as the blank. Pool the fractions containing Cas protein with an A_{260}/A_{280} ratio between 0.5 and 0.6 (Cas9 and Cas12 typically elute at the E2 and E3 fractions). Attach a 5 mL HiTrap Heparin HP column (GE Healthcare) to an HPLC system and equilibrate the column with buffer A (20 mM HEPES-NaOH and 300 mM NaCl). Load the pooled protein-containing fractions onto the column using a syringe of appropriate volume. Wash the column with 10 mL buffer A and elute the bound protein by applying a linear gradient of 0.3–2 M NaCl. Collect the eluate into a 96-well deep-well plate, distributing 500 μL per well. Analyze all peak fractions for the presence of Cas protein using SDS-PAGE. Pool the fractions containing Cas protein with an A_{260}/A_{280} ratio between 0.5 and 0.6. The pooled sample can be flash frozen in liquid nitrogen and stored at −80 °C until use.

2.3.1 Note 1

We do not remove the His_6 tag sequence from the purified Cas protein because protease treatment is time consuming and often promotes protein aggregation. The structurally flexible His6 tag is rarely problematic for cryo-EM single-particle analysis; however, if it is observed to impair the efficiency of in vitro complex reconstruction, it is advisable to use proteases like SUMO or TEV to remove it.

2.3.2 Note 2

If impurities are not adequately eliminated by ion exchange chromatography, gel filtration chromatography can be employed as an additional purification step. For this step, we typically use a Hiload 16/600 Superdex 200 pg column (GE Healthcare) equilibrated with buffer B (20 mM HEPES-NaOH and 500 mM NaCl).

3. Guide RNA preparation

The gRNA can be readily synthesized by in vitro transcription (IVT) using T7 RNA polymerase. The IVT template requires a double-stranded T7 promoter followed by a single- or double-stranded template sequence (Fig. 2). The T7 promoter should include two G nucleotides at the 3′ end, which are necessary for efficient transcription and will be incorporated at the 5′ end of the transcribed RNA. It is important to note that the gRNA polarity differs between Cas9 and Cas12: Cas9 gRNA has a 5′-spacer-repeat-3′ arrangement, whereas Cas12 gRNA has a 5′-repeat-spacer-3′ arrangement (Fig. 2). Consequently, the gRNA for Cas9 will contain a GG dinucleotide at the 5′ end of the spacer region, and the Cas12 gRNA will contain a GG dinucleotide at the 5′ end of the repeat region. The addition of the 5′ GG dinucleotide to the gRNA has little effect on the enzymatic activity of Cas9 and Cas12.

3.1 Template preparation for in vitro transcription of gRNA

We typically prepare a partially double-stranded DNA template, which consists of a forward strand containing the T7 promoter followed by a GG dinucleotide, and a reverse strand containing the full-length template

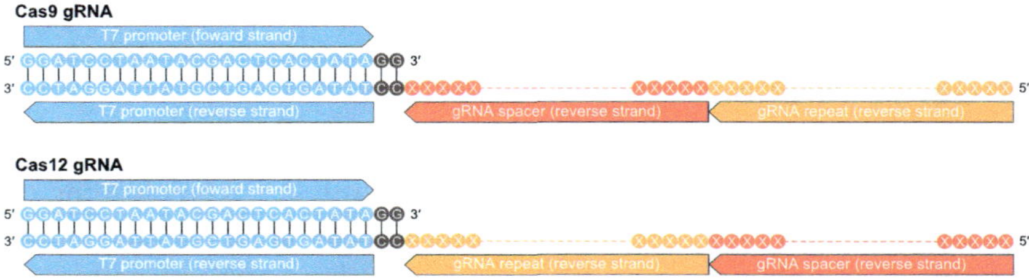

Fig. 2 Schematic of the gRNA template for in vitro transcription. Sequences corresponding to the T7 promoter, gRNA spacer region, and gRNA repeat region are represented as light blue, red, and orange, respectively.

sequence along with the T7 promoter (Fig. 2). Prior to IVT, dilute 0.1 nmol of the forward and reverse strands of DNA in 150 μL water, then anneal the strands by heating the solution at 95 °C for 3 min and cooling gradually to room temperature. The annealed template can be used directly for subsequent IVT.

A transcription template can also be prepared by PCR amplification of a single-stranded antisense template oligonucleotide. Forward primers are designed against the T7 promoter sequence, and reverse primers are designed against the 3′ end of the antisense template. For sufficient yield of the IVT template, perform the PCR at a scale of at least 50 μL with 35 cycles of amplification. The PCR mixture can be used directly for subsequent IVT; however, better yields will be obtained with purified PCR products.

3.2 In vitro transcription

All buffers should be sterilized by passing through a 0.22 μm polyethersulfone filter and should be prewarmed at 37 °C before use. To generate sufficient gRNA for cryo-EM analysis, IVT reactions can be scaled to 2.5 mL, yielding enough gRNA for approximately 2–5 rounds of cryo-EM screening.

For IVT, first prepare a 10 × IVT buffer solution (500 mM Tris-HCl, pH8.0, 400 mM KCl, 200 mM $MgCl_2$, 50 mM DTT, and 20 mM spermidine) and a 10 × NTP mix solution (50 mM NTP each). Set up a transcription reaction by mixing 250 μL 10 × IVT buffer, 250 μL 10 × NTP mix, 2.5 μL 40 U/μL RNase inhibitor, 5 μL 0.5 U/μL inorganic pyrophosphatase (Sigma-Aldrich), 40 mg/mL T7 RNA polymerase, and the prepared DNA template. Scale the total volume up to 2.5 mL with prewarmed water and incubate the reaction mixture at 37 °C for 2–5 h. Note that the use of inorganic pyrophosphatase is crucial to avoid magnesium pyrophosphate precipitation due to the release of inorganic pyrophosphate during NTP incorporation into the RNA transcripts.

3.3 Gel purification

The transcribed gRNA can be purified by denaturing urea polyacrylamide gel electrophoresis (Urea-PAGE). First, add 2.5 mL 2 × RNA loading dye (5% glycerol, 7 M urea, and trace bromophenol blue) to the sample. Perform electrophoresis on an 8% denaturing polyacrylamide gel (7 M urea) in 1 × Trix-Borate-EDTA (TBE) buffer (89 mM Tris-Borate, pH 8.3, and 2 mM EDTA). Stop the electrophoresis when the bromophenol blue tracking dye reaches the lower quarter of the gel, then visualize the RNA by ultraviolet shadowing. Carefully excise the band corresponding to the correct RNA product using a disposable scalpel. Transfer the excised gel slice to a 50 mL Falcon tube, add 10 mL 1 × TBE buffer, and rotate at 4 °C overnight to elute the RNA transcripts into the buffer.

After elution, centrifuge the gel suspension at 5000 × *g* for 5 min, then filter the supernatant through a 0.22 μm polyethersulfone filter. Add 1 mL 3 M sodium acetate (pH 5.2) to the filtered solution and mix thoroughly by tapping. Then, add 25 mL of 100% ethanol and mix well. Centrifuge the mixture at 10,000 × *g* for 30 min. Remove the supernatant, and wash the RNA pellet with 15 mL 70% ethanol. Air-dry the pellet and dissolve it in 400 μL water.

4. Complex reconstruction and purification

Structural analysis of Cas enzymes is typically performed with the enzyme in a ternary complex with gRNA and target DNA, providing mechanistic insights into the DNA binding and cleavage mediated by the Cas protein. The Cas9/Cas12–gRNA–DNA ternary complex can be reconstituted in vitro by incubating purified Cas protein with gRNA and a pair of complementary single-stranded DNA oligos in an appropriate buffer. After in vitro complex reconstruction, the ternary complex can be purified by size-exclusion chromatography.

4.1 Target DNA design

Target dsDNA can be prepared by annealing complementary oligos consisting of a PAM sequence and a sequence that is complementary to the gRNA. To ensure the NTS remains paired with the TS during Cas-mediated DNA unwinding, add 10 extra nucleotides of arbitrary complementary sequence at both termini (Fig. 3A). In addition, to prevent target DNA cleavage during the complex-reconstruction and purification processes, we use target DNA with phosphorothioate modifications (S-modification) around the cleavage site, rather than using a catalytically inactive variant of the Cas protein (Fig. 3A). The efficiency of the complex reconstruction process is enhanced by converting the guide region of the NTS into a poly-T sequence, which promotes efficient unwinding by the Cas protein (Fig. 3A).

4.2 Complex reconstruction

Anneal the gRNA by heating at 95 °C for 3 min and then cooling gradually to room temperature (Fig. 3B). In parallel, anneal the complementary

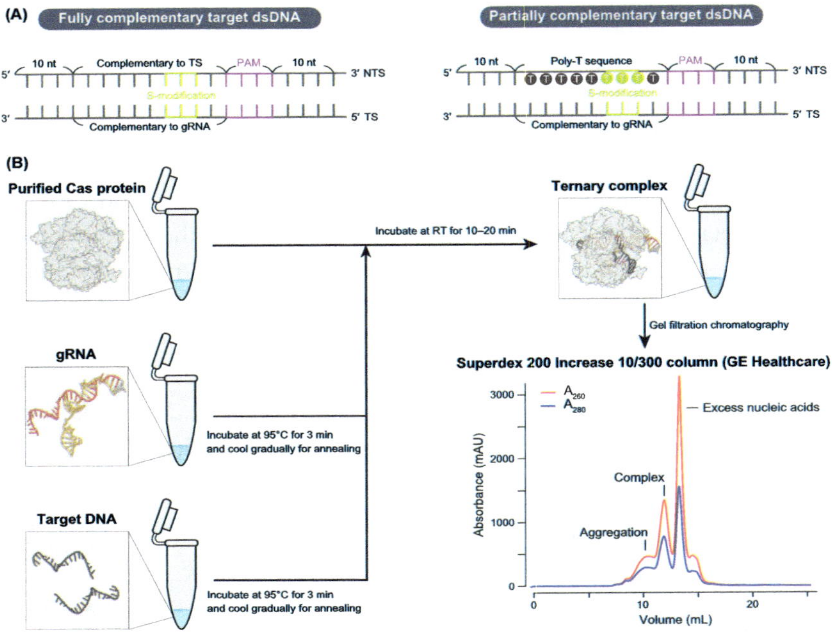

Fig. 3 Overview of in vitro complex reconstruction. (A) Target DNA design for in vitro complex reconstruction. The efficiency of complex reconstruction can be improved by replacing the guide region of the NTS with a poly-T sequence. (B) In vitro reconstruction of the Cas9/Cas12–gRNA–DNA ternary complex.

single-stranded DNA oligos by heating at 95 °C for 3 min and then cooling gradually to room temperature to form the target dsDNA (Fig. 3B). Mix 0.3 mg Cas protein with the annealed gRNA and the target dsDNA at a molar ratio of 1:1.5:2 in SEC buffer (20 mM HEPES-NaOH, pH7.6, 50 mM NaCl, 10 mM $MgCl_2$, 10 μM $ZnCl_2$, and 1 mM DTT), and incubate the mixture for 10–20 min at room temperature (Fig. 3B). Concentrate the assembled complex into a 400 μL volume using a 50,000 MWCO centrifugal concentrator.

4.3 Complex purification

Attach a Superdex 200 Increase 10/300 column (GE Healthcare) to an HPLC system and equilibrate the column with SEC buffer. Load the concentrated complex onto the column using a syringe of appropriate volume. Elute with 26 mL SEC buffer at a flow rate of 0.5 mL min^{-1}.

Collect the eluate into a 96-well deep-well plate, distributing 250 μL per well. The Cas9/Cas12–gRNA–DNA ternary complex typically elutes at a volume between 11 mL and 13 mL (Fig. 3B). Analyze the peak fractions using both SDS-PAGE and Urea-PAGE, and pool the fractions that contain Cas protein, gRNA, and target DNA.

4.4 Sample concentration

Instead of the protein concentration, we typically quantify the Cas9/Cas12–gRNA–DNA ternary complex concentration by measuring the absorbance at 260 nm. Concentrate the complex using a 50,000 MWCO centrifugal concentrator until the A_{260} reaches between 3 and 10. The concentrated sample can either be used directly for cryo-EM grid preparation or flash frozen in liquid nitrogen and stored at −80 °C until use.

5. Grid preparation for cryo-EM

The purified and concentrated sample is applied to a glow-discharged grid, blotted with filter paper to create a thin aqueous film, and plunge-frozen in liquid ethane—a process known as vitrification (Fig. 4). To achieve a high-resolution cryo-EM map, it is necessary to carefully optimize the vitrification conditions, including the choice of grid type, humidity, blotting force, and blotting time. For the Cas9/Cas12–gRNA–DNA ternary complex, we typically use Quantifoil R1.2/1.3 holey carbon on Au 300 mesh base grids that are glow-discharged with a plasma treatment device at 10 mA for 90 s. A 3 μL sample is then applied to the glow-discharged grid using a Vitrobot Mark IV vitrification system (FEI) at 4 °C, with a blotting force of 10, a blotting time of 4 s, and 100% humidity. Various preparation conditions, including glow-discharge settings, blotting forces, and blotting times, should be screened to achieve the optimal cryo-EM grid, with data collection focused on finding the best-performing grids. An ideal cryo-EM grid is characterized by particles that are evenly and richly distributed in a variety of orientations throughout the holes of the support film.

In our experience, particle orientation bias can present a major challenge in grid optimization for Cas proteins. Orientation bias is a phenomenon where particles adopt a limited range of unique views, which is not suitable for three-dimensional reconstruction (D'Imprima et al. 2019; Tan et al. 2017). This problem is mainly caused either by particle adsorption to the hydrophobic air-water interface or by the overall negative charge of the

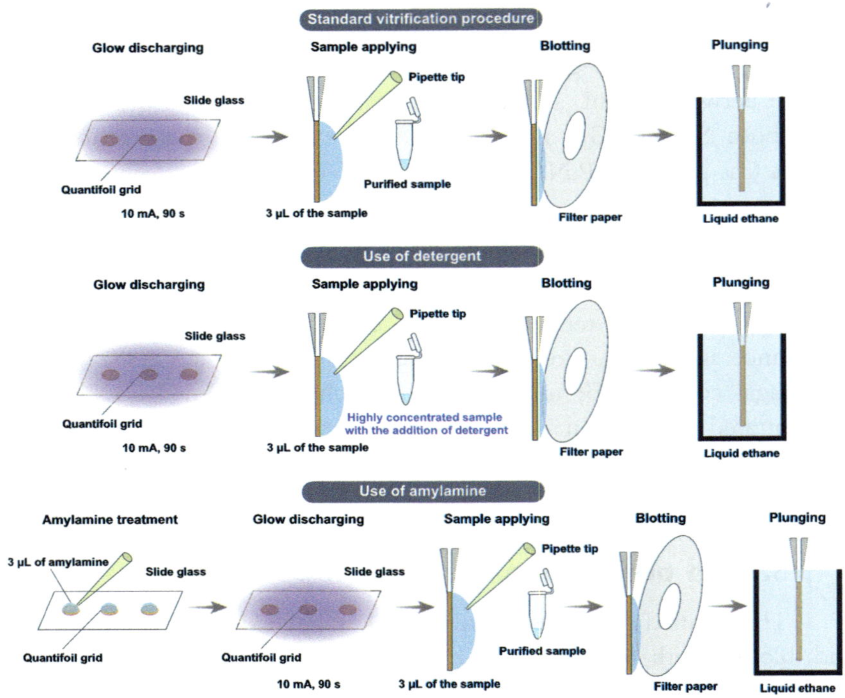

Fig. 4 Cryo-EM grid optimization strategies. Vitrification in the standard case, when using detergent, and when using amylamine.

glow-discharged grid surface, causing the proteins to align in a fixed orientation. Several strategies have been developed to overcome orientation bias, including addition of a mild detergent to the sample solution (Chen, Noble, Kang, & Darst, 2019), application of amylamine to the grid prior to glow discharge (Passmore and Russo, 2016; Fonseca and Morris, 2015), utilization of a solid support layer over the grid holes (Han et al. 2020; Naydenova, Peet, & Russo, 2019; Palovcak et al. 2018; Wang, Yanxin Liu, Sam Li, Yifan Cheng, & Agard, 2020), and data collection using a tilted stage (Tan et al. 2017). For Cas9 and Cas12, we recommend the use of detergent and amylamine to mitigate the orientation bias.

5.1 Use of detergent

The addition of a small amount of detergent can broaden the distribution of particle orientations by weakening the surface tension of the solution, thereby minimizing particle adsorption at the air-water interface. We typically use the zwitterionic detergent CHAPSO or the non-ionic detergent TWEEN 20 by mixing them with the sample solution at a final concentration of 0.25% or 0.005%, respectively. It is crucial to note that when using detergents, the sample concentration should be 5–10 times higher than the sample concentration used under detergent-free conditions (Fig. 4). When using detergents, we typically concentrate the complex using a 50,000 MWCO centrifugal concentrator until the A_{260} reaches between 30 and 60.

5.2 Use of amylamine

Another approach we frequently employ to address orientation bias is the application of amylamine on the grid. Prior to vitrification, the grid surface must be made hydrophilic to ensure even distribution of the sample. This is typically achieved by glow discharge, which makes the grid surface hydrophilic. When ambient air is used in glow discharge, the reactive species—primarily oxygen and water—give the grid an overall negative charge. This negatively charged surface can adsorb particles in a fixed orientation, leading to orientation bias. To address this, amylamine can be applied to the grid before glow discharge, giving the carbon surface a net positive charge, thereby reversing the electrostatic effects of preferred carbon-particle interactions to repel particles into the holes. We typically apply 3 μL amylamine to a grid placed on a glass slide and then glow discharge the grid with a plasma treatment device at 10 mA for 90 s (Fig. 4). An advantage of the use of amylamine is the lower particle concentration requirement compared with the use of detergent. We typically concentrate the complex using a 50,000 MWCO centrifugal concentrator until the A_{260} reaches between 3 and 10. It is important to note that amylamine treatment may cause denser ice formation in certain areas due to incomplete glow discharge. Therefore, it is advisable to screen grids prepared under different glow-discharge conditions to obtain the best quality grid.

6. Summary

Since the identification of SpCas9 in 2012, numerous functionally divergent CRISPR-Cas enzymes have been discovered. To date, four different subtypes of Cas9 (types II-A–D) and 17 different subtypes of Cas12 (types V-A–P and phage-encoded CRISPR-Casλ) have been functionally characterized, and the search for novel CRISPR-Cas systems is still ongoing. Single-particle analysis has greatly accelerated the structural elucidation of Cas9 and Cas12 and will continue to play a central role in the characterization of newly discovered Cas enzymes, providing a molecular basis for understanding their RNA-guided DNA cleavage mechanisms and a foundation for further engineering efforts. In this chapter, we provided detailed procedures to purify recombinant Cas proteins and gRNAs, reconstruct ternary complexes, and prepare cryo-EM grids. These protocols establish a general and robust experimental framework for the structural determination of Cas9 and Cas12 variants.

Acknowledgments

We would like to thank all past and present members of the Nureki laboratory who have contributed to the structural analysis of CRISPR-Cas. O.N. was supported by AMED Grant Numbers JP23fa627001 and JP19am0401005, the Platform Project for Supporting Drug Discovery and Life Science Research (Basis for Supporting Innovative Drug Discovery and Life Science Research (BINDS)) from AMED, under grant numbers JP23ama121002 (support number 3272, M.K.) and JP23ama121012 (support no. 4894, O.N.), and the Cabinet Office, Government of Japan, Public/Private R&D Investment Strategic Expansion Program (PRISM) Grant Number JPJ008000.

Author contributions

S.N.O. and O.N. wrote the manuscript.

Declaration of interests

O.N. is a co-founder, board member, and scientific advisor of Curreio.

References

Aliaga Goltsman, D. S., Alexander, L. M., Lin, J.-L., Fregoso Ocampo, R., Freeman, B., Lamothe, R. C., Perez Rivas, A., et al. (2022). Compact Cas9d and HEARO enzymes for genome editing discovered from uncultivated microbes. *Nature Communications, 13*(1), 7602.

Anders, C., Niewoehner, O., Duerst, A., & Jinek, M. (2014). Structural basis of PAM-dependent target DNA recognition by the Cas9 endonuclease. *Nature, 513*(7519), 569–573.

Bigelyte, G., Young, J. K., Karvelis, T., Budre, K., Zedaveinyte, R., Djukanovic, V., Van Ginkel, E., et al. (2021). Miniature type V-F CRISPR-Cas nucleases enable targeted DNA modification in cells. *Nature Communications, 12*(1), 6191.

Burstein, D., Harrington, L. B., Strutt, S. C., Probst, A. J., Anantharaman, K., Thomas, B. C., ... Banfield, J. F. (2017). New CRISPR-Cas systems from uncultivated microbes. *Nature, 542*(7640), 237–241.

Chen, J., Noble, A. J., Kang, J. Y., & Darst, S. A. (2019). Eliminating effects of particle adsorption to the air/water interface in single-particle cryo-electron microscopy: Bacterial RNA polymerase and CHAPSO (February) *Journal of Structural Biology: X, 1*. https://doi.org/10.1016/j.yjsbx.2019.100005.

Chen, J. S., Dagdas, Y. S., Kleinstiver, B. P., Welch, M. M., Sousa, A. A., Harrington, L. B., ... Doudna, J. A. (2017). Enhanced proofreading governs CRISPR-Cas9 targeting accuracy. *Nature, 550*(7676), 407–410.

Chen, W., Ma, J., Wu, Z., Wang, Z., Zhang, H., Fu, W., ... Ji, Q. (2023). Cas12n nucleases, early evolutionary intermediates of type V CRISPR, comprise a distinct family of miniature genome editors. *Molecular Cell*. https://doi.org/10.1016/j.molcel.2023.06.014.

Cong, L., Ran, F. A., Cox, D., Lin, S., Barretto, R., Habib, N., Hsu, P. D., et al. (2013). Multiplex genome engineering using CRISPR/Cas systems. *Science (New York, N. Y.), 339*(6121), 819–823.

D'Imprima, E., Floris, D., Joppe, M., Sánchez, R., Grininger, M., & Kühlbrandt, W. (2019). Protein denaturation at the air-water interface and how to prevent It. *ELife, 8*. https://doi.org/10.7554/eLife.42747.

Doudna, J. A. (2020). The promise and challenge of therapeutic genome editing. *Nature, 578*(7794), 229–236.

Edraki, A., Mir, A., Ibraheim, R., Gainetdinov, I., Yoon, Y., Song, C.-Q., Cao, Y., et al. (2019). A compact, high-accuracy Cas9 with a dinucleotide PAM for in vivo genome editing. *Molecular Cell, 73*(4), 714–726.e4.

Fonseca, P. C. A. D., & Morris, E. P. (2015). Cryo-EM reveals the conformation of a substrate analogue in the human 20S proteasome core. *Nature Communications, 6*(1), 7573.

Fu, Y., Foden, J. A., Khayter, C., Maeder, M. L., Reyon, D., Joung, J. K., & Sander, J. D. (2013). High-frequency off-target mutagenesis induced by CRISPR-Cas nucleases in human cells. *Nature Biotechnology, 31*(9), 822–826.

Gasiunas, G., Barrangou, R., Horvath, P., & Siksnys, V. (2012). Cas9-CrRNA ribonucleoprotein complex mediates specific DNA cleavage for adaptive immunity in bacteria. *Proceedings of the National Academy of Sciences of the United States of America, 109*(39), E2579–E2586.

Han, Y., Fan, X., Wang, H., Zhao, F., Tully, C. G., Kong, J., ... Yan, N. (2020). High-yield monolayer graphene grids for near-atomic resolution cryoelectron microscopy. *Proceedings of the National Academy of Sciences of the United States of America, 117*(2), 1009–1014.

Harrington, L. B., Burstein, D., Chen, J. S., Paez-Espino, D., Ma, E., Witte, I. P., ... Doudna, J. A. (2018). Programmed DNA destruction by miniature CRISPR-Cas14 enzymes. *Science (New York, N. Y.), 362*(6416), 839–842.

Hille, F., Richter, H., Wong, S. P., Bratovič, M., Ressel, S., & Charpentier, E. (2018). The biology of CRISPR-Cas: Backward and forward. *Cell, 172*(6), 1239–1259.

Hino, T., Omura, S. N., Nakagawa, R., Togashi, T., Takeda, S. N., Hiramoto, T., Tasaka, S., et al. (2023). An AsCas12f-based compact genome-editing tool derived by deep mutational scanning and structural analysis. *Cell, 0*(0), https://doi.org/10.1016/j.cell.2023.08.031.

Hsu, P. D., Scott, D. A., Weinstein, J. A., Ran, F. A., Konermann, S., Agarwala, V., Li, Y., et al. (2013). DNA targeting specificity of RNA-guided Cas9 nucleases. *Nature Biotechnology, 31*(9), 827–832.

Jinek, M., Chylinski, K., Fonfara, I., Hauer, M., Doudna, J. A., & Charpentier, E. (2012). A programmable dual-RNA-guided DNA endonuclease in adaptive bacterial immunity. *Science (New York, N. Y.), 337*(6096), 816–821.

Jinek, M., Jiang, F., Taylor, D. W., Sternberg, S. H., Kaya, E., Ma, E., Anders, C., et al. (2014). Structures of Cas9 endonucleases reveal RNA-mediated conformational activation. *Science (New York, N. Y.), 343*(6176), 1247997.

Kim, E., Koo, T., Park, S. W., Kim, D., Kim, K., Cho, H.-Y., Song, D. W., et al. (2017). In vivo genome editing with a small Cas9 orthologue derived from Campylobacter jejuni. *Nature Communications, 8*, 14500.

Kleinstiver, B. P., Pattanayak, V., Prew, M. S., Tsai, S. Q., Nguyen, N. T., Zheng, Z., & Joung, J. K. (2016). High-fidelity CRISPR-Cas9 nucleases with no detectable genome-wide off-target effects. *Nature, 529*(7587), 490–495.

Koonin, E. V., Gootenberg, J. S., & Abudayyeh, O. O. (2023). Discovery of diverse CRISPR-Cas systems and expansion of the genome engineering toolbox. *Biochemistry*. https://doi.org/10.1021/acs.biochem.3c00159.

Kurihara, N., Nakagawa, R., Hirano, H., Okazaki, S., Tomita, A., Kobayashi, K., Kusakizako, T., et al. (2022). Structure of the type V-C CRISPR-Cas effector enzyme. *Molecular Cell*. https://doi.org/10.1016/j.molcel.2022.03.006.

Kuscu, C., Arslan, S., Singh, R., Thorpe, J., & Adli, M. (2014). Genome-wide analysis reveals characteristics of off-target sites bound by the cas9 endonuclease. *Nature Biotechnology, 32*(7), 677–683.

Nakagawa, R., Ishiguro, S., Okazaki, S., Mori, H., Tanaka, M., Aburatani, H., ... Nureki, O. (2022). Engineered Campylobacter jejuni Cas9 variant with enhanced activity and broader targeting range. *Communications Biology, 5*(1), 211.

Naydenova, K., Peet, M. J., & Russo, C. J. (2019). Multifunctional graphene supports for electron cryomicroscopy. *Proceedings of the National Academy of Sciences of the United States of America, 116*(24), 11718–11724.

Nishimasu, H., Ran, F. A., Hsu, P. D., Konermann, S., Shehata, S. I., Dohmae, N., ... Nureki, O. (2014). Crystal structure of Cas9 in complex with guide RNA and target DNA. *Cell, 156*(5), 935–949.

Nishimasu, H., Shi, X., Ishiguro, S., Gao, L., Hirano, S., Okazaki, S., Noda, T., et al. (2018). Engineered CRISPR-Cas9 nuclease with expanded targeting space. *Science (New York, N. Y.), 361*(6408), 1259–1262.

Omura, S. N., Nakagawa, R., Südfeld, C., Villegas Warren, R., Wu, W. Y., Hirano, H., Laffeber, C., et al. (2023). Mechanistic and evolutionary insights into a type V-M CRISPR-Cas effector enzyme. *Nature Structural & Molecular Biology*. https://doi.org/10.1038/s41594-023-01042-3.

Pacesa, M., Lin, C.-H., Cléry, A., Saha, A., Arantes, P. R., Bargsten, K., Irby, M. J., et al. (2022). Structural basis for Cas9 off-target activity. *Cell, 185*(22), 4067–4081.e21.

Palovcak, E., Wang, F., Zheng, S. Q., Yu, Z., Li, S., Betegon, M., ... Cheng, Y. (2018). A simple and robust procedure for preparing graphene-oxide cryo-EM grids. *Journal of Structural Biology, 204*(1), 80–84.

Passmore, L. A., & Russo, C. J. (2016). Specimen preparation for high-resolution cryo-EM. *Methods in Enzymology, 579*, 51–86 (June).

Pattanayak, V., Lin, S., Guilinger, J. P., Ma, E., Doudna, J. A., & Liu, D. R. (2013). High-throughput profiling of off-target DNA cleavage reveals RNA-programmed Cas9 nuclease specificity. *Nature Biotechnology, 31*(9), 839–843.

Ran, F. A., Cong, L., Yan, W. X., Scott, D. A., Gootenberg, J. S., Kriz, A. J., Zetsche, B., et al. (2015). In vivo genome editing using Staphylococcus aureus Cas9. *Nature, 520*(7546), 186–191.

Slaymaker, I. M., Gao, L., Zetsche, B., Scott, D. A., Yan, W. X., & Zhang, F. (2016). Rationally engineered Cas9 nucleases with improved specificity. *Science (New York, N. Y.), 351*(6268), 84–88.

Swarts, D. C., & Jinek, M. (2019). Mechanistic insights into the cis- and trans-acting DNase activities of Cas12a. *Molecular Cell, 73*(3), 589–600.e4.

Swarts, D. C., van der Oost, J., & Jinek, M. (2017). Structural basis for guide RNA processing and seed-dependent DNA targeting by CRISPR-Cas12a. *Molecular Cell, 66*(2), 221–233.e4.

Takeda, S. N., Nakagawa, R., Okazaki, S., Hirano, H., Kobayashi, K., Kusakizako, T., ... Nureki, O. (2021). Structure of the miniature type V-F CRISPR-Cas effector enzyme. *Molecular Cell, 81*(3), 558–570.e3.

Tan, Y. Z., Baldwin, P. R., Davis, J. H., Williamson, J. R., Potter, C. S., Carragher, B., & Lyumkis, D. (2017). Addressing preferred specimen orientation in single-particle Cryo-EM through tilting. *Nature Methods, 14*(8), 793–796.

Vakulskas, C. A., Dever, D. P., Rettig, G. R., Turk, R., Jacobi, A. M., Collingwood, M. A., Bode, N. M., et al. (2018). A high-fidelity Cas9 Mutant delivered as a ribonucleoprotein complex enables efficient gene editing in human hematopoietic stem and progenitor cells. *Nature Medicine, 24*(8), 1216–1224.

Walton, R. T., Christie, K. A., Whittaker, M. N., & Kleinstiver, B. P. (2020). Unconstrained genome targeting with near-PAMless engineered CRISPR-Cas9 variants. *Science (New York, N. Y.), 368*(6488), 290–296.

Wang, D., Zhang, F., & Gao, G. (2020). CRISPR-based therapeutic genome editing: Strategies and in vivo delivery by AAV vectors. *Cell, 181*(1), 136–150.

Wang, F., Yanxin Liu, Z. Y., Sam Li, S. F., Yifan Cheng, & Agard, D. A. (2020). General and robust covalently linked graphene oxide affinity grids for high-resolution Cryo-EM. *Proceedings of the National Academy of Sciences of the United States of America, 117*(39), 24269–24273.

Wu, X., Scott, D. A., Kriz, A. J., Chiu, A. C., Hsu, P. D., Dadon, D. B., Cheng, A. W., et al. (2014). Genome-wide binding of the CRISPR endonuclease Cas9 in mammalian cells. *Nature Biotechnology, 32*(7), 670–676.

Wu, Z., Zhang, Y., Yu, H., Pan, D., Wang, Y., Wang, Y., Li, F., et al. (2021). Programmed genome editing by a miniature CRISPR-Cas12f nuclease. *Nature Chemical Biology, 17*(11), 1132–1138.

Zetsche, B., Gootenberg, J. S., Abudayyeh, O. O., Slaymaker, I. M., Makarova, K. S., Essletzbichler, P., Volz, S. E., et al. (2015). Cpf1 is a single RNA-guided endonuclease of a class 2 CRISPR-Cas system. *Cell, 163*(3), 759–771.

Zhang, F. (2019). Development of CRISPR-Cas systems for genome editing and beyond. *Quarterly Reviews of Biophysics, 52*, e6. https://doi.org/10.1017/s0033583519000052.

CHAPTER THREE

Visualizing the conformational landscape of CRISPR-Cas9 through kinetics-informed structural studies

Grace N. Hibshman[a,b,*] and David W. Taylor[a,b,*]
[a]Interdisciplinary Life Sciences Graduate Programs, University of Texas at Austin, Austin, TX, United States
[b]Department of Molecular Biosciences, University of Texas at Austin, Austin, TX, United States
*Corresponding authors. e-mail address: grace.hibshman@utexas.edu; dtaylor@utexas.edu

Contents

1. Introduction 42
2. Kinetic characterization 44
   2.1 Purification of Cas9 44
   2.2 gRNA preparation 44
   2.3 DNA preparation 44
   2.4 Reconstitution of active Cas9-gRNA complex 46
   2.5 Cleavage assay 46
   2.6 Stopped-flow measurements of DNA unwinding 46
3. Cryo-EM structural studies 47
   3.1 Sample Preparation 47
   3.2 Cryo-EM Data Collection and Processing Pipeline 48
4. Alternative Approaches 51
5. Conclusion 52
Acknowledgements 52
References 52

Abstract

CRISPR-Cas9 has transformed genome editing through its programmability and versatility. Its DNA cleavage activity involves dynamic conformational changes during gRNA binding, DNA recognition, R-loop formation, and endonuclease activation. Understanding these molecular transitions is critical for improving the specificity and efficiency of Cas9, but this remains challenging precisely due to these rapid structural rearrangements. Early structural studies provided foundational insights but were limited to static states under catalytically inactive conditions. Cryo-EM has since enabled visualization of the dynamic nature of active Cas9, by enriching for specific conformations. This chapter introduces a kinetics-informed cryo-EM approach to capture the stepwise activation of Cas9 in real time. With

Methods in Enzymology, Volume 712
ISSN 0076-6879, https://doi.org/10.1016/bs.mie.2025.01.004

thorough kinetic analyses, such as stopped-flow measurements of R-loop formation, we describe how to identify optimal timepoints to visualize key conformational states with cryo-EM. Integration of kinetic and structural data enables precise mapping of the conformational landscape of Cas9 and other dynamic enzymes, advancing our understanding of their molecular mechanisms and providing a framework for engineering enhanced variants.

1. Introduction

CRISPR-Cas systems exhibit sophisticated adaptive immune mechanisms in bacteria and archaea to identify and defend against invading genetic elements such as phages and plasmids (Sorek, Lawrence, & Wiedenheft, 2013). Among these systems, the Type II CRISPR-Cas9 system has been extensively studied due to its simplicity and versatility (Jinek et al., 2012). Cas9, an RNA-guided endonuclease, functions by cleaving DNA sequences complementary to a guide RNA (gRNA). This activity depends on the recognition of a protospacer adjacent motif (PAM) and subsequent base pairing between the gRNA and the target DNA (Anders, Niewoehner, Duerst, & Jinek, 2014).

The activation mechanism of Cas9 involves several discrete steps including binding to gRNA to form a ribonucleoprotein complex, recognizing the PAM through protein-nucleic acid interactions, and unwinding the DNA to facilitate gRNA-DNA hybridization, forming an R-loop. This process triggers conformational activation of Cas9, enabling its two endonuclease domains, HNH and RuvC, to rearrange and cleave the target and non-target DNA strands, respectively. Each stage of this mechanism is associated with substantial conformational transitions, which are essential for target discrimination and catalytic activity.

Cas9 has been widely adapted as a genome editing tool, with applications spanning gene editing, base editing, epigenetic modulation, and molecular diagnostics (Anzalone, Koblan, & Liu, 2020; Hilton et al., 2015; Li et al., 2023). The programmable nature of the gRNA has made it a powerful platform for genome engineering. However, unintended cleavage or off-target activity has precipitated the need to enhance Cas9 specificity and efficiency (Doudna, 2020; Fu et al., 2013). A deep understanding of the conformational landscape of Cas9 could help to address these limitations and optimize its therapeutic and biotechnological applications.

Identifying the molecular composition of Cas9 through structural characterization posed a challenge due to the rapid and dynamic nature of Cas9 activation. Initial structural studies employed x-ray crystallography and clever tactics to trap Cas9 in specific states to achieve the best resolution possible. These tactics included using dead enzyme where the catalytic residues were mutated to alanine (Nishimasu et al., 2014), replacing magnesium with manganese to prevent cleavage (Jiang et al., 2016), and crosslinking the target DNA to Cas9 (Jinek et al., 2014). The seminal structures revealed the bilobed architecture of Cas9, PAM recognition residues, and the formation of an R-loop, all steps preceding cleavage. A breakthrough in understanding Cas9 came from visualizing catalytically active Cas9 with cryo-electron microscopy (cryo-EM) (Zhu et al., 2019). The power of cryo-EM to capture multiple conformational states enabled visualization of the large-scale conformational changes required for Cas9 activation for the first time. Since then, there have been an explosion of Cas9 cryo-EM structures with carefully designed substrates capturing the conformational landscape of Cas9 in fine detail (Bravo et al., 2022; Cofsky, Soczek, Knott, Nogales, & Doudna, 2022; Pacesa et al., 2022a, 2022b). Yet these methods synthetically enrich for a particular Cas9 conformation along the activation pathway, potentially missing discrete conformational transitions.

Integrating kinetic analysis with cryo-EM provides a powerful framework for elucidating a wholistic view of the conformational landscape of Cas9 and Cas9 variants. One important variant is SpRY-Cas9 (SpRY), a near-PAMless version of the enzyme that relaxes the strict PAM recognition requirement of wild-type Cas9 (Christie, et al., 2022; Walton, Christie, Whittaker, & Kleinstiver, 2020). This property expands the targeting scope and enables the editing of previously inaccessible genomic sites, making SpRY a compelling example for applying kinetics-informed cryo-EM data processing frameworks. Specifically, kinetic studies using the stopped-flow method facilitate characterization of fast molecular processes such as R-loop formation. Rigorous kinetic characterization of SpRY uncovers a temporal map of functional transitions, which can inform cryo-EM sample preparation to enrich for conformational heterogeneity. Here, we detail how to perform kinetic analyses to inform cryo-EM structural determination, ultimately unveiling the conformational landscape of SpRY in near-atomic detail. This approach offers a comprehensive understanding of the activation mechanism of SpRY, providing a molecular blueprint for advancing its application in biotechnology and beyond.

2. Kinetic characterization

2.1 Purification of Cas9

Purification of Cas9 has been previously described in detail (Liu et al., 2020). Cas9 variants, including SpRY can be purified in the same manner as Cas9. The experiments described in this chapter were performed using SpRY, although these methods may be applied to other Cas9 variants, or DNA unwinding enzymes.

2.2 gRNA preparation

Customizable, full-length gRNA for SpCas9 is now commercially available, and no longer requires in vitro transcription (IVT), although this method is still a valid way to produce gRNA. IVT for gRNA production is described in detail by Anders and Jinek (2014). The experiments outlined in this chapter were designed to target lambda phage DNA, although these methods may be adapted to any sequence.

1. Purchase gRNA comprised of the SpCas9 scaffold sequence and a 20 bp programmable spacer sequence from Synthego.
2. In an RNase-free environment, reconstitute the synthetic gRNA to a concentration of 100 μM using nuclease-free or DEPC-treated water.
3. Heat the reconstituted gRNA to 95 °C for 5 min, then allow it to gradually cool to room temperature to enable proper RNA folding.

2.3 DNA preparation

1. Purchase PAGE-purified 55-nt DNA oligonucleotides for the TS and NTS from Integrated DNA Technologies. Order unlabeled TS and NTS, in addition to TS labeled with 6-FAM for cleavage measurements. Purchase NTS labeled with fluorescent tricyclic cytosine (tC°) analog at a PAM-proximal position and PAM-distal position where a C exists in the sequence from Bio-Synthesis, Inc. We chose position 1 and 16 for our DNA unwinding experiments (Fig. 1A). Alternatively, 2-aminopurine (2-AP) may be used to measure DNA unwinding at positions where an A exists as opposed to tC° replacing a C. Either tC° or 2-AP should be used to measure unwinding, not a combination of both, as the fluorescent read outs cannot be easily reconciled.
2. Reconstitute the TS and NTS oligonucleotides with nuclease-free water. Measure the concentrations of the reconstituted oligonucleotides.
3. Anneal unlabeled TS and NTS at a 1:1 molar ratio for cryo-EM and labeled oligo and unlabeled oligo at a 1:1.15 molar ratio for kinetic

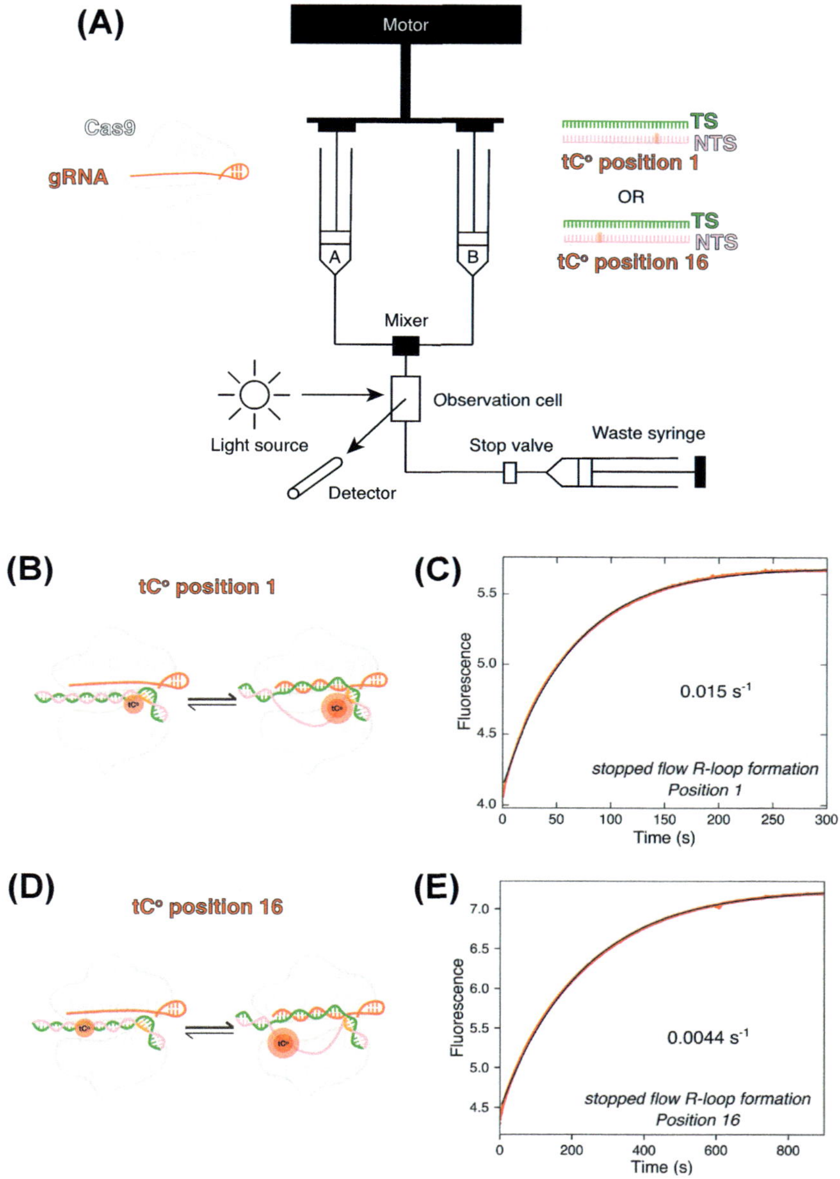

Fig. 1 The rate of DNA unwinding measured by stopped-flow. (A) Stopped-flow schematic for measurement of Cas9 DNA unwinding where syringe A contains Cas9-gRNA and syringe B contains DNA with tC° in either position 1 or 16 of the NTS. (B) Schematic representation of tC° fluorescence upon DNA unwinding at position 1 of the NTS. (C) Stopped-flow measurement of DNA unwinding (R-loop formation) at position 1. (D) Schematic representation of tC° fluorescence upon DNA unwinding at position 16 of the NTS. (E) Stopped-flow measurement of DNA unwinding at position 16. *(C and E) Adapted from Hibshman et al. (2024).*

experiments in annealing buffer (10 mM Tris-HCl pH 8, 50 mM NaCl, 1 mM EDTA), heating to 95 °C for 5 min, then allow it to gradually cool to room temperature over the course of one hour.

2.4 Reconstitution of active Cas9-gRNA complex

1. Hard spin purified Cas9 at 4 °C for 10 min to pellet any aggregates.
2. Mix Cas9 with heat annealed gRNA at a 1:1.5 molar ratio in reaction buffer (20 mM Tris-Cl, pH 7.5, 100 mM KCl, 5 % glycerol, and 5 mM DTT).
3. Incubate at room temperature for 10 min.

2.5 Cleavage assay

Perform DNA cleavage assays as previously described in detail (Bravo et al., 2022; Liu et al., 2020). Briefly, to measure substrate-initiated cleavage (limited by the rate of R-loop formation), simultaneously mix Cas9-gRNA (25 nM active site concentration) with 10 nM 6-FAM labeled DNA in the presence of Mg^{2+}. Quench the reaction at various timepoints with 0.3 M EDTA. Resolve and quantify the reaction products using an Applied Biosystems DNA sequencer (ABI 3130xl) (Dangerfield, Huang, & Johnson, 2021). Fit data using a single- or double-exponential equation. The same steps can be followed to measure pre-formed R-loop cleavage, the difference being rather than initiating the reaction with DNA, Mg^{2+} instead initiates the reaction. In this scenario, Cas9-gRNA and 6-FAM labeled DNA are pre-incubated in the absence of Mg^{2+} enabling R-loop formation, but not DNA cleavage. The reaction is then initiated by addition of Mg^{2+}, and subsequently quenched and quantified as described for substrate-initiated cleavage.

2.6 Stopped-flow measurements of DNA unwinding

We can directly measure the rate of DNA unwinding during the process of R-loop formation with the tC° labeled DNA. The tC° fluorophore is quenched by base stacking. Therefore, as the DNA duplex unwinds, we observe an increase in fluorescence as the tC°-labeled nucleotide is displaced (Fig. 1B–E). The stopped-flow method enables measurement of increasing fluorescence as depicted in Fig. 1.

1. Prepare various concentrations of tC°-labeled DNA duplex in reaction buffer ranging from 200-500 nM. Reconstitute 500 nM Cas9-gRNA complex. The final concentrations of reactants are 250 nM Cas9-gRNA and 100–250 nM variable concentrations of DNA.

2. Load syringe A of an AutoSF-120 stopped-flow instrument (KinTek Corp., Austin, TX, USA) with Cas9-gRNA and syringe B with tC°-labeled DNA (Fig. 1A).
3. Excite the fluorophore at 367 nm and monitor the change in fluorescence emission over time using a 445 nm single band-pass filter with a 20 nm bandwidth (Semrock).
4. Fit the data to either a single- or double-exponential equation to obtain the observed rate of DNA unwinding. DNA unwinding data for SpRY were fit to single-exponential equations yielding observed rates of $0.015\,s^{-1}$ and $0.0044\,s^{-1}$ for positions 1 and 16, respectively (Fig. 1C and E). It should be noted that this method is able to measure rates of up to $10\,s^{-1}$ for SpCas9.

3. Cryo-EM structural studies

3.1 Sample Preparation

To capture Cas9 during nuclease activation, rigorous kinetic analyses should be performed prior to cryo-EM sample preparation. These data will inform exactly when to quench the cleavage reaction. For SpRY, we chose the timepoint of one minute to enrich for heterogeneity in our cryo-EM sample (Fig. 2A). Under these conditions, we aimed to capture multiple intermediates during the process of DNA interrogation.

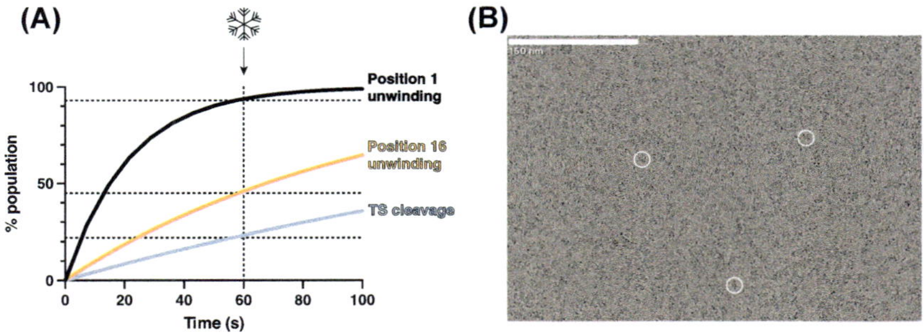

Fig. 2 Kinetics-informed cryo-EM preparation. (A) Overlay of unwinding and cleavage data. A one-minute timepoint was chosen to freeze cryo-EM grids, where ~90 % of the DNA is unwound at position 1, ~50 % of the DNA is unwound at position 16, and only ~10 % of the DNA is cleaved. (B) Representative micrograph highlighting representative particle picks from cryo-EM grid frozen after a one-minute incubation. *Created using data from Hibshman et al. (2024).*

1. Given previously performed DNA unwinding and cleavage assays, determine a timepoint that will optimize for your desired Cas9 conformation(s).
2. Initiate the cleavage reaction by mixing the preformed Cas9-gRNA complex with annealed DNA substrate at a 1:1 molar ratio. The final concentration of enzyme should be 10 μM or greater, especially if multiple conformations are expected at the chosen timepoint. The reaction should be performed at the same temperature as the previous kinetic experiments (e.g. 37 °C).
3. Apply 2.5 μL of the reaction sample to a glow-discharged holey carbon Quantifoil 1.2/1.3 grid.
4. Quench the reaction at your chosen timepoint (e.g. 1 min) by plunge-freezing the grid in liquid ethane using a Mark IV Vitrobot (Thermo Fisher Scientific) set to 4 °C and 100 % humidity with a blot force of 0 and blot time of 6 s, or otherwise optimized blotting conditions.
5. Store the grids in liquid nitrogen prior to imaging.

3.2 Cryo-EM Data Collection and Processing Pipeline

Cryo-EM datasets are collected on an FEI Titan Krios cryo-electron microscope equipped with a Gatan Biocontinuum Imaging Filter and a Gatan K3 Summit direct electron detector. Images are recorded with SerialEM v4.1 (Mastronarde, 2005). Datasets are collected with a defocus range of −1.5 to −2.5 μm. For SpRY, we collected 10,151 micrographs.

With concentrated, well-distributed particles (Figs. 2B and 3A), multiple rounds of 3D variability and 3D classification can be used to isolate distinct conformational states from a single cryo-EM dataset. An example workflow is depicted in Fig. 3 for SpRY, and the resulting reconstructions highlighted in Fig. 4.

1. Perform motion correction, CTF estimation, and particle picking on-the-fly using cryoSPARC Live (Punjani, 2021). Using a previous Cas9 reconstruction to create templates for templated particle picking typically yields the best Cas9 particles. Alternatively, blob picker can be used with a minimum particle diameter set to 100 Å, and a maximum particle diameter set to 180 Å. For the SpRY dataset, template picker yielded 9,654,609 picks from which we extracted 7,971,106 particles.
2. 2D Classification: Subject the particles to 2D classification with a batch size per class of 200, 40 online EM iterations, and 2 final full iterations, without forcing max over poses/shifts (Fig. 3B).

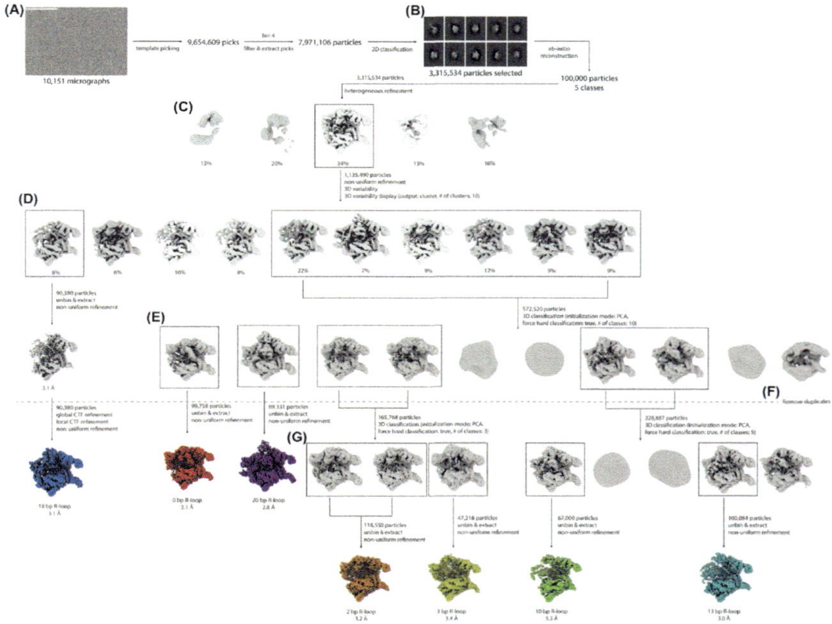

Fig. 3 Data processing flowchart of a single cryo-EM dataset yielding seven distinct structures. *Created using a dataset from Hibshman et al. (2024).*

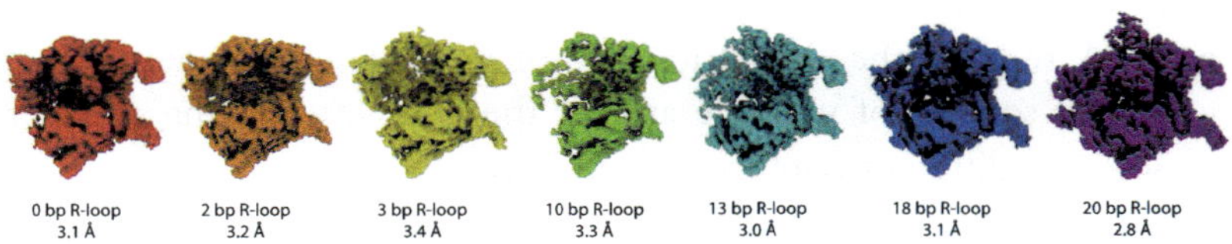

Fig. 4 Final cryo-EM reconstructions of R-loop intermediates isolated from a single dataset. *Structures originally published in Hibshman et al. (2024).*

3. Select the 2D classes that appear bilobed, have other features of the enzyme, and/or appear to contain high-resolution structural information. Avoid classes that look bumpy with lots of noise. We selected 3,315,534 particles from 2D classification, making sure to include views from as many angles as possible.
4. Ab Initio Reconstruction: Process 100,000 particles from the selected 2D classified particles via ab initio reconstruction. The ideal number of ab initio classes varies given the quantity and quality of particles. We chose 5 ab initio classes for this dataset.

5. Heterogeneous Refinement: Using the initial volumes from ab initio reconstruction, perform heterogeneous refinement on all the particles selected from 2D classification (Fig. 3C).
6. Consensus Reconstruction: One dominant volume representing the expected particle should emerge from heterogeneous refinement. Use this set of particles and the corresponding volume to get a consensus reconstruction via non-uniform refinement.
7. 3D Variability Analysis: Use the particles and mask from the non-uniform refinement to perform 3D variability analysis (Fig. 3D). Output the 3D variability as 10 clusters with a low-pass filter resolution of 5 Å. We subjected 1,135,490 particles to 3D variability analysis.
8. Analyze Clusters & 3D Classification: Analyze the variation in volumes for each cluster by downloading the volume and toggling the threshold, paying close attention to domain occupancy and base pairing between the gRNA and TS. Group the clusters based on the number of base pairs formed. At this stage, we could distinguish the 18 bp structure, so we extracted unbinned particles and executed non-uniform refinement to reconstruct the final 18 bp final structure. If it is difficult to distinguish between clusters, group them and subject this group of particles to 3D classification using principal component analysis (PCA) as the initialization mode, forcing hard classification, and outputting 10 classes (Fig. 3F).
   a. PCA reduces the dimensionality of the dataset by identifying the largest sources of variance among the particles, making it easier to separate different conformational states.
   b. Hard classification assigns each particle strictly to one class, rather than distributing its probability across multiple classes. This helps in isolating particles that represent distinct conformations.

   Be sure to discard particles from clusters with any sign of preferred orientation before performing 3D classification. We subjected 572,520 particles to 3D classification under these conditions.
9. Remove duplicate particles (Fig. 3F).
10. Refine or Re-Classify: Again, analyze the variation in volumes for each class. If a class is a unique state with a distinct number of base pairs formed, extract unbinned particles and perform non-uniform refinement to achieve the final reconstruction. If unable to decipher a single state, group similar classes and perform another round of 3D classification using PCA as the initialization mode, forcing hard classification, and outputting an appropriate number of classes (Fig. 3G). For example, we

had trouble distinguishing the 2 and 3 bp structures, so we subjected the corresponding 165,768 particles to an additional round of 3D classification. We discovered that two of the three clusters represented the 2 bp intermediate, and one cluster represented the 3 bp intermediate. Therefore, we combined the two clusters representing the 2 bp structure and kept the 3 bp particles separate for subsequent refinements.

11. Final Refinements: Inspect the volumes from the second round of 3D classification, group them accordingly to how many base pairs are formed, extract unbinned particles, and execute non-uniform refinement. This method resulted in final reconstructions of SpRY with 0, 2, 3, 10, 13, 18, and 20 bp of R-loop formed (Fig. 4).

4. Alternative Approaches

Recent advancements in cryo-EM data processing provide alternative strategies to those described above. One such approach involves using RELION, a widely adopted software package that facilitates maximum-likelihood-based 3D classification and refinement (Scheres, 2012, 2016). The robust RELION algorithms for particle alignment and classification allow for similar extraction of discrete conformational states, albeit with different computational strategies compared to cryoSPARC. For instance, Bayesian polishing and multi-body refinement in RELION have been shown to improve the resolution of flexible regions, offering an attractive option for studying transient intermediates in dynamic systems.

Another promising development is the use of deep learning-based methods such as cryoDRGN (Zhong, Bepler, Berger, & Davis, 2021). This program leverages neural networks to model continuous heterogeneity in cryo-EM datasets, offering a novel way to visualize the full spectrum of conformational states rather than discrete classes. CryoDRGN has been demonstrated to provide insights into subtle conformational variations that traditional classification methods might overlook.

Additionally, time-resolved cryo-EM (TRCEM) enables rapid capture of biomolecular intermediates on the 10–1000 ms timescale (Feng & Frank, 2024). By integrating a polydimethylsiloxane-based microfluidics chip assembly for fast mixing and sample deposition, TRCEM offers a way to observe pre-equilibrium states that are otherwise inaccessible by conventional blotting methods. This approach, successfully applied to ribosome recycling, addresses challenges like sample adsorption and poor mixing, and holds promise for revealing critical transitions in other dynamic systems. This method has set a benchmark for capturing fast structural transitions, and its integration with modern processing tools like RELION and cryoDRGN represents a powerful convergence of experimental and computational innovation in the field.

5. Conclusion

Combining kinetic studies with cryo-EM offers a transformative approach to uncovering the intricate conformational landscape of Cas9. By leveraging real-time kinetics, such as stopped-flow measurements of R-loop formation, we can precisely time cryo-EM sample preparation to capture transient intermediates. Recently, this method has been employed to extract 25 distinct Cas9 conformations from a single cryo-EM dataset (Kiernan & Taylor, 2024). This kinetics-informed strategy enriches our understanding of the dynamic activation pathway of Cas9, revealing structural transitions that underpin its specificity and catalytic efficiency. The methods described in this chapter provide a comprehensive framework for dissecting Cas9 function at atomic resolution, paving the way for the rational design of enhanced Cas9 variants with improved precision for therapeutic and biotechnological applications. This synergistic approach exemplifies the power of combining structural and kinetic insights to unravel the mechanisms of complex biomolecular machines.

Acknowledgements

We thank J.P.K. Bravo, K.A. Kiernan, T.L. Dangerfield, and K.A. Johnson for helpful discussions and insights incorporated within this chapter.

References

Anders, C., & Jinek, M. (2014). In vitro enzymology of Cas9. *Methods in Enzymology, 546*, 1–20.

Anders, C., Niewoehner, O., Duerst, A., & Jinek, M. (2014). Structural basis of PAM-dependent target DNA recognition by the Cas9 endonuclease. *Nature, 513*, 569–573.

Anzalone, A. V., Koblan, L. W., & Liu, D. R. (2020). Genome editing with CRISPR–Cas nucleases, base editors, transposases and prime editors. *Nature Biotechnology, 38*(7), 824–844.

Bravo, J. P. K., et al. (2022). Structural basis for mismatch surveillance by CRISPR–Cas9. *Nature, 603*, 343–347.

Christie, K. A., Guo, J. A., Silverstein, R. A., Doll, R. M., Mabuchi, M., Stutzman, H. E., ... Kleinstiver, B. P. (2022). Precise DNA cleavage using CRISPR-SpRYgests. *Nature Biotechnology, 41*, 409–416.

Cofsky, J. C., Soczek, K. M., Knott, G. J., Nogales, E., & Doudna, J. A. (2022). CRISPR–Cas9 bends and twists DNA to read its sequence. *Nature Structural & Molecular Biology, 29*(4), 395–402.

Dangerfield, T. L., Huang, N. Z., & Johnson, K. A. (2021). High throughput quantification of short nucleic acid samples by capillary electrophoresis with automated data processing. *Analytical Biochemistry, 629*, 114239.
Doudna, J. A. (2020). The promise and challenge of therapeutic genome editing. *Nature, 578*(7794), 229–236.
Fu, Y., et al. (2013). High-frequency off-target mutagenesis induced by CRISPR-Cas nucleases in human cells. *Nature Biotechnology, 31*(9), 822–826.
Feng, X., & Frank, J. (2024). Time-resolved cryo-EM (TRCEM) sample preparation using a PDMS-based microfluidic chip assembly. *BioRxiv, 12*(8), 627437.
Hibshman, G. N., et al. (2024). Unraveling the mechanisms of PAMless DNA interrogation by SpRY-Cas9. *Nature Communications, 15*(1), 1–15.
Hilton, I. B., et al. (2015). Epigenome editing by a CRISPR-Cas9-based acetyltransferase activates genes from promoters and enhancers. *Nature Biotechnology, 33*(5), 510–517.
Jiang, F., et al. (2016). Structures of a CRISPR-Cas9 R-loop complex primed for DNA cleavage. *Science (1979), 351*, 867–871.
Jinek, M., et al. (2012). A programmable dual-RNA-guided DNA endonuclease in adaptive bacterial immunity. *Science (1979), 337*, 816–821.
Jinek, M., et al. (2014). Structures of Cas9 endonucleases reveal RNA-mediated conformational activation. *Science (New York, N. Y.), 343, 1979.*
Kiernan, K. A., & Taylor, D. W. (2024). Visualization of a multi-turnover Cas9 after product release. *BioRxiv, 11*(25), 625307.
Li, T., et al. (2023). CRISPR/Cas9 therapeutics: Progress and prospects. *Signal Transduction and Targeted Therapy, 8*(1), 1–23.
Liu, M.-S., et al. (2020). Engineered CRISPR/Cas9 enzymes improve discrimination by slowing DNA cleavage to allow release of off-target DNA. *Nature Communications, 11*(1), 1–13.
Mastronarde, D. N. (2005). Automated electron microscope tomography using robust prediction of specimen movements. *Journal of Structural Biology, 152*, 36–51.
Nishimasu, H., et al. (2014). Crystal structure of Cas9 in complex with guide RNA and target DNA. *Cell, 156*, 935–949.
Pacesa, M., et al. (2022a). R-loop formation and conformational activation mechanisms of Cas9. *Nature, 609*, 191–196.
Pacesa, M., et al. (2022b). Structural basis for Cas9 off-target activity. *Cell, 185*, 4067–4081.e21.
Punjani, A. (2021). Real-time cryo-EM structure determination. *Microscopy and Microanalysis, 27*, 1156–1157.
Scheres, S. H. W. (2012). RELION: Implementation of a Bayesian approach to cryo-EM structure determination. *Journal of Structural Biology, 180*, 519–530.
Scheres, S. H. W. (2016). Processing of structurally heterogeneous Cryo-EM data in RELION. *Methods in Enzymology, 579*, 125–157.
Sorek, R., Lawrence, C. M., & Wiedenheft, B. (2013). CRISPR-mediated adaptive immune systems in bacteria and archaea. *Annual Review of Biochemistry, 82*, 237–266.
Walton, R. T., Christie, K. A., Whittaker, M. N., & Kleinstiver, B. P. (2020). Unconstrained genome targeting with near-PAMless engineered CRISPR-Cas9 variants. *Science, 368*(6488), 290–296.
Zhu, X., Clarke, R., Puppala, A. K., Chittori, S., Merk, A., Merrill, B. J., ... Subramaniam, S. (2019). Cryo-EM structures reveal coordinated domain motions that govern DNA cleavage by Cas9. *Nature Structural & Molecular Biology, 26*(8), 679–685.
Zhong, E. D., Bepler, T., Berger, B., & Davis, J. H. (2021). CryoDRGN: reconstruction of heterogeneous cryo-EM structures using neural networks. *Nature Methods, 18*(2), 176–185.

CHAPTER FOUR

Biochemical reconstitution of a type I-B CRISPR-associated transposon

Shukun Wang[a] and Leifu Chang[a,b,*]
[a]Department of Biological Sciences, Purdue University, West Lafayette, IN, United States
[b]Purdue Institute for Cancer Research, Purdue University, West Lafayette, IN, United States
*Corresponding author. e-mail address: lchang18@purdue.edu

Contents

1. Introduction 56
2. Plasmid preparation 58
   2.1 pCascade 58
   2.2 TnsAB 63
   2.3 TnsC 63
   2.4 TniQ 63
   2.5 pDonor and pTarget 63
3. Purification of Cascade 64
   3.1 Transformation 64
   3.2 Inoculation 65
   3.3 Induction 65
   3.4 Cell collection 66
   3.5 Immobilized metal affinity chromatography using Ni-NTA beads 66
   3.6 Ion exchange chromatography 67
   3.7 Size exclusion chromatography 68
4. Purification of TnsAB and TnsC 68
   4.1 Transformation, inoculation, induction 69
   4.2 Cell collection and affinity chromatography using Streptactin beads 69
   4.3 Size exclusion chromatography 70
5. Purification of TniQ-TnsC complex 70
   5.1 Transformation 71
   5.2 Inoculation 71
   5.3 Induction 71
   5.4 Cell collection and Strep-tag affinity chromatography 71
   5.5 Size exclusion chromatography 72
6. *In vitro* transposition assay 72
   6.1 Qualitative detection by PCR 73
   6.2 Quantitative detection by qPCR 74
   6.3 Qualitative detection by transformation 76
7. Conclusion 77

Methods in Enzymology, Volume 712
ISSN 0076-6879, https://doi.org/10.1016/bs.mie.2025.01.042

Acknowledgments 77
Author contributions 77
References 77

Abstract

CRISPR-associated transposons (CASTs) are potential gene editing tools because of their RNA-guided DNA insertion activity. It is essential to understand the mechanisms underlying the transposition for the application of CASTs. Here, we provide protocols for the biochemical reconstitution of a type I-B CAST for RNA-guided transposition. The procedures may be applicable to other types of CASTs and facilitate the mechanism studies of various CASTs.

1. Introduction

CRISPR-associated transposons (CASTs) are mostly Tn7-like transposons that have co-opted CRISPR-Cas systems to facilitate DNA transposition (Faure et al., 2019; Peters, Makarova, Shmakov, & Koonin, 2017). The cargo capacity of native CASTs can reach up to 43 kb in certain cases (Hsieh & Peters, 2023). In contrast to the widely used CRISPR-Cas9 system in gene editing, CAST systems insert DNA independent of host repair pathways and avoid creating exposed double-strand breaks (Lampe et al., 2024; Tou, Orr, & Kleinstiver, 2023), making them promising tools for programmable gene insertion into the genome.

Various CASTs have been reported, hypothetically resulting from independent capture of different types of CRISPR-Cas systems by Tn7-like transposons. To date, four types of CASTs have been experimentally characterized: types V-K (Strecker et al., 2019), I-B (Saito et al., 2021) (divided into I-B1 and I-B2), I-F (Klompe, Vo, Halpin-Healy, & Sternberg, 2019), and I-D (Faure et al., 2023; Hsieh & Peters, 2023). Except for the I-D CAST from *Cyanothece sp (*Faure et al., 2023*),* all the CRISPR modules in other CASTs are nuclease deficient. In these systems, the guide RNA and Cas protein(s) form an effector complex to recognize target DNA that is complementary to the guide RNA. The transposase TnsB recognizes the inverted repeat sequences of transposon ends, cleaves the 3′ ends of the transposon, and rejoins them to the target DNA. TnsA, an endonuclease-like protein, is recruited by TnsB to cut the 5′ end of the transposon to facilitate complete excision of the transposon. Recruitment of TnsAB-transposon ends to the target DNA requires TniQ, a zinc-binding protein, and TnsC, an AAA+ ATPase (Fig. 1A). Structural studies

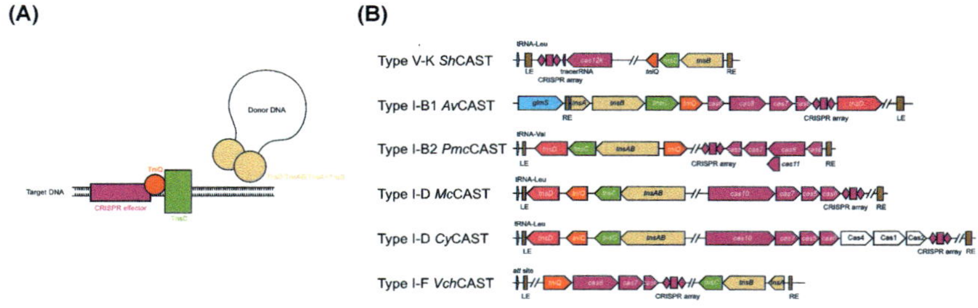

Fig. 1 Diversity of CRISPR-associated transposons. (A) A general illustration of the RNA-guided transposition. (B) Gene loci of various CRISPR-associated transposons. Abbreviations: *Sh, Scytonema hofmanni; Av, Anabaena variabilis; Pmc, Peltigera membranacea cyanobiont; Mc, Myxacorys californica; Cy, Cyanothece sp.; Vch, Vibrio cholerae.*

of various CAST systems have revealed significant diversity in the interactions between Tns proteins and Cas proteins (Halpin-Healy, Klompe, Sternberg, & Fernandez, 2020; Hoffmann et al., 2022; Jia, Xie, de la Cruz, Eng, & Patel, 2020; Li, Zhang, Xiao, & Chang, 2020; Park et al., 2021; Park et al., 2023a, 2023b; Querques, Schmitz, Oberli, Chanez, & Jinek, 2021; Schmitz, Querques, Oberli, Chanez, & Jinek, 2022; Wang, Xu, & Yang, 2020; Wang, Gabel, Siddique, Klose, & Chang, 2023; Xiao et al., 2021). For example, in type I-F CAST, a TniQ dimer binds to the effector before target recognition (Halpin-Healy et al., 2020; Jia et al., 2020; Li et al., 2020; Wang et al., 2020), whereas in type V-K and type I-B2, target DNA recognition is required for TniQ to bind the effector (Park et al., 2023a; Schmitz et al., 2022; Wang et al., 2023). The diversity of different CASTs creates a versatile toolbox that may be harnessed for application under different conditions and provides insights for the rational design of gene editing tools.

To harness CASTs for genome editing, it is important to understand the molecular mechanism underlying the targeted DNA insertion. Biochemical reconstitution of CASTs is a critical technique for elucidating the mechanisms of CASTs and establishing a fundamental basis for their applications. Compared to cellular experiments, biochemically reconstituted systems are more defined, enabling precise control of the experimental conditions. Precise control is invaluable for future studies aimed at optimizing these systems, such as enhancing efficiency through protein engineering. Furthermore, biochemical reconstitution aids in identifying the host factors or unknown components essential for transposition. Notable discoveries include ribosomal protein S15 in type V-K CAST system (Schmitz et al., 2022) and Cas11 in type I-B2 CAST system (Wang et al., 2023).

Biochemical reconstitution of CASTs is considerably challenging due to the involvement of multiple components. The type V-K CAST was first biochemically reconstituted using purified components due to its relatively compact nature with a single effector protein, Cas12k, in the CRISPR module (Strecker et al., 2019). However, type I systems are more complex (Fig. 1B). Among type I CASTs, only the type I-B2 system has been biochemically reconstituted (Wang et al., 2023). The type I-B2 CAST from *Peltigera membranacea cyanobiont* 210 A (*Pmc*CAST) utilizes both RNA-guided and TnsD-guided pathways, employing the same core machinery TnsABC for donor DNA recognition and insertion but using different target selectors (Saito et al., 2021). Here, we will focus on the RNA-guided transposition pathway, which requires proteins Cas5, Cas6, Cas7, Cas8, Cas11, TnsAB, TnsC, TniQ and one CRISPR RNA (crRNA) (Saito et al., 2021; Wang et al., 2023). The Cas proteins assemble with crRNA to form a multi-subunit surveillance complex known as the Cascade complex (Jiang & Doudna, 2015). This Cascade complex recognizes a protospacer adjacent motif (PAM) sequence "ATG" and facilitates the formation of a crRNA-DNA duplex via base pairing (Saito et al., 2021). Transposase TnsAB, which is a natural fusion protein of TnsA and TnsB, recognizes and excises the transposon from both ends. Subsequently, the excised transposon is inserted 47–58 bp downstream of the target site complimentary to crRNA, a process mediated by TnsC and TniQ (Saito et al., 2021).

In this chapter, we will discuss the reconstitution method for the type I-B2 *Pmc*CAST system, covering plasmid construction, protein expression and purification, and resultant DNA detection. These step-by-step methods are designed to help readers establish the transposition assay in a standard biochemistry laboratory setting.

2. Plasmid preparation

Four plasmids were used for the expression and purification of proteins in *E. coli*: pCascade, pTnsAB, pTnsC, and pTniQ (Fig. 2A).

2.1 pCascade

The methods used to assemble the Cascade complex in different type I CRISPR-Cas systems can be categorized into two strategies (Table 1).

The first strategy involves co-expressing all the Cas proteins and the pre-crRNA in the cell, allowing the complex to assemble in the cell before

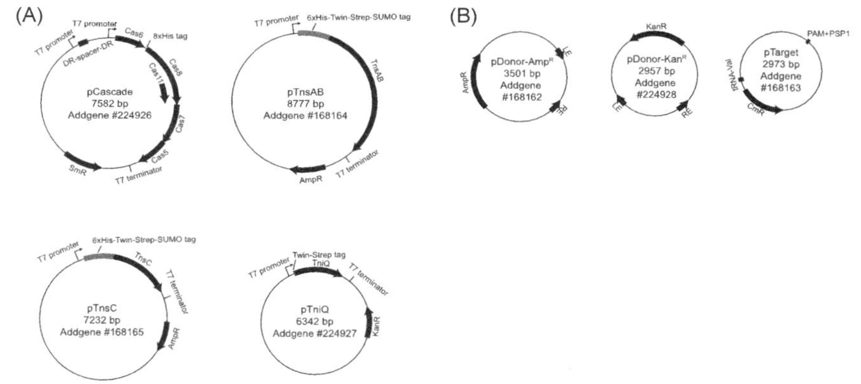

Fig. 2 Plasmid maps for the plasmids used in this chapter. (A) Plasmid maps for the protein expression. (B) Plasmid maps for pDonor and pTarget.

purification. Typically, the genes for the proteins and CRISPR array are cloned into two to three expression plasmids, with one affinity tag fused to a Cas protein for purification (Hochstrasser, Taylor, Kornfeld, Nogales, & Doudna, 2016; Hu et al., 2022; Lu et al., 2024; McBride et al., 2020; Wiedenheft et al., 2011; Xiao et al., 2017). For example, in the type I-A CRISPR-Cas system, three plasmids are co-transformed into *E. coli* BL21(DE3). One plasmid encodes Cas8 with an N-terminal Twin-strep tag, another encodes Cas11, Cas7, and Cas5, and the third encodes the CRISPR array and Cas6. The complex is then purified using affinity chromatography followed by size-exclusion chromatography (SEC) (Hu et al., 2022).

For the second strategy, the Cas proteins are expressed and purified individually, and Cascade complex is assembled subsequently by mixing the purified proteins with RNA precursor or crRNA (Plagens et al., 2014; Richter, Rompf, Wiegel, Rau, & Randau, 2017; Shangguan, Graham, Sundaramoorthy, & White, 2022). The complex is then purified using SEC. One disadvantage of this method is that some Cas proteins may not be soluble when expressed and purified alone, which require refolding before the assembly (Plagens et al., 2014).

Here we present the method for the co-expression and purification of the *Pmc*Cascade. The pCascade plasmid was constructed by deleting the transposon proteins from a previously constructed pHelper plasmid from Feng Zhang Lab (Saito et al., 2021). The *Pmc*Cascade proteins and crRNA array were inserted into the pCDFDuet-1 plasmid backbone under different T7 promoters. An N-terminal 8xHis tag was cloned on Cas8 for purification (Wang et al., 2023) (Fig. 2A).

Table 1 Summary of the methods for the purification of Cascade in different type I CRISPR-Cas systems.

Subtypes	Origin	Codon optimization	Tag	Expression and purification method	References
I-A	*Thermoprotus tenax*	No	C-terminal His-SUMO tag on Cas7	*In vitro* assembly Csa 5 and Cas7 were purified as soluble proteins, whereas Cas5a, Cas3′, Cas3″ and Cas8a2 were insoluble. The cascade complex (without crRNA) was assembled through co-folding, heat treatment, and size-exclusion chromatography, then completed by incubating with crRNA.	Plagens et al. (2014)
I-A	*Pyrococcus furiosus DSM 3638*	No	N-terminal Twin-strep tag on Cas8a	*In vivo* assembly pCDFDuet-Twin-Strep-Cas8a, pETDuet-Cas11-Cas7-Cas5 and pcrRNA-Cas6 were co-transformed into *E.coli* BL21(DE3). The Cascade complex was purified via streptavidin affinity chromatograph and size-exclusion chromatograph.	Hu et al. (2022)
I-B	*Clostridium thermocellum*	Yes	N-terminal His tag on Cas5 and Cas7; C-terminal His tag on Cas8b and Cas6b	*In vitro* assembly Individual Cas proteins were expressed in *E. coli* Rosetta(DE)pLysS and purified via nickel affinity, cation-exchange, and size exclusion chromatography. The Cascade complex was then assembled in vitro by incubation with RNA precursor and further purified by size exclusion chromatography.	Richter et al. (2017)

I-B	*Synechocystis sp. PCC 6714*	No	N-terminal Strep tag on Cas8b	*In vivo* assembly pRSFDuet-Cas6b-Cas11b, pCDFDuet-Cas8b-Cas7b-Cas5b, and PUC19-CRISPR array were co-transformed into *E. coli* BL21(DE3). The complex was purified via Strep-Tactin affinity chromatograph and size-exclusion chromatograph.	Lu et al. (2024)
I-C	*Desulfovibrio vulgaris str. Hildenborough*	No	N-terminal His-MBP tag on Cas5c	*In vivo* assembly pHMGWA-Cas5C-Cas8c-Cas7 and pACYCDuet-CRISPR array were co-transformed into *E. coli* for co-expression of the proteins and pre-crRNA. The cascade complex was purified by nickle affinity chromatograph and size-exclusion chromatograph.	Hochstrasser et al. (2016)
I-D	*Synechocystis sp. PCC 6803*	No	N-terminal His tag on Cas10d	*In vivo* assembly pQE−80LoriT-Cas10d-Cas7d-Cas5d and pACYCDuet-Cas6d-CRISPR array were co-transformed into *E. coli* BL21(DE3). The cascade complex was purified by Nickle affinity chromatograph and size exclusion chromatograph.	McBride et al. (2020)

(continued)

Table 1 Summary of the methods for the purification of Cascade in different type I CRISPR-Cas systems. (*cont'd*)

Subtypes	Origin	Codon optimization	Tag	Expression and purification method	References
I-E	*Thermobifida fusca*	No	N-terminal Twin-strep-SUMO tag on Cse1	*In vivo* assembly pTwin-Strep-SUMO-pET19b-Cse1, pCDFDuet-Cse2-Cas7-Cas5e-Cas6e, and pRSFDuet-CRISPR array were co-transformed into *E. Coli* BL21star (DE3). The cascade complex was purified by Strep-Tactin affinity chromatograph and size exclusion chromatograph.	Xiao et al. (2017)
I-F	*Pseudomonas aeruginosa*	No	N-terminal His tag on Csy4	*In vivo* assembly Csy proteins and the synthetic CRISPR RNA were coexpressed in *Rosetta* 2 (DE3) Escherichia coli. The cascade complex was purified by Nickle affinity chromatograph and size exclusion chromatograph.	Wiedenheft et al. (2011)
I-G	*Thioalkalivibrio sulfidiphilus*	Yes	N-terminal His Tag on Cas8g, Cas7 and Cas3	*In vitro* assembly *Cas8g*, *Cas7*, and *Cas3* were cloned into expression plasmid pEV5HisTEV individually. Csb2 was cloned into pET-Duet. Csb2 and Cas7 were co-expressed and co-purified. Cas8g and Cas3 were purified separately. The cascade complex was assembled by mixing the purified proteins in the presence of crRNA, followed by size exclusion chromatograph.	Shangguan et al. (2022)

2.2 TnsAB

The plasmid for the expression and purification of TnsAB was constructed by the Feng Zhang lab (Addgene: 168164) (Saito et al., 2021). The gene for TnsAB was inserted into the ColE1 plasmid backbone, with 6xHis, Twin Strep, SUMO tags on the N terminus to aid in protein purification (Fig. 2A).

2.3 TnsC

The construction of pTnsC (Addgene: 168165) is identical to that of TnsAB except the gene of TnsC.

2.4 TniQ

We initially cloned the gene of TniQ into multiple different plasmids with different tags to test the expression and purification. However, all the constructs we tested including His-TniQ, TniQ-His, Strep-TniQ and His-Strep-TniQ failed to express soluble TniQ. Finally, we purified TniQ by co-transforming one plasmid that is compatible with the pTnsC for co-expression and co-purification. In that construct, the gene for TniQ was inserted into the pET StrepII TEV LIC cloning vector with a Twin-Step-TEV tag on the N terminus (Fig. 2A).

2.5 pDonor and pTarget

For the transposition assay, three additional plasmids were used: pDonor_Amp^R, pDonor_Kan^R, and pTarget (Fig. 2B). All the plasmids were initially constructed by the Feng Zhang lab (Saito et al., 2021). The pDonor_Amp^R plasmid contains one left end (LE, 132 bp), and one right end (RE, 216 bp), with a cargo gene of 417 bp between them. This plasmid was used for PCR and qPCR to detect the product. The pDonor_Kan^R plasmid contains a Kanamycin resistance gene between LE and RE, used to detect the transposition product by transformation into *E. coli* cells. Successful transposition will confer kanamycin resistance to the target plasmid, which can be purified by transforming to Top 10 competent cell with kanamycin selection. The pTarget plasmid contains a target DNA fragment with the PAM "ATG" sequence and a 35 bp protospacer complementary to the spacer region of crRNA.

The plasmids from Addgene are transported in the format of stab-cultured bacteria. Once received, they need to be cultured, and the plasmids can be extracted from the cultured bacteria. The following protocol details the preparation of the plasmids from stab-cultured bacteria:

1. Prepare agar plates with 50 μg/mL spectinomycin, 100 μg/mL ampicillin, 50 μg/mL kanamycin or 50 μg/mL chloramphenicol.
2. Streak the plates using the received stab-cultured bacteria from Addgene. Ensure the correct antibiotic is used for each plasmid. Specifically, pCascade contains a spectinomycin resistance gene; pTnsAB, pTnsC and pDonor_Amp^R contain an ampicillin resistance gene; pDonor_Kan^R and pTniQ contain a kanamycin resistance gene; and pTarget contains a chloramphenicol resistance gene (Fig. 2B).
3. Incubate the plates at 37 °C overnight.
4. Pick a single colony from each plate and inoculate it in 5 mL LB broth containing appropriate antibiotic. Shake the broth at 37 °C and 180 rpm overnight.
5. Transfer 500 μL of the bacterial culture into a 1.5 mL microcentrifuge tube and mix with 500 μL of sterile 50 % glycerol. Store the glycerol stock at −80 °C for long-term preservation.
6. Use the remaining bacterial culture to extract the plasmid using a commercial miniprep kit (QIAprep Spin Miniprep Kit or IBI High-Speed Mini Plasmid Kit).
7. Measure the DNA concentration, label the samples, and store them in a −20 °C freezer.

3. Purification of Cascade

The following protocol outlines the detailed procedures to purify the Cascade complex in type I-B2 CAST system. The complex was assembled in the cell by co-expressing Cas5, Cas6, Cas7, Cas8, Cas11, and pre-crRNA. Within the cell, pre-crRNA is processed by Cas6 into mature crRNA (Jiang & Doudna, 2015), facilitating complex assembly. The entire experiment typically spans four days, with the first three days dedicated to cultivation and expression, and the fourth day to protein purification. Three steps of purification were conducted including Ni-NTA chromatography, ion exchange chromatography, and SEC. Fig. 3A shows a representative result after each step of purification (Wang et al., 2023).

3.1 Transformation

a. Prepare LB agar plate with 50 μg/mL Spectinomycin and 50 μg/mL Chloramphenicol. Spectinomycin is added to maintain the Cascade

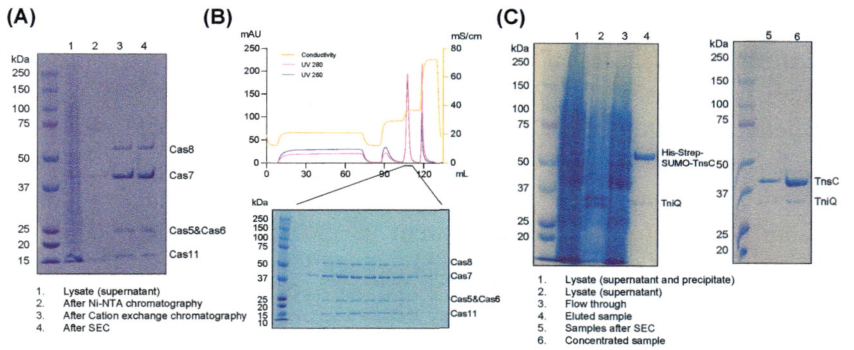

Fig. 3 SDS-PAGE gels for the purification of Cascade and TnsC-TniQ complex. (A) SDS-PAGE gel for the purification of Cascade. (B) The anion exchange chromatography profile of Cascade. The SDS-PAGE gel is on the bottom of the profile. (C) SDS-PAGE gels for the purification of TnsC-TniQ complex.

expression plasmid. Chloramphenicol is added to keep the plasmid for rare tRNAs in Rosetta™(DE3)pLysS Competent cells.

b. Transform the pCascade plasmid to the Rosetta™(DE3)pLysS Competent Cells (Sigma-Aldrich, 70956) with the traditional heat-shock method.

c. Incubate the plate at 37 °C overnight

3.2 Inoculation

a. Take out the plate and save it at 4 °C fridge until use.

b. Prepare 50 mL LB broth in a 250 mL flask and cool it down to room temperature before adding 50 μg/mL Spectinomycin and 50 μg/mL Chloramphenicol.

c. Pick a single colony and inoculate it in LB with antibiotics.

d. Shake the flask at 37 °C and 180 rpm overnight.

e. Prepare 4 flasks of 750 mL TB broth using 2 L flasks for overexpression next day. Alternatively, these can be prepared the next day in the morning, but it takes time to cool down for inoculation.

3.3 Induction

a. Add 50 μg/mL Spectinomycin to TB broth.

b. Inoculate 7.5 mL of bacteria in each flask and shake the flasks at 37 °C and 180 rpm until OD_{600} reaches around 0.6.

c. Cool down the medium to 16 °C, which takes about 30 min.

d. Add 0.25 mM IPTG and incubate the flasks at 16 °C and 180 rpm overnight.

3.4 Cell collection

a. Centrifuge the bacterial culture at 6000 rpm for 10 min at 16 °C.

b. After centrifugation, combine the pellet into two 50 mL falcon tubes. Briefly centrifuge the falcon tubes at 4 °C to ensure the pellets are collected at the bottom of the tubes.

c. If purification will be done on another day, freeze the pellets quickly using liquid nitrogen and store them at −20 °C or −80 °C.

3.5 Immobilized metal affinity chromatography using Ni-NTA beads

a. Lysis buffer (200 mL): 50 mM Tris 7.5, 500 mM NaCl, 5 % glycerol, 10 mM imidazole, and two pills of protease inhibitor tablet (Thermo Scientific™ Pierce™ Protease Inhibitor Tablets, EDTA-free, PIA32965 or cOmplete™, Mini, EDTA-free Protease Inhibitor Cocktail, 11836170001).
Wash buffer 1 (500 mL): 50 mM Tris 7.5, 500 mM NaCl, 5 % glycerol, 10 mM imidazole.
Wash buffer 2 (200 mL): 50 mM Tris 7.5, 500 mM NaCl, 5 % glycerol, 50 mM imidazole.
Elution buffer (20 mL): 50 mM Tris 7.5, 500 mM NaCl, 5 % glycerol, 250 mM imidazole.

b. Resuspend the bacteria pellet with 150–200 mL lysis buffer.

c. Place the resuspended sample on ice.

d. Sonicate the resuspension on an ice water bath using a sonicator (Fisher Scientific, FB505, 500 WATTS, 20 kHz) at 67 % Ampl, work 3 s, rest 6 s until the sample is clear (approximately 15 to 20 min for every 35 mL sample).

e. Clarify the sample by centrifugation at 4 °C, 20,000 rpm for 30 min.

f. Collect the supernatant in a beaker and place it on ice. Take 20 μL supernatant for SDS-PAGE analysis (Fig. 3A, lane 1).

g. Resuspend the precipitate with the same volume of washing buffer 1.

h. Equilibrate 1 mL Ni-NTA beads with 5 column volume (CV) of H_2O, and then 5 CV of washing buffer 1.

i. Incubate the supernatant with pre-equilibrated Ni-NTA beads at 4 °C for 30 min.

j. Load the sample-beads mixture into an empty gravity column.

k. Wash the beads with 5 CV of the washing buffer 1 and 10 CV of the washing buffer 2.

l. Close the column outlet and add 10 mL of the elution buffer.

m. Resuspend the beads and allow them to settle down.
n. Collect the eluted sample. Collect 20 μL of the eluted sample for SDS-PAGE analysis (Fig. 3A, lane 2).

3.6 Ion exchange chromatography

Cascade is further purified using anion exchange chromatography (Cytiva, HiTrap Q HP). The principle of anion exchange chromatography is based on the interactions between the negatively charged molecules and the positively charged functional groups on the stationary phase. As the salt concentration in the elution buffer increases, the interaction between the negatively charged molecules and the positively charged beads decreases, allowing for the separation of molecules based on their net charge.

a. Prepare buffers for ion exchange chromatography
Buffer A: 50 mM Tris 7.5, 100 mM NaCl, 5 % glycerol, 1 mM DTT.
Buffer B: 50 mM Tris 7.5, 1000 mM NaCl, 5 % glycerol, 1 mM DTT.
b. Dilute the sample with buffer A to decrease the salt concentration to around 200 mM NaCl by mixing one volume of the sample with 3 vol of buffer A. This step should be taken quickly but gently to avoid potential protein denaturation.
c. Open the AKTA protein purification system and place the pump A into buffer A, pump B into buffer B, and the sample pump tube into buffer A.
d. Wash pump A with buffer A, pump B with buffer B, sample pump with buffer A, and system with buffer A.
e. Equilibrate the 5 mL prepacked HiTrap Q HP anion exchange column with buffer A for 5 CV until the baseline is steady to zero.
f. Place the sample pump tube into the sample and load the sample to the column.
g. Wash the column with buffer A until the reads for A260 and A280 are stable to zero.
h. Set the salt concentration of the buffer to around 316 mM NaCl (~27 mS/cm) by increasing the percentage of buffer B to 24 %. The impurity will be washed off the beads under this condition.
i. Set the salt concentration of the buffer to around 424 mM NaCl (~35 mS/cm) by increasing the percentage of buffer B to 36 %. The cascade complex will be eluted under this condition.
j. Set the salt concentration to 1000 mM NaCl by increasing the buffer B to 100 %. The remaining impurities are eluted off the beads.

k. Run an SDS-PAGE gel to check the presence of Cascade (Fig. 3B).
l. Concentrate the fractions from the peak with Cascade. Take 20 μL of the sample for SDS-PAGE analysis (Fig. 3A, lane 3).
m. Clean the column by sequentially washing with 3 CV of buffer B, 3 CV of buffer A, and H_2O until the read of conductivity is zero.

3.7 Size exclusion chromatography

Size exclusion chromatography separates molecules based on their molecular weight. Given that the molecular weight of the Cascade complex is 409.5 kDa, the Superdex 200 Increase 10/300 GL column is an appropriate choice to further purify the complex and confirm its assembly.
a. Prepare SEC buffer containing 50 mM Tris 7.5, 150 mM NaCl, 5 % glycerol, 1 mM DTT.
b. Equilibrate the SEC column for at least one CV.
c. Concentrate the sample from ion exchange to 1 mL with Millipore Amicon Ultra Centrifugal Filter, 100 kDa cutoff.
d. Load the sample to SEC column and collect the fractions after SEC.
e. Run an SDS-PAGE gel to check the presence and purity of the Cascade complex.
f. Concentrate the fractions with Cascade (Fig. 3A, lane 4).
g. Measure the concentration of Cascade using Bradford protein assay. A typical yield out of 3 L of bacteria broth is around 200 ng (e.g. 4 mg/mL at the final volume of 50 μL).
h. Aliquot and freeze the sample with liquid nitrogen before storing at −80 °C freezer.

Note: It is important to mention that the plasmid for Cascade expression needs to be transformed freshly every time. From our experience, cells stroked from the glycerol stock grew fewer colonies, and the colony picked from the plate barely grow in LB broth after overnight cultivation.

4. Purification of TnsAB and TnsC

Several TnsA, TnsB and TnsC in various CAST or Tn7 systems have been successfully purified (Hoffmann et al., 2022; Park et al., 2021; Querques et al., 2021; Saito et al., 2021; Strecker et al., 2019; Xiao et al., 2021). However, due to the distinct properties of transposon proteins across different systems, not all Tns proteins in CASTs have been successfully purified to date. For instance, the purification of TnsA and TnsB from type I-F CAST and type I-D CAST remains unreported.

Here, we discuss the detailed procedure to purify TnsAB and TnsC in type I-B2 CAST system. The purification method was established by the Feng Zhang lab (Saito et al., 2021), using a two-step process: Strep affinity chromatography followed by SEC. The experiment typically takes five days including three days for cell cultivation and two days for protein purification.

4.1 Transformation, inoculation, induction

The bacterial cultivation procedures are similar to those used for Cascade, except for the cells and antibiotics. TnsAB and TnsC are expressed in BL21-CodonPlus (DE3)-RIPL cells (Agilent #230280). These cells contain extra copies of rare tRNAs encoded in two extra plasmids, making them suitable for expressing heterologous proteins in *E. coli*. To retain the plasmids, 50 μg/mL spectinomycin and 50 μg/mL chloramphenicol are added to the LB medium. 50 μg/mL ampicillin is also added in the overnight growth medium to maintain the TnsAB or TnsC plasmid. For protein expression in TB broth, only 50 μg/mL of ampicillin is required.

4.2 Cell collection and affinity chromatography using Streptactin beads

The Twin-Strep tag on N terminus of TnsAB and TnsC binds specifically to the Streptactin beads and can be eluted by buffers containing 2.5 mM dethiobiotin.

a. Prepare buffers
Lysis buffer (200 mL): 50 mM Tris 8.0, 500 mM NaCl, 10 % glycerol, 1 mM DTT, and 1 mM PMSF.
Wash buffer (500 mL): 50 mM Tris 8.0, 500 mM NaCl, 10 % glycerol, 1 mM DTT.
Elution buffer (20 mL): 50 mM Tris 8.0, 500 mM NaCl, 10 % glycerol, 1 mM DTT, and 2.5 mM dethiobiotin.

b. Resuspend the pellets from 1.5 L TB broth with 150–200 mL lysis buffer.

c. Sonicate 15 min for every 35 mL resuspension until the sample is clear. The parameters for the sonication are the same as those for Cascade purification. Take 20 μL of the sample for SDS-PAGE analysis.

d. Clarify the resuspension by centrifugation at 4 °C, 20000 rpm, for 30 min. Collect the supernatant in a beaker, filter to remove any cell debris, and place it on the ice. Take 20 μL of the supernatant for SDS-PAGE analysis.

e. Equilibrate 1 mL Streptavidin beads with 5 CV of washing buffer.

f. Incubate the supernatant with pre-equilibrated Streptactin beads on a rocker at 4 °C for 1 h.

g. Load the mixture to an empty gravity column and collect the flow through for SDS-PAGE analysis.

h. Wash the beads with washing buffer for 10 CV. Collect 20 μL of the sample for SDS-PAGE analysis.

i. Elute the sample with 5 mL elution buffer.

j. Collect the eluted sample and repeat the elution step with another 5 mL elution buffer.

k. Run an SDS-PAGE gel to check the purity.

l. Add SUMO protease to the eluted sample at the 1:100 ratio to cleave the His-Twin-Strep-SUMO tag.

m. Leave the sample at 4 °C cold room overnight

4.3 Size exclusion chromatography

a. Prepare the SEC buffer containing 50 mM Tris 7.5, 500 mM NaCl, 5 % glycerol, 1 mM DTT.

b. Equilibrate the S200 column with the SEC buffer for at least one CV.

c. Concentrate the sample to 1 mL using Millipore Amicon Ultra Centrifugal Filter (TnsAB, 50 kDa cutoff; TnsC, 30 kDa cutoff).

d. Load the sample to the column and collect the fractions through the column.

e. Run an SDS-PAGE to confirm the presence and purity of protein.

f. Combine all the fractions with the target protein and measure the concentration using NanoDrop.

g. Aliquot and freeze the sample with liquid nitrogen before storing it at −80 °C freezer.

5. Purification of TniQ-TnsC complex

In our research on type I-B2, we observed that TniQ is unstable and tends to interact with GroEL when purified alone. To address this issue, we co-purified TniQ with TnsC to stabilize the TniQ (Wang et al., 2023). However, a later study in the Feng Zhang lab demonstrated that TniQ in type I-B2 can be purified alone using a method similar to that employed for

TnsAB and TnsC (Faure et al., 2023). Here, we focus on the co-expression and purification of TniQ-TnsC complex. Two steps of purification were conducted including Strep-tag affinity chromatography and SEC.

5.1 Transformation

a. Prepare one LB agar plate with 50 μg/mL spectinomycin, 100 μg/mL ampicillin, 50 μg/mL kanamycin, and 50 μg/mL chloramphenicol.
b. Co-transform 50 ng of the pTnsC and 50 ng of the pTniQ plasmids to 20 μL BL21-CodonPlus (DE3)-RIPL Competent Cells. A lower volume of competent cells helps to increase the co-transformed cells.
c. Incubate the plate at 4 °C overnight.

5.2 Inoculation

a. Prepare 50 mL LB broth with 50 μg/mL spectinomycin, 100 μg/mL ampicillin, 50 μg/mL Kanamycin and 50 μg/mL chloramphenicol.
b. Pick a single colony from the plate and inoculate to the LB broth
c. Shake the LB broth at 37 °C, 180 rpm overnight
d. Prepare 4 flasks of 750 mL TB broth for the experiment the next day.

5.3 Induction

a. Inoculate 7.5 mL of the bacteria to 750 mL TB broth with 100 μg/mL ampicillin and 50 μg/mL kanamycin.
b. Shake the flasks at 37 °C, 180 rpm until the OD600 value reaches around 0.8
c. Add 0.5 mM IPTG and keep culturing overnight at 16 °C

5.4 Cell collection and Strep-tag affinity chromatography

The steps for cell collection and Strep-tag affinity chromatography are the same as these for TnsAB and TnsC, except for the buffers. Since the buffers contain 1000 mM NaCl which affects the activity of SUMO protease, the eluted sample is dialyzed against a low salt buffer during tag cleavage. A representative result of purification was shown in Fig. 3C, lane 1–4 (Wang et al., 2023).

The compositions of the buffers are listed below.

Lysis buffer: 50 mM Tris 8.0, 1000 mM NaCl, 10 % glycerol, 5 mM EDTA, 1 mM DTT, protease inhibitor.

Wash buffer: 50 mM Tris 8.0, 1000 mM NaCl, 10 % glycerol, 5 mM EDTA, 1 mM DTT.

Elution buffer: 50 mM Tris 8.0, 1000 mM NaCl, 10 % glycerol, 5 mM EDTA, 1 mM DTT, 2.5 mM desthiobiotin.

Dialysis buffer: 50 mM Tris 8.0, 500 mM NaCl, 10 % glycerol, 1 mM DTT.

5.5 Size exclusion chromatography

a. Prepare SEC buffer that contains 50 mM Tris 7.5, 500 mM NaCl, 5 % glycerol, 1 mM DTT.
b. Equilibrate the SEC column for at least one column volume.
c. Concentrate the sample to 1 mL using Millipore Amicon Ultra Centrifugal Filter with 50 kDa cutoff.
d. Load the sample to SEC column and collect the fractions after SEC.
e. Run an SDS-PAGE gel to check the samples.
f. Concentrate the fractions with TniQ-TnsC complex (Fig. 3C, lane 5–6) (Wang et al., 2023).
g. Aliquot and flash freeze the sample with liquid nitrogen. Store the tubes in −80 °C freezer.

6. *In vitro* transposition assay

Transposition activity was assessed through an *in vitro* transposition assay, in which purified proteins were incubated with donor and target plasmids. Following incubation, the reaction products consisted of a mixture containing intact donor and target plasmids, target plasmids with simple insertions, target plasmids with cointegrations, and various potential off-target insertions. Simple insertion refers to the direct integration of the transposon fragment while cointegrate includes the insertion of both the transposon fragment and the donor plasmid backbone. Although direct PCR and qPCR can detect on-target insertions, they are unable to detect off-target insertions or differentiate between simple insertions and cointegrates. To further analyze off-target insertions and elucidate individual transposition events, transformation can be done to isolate and purify the transposition products.

The way to prepare the sample for all three detection ways are similar, except for the donor plasmid used. For transformation, a special donor plasmid is required, which will be discussed later in the transformation section. Below is the detailed protocol to prepare the sample.

a. Prepare the dilution buffer and 5x transposition buffer
Dilution buffer: 25 mM Tris-HCl (pH 8.0), 500 mM NaCl, 1 mM EDTA, 1 mM DTT, and 25 % glycerol.
5x transposition buffer: 130 mM HEPES (pH 7.5), 21 mM Tris-HCl (pH 8.0), 10.5 mM DTT, 0.25 mM EDTA, 1 mM $MgCl_2$, 140 mM NaCl, 105 mM KCl, 6.75 % glycerol, 250 μg/mL BSA, and 10 mM ATP.

b. Measure the protein concentrations in molarity. For TnsAB and TnsC-TniQ, the concentrations are measured using Nanodrop and then converted to molarity according to the molecular weight and extinction coefficients. For Cascade, the concentration measured by Nanodrop is not accurate because of the interference of the crRNA in Cascade complex. Therefore, Braford protein assay is used to measure concentration. The molarity is converted through the total molecular weight of the protein in the cascade complex.
c. Dilute the proteins to 5 μM with dilution buffer.
d. Add 200 nM of TnsAB, 200 nM of TnsC-TniQ, 200 nM of Cascade,100 ng of pTarget, 100 ng of pDonor_Amp^R or pDonor_Kan^R and H_2O to the transposition buffer for a total volume of 20 μL.
e. Incubate the mixture at 30 °C for 30 min
f. Add 25 mM $MgOAc_2$ and incubate at 30 °C for another 2 h.
g. Take 2 μL of the final products for qPCR or direct PCR readout. For transformation, 10–15 μL of the final products is typically used.

6.1 Qualitative detection by PCR

Direct PCR is a simple method to test whether the transposition occurred. A forward primer is designed on the target plasmid upstream of the insertion site, and a reverse primer is on the LE on the donor plasmid. Alternatively, the forward primer can be designed on the RE on the donor plasmid, with the reverse primer downstream of the insertion site on the target plasmid (Fig. 4A). It should be noted that this strategy is designed to detect the LE-RE insertion orientation, which is preferred in *Pmc*CAST. To detect the RE-LE insertion, another set of primers should be designed. The PCR reaction is prepared as a standard PCR mixture. 0.5 μM of each primer and in vitro transposition reaction product were mixed to a final volume of 20 μL for PCR reactions with Phusion Hot Start II High-Fidelity PCR Master Mix (Thermo Scientific, F565L). The cycling conditions are as follows: 1 cycle, 98 °C, 3 min; 35 cycles, 98 °C, 10 s, 69 °C, 30 s, 72 °C, 10 s; 1 cycle, 72 °C, 1 min. After electrophoresis on an agarose gel, a clear band at the expected position indicates that transposition has occurred.

Tips: Different brands of DNA polymerase or different primer sets may require optimization of the annealing temperature to avoid non-specific PCR products. After visualizing the PCR products on an agarose gel, the product can be purified using a spin column or gel extraction for Sanger sequencing, using one of the primers as sequencing primer. Because the

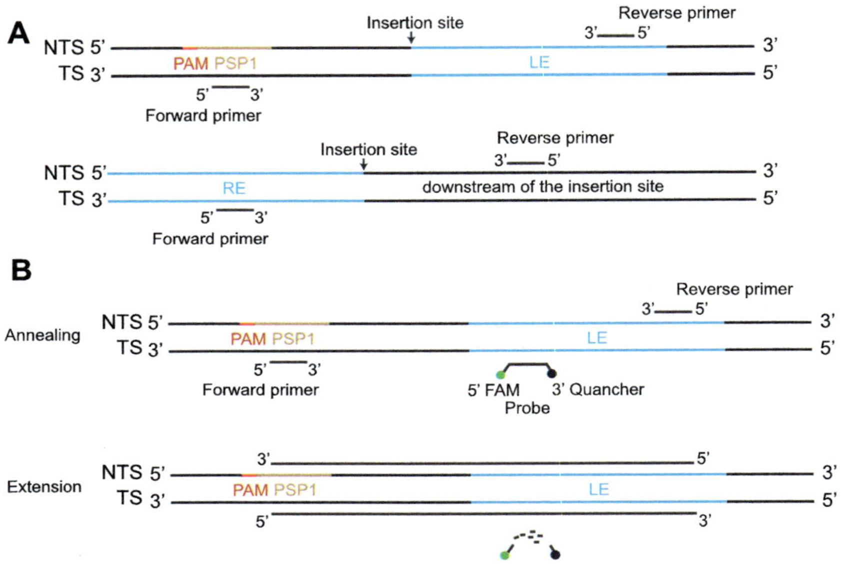

Fig. 4 Illustration for the principle of PCR methods in the transposition assay. (A) The primers designed for direct PCR. (B) The principle of qPCR and the design of primers and probe of qPCR.

insertion rate is not 100 % (typically < 1 % insertion efficiency), the PCR product is a mixture of DNA with and without insertion. Therefore, the sequencing result will be noisy around the expected insertion site indicating a successful transposition.

6.2 Quantitative detection by qPCR

Quantitative PCR (qPCR) can be used to assess the copy number of plasmids in the reaction. The principle of qPCR using TaqMan technology is to use a probe that generates quantitative fluorescence signals corresponding to the number of amplicons in the reaction (Fig. 4B). The probe is a small DNA fragment with a 5′ FAM (fluorescent reporter) modification and a 3′ quencher modification. When intact, the probe does not emit fluorescence and can anneal to the template like a normal primer. In this experiment, the probe is designed to anneal at the LE of the transposon. At the start of the qPCR reaction, the temperature is raised to 95 °C to denature the template, breaking all secondary structures. As the temperature decreases, the probe and two primers anneal to their corresponding sites. The forward primer binds upstream of the insertion site, while the reverse primer and probe anneal at the LE. When the polymerase begins to extend the primers, its

5′ to 3′ exonuclease activity cleaves the probe, separating the fluorescent reporter from the quencher and generating a fluorescent signal. The fluorescence signal is proportional to the numbers of the amplicon present. By comparing the Ct value of the sample to a standard curve of Ct values from known standards under identical conditions, the copy number of the target sequence in the sample can be estimated.

The first step in qPCR is to make two standard curves: one for the target plasmid and one for the inserted plasmid. The template for the inserted plasmid standard curve is a previously sequenced plasmid containing the on-target insertion. Detailed instructions for creating the inserted plasmid standard curve are provided below. The protocol for the target plasmid follows a similar procedure, with adjustments in the cycling conditions.

a. Measure the concentration of the plasmid (ng/μL) using a NanoDrop spectrophotometer.
b. Calculate the molecular weight of the plasmid using the sequence and an online web server (e.g., https://www.bioinformatics.org/sms2/dna_mw.html).
c. Use an online DNA copy number calculator (e.g., provided by Thermo Fisher Scientific) to convert the concentration of plasmid from ng/μL to copies/μL.
d. Dilute the stock plasmid to create serial dilutions of 10^9, 10^8, 10^7, 10^6, 10^5, 10^4, 10^3, and 10^2 copies/μL
e. Prepare the qPCR reaction mixture, including two primers (900 nM each), one probe (250 nM), and TaqMan Fast Advanced Master Mix (Applied Biosystems, 4444557). Prepare enough mixture to run triplicates for each concentration. Carefully aliquot 18 μL of the reaction mixture into each well of a 96-well plate.
f. Add 2 μL of each diluted stock to the corresponding wells.
g. Centrifuge the plate briefly to ensure the reagents are properly mixed and settled at the bottom of the wells.
h. Set up the Bio-Rad CFX Connect Real-Time PCR Detection system with the following cycling conditions:
1 cycle: 50 °C for 2 min
1 cycle: 95 °C for 1.5 min
39 cycles: 95 °C for 10 s, 56 °C for 30 s
i. Obtain the Ct value for each reaction using Bio-Rad CFX Maestro software (or the appropriate software for your qPCR system).

j. Plot the Ct values against the log (number of copies/μL) and calculate the R (Faure et al., 2019) value using GraphPad Prism 8 or Excel. A good R (Faure et al., 2019) value should be higher than 0.99.

k. Calculate the efficiency of the qPCR reaction using the slope of the standard curve. An ideal efficiency should be between 90 % and 110 %, corresponding to a slope between −3.1 and −3.6.

To test the amount of inserted plasmid in products after *in vitro* transposition, 2 μL of the product, 900 nM forward primer, 900 nM reverse primer, and 250 nM of LE probe were mixed. Thermal cycling conditions are as follows: 1 cycle, 50 °C, 2 min; 1 cycle, 95 °C, 1.5 min; 39 cycles, 95 °C, 10 s, 56 °C, 30 s; End. Following the run, retrieve the Ct values using the Bio-Rad CFX Maestro software. The copy number of the inserted plasmid can be determined by plotting these Ct values against the standard curve.

To test the amount of target plasmid in products after *in vitro* transposition, the samples were diluted 100 times to make the concentration within the range of the standard curve. 2 μL of the diluted samples, 900 nM forward primer, 900 nM reverse primer, and 250 nM of target plasmid probe were mixed. Thermal cycling conditions were as follows: 1 cycle, 50 °C, 2 min; 1 cycle, 95 °C, 1.5 min; 39 cycles, 95 °C, 10 s, 59 °C, 30 s; End. After the run, obtain the Ct values using the Bio-Rad CFX Maestro software. Determine the copy number of the diluted target plasmid by plotting these Ct values against the standard curve. To obtain the original copy number of the target plasmid, multiply the determined copy number by 100. The insertion frequency was calculated as the copy number of inserted plasmid divided by the copy number of target plasmid.

6.3 Qualitative detection by transformation

The pDonor_Kan^R plasmid is used in this experiment. As mentioned above in the plasmid construction section, pDonor_Kan^R has the LE and RE of the *Pmc*CAST with a kanamycin resistance gene in between. The Top10 cells transformed with the transposition products will be resistant to Chloramphenicol and Kanamycin. It is important to note that co-transformant of the pDonor_Kan^R and pTarget will not survive, because pDonor_Kan^R cannot be maintained in Top10 cells due to the lack of replication protein π in the host cell. After transformation, single colonies can be picked, and the products of individual events can be obtained by extracting the plasmids. The exact insertion site and the off-taget insertion can be confirmed by Sanger sequencing and whole plasmid sequencing.

Restriction enzyme digestion is performed to determine whether the resultant transposition product is a simple insertion or a cointegration event. In our experiment, we used *Bam*HI restriction enzyme, which results in different patterns when testing simple insertion and cointegrate (Wang et al., 2023).

7. Conclusion

The recently discovered CAST systems represent a promising tool for gene knock-in applications. The ability to measure the activity of these systems *in vitro* is essential for the molecular mechanism studies of the systems. This chapter provides detailed protocols for reconstituting the type I-B2 *Pmc*CAST system. The type I CAST systems constitute the majority of CAST systems identified so far, including type I-B (Saito et al., 2021), I-F (Klompe et al., 2019), I-D (Faure et al., 2023; Hsieh & Peters, 2023). Biochemical reconstitution of the other two systems have not been reported yet, which limits the mechanistic study of these systems. The methods described in this chapter could be applied to reconstitute other types of CAST systems. The mechanistic insights gained from the reconstituted system will be critical for revealing the full potential of CASTs for future therapeutic applications.

Acknowledgments

This work was supported by the NIH grant R01GM138675 and NSF CAREER 2339799 to L.C.

Author contributions

S.W. prepared the first draft of the manuscript. L.C. provided revisions and edits.

References

Faure, G., Saito, M., Benler, S., Peng, I., Wolf, Y. I., Strecker, J., ... Makarova, K. S. (2023). Modularity and diversity of target selectors in Tn7 transposons. *Molecular Cell, 83*, 2122–2136.e2110. https://doi.org/10.1016/j.molcel.2023.05.013.

Faure, G., Shmakov, S. A., Yan, W. X., Cheng, D. R., Scott, D. A., Peters, J. E., ... Koonin, E. V. (2019). CRISPR-Cas in mobile genetic elements: Counter-defence and beyond. *Nature Reviews. Microbiology, 17*, 513–525. https://doi.org/10.1038/s41579-019-0204-7.

Halpin-Healy, T. S., Klompe, S. E., Sternberg, S. H., & Fernandez, I. S. (2020). Structural basis of DNA targeting by a transposon-encoded CRISPR-Cas system. *Nature, 577*, 271–274. https://doi.org/10.1038/s41586-019-1849-0.

Hochstrasser, M. L., Taylor, D. W., Kornfeld, J. E., Nogales, E., & Doudna, J. A. (2016). DNA targeting by a minimal CRISPR RNA-guided cascade. *Molecular Cell, 63*, 840–851. https://doi.org/10.1016/j.molcel.2016.07.027.

Hoffmann, F. T., Kim, M., Beh, L. Y., Wang, J., Vo, P. L. H., Gelsinger, D. R., ... Sternberg, S. H. (2022). Selective TnsC recruitment enhances the fidelity of RNA-guided transposition. *Nature, 609*, 384–393. https://doi.org/10.1038/s41586-022-05059-4.

Hsieh, S. C., & Peters, J. E. (2023). Discovery and characterization of novel type I-D CRISPR-guided transposons identified among diverse Tn7-like elements in cyanobacteria. *Nucleic Acids Research, 51*, 765–782. https://doi.org/10.1093/nar/gkac1216.

Hu, C. Y., Ni, D. C., Nam, K. H., Majumdar, S., McLean, J., Stahlberg, H., ... Ke, A. L. (2022). Allosteric control of type I-A CRISPR-Cas3 complexes and establishment as effective nucleic acid detection and human genome editing tools. *Molecular Cell, 82*, 2754. https://doi.org/10.1016/j.molcel.2022.06.007.

Jia, N., Xie, W., de la Cruz, M. J., Eng, E. T., & Patel, D. J. (2020). Structure-function insights into the initial step of DNA integration by a CRISPR-Cas-transposon complex. *Cell Research, 30*, 182–184. https://doi.org/10.1038/s41422-019-0272-2.

Jiang, F., & Doudna, J. A. (2015). The structural biology of CRISPR-Cas systems. *Current Opinion in Structural Biology, 30*, 100–111. https://doi.org/10.1016/j.sbi.2015.02.002.

Klompe, S. E., Vo, P. L. H., Halpin-Healy, T. S., & Sternberg, S. H. (2019). Transposon-encoded CRISPR-Cas systems direct RNA-guided DNA integration. *Nature, 571*, 219–225. https://doi.org/10.1038/s41586-019-1323-z.

Lampe, G. D., King, R. T., Halpin-Healy, T. S., Klompe, S. E., Hogan, M. I., Vo, P. L. H., ... Sternberg, S. H. (2024). Targeted DNA integration in human cells without double-strand breaks using CRISPR-associated transposases. *Nature Biotechnology, 42*, 87–98. https://doi.org/10.1038/s41587-023-01748-1.

Li, Z., Zhang, H., Xiao, R., & Chang, L. (2020). Cryo-EM structure of a type I-F CRISPR RNA guided surveillance complex bound to transposition protein TniQ. *Cell Research, 30*, 179–181. https://doi.org/10.1038/s41422-019-0268-y.

Lu, M., Yu, C., Zhang, Y., Ju, W., Ye, Z., Hua, C., ... Xiao, Y. (2024). Structure and genome editing of type I-B CRISPR-Cas. *Nature Communications, 15*, 4126. https://doi.org/10.1038/s41467-024-48598-2.

McBride, T. M., Schwartz, E. A., Kumar, A., Taylor, D. W., Fineran, P. C., & Fagerlund, R. D. (2020). Diverse CRISPR-Cas complexes require independent translation of small and large subunits from a single gene. *Molecular Cell, 80*, 971–979.e977. https://doi.org/10.1016/j.molcel.2020.11.003.

Park, J. U., Petassi, M. T., Hsieh, S. C., Mehrotra, E., Schuler, G., Budhathoki, J., ... Peters, J. E. (2023b). Multiple adaptations underly co-option of a CRISPR surveillance complex for RNA-guided DNA transposition. *Molecular Cell, 83*, 1827–1838.e1826. https://doi.org/10.1016/j.molcel.2023.05.005.

Park, J. U., Tsai, A. W., Mehrotra, E., Petassi, M. T., Hsieh, S. C., Ke, A., ... Kellogg, E. H. (2021). Structural basis for target site selection in RNA-guided DNA transposition systems. *Science (New York, N. Y.), 373*, 768–774. https://doi.org/10.1126/science.abi8976.

Park, J. U., Tsai, A. W., Rizo, A. N., Truong, V. H., Wellner, T. X., Schargel, R. D., & Kellogg, E. H. (2023a). Structures of the holo CRISPR RNA-guided transposon integration complex. *Nature, 613*, 775–782. https://doi.org/10.1038/s41586-022-05573-5.

Peters, J. E., Makarova, K. S., Shmakov, S., & Koonin, E. V. (2017). Recruitment of CRISPR-Cas systems by Tn7-like transposons. *Proceedings of the National Academy of Sciences of the United States of America, 114*, E7358–E7366. https://doi.org/10.1073/pnas.1709035114.

Plagens, A., Tripp, V., Daume, M., Sharma, K., Klingl, A., Hrle, A., ... Randau, L. (2014). In vitro assembly and activity of an archaeal CRISPR-Cas type I-A Cascade interference complex. *Nucleic Acids Research, 42*, 5125–5138. https://doi.org/10.1093/nar/gku120.

Querques, I., Schmitz, M., Oberli, S., Chanez, C., & Jinek, M. (2021). Target site selection and remodelling by type V CRISPR-transposon systems. *Nature, 599*, 497–502. https://doi.org/10.1038/s41586-021-04030-z.

Richter, H., Rompf, J., Wiegel, J., Rau, K., & Randau, L. (2017). Fragmentation of the CRISPR-Cas Type I-B signature protein Cas8b. *Biochimica et Biophysica Acta (BBA) - General Subjects, 1861*, 2993–3000. https://doi.org/10.1016/j.bbagen.2017.02.026.

Saito, M., Ladha, A., Strecker, J., Faure, G., Neumann, E., Altae-Tran, H., ... Zhang, F. (2021). Dual modes of CRISPR-associated transposon homing. *Cell, 184*, 2441–2453.e2418. https://doi.org/10.1016/j.cell.2021.03.006.

Schmitz, M., Querques, I., Oberli, S., Chanez, C., & Jinek, M. (2022). Structural basis for the assembly of the type V CRISPR-associated transposon complex. *Cell, 185*, 4999–5010.e4917. https://doi.org/10.1016/j.cell.2022.11.009.

Shangguan, Q., Graham, S., Sundaramoorthy, R., & White, M. F. (2022). Structure and mechanism of the type I-G CRISPR effector. *Nucleic Acids Research, 50*, 11214–11228. https://doi.org/10.1093/nar/gkac925.

Strecker, J., Ladha, A., Gardner, Z., Schmid-Burgk, J. L., Makarova, K. S., Koonin, E. V., & Zhang, F. (2019). RNA-guided DNA insertion with CRISPR-associated transposases. *Science (New York, N. Y.), 365*, 48–53. https://doi.org/10.1126/science.aax9181.

Tou, C. J., Orr, B., & Kleinstiver, B. P. (2023). Precise cut-and-paste DNA insertion using engineered type V-K CRISPR-associated transposases. *Nature Biotechnology, 41*, 968–979. https://doi.org/10.1038/s41587-022-01574-x.

Wang, B., Xu, W., & Yang, H. (2020). Structural basis of a Tn7-like transposase recruitment and DNA loading to CRISPR-Cas surveillance complex. *Cell Research, 30*, 185–187. https://doi.org/10.1038/s41422-020-0274-0.

Wang, S., Gabel, C., Siddique, R., Klose, T., & Chang, L. (2023). Molecular mechanism for Tn7-like transposon recruitment by a type I-B CRISPR effector. *Cell, 186*, 4204–4215.e4219. https://doi.org/10.1016/j.cell.2023.07.010.

Wiedenheft, B., van Duijn, E., Bultema, J. B., Waghmare, S. P., Zhou, K., Barendregt, A., ... Doudna, J. A. (2011). RNA-guided complex from a bacterial immune system enhances target recognition through seed sequence interactions. *Proceedings of the National Academy of Sciences of the United States of America, 108*, 10092–10097. https://doi.org/10.1073/pnas.1102716108.

Xiao, R., Wang, S., Han, R., Li, Z., Gabel, C., Mukherjee, I. A., & Chang, L. (2021). Structural basis of target DNA recognition by CRISPR-Cas12k for RNA-guided DNA transposition. *Molecular Cell, 81*, 4457–4466.e4455. https://doi.org/10.1016/j.molcel.2021.07.043.

Xiao, Y., Luo, M., Hayes, R. P., Kim, J., Ng, S., Ding, F., ... Ke, A. (2017). Structure basis for directional R-loop formation and substrate handover mechanisms in type I CRISPR-Cas system. *Cell, 170*, 48–60.e11. https://doi.org/10.1016/j.cell.2017.06.012.

CHAPTER FIVE

Determining the biochemical function of type IV CRISPR ribonucleoprotein complexes and accessory proteins

Andrew A. Williams, Olivine Redman, Hannah Domgaard, Matthew J. Armbrust, and Ryan N. Jackson*
Department of Chemistry and Biochemistry, Utah State University, Logan, UT, United States
*Corresponding author. e-mail address: ryan.jackson@usu.edu

Contents

1. Introduction 82
2. Plasmid clearance assay to assess in vivo type IV-A immunity 85
   2.1 Plasmid clearance assay materials and equipment 87
   2.2 Plasmid clearance assay protocol day 1 – transform the type IV-A system 87
   2.3 Plasmid clearance assay protocol day 2 – make a small liquid starter culture 89
   2.4 Plasmid clearance assay protocol day 3 – make cells competent 89
   2.5 Plasmid clearance assay protocol day 3 or 4 – perform plasmid curing assay 90
3. Expressing and purifying the type IV-A Csf ribonucleoprotein complex 91
   3.1 Csf complex expression materials and equipment 91
   3.2 Csf complex expression day 1 – bacterial transformation 91
   3.3 Csf complex expression day 2 – small culture growth 92
   3.4 Csf complex expression day 3 – large culture growth 94
   3.5 Csf complex expression day 4 – harvesting 94
   3.6 Csf complex purification materials and equipment 94
   3.7 Csf complex purification – lysis and strep-affinity chromatography 95
   3.8 Csf complex purification – concentrate and size-exclusion chromatography 96
4. Expressing and purifying the CasDinG helicase 96
   4.1 CasDinG expression materials and equipment 97
   4.2 CasDinG expression days 1 and 2 – bacterial transformation and liquid culture 97
   4.3 CasDinG expression day 3 – large culture growth 98
   4.4 CasDinG expression day 3 – harvesting cells 98
   4.5 CasDinG purification materials and equipment 98
   4.6 CasDinG purification – lysis and strep-affinity chromatography 99
   4.7 CasDinG purification – desalting and heparin chromatography 100
   4.8 CasDinG purification – concentration and size-exclusion chromatography 100

Methods in Enzymology, Volume 712
ISSN 0076-6879, https://doi.org/10.1016/bs.mie.2025.01.039

5. Purification of type IV-A Csf complex bound to DNA targets and the CasDinG helicase 101
  5.1 Materials for SEC of Csf complexes bound to target DNA and CasDinG 101
  5.2 SEC of Csf complex bound to target DNA - preparation 104
  5.3 SEC of Csf complex bound to target DNA – protocol 104
  5.4 SEC of CasDinG recruited to R-loop-bound Csf complexes - preparation 105
  5.5 SEC of CasDinG recruited to R-loop-bound Csf complexes – protocol 105
6. Determining Csf complex or CasDinG affinity to nucleic acid targets with fluorescence anisotropy 106
  6.1 Materials and equipment 107
  6.2 Set up and perform fluorescence anisotropy 107
  6.3 Fluorescence anisotropy troubleshooting 110
7. CasDinG helicase assay 110
  7.1 Materials 111
  7.2 Duplex substrate annealing 111
  7.3 Prepare controls 112
  7.4 Time course protocol 112
8. Conclusion 114
References 115

Abstract

Type IV CRISPR systems are phylogenetically diverse and poorly understood. However, recently, major strides have been made toward understanding type IV-A systems. In type IV-A systems, a multi-subunit ribonucleoprotein complex, called the Csf complex, uses a CRISPR-derived guide to bind double-stranded DNA, forming an R-loop to which a helicase called CRISPR-associated DinG (CasDinG) is recruited. It is proposed that the ATP-dependent helicase activity of CasDinG then unwinds duplex DNA near the targeting site, impairing RNA transcription, and gene expression. Here we describe methods used to investigate the type IV-A system from *Pseudomonas aeruginosa strain 83* including a plasmid clearance assay, expression and purification of type IV ribonucleoprotein complexes and proteins, nucleic acid binding assays, and CasDinG helicase assays. These methods provide a foundation for future work aimed at understanding these enigmatic systems.

1. Introduction

Multi-subunit type IV CRISPR systems are the least understood of the six CRISPR types (Makarova et al., 2020; Moya-Beltrán et al., 2021; Pinilla-Redondo et al., 2020; Taylor et al., 2021). Two schemes for defining the broadly diverse type IV subtypes have been described; one scheme separates the systems into three subtypes (Moya-Beltrán et al., 2021), and another into five (IV-A to IV-C or to IV-E) (Pinilla-Redondo et al., 2020).

No matter the scheme, each subtype is distinguished by a signature accessory or fusion protein (e.g. IV-A systems encode a CRISPR-associated DinG (CasDinG) helicase, IV-B systems encode a CysH-like protein, and IV-C encode a Cas10-like protein) (Makarova et al., 2020; Moya-Beltrán et al., 2021; Taylor et al., 2021). Additionally, subtypes have been further classified into distinct subtype variants (e.g. IV-A1, IV-A2, and IV-A3) using sequence differences of a highly diverged Cas7-like protein, called Csf2 (Pinilla-Redondo et al., 2020).

Type IV systems are distinct from most other CRISPR-Cas systems in that (i) instead of being encoded on bacterial chromosomes they are primarily encoded on plasmids, megaplasmids, integrative and conjugative elements (ICEs), and prophages, and (ii) instead of targeting primarily bacteriophage sequences, type IV systems appear to generally target non-bacteriophage mobile genetic elements (MGEs), like plasmids, megaplasmids, and conjugational transfer genes (Newire, Aydin, Juma, Enne, & Roberts, 2020; Pinilla-Redondo et al., 2020). From these observations, it has been hypothesized that type IV systems are more involved in interplasmid competition than defense against viral pathogens. While many type IV systems are associated with a CRISPR array, most lack genes encoding the protein machinery (i.e. Cas1 and Cas2) that acquire immunity to new targets by adding "spacers" in between the repeats of the CRISPR array (Pinilla-Redondo et al., 2020), relying on other CRISPR systems for in *trans* spacer acquisition (Benz et al., 2024).

Although a complete functional understanding of all type IV subtypes is lacking, recent studies have revealed a model describing the mechanism of type IV-A (Čepaitė et al., 2024; Cui et al., 2023; Guo et al., 2022; Domgaard et al., 2022). Type IV-A systems encode five proteins: Csf1, Csf2, Csf3, Csf4 (CasDinG), and Csf5 (Cas6) that are each essential to mediate plasmid and phage interference (Altae-Tran et al., 2023; Čepaitė et al., 2024; Crowley et al., 2019; Domgaard et al., 2023; Guo et al., 2022; Özcan et al., 2019; Sanchez-Londono et al., 2024) Like other class I systems, the CRISPR array in the type IV-A system is expressed as a long pre-crRNA that is processed by a Cas6-like protein (e.g. Csf5) (Özcan et al., 2019; Taylor et al., 2019). After cleaving the crRNA the protein remains bound to the mature crRNA, likely serving as a scaffold for assembly of the multi-subunit ribonucleoprotein complex. The Csf1, Csf2, Csf3, and Csf5 proteins bind the crRNA with unequal stoichiometry to form a multi-subunit ribonucleoprotein complex (1x-Csf1:5x-Csf2:1x-Csf3:1x-Csf5), called the Csf complex (Čepaitė et al., 2024; Cui et al., 2023). Once formed, the Csf ribonucleoprotein complex binds dsDNA targets that meet two criteria; (i) the sequence of the dsDNA target is complementary to the crRNA guide, and (ii) the target sequence is flanked on the 5′-side of the non-target strand by a small "non-self" sequence, called a protospacer adjacent motif (PAM) (Benz et al., 2024; Domgaard et al., 2023; Guo et al., 2022). Binding of the Csf complex crRNA guide to the dsDNA target forms an R-loop, to which the CasDinG protein is recruited (Čepaitė et al., 2024; Cui et al., 2023). Once bound to the R-loop, it is proposed CasDinG hydrolyzes ATP to move the helicase in a 5′−3′ direction away from the target, distorting duplex DNA and displacing bound RNAs (Čepaitė et al., 2024; Domgaard et al., 2023; Guo et al., 2022; Sanchez-Londono et al., 2024). Recently it was reported that a CasDinG helicase fused to an HNH domain degrades the dsDNA target, similar to the Cas3 helicase of type I multi-subunit systems (Altae-Tran et al., 2023; Mulepati & Bailey, 2013; Sinkunas et al., 2011; van Erp et al., 2015). However, the majority of type IV-A systems do not encode a recognizable nuclease domain on CasDinG or the Csf complex, and cell-based and biochemical data indicate these type IV-A systems silence transcription near type IV targeting sites rather than cleave DNA targets (Čepaitė et al., 2024; Domgaard et al., 2023; Guo et al., 2022; Sanchez-Londono et al., 2024).

CasDinG is the signature protein of type IV-A systems and is an accessory protein to the Csf complex (Makarova et al., 2020). CasDinG proteins are superfamily 2 helicases that form a phylogenetically distinct clade from related DinG (Damage inducible gene G) helicases involved in recombination and repair, indicating CasDinG and DinG helicases have distinct biological roles (Taylor et al., 2021). CasDinG is an ATP- and metal ion-dependent helicase that unwinds DNA-DNA and DNA-RNA duplexes with a 5′−3′ polarity (Cui et al., 2023; Domgaard et al., 2023). It is hypothesized that after the Csf complex loads CasDinG onto the R-loop the 5′−3′ helicase activity of CasDinG drives the transcription interference mechanism observed in vivo by distorting duplex DNA and disrupting RNA transcription (Čepaitė et al., 2024; Crowley et al., 2019; Domgaard et al., 2023; Guo et al., 2022; Sanchez-Londono et al., 2024). CasDinG's activity is essential to type IV-A immunity, and it has structural elements vital to its function, including a helicase core and three accessory domains; the N-terminal domain, a vestigial FeS domain, and an arch domain (Čepaitė et al., 2024; Cui et al., 2023; Domgaard et al., 2023; Hallmark et al., 2024). While it is clear each accessory domain is essential to CasDinG activity and type IV-A mediated transcriptional silencing, their roles in the mechanism remain largely unknown. In particular, the role of the N-terminal domain remains especially elusive (Domgaard et al., 2023). Further biochemical and structural studies of the type IV systems are needed to better understand the essential mechanisms of the CasDinG accessory domains and the mechanism of type IV-A mediated transcriptional silencing. Additionally, very little cellular, biochemical, or structural work has been done to determine the function of the other type IV subtypes (Zhou et al., 2021).

To support future efforts aimed at understanding type IV CRISPR systems, here we summarize protocols necessary for testing the in vivo and in vitro activity of a type IV-A system from *Pseudomonas aeruginosa*. We detail the methodology for conducting cell-based plasmid clearance assays and provide protocols to express and purify the type IV-A Csf ribonucleoprotein complex and CasDinG separately, and in complex with a dsDNA target. Additionally, we detail methods for testing the DNA binding and helicase activities of CasDinG with fluorescence polarization anisotropy (FP anisotropy) and helicase assays.

2. Plasmid clearance assay to assess in vivo type IV-A immunity

Unlike CRISPR systems that cleave DNA targets (e.g. Cas9, and type I-E), the general mechanism of type IV-A systems from *Pseudomonas* appears to be transcriptional silencing (Guo et al., 2022; Sanchez-Londono et al., 2024). This means that in vitro DNA cleavage assays cannot be used to measure the activity of recombinantly expressed and purified type IV-A systems. Therefore, in vivo methods must be used instead to evaluate immune system activity and measure the influence of introduced mutations on system function. CRISPR immunity against DNA is often studied using a plasmid clearance assay, where a cell is first immunized against a specific target sequence by transfecting the cell with a plasmid containing a CRISPR system programmed against a target sequence (Liao et al., 2019). The cell containing the CRISPR system is subsequently challenged by transforming a plasmid containing the target sequence and an antibiotic-resistance gene. Plating on selection media kills the cells that effectively clear the target plasmid or silence antibiotic-resistance, allowing for a comparison of targeting vs non-targeting conditions. In our version of a plasmid clearance assay, we transplanted the type IV-A system from

Pseudomonas aeruginosa strain 83 (*Pa83*) into *E. coli* HMS174(DE3) on two plasmids. Two versions of the immune system were made, one with a CRISPR programmed to target a DNA sequence found on a third plasmid, and another programmed with a non-targeting guide. Both systems were then challenged with the plasmid containing the target sequence and surviving colonies were counted to observe immunity (Fig. 1). We found that the *E. coli* HMS174(DE3) provided the clearest results for plasmid clearance assays and we highly recommend beginning plasmid clearance assays with this cell line.

The genes of the *Pa83* type IV-A system were codon optimized for *E. coli* and ordered from Twist Bioscience. Genes *csf1, csf2,* and *csf5/cas6* were inserted into the pCDFDuet expression vector (Novagen) under T7 control, and genes *csf3, casdinG,* and a single spacer CRISPR array were inserted into pACYC (Novagen), also under a T7 promoter. Of particular note is the target spacer sequence, which has a high impact on observed targeting. We see the best targeting when the crRNA guide sequence (CRISPR spacer) base pairs with the coding strand of the targeted gene. When the Csf complex binds the coding strand, CasDinG loads onto the template strand in an orientation where the 5′−3′ helicase activity of CasDinG will move the helicase upstream of the target site towards the promoter (Domgaard et al., 2023; Guo et al., 2022). Because the IV-A

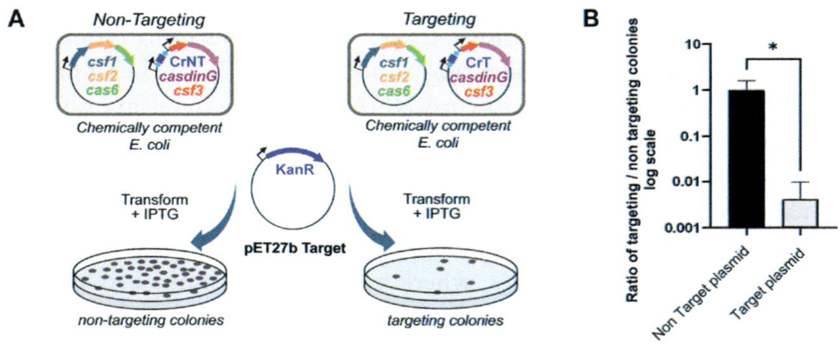

Fig. 1 Plasmid clearance assay. (A) Diagram of the plasmid clearance assay measuring immunity with a target plasmid (pET27b). The rounded rectangles represent chemically competent bacteria expressing type IV-A immune systems (non-targeting or targeting) from plasmids (circles). Colored arrows indicate protein-coding regions, the CRISPR is represented by small boxes, and small arrows represent promoters. (B) Representative results from a plasmid clearance assay in cells with recombinant type IV-A systems, shown in log scale. Bars represent the average ratio of target to non-target colonies in a triplicate experiment, with error bars showing standard deviation.

system represses transcription, the targeted gene must be essential for plasmid maintenance to observe a targeting phenotype. For example, a strong target would be an antibiotic resistance gene under conditions of antibiotic selective pressure. We designed our system to target a sequence within the kanamycin resistance gene located on the plasmid pET27b_CA01_GCTAG, and refer to the plasmid harboring this targeting CRISPR as "CrT". We also created a plasmid harboring a non-targeting CRISPR array ("CrNT") with the same pACYC backbone containing the *csf3* and *casdinG* genes. Below we explain the materials, equipment, and protocols we follow to execute a type IV plasmid clearance assay.

2.1 Plasmid clearance assay materials and equipment

- *E. coli* HMS174(DE3) chemically competent cells
- 1000x streptomycin (50 mg/mL)
- 1000x chloramphenicol (25 mg/mL)
- 1000x kanamycin (50 mg/mL)
- LB liquid media
- LB agar (1.5 %) solid media
- Purified plasmid (expression and target vectors)
- Competent cell buffer 1 (100 mM RbCl, 50 mM $MnCl_2$, 30 mM potassium acetate, 15 % m/v glycerol)
- Competent cell buffer 2 (10 mM RbCl, 10 mM MOPS, 75 mM $CaCl_2$, 15 % m/v glycerol)
- IPTG stock (1 M)
- 1.5 mL Microcentrifuge tubes
- 15 mL sterile conical tubes
- Petri plates
- 42 °C water bath or heat block
- 37 °C incubator
- Swing-bucket rotor centrifuge
- Sterile glass beads

2.2 Plasmid clearance assay protocol day 1 – transform the type IV-A system

1. Thaw two tubes of chemically competent HMS174(DE3) cells on ice. One of these tubes will be transformed with 100 ng each of pCDF_csf1-csf2-cas6 and pACYC_casdinG-csf3-CrT (targeting system), and the other will be transformed with 100 ng each of pCDF_Pa_csf1_csf2_cas6 and pACYC_casdinG-csf3-CrNT (non-targeting system) (Table 1).

Table 1 Plasmids used in the plasmid clearance assay.

Plasmid name	Plasmid description	Link
pACYC_casdinG-csf3-CrT	Plasmid containing type IV-A1 system with targeting spacer for in vivo plasmid clearance assay	https://benchling.com/s/seq-PqtrRfi7JJk5cnwvLvjt?m=slm-wmVZypwPz2wrnaxPHi2K
pACYC_casdinG-csf3-CrNT	Plasmid containing type IV-A1 system with non-targeting spacer for in vivo plasmid clearance assay	https://benchling.com/s/seq-PqtrRfi7JJk5cnwvLvjt?m=slm-wmVZypwPz2wrnaxPHi2K
pCDF_Pa_csf1_csf2_cas6	Plasmid containing type IV-A1 system for in vivo plasmid clearance assay	https://benchling.com/s/seq-vCC4NsYDR0A102mbjuA2?m=slm-EYISoKlhUzfGonZnjvMX
pET27b_CA01_GCTAG	Target plasmid for plasmid clearance assay	https://benchling.com/s/seq-HLKXXq9t26FTABYymGYd?m=slm-ebrayuypiVO1XJe1FHXb

2. Add DNA to each tube and let tubes incubate on ice for 30 min.
3. Heat shock at 42 °C for 30 s, followed by another 2-minute incubation on ice.
4. Add 400 μL of recovery LB media without selection antibiotics to each tube and recover at 37 °C for 1 h.
5. Plate 200 μL of each transformation onto an LB agar plate containing 50 μg/mL streptomycin and 25 μg/mL chloramphenicol, and incubate at 37 °C overnight.

2.3 Plasmid clearance assay protocol day 2 – make a small liquid starter culture

1. Pick three well-isolated colonies from each plate (3 targeting systems and 3 non-targeting systems) and use each to inoculate a respective 5 mL overnight LB culture with antibiotic selection (6 total 5 mL overnights). Each colony will provide one biological replicate for a triplicate set of targeting or non-targeting systems.

2.4 Plasmid clearance assay protocol day 3 – make cells competent

1. Prepare six 50 mL conical tubes each with 13 mL of LB media with 50 μg/mL streptomycin and 25 μg/mL chloramphenicol.
2. Add 100 μL of each overnight growth to a respective tube of LB media with antibiotics, making sure to label tubes well.
3. Let liquid cultures grow until an OD_{600} of 0.25-0.27 is reached.
4. Once the proper OD_{600} is reached, add IPTG to a concentration of 0.1 mM and let cultures grow for another 45 min at 37 °C.
5. Incubate tubes on ice for 15 min to chill.
6. Spin down cell pellets in a swing-bucket rotor at 2700 × *g* for 10 min at 4 °C.
7. Discard the supernatant and resuspend cells in ice-cold 4.2 mL of Competent cell buffer 1 (see Section 2.1). Let sit on ice for 15 min.
8. Centrifuge cells at 580 × *g* for 15 min at 4 °C.
9. Discard supernatant and resuspend cells in 0.5 mL of ice-cold RF2 buffer (see Section 3.1).
10. Pipette resuspended cells as 100 μL aliquots into 1.5 mL micro-centrifuge tubes for transformation.
11. Flash freeze all aliquots and store them at −80 °C, for future use. If wanting to perform assay on same day set aside 1 aliquot from each set of the 6 overnight growth cells.

Note: more aliquots can be set aside at this point for experimental replicates and examination of competency if desired. We recommend performing the assay within one week of preparing competent cells, but optimally the next day.

2.5 Plasmid clearance assay protocol day 3 or 4 – perform plasmid curing assay

1. Add 50 ng of target plasmid to each of the six aliquots that were set aside and incubate on ice for 30 min.
2. Heat shock tubes at 42 °C for 30 s and return to ice for two minutes.
3. To each tube add 400 μL LB supplemented with 0.1 mM IPTG. Place in a shaking 37 °C incubator for 40 min.
4. Prepare 2 LB agar plates per transformation (12 total), each supplemented with 0.1 mM IPTG, 50 μg/mL streptomycin, 50 μg/mL kanamycin, and 25 μg/mL chloramphenicol and keep warm in 37 °C.
5. Prepare 6 tubes with 300 μL LB at 0.1 mM IPTG.
6. Dilute 100 μL of each transformation into the tubes prepared with 300 μL of LB and 0.1 mM IPTG. This splits the transformation reaction into 4/5 (undiluted) and 1/5 (diluted) parts for each replicate but keeps each portion at a constant total volume of 400 μL of LB media.
7. Plate each 400 μL part (4/5 and 1/5) entirely onto one of the warmed LB agar plates supplemented with antibiotics and IPTG. Spread the culture with sterile glass beads or a spreader and let dry. Once dry, cover and place each plate upside down in a 37 °C incubator for 18 h.
8. The next day count the colonies on each plate. The transformation was divided into parts so that if the 4/5 plate has too many colonies to accurately count, the 1/5 plate can be counted in its place.
9. Assess targeting efficiency by calculating the ratio of # targeting colonies / # non-targeting colonies. A lower ratio of T:NT indicates more efficient immunity against the target plasmid.
10. To control for any differences between the competency of the targeting and non-targeting systems measure competency by thawing 1 tube from each biological replicate of chemically competent cells (step 11 of 3.4) and transform the cells with a non-target control plasmid that doesn't contain a kanamycin resistance gene and a ColE1 origin (e.g. 10 ng of pUC19 plasmid) (steps 1-9 of 2.5). (Fig. 1B).

3. Expressing and purifying the type IV-A Csf ribonucleoprotein complex

The components of the type IV-A Csf ribonucleoprotein complex are expressed by two plasmids in *E. coli* strain HMS174(DE3). Codon-optimized gene sequences from the *Pa83* type IV system were ordered from Twist Bioscience and were inserted by restriction cloning into the expression vectors pCDFDuet (Novagen) and pACYCDuet (Novagen) under T7 promoters. Genes *csf1, csf2,* and *csf5* were inserted into a pCDFDuet (Novagen) backbone, while the CRISPR and *csf3* gene were inserted into pACYCDuet (Novagen) plasmid (see Table 1). Sequence to encode an N-terminal 6x-His tag was added to the *csf1* gene and an N-terminal Strep-tag ®II tag to the *csf2* to allow for purification by either nickel or strep affinity chromatography.

The purification protocol of the Csf complex consists of two chromatography steps, a Strep-affinity step and a size exclusion chromatography step. The Strep-affinity protocol is based on the manufacturer recommendations (IBA Life Sciences) for Strep-Tactin®XT 4Flow® resin with minor adjustments and additional washes (Fig. 2).

3.1 Csf complex expression materials and equipment

- *E. coli* HMS174(DE3) chemically competent cells (Novagen)
- LB media
- LB Agar solid media
- 1000x streptomycin (50 mg/mL)
- 1000x chloramphenicol (25 mg/mL)
- Purified expression plasmids
- 42 °C water bath or heat block
- Sterile glass beads or cell spreaders
- 2.8 L baffled culture flasks
- Swing-bucket centrifuge
- 50 mL conical tubes
- 1 L centrifuge bottles
- 600_{nm} spectrophotometer

3.2 Csf complex expression day 1 – bacterial transformation

1. Thaw one 50 μL vial of chemically competent HMS174(DE3) cells on ice and add ~100 ng of each expression plasmid (pACYC-PaCR83-csf3, pCDF_Pa_N-His-csf1_N-Strep-csf2_cas6) to the cells (Table 2). Gently flick the tube to mix, and let the cells incubate on ice for 20–30 min.

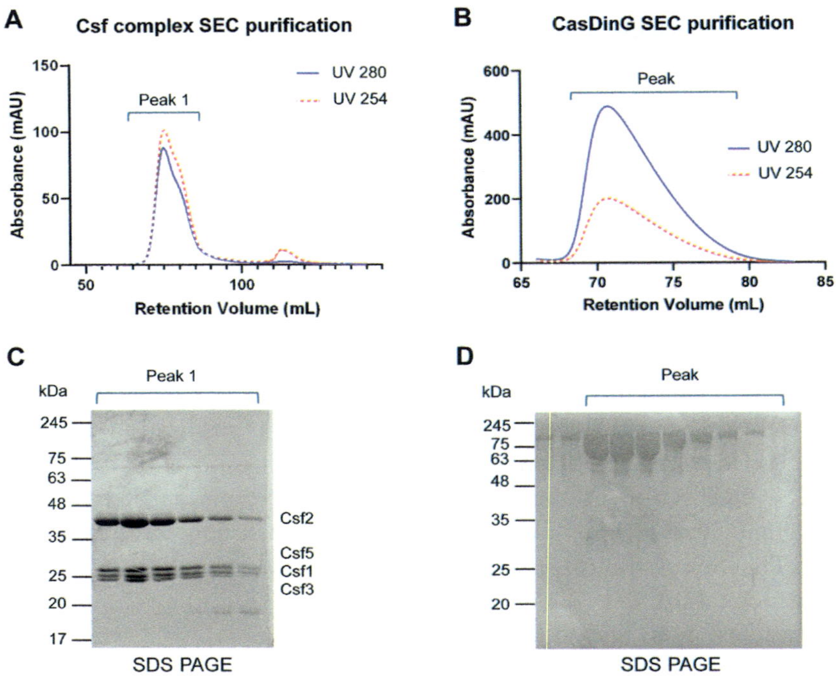

Fig. 2 Purification of the Csf ribonucleoprotein complex and CasDinG. (A, B) Representative SEC chromatograms of type IV-A Csf complex and CasDinG protein. UV absorbances at 280_{nm} measuring protein and 254_{nm} measuring nucleic acid are shown. (C, D) Representative SDS-PAGE gels of collected elutions after size exclusion chromatography for the Csf complex and CasDinG.

2. Heat-shock the cells in a 42 °C water bath for 30 s followed by a 1–2 min incubation back on ice.
3. Add 450 μL of LB media to the cells and recover at 37 °C with shaking for 1 h.
4. Plate 250 μL (1/2) and 50 μL (1/10) volumes of the transformed cell reaction onto LB agar plates containing 50 μg/mL streptomycin and 25 μg/mL chloramphenicol. Spread the cells around the plate with sterile beads or cell spreaders, and let excess LB evaporate before placing plates in a 37 °C incubator to grow overnight.

3.3 Csf complex expression day 2 – small culture growth

1. Examine the transformation plates from the previous day and pick one colony with a sterile tool/pipette tip to inoculate a starter culture. Add this colony to a 250 mL baffled flask containing 150 mL liquid LB

Table 2 Plasmids used in the expression of the IV-A Csf complex and CasDinG.

Plasmid name	Plasmid description	Link
pACYC-PaCR83-csf3	Expression of the type IV-A1 Csf complex containing the CRISPR and csf3	https://benchling.com/s/seq-XmOW3BbbE3HwcfPRcwX6?m=slm-uZDdDwgQY35SHEJoDlzd
pCDF_Pa_N-His-csf1_N-Strep-csf2_cas6	Expression of the type IV-A1 Csf complex containing N-His-csf1, N-strep-csf2, and cas6	https://benchling.com/s/seq-xZMVfUgpOLAwFBXQkVAS?m=slm-rgeAmepDJOy0T7O6JKf3
2R-T_CasDinG	Expression of N-strep-casdinG	https://benchling.com/s/seq-jy1WVocbqxq03DUMmmS4?m=slm-m5s5NixiWQ0F4WNuBRb8

culture with 50 μg/mL streptomycin and 25 μg/mL chloramphenicol. Place the flask into a shaking incubator at 37 °C overnight.

Optionally, cells from a previously prepared glycerol stock can be used to start the overnight culture (see 3.4.1 below).

2. Prepare 4 liters of sterile LB in 2.8-liter baffled flasks for use the following day.

3.4 Csf complex expression day 3 – large culture growth

1. Add 25 mL of overnight starter culture to each 2.8 L flask containing 1 L of sterile LB along with 50 μg/mL streptomycin and 25 μg/mL chloramphenicol.

   Optionally, a glycerol stock may be prepared from this small overnight growth by adding sterile glycerol to a concentration of 25 % v/v and storing it at −80 °C. This glycerol stock can then be used in place of a fresh colony to start new overnight starter growths (skipping the need for step 3.2). We have successfully used glycerol stocks up to 6 months old for this purpose.
2. Place flasks into a 37 °C incubator shaking at 200 rpm and let incubate until cells have reached an OD_{600} of 0.8–1.0.
3. Remove flasks from the incubator and place at room temperature to cool down for 30 min, and during that time set the incubator to cool to 18 °C.
4. Add IPTG to each flask to a final concentration of 0.5 mM and return to the incubator to grow for another 16–18 h at 18 °C.

3.5 Csf complex expression day 4 – harvesting

1. Harvest cells by centrifugation at ~2700 × *g* for 30 min and recover cells with a spatula.
2. Store cells in a 50 mL conical tube at −80 °C for future use, or immediately proceed to lysis and purification.

3.6 Csf complex purification materials and equipment

- Lysis buffer (500 mM NaCl, 100 mM Tris-HCl pH 8.0, 2 mM $MgCl_2$, 1 mM TCEP)
- High-salt wash buffer (1000 mM NaCl, 100 mM Tris-HCl pH 8.0, 2 mM $MgCl_2$, 1 mM TCEP)
- Strep elution buffer (150 mM NaCl, 100 mM Tris-HCl pH 8.0, 2 mM $MgCl_2$, 50 mM biotin, 1 mM TCEP)
- IV-A SEC Buffer (150 mM NaCl, 50 mM Tris-HCl pH 8.0, 2 mM $MgCl_2$, 1 mM TCEP)

- 5000x Aprotinin solution (0.5 mg/mL dissolved in H_2O)
- 1000x Pepstatin A solution (0.7 mg/mL dissolved in methanol)
- 1000x Leupeptin solution (0.5 mg/mL dissolved in ddH_2O)
- 100x AEBSF solution (100 mM dissolved in ddH_2O)
- Ultrasonic Cell Homogenizer
- Fixed-angle rotor centrifuge
- 50 mL centrifuge bottles
- Strep-Tactin®XT 4Flow® high-capacity resin
- Protein chromatography gravity flow column
- 15 mL 100 kDa MWCO spin-concentrator (e.g. Sartorius Vivaspin)
- Swing-bucket centrifuge
- ÄKTA Pure Protein FPLC system (Cytiva)
- HiLoad 16/60 Superose 6 pg SEC column (Cytiva)
- 0.2 μm syringe filters with luer-lock syringes
- 14 % polyacrylamide SDS-PAGE gels

3.7 Csf complex purification – lysis and strep-affinity chromatography

1. Resuspend cell pellet in 50 mL of ice-cold lysis buffer (**see 3.6**) supplemented with 0.5 μg/mL aprotinin, 0.5 μg/mL leupeptin, 0.7 μg/mL pepstatin A, and 1 mM AEBSF.
2. Lyse cells by sonication using a homogenizer at a power setting of 6 and a duty cycle of 60 %. Cells should be lysed in a metal container on ice to keep chilled in intervals of 30 s on and a 30-second rest. Continue sonicating in cycles until the suspension visibly darkens and becomes less viscous, indicating total lysis.
3. Clarify lysate by centrifugation in 50 mL centrifuge bottles at 4 °C at 32,000 × *g* for 25 min. Pour off clarified lysate into a new tube on ice.
4. Immediately proceed to affinity chromatography. All further steps take place in a 4 °C environment.
5. Place lysate over 4 mL of freshly regenerated and equilibrated Strep-Tactin®XT 4Flow® high-capacity resin in a 50 mL gravity flow column and incubate for 1 h at 4 °C with gentle shaking.
6. Elute lysate from the resin before washing the resin with a further 300 mL of lysis buffer under gravity flow. Collect lysate elution and lysis buffer washes.
7. Perform a second wash with 150 mL of high-salt wash buffer (Section 3.6) under gravity flow. Collect the flow through.

8. Add 1 column volume (4 mL) of elution buffer (Section 3.6), let sit on the resin with gentle shaking for 10 min, then elute into a tube.
9. Perform five more elutions of 5 mL with elution buffer, letting the elution buffer sit over the resin with gentle shaking for 10 min each time before elution.
10. Assess purity and yield by SDS-PAGE analysis using a 14% polyacrylamide gel. We run samples collected at each step of affinity chromatography including the cell pellet, lysate, washes, and elutions to confirm the protein is in the elution buffers.

3.8 Csf complex purification – concentrate and size-exclusion chromatography

1. Pool elutions from Strep affinity and equilibrate a 15 mL 100 kDa MWCO spin-concentrator with lysis buffer.
2. Filter pooled elutions with a 0.2 μm syringe filter and proceed to concentrate as per the manufacturer's instructions by centrifugation at 2000 × *g* for 20-minute intervals at 4 °C, mixing the solution by pipetting between runs. Concentrate until the final volume of protein is below 1 mL.
3. Load protein onto a HiLoad 16/600 Superose 6 pg SEC column (Cytiva), and elute in SEC buffer at a flow rate of 1 mL/min and with 1 mL fractions during the resulting peaks. SEC chromatography was performed on an ÄKTA Pure FPLC system.
4. Analyze peak fractions by SDS-PAGE to confirm protein purity, and assess concentration by 280 absorbance (e.g. NanoDrop) using an estimated extinction coefficient of the full Csf complex (e.g. 581,840 M^{-1} cm^{-1}) assuming a stoichiometric ratio of 1:5:1:1:1 Csf1:Csf2:Csf3:Csf5:crRNA and considering differences caused by changes in the crRNA target sequence.
5. Syringe filter and concentrate the purified Csf complex over a new 100 kDa MWCO spin concentrator to the desired concentration. We routinely concentrate purified Csf complex to 10 μM without issue.
6. Aliquot 50–100 μL of concentrated protein into PCR tubes and flash freeze by plunging into liquid nitrogen before storing the purified protein at −80 °C.

4. Expressing and purifying the CasDinG helicase

CasDinG from *Pa83* is expressed in HMS174(DE3) *E. coli* cells (Novagen) using a pET expression system under a T7 promoter. The codon-optimized *casdinG* gene was ordered from Twist Bioscience and inserted into a 2R-T LIC expression vector which contains an N-terminal Strep-tag ®II tag and TEV cleavage site upstream of the *casdinG* gene (Addgene plasmid #29717).

CasDinG is purified by column chromatography in 3 stages – strep affinity, heparin ion exchange, and size-exclusion chromatography. The strep-affinity protocol is based on the manufacturer recommendations (IBA Life Sciences) for the Strep-Tactin®XT 4Flow® resin with minor adjustments and additional washes (Fig. 2).

4.1 CasDinG expression materials and equipment

- *E. coli* HMS174(DE3) chemically competent cells (Novagen)
- LB media
- LB Agar solid media
- 1000x ampicillin (100 mg/mL)
- 1000x metals mix (0.1 M $FeCl_3$, 1 M $CaCl_2$, 1 M $MnCl_2$, 1 M $ZnSO_4$, 0.2 M $CoCl_2$, 0.1 M $CuCl_2$, 0.2 M $NiCl_2$, 0.1 M Na_2MoO_4, 0.1 M Na_2SeO_3, 0.1 M H_3BO_3) (Studier, 2005)
- 1000x $MgSO_4$ (1 M)
- Purified CasDinG expression plasmid
- 42 °C water bath or heat block
- Sterile glass beads
- 2.8 L baffled culture flasks
- Swing-bucket centrifuge
- 50 mL conical tubes
- 1 L centrifuge bottles
- 600 nm spectrophotometer

4.2 CasDinG expression days 1 and 2 – bacterial transformation and liquid culture

1. Transform ~100 ng 2R-T_casdinG expression vector (Table 2) into HMS174(DE3) chemically competent cells as described above for transforming the Csf complex (Section 4.1) but using LB agar plates containing 100 μg/mL ampicillin instead of other antibiotics.
2. Pick a colony and add it to an overnight 150 mL liquid culture of LB media with 100 μg/mL ampicillin in a 250 mL baffled flask.
3. Prepare 4 L of sterile LB in four 2.8-liter baffled flasks (1 L per flask) for use the following day.

4.3 CasDinG expression day 3 – large culture growth

1. Add 25 mL of overnight starter culture to each flask of sterile LB with 100 μg/mL ampicillin. Further, supplement each flask with 1 mM $MgSO_4$ and 1 mL of 1000x metals mix (see 4.1).
2. Place flasks into a 37 °C incubator shaking at 200 rpm and let incubate until cells have reached an OD_{600} of 0.8–1.0.
3. Remove flasks from the incubator and place them at room temperature to cool down for 30 min, and during that time set the incubator to cool to 18 °C.
4. Add IPTG to each flask to a final concentration of 0.5 mM and return to the incubator to grow for 5 h.

4.4 CasDinG expression day 3 – harvesting cells

1. Harvest cells by centrifugation at ~2700 × *g* for 30 min. Pour off the media and recover pelleted cells with a spatula.
2. Store cells in a 50 mL conical tube at −80 °C for future use, or immediately proceed to lysis and purification.

4.5 CasDinG purification materials and equipment

- Lysis buffer (500 mM NaCl, 100 mM Tris-HCl pH 8.0, 2 mM $MgCl_2$, 1 mM TCEP)
- High-salt wash buffer (1000 mM NaCl, 100 mM Tris-HCl pH 8.0, 2 mM $MgCl_2$, 1 mM TCEP)
- Strep elution buffer (150 mM NaCl, 100 mM Tris-HCl pH 8.0, 2 mM $MgCl_2$, 50 mM biotin, 1 mM TCEP)
- Low-salt buffer (20 mM NaCl, 100 mM Tris-HCl pH 8.0, 2 mM $MgCl_2$, 1 mM TCEP)
- 5000x Aprotinin solution (0.5 mg/mL dissolved in H_2O)
- 1000x Pepstatin A solution (0.7 mg/mL dissolved in methanol)
- 1000x Leupeptin solution (0.5 mg/mL dissolved in ddH_2O)
- 100x AEBSF solution (100 mM dissolved in ddH_2O)
- Ultrasonic Cell Homogenizer
- Fixed-angle rotor centrifuge
- 50 mL centrifuge bottles
- Strep-Tactin®XT 4Flow® high-capacity resin (IBA)
- Protein chromatography gravity flow column
- 5 mL HiTrap Heparin HP column (Cytiva)
- HiPrep 26/10 Desalting column (Cytiva)
- 15 mL 100 kDa MWCO spin-concentrator (e.g. Sartorius Vivaspin)

- 4 mL 30 kDa MWCO spin-concentrator(e.g. Sartorius Vivaspin)
- Swing-bucket centrifuge
- Protein FPLC system
- HiLoad 26/600 200 pg SEC column (Cytiva)
- 0.2 μm syringe filters with luer-lock syringes
- 12 % polyacrylamide SDS-PAGE gels

4.6 CasDinG purification – lysis and strep-affinity chromatography

1. Resuspend cell pellet in 50 mL of ice-cold lysis buffer supplemented with 0.5 μg/mL aprotinin, 0.5 μg/mL leupeptin, 0.7 μg/mL pepstatin A, and 1 mM AEBSF.
2. Lyse cells by sonication using a homogenizer at a power setting of 6 and a duty cycle of 60 %. Cells should be lysed in a metal container on ice to keep chilled in intervals of 30 s and then given a 30-second rest. Continue sonicating in cycles until the suspension visibly darkens and becomes thinner in texture, indicating total lysis.
3. Clarify lysate by centrifugation in 50 mL tubes at 32,000 × *g* for 25 min. Pour off clarified lysate into a new tube on ice.
4. Immediately proceed to affinity chromatography. All further steps take place in a 4 °C environment.
5. Place lysate over 4 mL of freshly regenerated and equilibrated Strep-Tactin®XT 4Flow® high-capacity resin in a gravity flow column and let incubate for 1 h with gentle shaking.
6. Filter lysate from resin before washing the resin with a further 250 mL of lysis buffer (Section 4.5) under gravity flow. Collect and label flow through.
7. Perform a second wash with 100 mL of high-salt wash buffer (Section 4.5) under gravity flow. Collect and label all washes.
8. Add 1 column volume (4 mL) of elution buffer and let sit on the column for 10 min. Elute the resin once by gravity flow, collecting the elution in a tube.
9. Perform 5 more elutions with 5 mL each, letting the elution buffer sit on the resin with gentle shaking for 10 min before eluting.
10. Assess purity and yield by SDS-PAGE analysis using a 12 % polyacrylamide gel. We run samples collected at each step of affinity chromatography including the cell pellet, lysate, washes, and elutions to confirm the protein is in the elution buffers.

4.7 CasDinG purification – desalting and heparin chromatography

1. Pool Strep elutions together and pass through a 0.2 μm syringe filter to remove aggregated proteins. All further chromatography steps were performed on an ÄKTA Pure FPLC system.
2. Equilibrate a HiPrep 26/10 Desalting column (Cytiva) with low-salt buffer before loading CasDinG elutions. Desalt in two steps, equilibrating the column between runs with Low-salt buffer (Section 4.5).
3. If desalted fractions appear to contain precipitate, immediately filter through a 0.2 μm syringe filter and proceed quickly to heparin chromatography.
4. Equilibrate a 5 mL HiTrap Heparin HP column (Cytiva) with low-salt buffer.
5. Load the desalted fractions of CasDinG onto the column at a flow rate of 5 mL/min. Wash the column with additional low-salt buffer until the UV 280 $_{nm}$ signal returns to baseline.
6. Reverse the flow direction of the column to up-flow, and wash with low-salt buffer until the UV 280 $_{nm}$ signal hits a baseline.
7. Elute CasDinG from the column with lysis buffer (Section 4.5) at a flow rate of 0.5 mL/min. The combination of reversed direction flow and reduced flow rate results in a narrow elution peak with concentrated protein.
8. Collect and pool peak fractions.

4.8 CasDinG purification – concentration and size-exclusion chromatography

1. Equilibrate a HiLoad 26/600 Superdex 200 pg SEC column (Cytiva) with type IV-A SEC buffer (Section 3.6).
2. Pool elutions from heparin and equilibrate a 4 mL 30 kDa MWCO spin-concentrator with lysis buffer.
3. Concentrate as per the manufacturer's instructions by centrifugation at 4000 × *g* for 20-minute intervals at 4 °C, mixing by pipetting concentrated protein up and down between runs. Concentrate until the final volume of protein is 1-2 mL.
4. Load protein onto a HiLoad 26/600 Superdex 200 pg SEC column (Cytiva) equilibrated with type IV-A SEC buffer, and elute in SEC buffer at a flow rate of 1 mL/min and with 1 mL fractions during the resulting peaks. SEC chromatography was performed on an ÄKTA Pure FPLC system.

5. Analyze peak fractions by SDS-PAGE to confirm protein purity and assess concentration by absorbance at 280 $_{nm}$ with a NanoDrop.
6. Syringe filter and concentrate the purified CasDinG over a new 30 kDa MWCO spin concentrator to the desired concentration. We routinely bring our purified protein to over 60 μM without issue.
7. Aliquot protein and store at 4 °C.

Note: Do not freeze CasDinG, freezing has been observed to eliminate enzymatic activity (Domgaard et al., 2023). Protein kept at 4 °C has been observed to retain activity for several months, however, the N-terminal domain of CasDinG has been demonstrated to slowly proteolyze during storage, especially if the temperature fluctuates above 4 °C (Hallmark et al., 2024). As such, it is recommended to periodically assess the protein by SDS-PAGE to observe proteolysis as well as by activity assays (ATPase, helicase, binding) when relevant. Additionally, we have observed that the addition of glycerol to purification, storage, or reaction buffers impairs CasDinG activity (Domgaard et al., 2023).

5. Purification of type IV-A Csf complex bound to DNA targets and the CasDinG helicase

The type IV-A system functions by binding to a dsDNA target and then recruiting the accessory helicase CasDinG to the target site. Once loaded, CasDinG translocates along the DNA strand and silences gene transcription, purportedly by displacing RNA transcripts or disrupting DNA structure (Čepaitė et al., 2024; Guo et al., 2022; Sanchez-Londono et al., 2024). We recently isolated the Csf complex bound to an intact dsDNA target and CasDinG recruited to Csf-bound to an intact R-loop using size exclusion chromatography (SEC) (Fig. 3). This method provides a framework for examining structural features involved in Csf-mediated DNA binding or CasDinG recruitment to Csf-bound R-loops. The following protocols detail how to perform SEC to observe binding and isolate the Csf-DNA target and Csf-R-loop-CasDinG complexes.

5.1 Materials for SEC of Csf complexes bound to target DNA and CasDinG

- Purified IV-A1 Csf Complex (10 μM)
- DNA Oligo 5, 100 μM (Table 3)

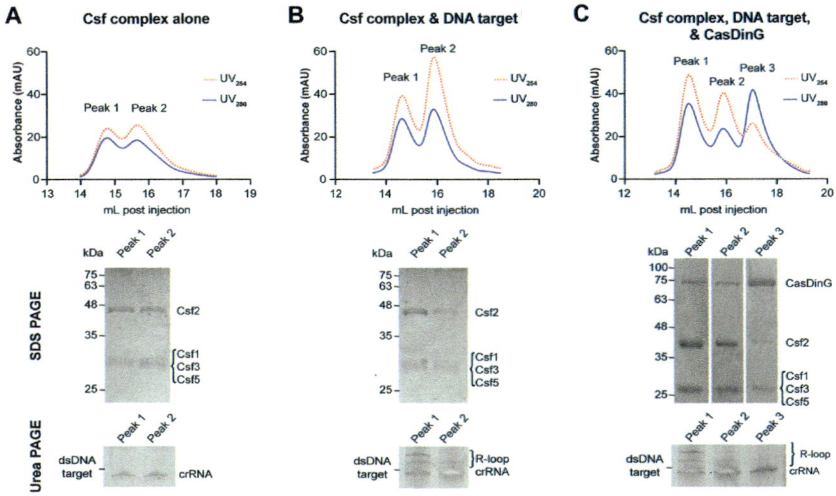

Fig. 3 **Purification of Csf complex bound to DNA and CasDinG**. (A–C). Representative size exclusion chromatograms and gels of the IV-A Csf complex, Csf complex bound to a dsDNA target, and Csf complex bound to a dsDNA target and CasDinG. (A) SEC of the Csf Complex alone. Top: SEC chromatogram, Middle: SDS PAGE, Bottom: Urea PAGE. (B) SEC of the IV-A Csf Complex bound to dsDNA. Top: SEC chromatogram, Middle: SDS PAGE, Bottom: Urea PAGE. (C) SEC of the IV-A Csf Complex bound to dsDNA and CasDinG. Top: SEC chromatogram, Middle: SDS PAGE, Bottom: Urea PAGE.

- Purified CasDinG (50 μM)
- IV-A Size Exclusion Buffer (**See 3.6**)
- Nuclease-free water, DEPC-treated
- 1.5 mL microcentrifuge tubes
- 1 mL syringe
- 14% SDS PAGE gel
- Acid phenol:chloroform (Ambion)
- 50% glycerol
- 15% Urea PAGE gel
- 10x TBE buffer (108 g Tris base, 55.2 g boric acid, 7.44 g EDTA, dissolve in 1 L ddH_2O)
- Superose 6 10/300 increase GL sizing column (Cytiva)
- 100 kDa or 50 kDa MWCO Concentrator
- ÄKTA Pure pump system (Cytiva)
- Incubator or thermocycler set to 37 °C

Table 3 DNA oligo sequences used in biochemical analysis of Csf complexes and CasDinG.

Name	Description	DNA sequence from 5′ to 3′
DNA Oligo 1	Fluorescence polarization anisotropy 5′-FAM labeled oligo	FAM-TGTGGAATGCTACAGGCGTTGTAGTTTGTACTG GTGACGACAGGTGTTGCTCCAGAAAGGTGAGTTCTT CTTGTTGTT
DNA Oligo 2	5′-FAM labeled short oligo for CasDinG Helicase assay.	FAM-TCGTCACCAGTACAAAC
DNA Oligo 3	Quenching oligo for CasDinG helicase assay	TCGTCACCAGTACAAAC
DNA Oligo 4	Complementary 5′ overhang oligo to Oligo 2 for CasDinG helicase assay	TTTTTTTTTTTTTTTGTTTGTACTGGTGACGA
DNA Oligo 5	dsDNA target oligo for the type IV-A Csf complex	AACTCACGCTAGTGGAGCAACACCTGAAGGAA GGCTTGATGAGCAAGTGCGGCA (Top strand)

5.2 SEC of Csf complex bound to target DNA - preparation

1. Prepare an incubator or thermal cycler and set at 37 °C
2. Equilibrate the Cytiva Superose 6 10/300 GL Increase Size Exclusion Column with 2 column volumes of type IV-A SEC buffer (Section 3.6) and equilibrate the 500 μL injection loop with 5 mL of SEC buffer.

5.3 SEC of Csf complex bound to target DNA – protocol

1. Thaw frozen aliquots of complex or take freshly prepped complex and place them on ice. We generally use at least 1.25 nanomoles of protein. Also, thaw or prepare at least 1.5 nanomoles of target DNA oligo.
2. Add the IV-A complex and DNA Oligo 5 (Table 3) in SEC buffer to a final volume of 500 μL according to the conditions described in Table 4. The complex:DNA Oligo 5 ratio should be 1:1.2, with a slight molar excess of DNA.

   The total concentration of the complex can be modified, however, the maximum concentration at which peak separation can be achieved remains to be determined empirically.
3. Incubate the Csf complex and DNA Oligo 5 at 37 °C for 30 min
4. Use a 1 mL syringe to aspirate the mixture and inject it directly into a previously prepared 500 μL injection loop.
5. Load the sample from the injection loop onto the column and empty the injection loop with 2.5 mL buffer
6. Elute the size exclusion column with 1.5 column volumes, while monitoring absorbance at UV 280 $_{nm}$ and 254 $_{nm}$ collecting any absorbance peaks (Fig. 3).

   We monitor absorbance at both UV 254 nm (nucleic acid) and 280 $_{nm}$ (protein) to measure the ratio of nucleic acid:protein

Table 4 SEC complex formation conditions.

Components	[Initial]	[Final]	Volume
IV-A SEC Buffer			310 or 360 μL
IV-A Csf Complex	10 μM	2.5 μM	125 μL
DNA substrate	100 μM	3 μM	15 μL
CasDinG	50 μM	5 μM	50 μL
Total			500 μL

Note: For generating a Csf complex bound only to the target DNA (DNA Oligo 5), omit CasDinG and replace its volume with IV-A SEC Buffer. ***When generating a complex bound to CasDinG add CasDinG AFTER a 30-minute incubation of the Csf complex and DNA alone.

absorbance peaks. A 254:280 ratio greater than 1.5 indicates a Csf complex bound to a DNA duplex.

7. Take samples of each peak (there are usually 2) and run on a 14 % SDS PAGE gel at 200 V for 65 min.
8. Take 50 μL samples of each peak and combine with 50 μL acid phenol:chloroform. Mix by shaking and then centrifuge briefly to obtain phase separation.
9. Pipette 10 μL of the aqueous phase into a new PCR tube containing 2 μL 50 % glycerol.

   Note: addition of formaldehyde and a heat treatment can be used at this step to completely denature nucleic acid, if desired.
10. Run 5-10 μL of the aqueous portion of the acid phenol:chloroform extraction on a 15 % Urea PAGE gel at 150 V for 1.5 h in 1x TBE buffer to visualize crRNA and bound dsDNA. Visualize with SYBR Gold or equivalent (Fig. 3).

    Note: Better resolution is also obtained by washing out the wells with a micropipette before loading samples and pre-running the Urea PAGE gel at 100 V for one hour.
11. Pool fractions containing complex confirmed by gel analysis and concentrate to the desired concentration with either a 100 kDa or 50 kDa MWCO concentrator. Quantify using the following extinction coefficient and molecular weight: 1,462,216 mol^{-1} cm^{-1}, 319,710 Da if using the same DNA sequences described in Table 3.

    Note: the extinction coefficient will change depending on the sequence of the crRNA guide and bound DNA target.
12. We often concentrate Csf-R-loop to 50–100 μM. Concentrated aliquots can be kept at 4 °C or flash frozen and kept at −80 °C

5.4 SEC of CasDinG recruited to R-loop-bound Csf complexes - preparation

1. Prepare an incubator or thermal cycler and set it at 37 °C
2. Equilibrate a Cytiva Superose 6 10/300 GL Increase size exclusion column with 2 column volumes of type IV-A size exclusion buffer (Section 3.6) and equilibrate the 500 μL injection loop with 5 mL of size exclusion buffer.

5.5 SEC of CasDinG recruited to R-loop-bound Csf complexes - protocol

1. Thaw a frozen aliquot of complex or take freshly prepped complex and place it on ice (at least 1.25 nanomoles). Also, thaw the target DNA

oligo (at least 1.5 nanomoles). Remove an aliquot of CasDinG and place it on ice as well (at least 2.5 nanomoles).

2. As in protocol 5.3, add the IV-A complex and DNA Oligo 4 as prescribed in Table 4, but leave volume for the addition of CasDinG. The complex:target DNA oligo ratio should be 1:1.2, with a slight molar excess of DNA.
3. Incubate at 37 °C for 30 min.
4. After the Csf complex and the DNA have incubated for 30 min add purified CasDinG at a 2:1 ratio CasDinG:Csf complex (5 μM) and add SEC buffer to bring to a total volume of 500 μL. Mix by pipetting up and down (Table 4).
5. Incubate at 37 °C for 30 min
6. Use the 1 mL syringe to aspirate the Csf complex-R-loop-CasDinG mixture and inject it into a 500 μL injection loop.
7. Load the sample onto the column from the injection loop and empty the loop with 2.5 mL of additional SEC buffer.
8. Following steps 6–11 from Section 5.3 to elute protein off of SEC column, run samples on gels, and concentrate protein (Fig. 3).
9. Keep concentrated aliquots at 4 °C as freezing samples tends to impair CasDinG activity.

6. Determining Csf complex or CasDinG affinity to nucleic acid targets with fluorescence anisotropy

CasDinG and the type IV-A Csf complex both demonstrate an affinity for nucleic acid (Cui et al., 2023; Domgaard et al., 2023). Fluorescence anisotropy provides a fast and easy method to measure this affinity to a variety of fluorescently labeled substrates. The method relies on a difference in tumbling rates between a relatively small fluorescent probe (e.g. a FAM-labeled nucleic acid) when unbound and bound to a much larger molecule (e.g. the protein of interest). Measurements can be taken on a standard plate reader equipped with bandpass filters for fluorescence detection and polarizers to determine changes in the anisotropy of the fluorescence response from a polarized excitation source (Fig. 4).

Nucleic acid probes with a 5′ or 3′ FAM label were ordered from IDT and subsequently annealed and gel-extracted from native PAGE gels in the case of double-stranded probes. Varying concentrations of protein are added to a constant concentration of a FAM-labeled nucleic acid probe in

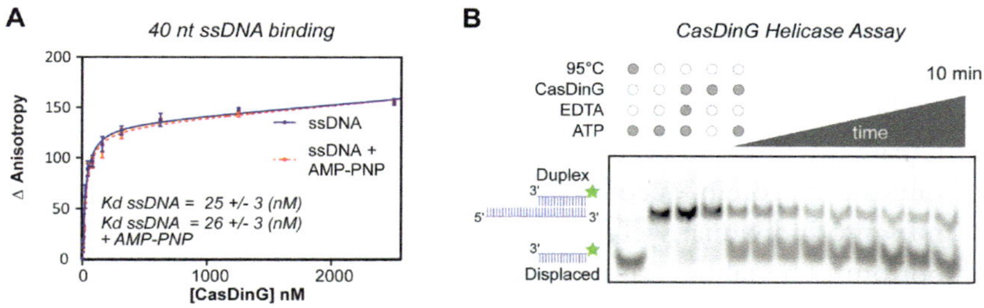

Fig. 4 Fluorescence anisotropy binding and helicase assays. (A) Representative fluorescence anisotropy binding assay with CasDinG. Higher values of delta anisotropy correlate with greater amount of bound FAM-labeled nucleic acid substrate. Isotherms of binding to a FAM-labeled ssDNA substrate in the presence of AMP-PNP (dashed line) and without NTP (solid). (B) Representative helicase assay of CasDinG with a dsDNA duplex with a 5′ overhang. Controls and 10 min time course reactions were run on a 15% TBE native gel and imaged to detect the fluorescent labeled nucleic acid (these data were previously presented in (Domgaard et al., 2023)).

multiple wells of a black bottom 96-well plate. The fluorescence values at perpendicular angles to the plate are measured by the plate reader, and the change in anisotropy is calculated by the plate reader software for each condition. The plate reader used for these experiments was a BioTek Synergy H4 Hybrid Multi-Mode Microplate Reader which is controlled by BioTek Gen5 software.

6.1 Materials and equipment

- 4x binding buffer (400 mM Tris pH 8.0, 4 mM TCEP, 20 mM $MgCl_2$)
- Concentrated protein stock (70 μM for CasDinG and preferably at least 15 μM for the Csf complex)
- DNA Oligo 1 stock (100 nM)
- ATP or other NTP stock (if desired)
- DEPC-treated water
- 96-well assay plates (Corning Costar #3915: Black, Flat Bottom, Non-Treated, Polystyrene)
- 12-tube PCR strips
- Multichannel pipette (12 channel)
- BioTek Synergy H4 Hybrid Multi-Mode Microplate Reader

6.2 Set up and perform fluorescence anisotropy

1. Gather the following reagents and place on ice:
   - a. 140 μL protein stock (Csf complex or CasDinG)
   - b. 4x binding buffer

   c. 100 nM (10x) FAM labeled DNA oligo 1
   d. 10 mM NTP stock (if desired)
   e. DEPC-treated water
2. Create the master mix for the assay, which contains all the components minus the protein, according to Table 5. Aliquot 60 μL of this master mix into individual wells of three rows of a 96-well plate with a black bottom (The table accounts for a slight excess of master mix, so there will be some left in the prepared mixture). Cover with foil until adding the protein in step 7 to prevent photobleaching of the fluorophore.
3. Create a stock of 1x binding buffer by diluting the 4x solution into ddH_2O. 2 mL is sufficient.
4. Place a 12-tube PCR strip into a chilled rack for the protein dilution series.
5. Prepare the triplicate series of 2-fold dilutions by first pipetting 70 μL of 1x binding buffer into the last 11 tubes in the chilled 12-tube PCR strips.
6. Pipette 140 μL of the protein stock solution in the first tube, and dilute it through the two fold dilution series by taking 70 μL from the first tube and mix it with the 70 μL of 1x binding buffer in the subsequent tube. Continue this pattern **stopping at tube 11.** Tube 12 should **only** contain 1x binding buffer to serve as a background control during data processing. Exchange pipette tips between each dilution.
7. Using a 12-channel pipette, add 20 μL of the dilution series into the rows of the previously prepared well plate containing 60 μL of master

Table 5 FP Anisotropy reaction conditions and master mix parameters.

Components	Initial	Final	Volume per well	Volume for master mix
Protein from 1x dilution series			20 μL	0 μL
FAM-labeled oligo	100 nM	10 nM	8 μL	345.6 μL
4x Binding buffer	4x	1x	20 μL	864 μL
DEPC treated H_2O			32 μL	1382.4 μL
Total			80 μL	2592 μL

Note: If it is desired to perform this assay with an ATP analog present, use a 10 mM stock and a final reaction concentration of 1 mM. The volume per reaction will be 8 μL and 345.6 μL in the master mix. Subtract the same volume from the volume for DEPC treated H_2O.

mix and gently mix by pipetting up and down. Repeat this step 2 more times, creating triplicate sets of each reaction containing a distinct protein concentration. Avoid making bubbles in the solution during mixing as they will interfere with reading the samples in the plate reader.

Note: Should any bubbles form by mixing, the well plate can be spun briefly in a centrifuge to remove them.

8. Allow the assembled well plate to incubate at room temperature for one hour for the reaction to reach binding equilibrium.
9. Using the plate reader software package (e.g. Gen5) define a protocol of which excitation and emission wavelengths to use and data to be collected. Set the protocol to define wells at runtime. We use the following parameters:
   a. Gain – use automatic gain adjustment.
   b. Excitation – 485/20 nm
   c. Emission – 528/20 nm
   d. Optics Position – Top 505 nm
   e. Read height – 4.00 mm
10. Define the data output in the "data reduction" window to include the following:
    a. Parallel Intensity
    b. Perpendicular Intensity
    c. Fluorescence Polarization
    d. Anisotropy
11. To correct for unequal passing of light between the parallel and perpendicular polarizing filters it may be needed to specify a correction value in the software. This value must be determined for the specific instrument being used, instructions for how to calculate it are usually found in the instruction manual of the instrument. For the machine we used, the correction term was called a G factor, and a value of 0.94 was utilized.
12. Once all parameters are programmed, remove the cover from the plate containing the reactions, insert the 96-well plate into the reader, and define the wells to be measured. Run the protocol and export the generated data into a spreadsheet.
13. Calculate a ΔAnisotropy (Δr) value for each well by subtracting the control well (the last well in each dilution series with only buffer) from each measured anisotropy value in that set.
14. Transfer the triplicate data into preferred statistics software and perform a non-linear regression fit to determine K_D.

6.3 Fluorescence anisotropy troubleshooting

If the experiment is successful, the data when plotted should start near 0 ΔAnisotropy for the low concentrations of ligand and then increase until a saturation point is reached. If a saturation point is not reached then higher concentrations of ligand must be used.

If the calculated K_D value is near or below 10 nM, which is the concentration of the probe, then the calculated affinity may be in error. A reduced probe concentration may be required.

If the measured data does not fit nicely to a hyperbolic binding curve then the possibility of quenching must be considered. It is possible for resonance energy to transfer from the fluorophore to residues on the protein, resulting in ΔAnisotropy that gets closer to zero or becomes more negative as the concentration of protein increases. A putative solution to this problem is to change the position of the fluorophore on the nucleic acid substrate, and thus change how the probe is oriented in relation to the protein, either by changing the length of the probe, moving it from one strand to the other, or moving it from one end of the strand to the other.

7. CasDinG helicase assay

CasDinG is a 5′−3′ helicase that can unwind DNA-DNA or DNA-RNA hybrids (Domgaard et al., 2023). Here we describe an EMSA method for measuring CasDinG helicase activity over time. CasDinG is mixed with a DNA oligonucleotide duplex, where one strand is labeled with a fluorophore and the other is unlabeled. Usually, the smaller oligonucleotide is the labeled strand. ATP is added to start the reaction. At prescribed time points, a sample of the reaction is taken and quenched in a buffer that stops the helicase activity with EDTA and SDS but does not denature the duplex. A high concentration of a single stranded oligonucleotide complementary to non-labeled strand (Oligo 3 – Table 3) is added to the reaction to disallow the displaced FAM-labeled strands to reanneal. After all time points of the reaction are collected, the samples are run on a TBE native gel to visualize unwinding (Fig. 4). The original duplex containing the FAM-labeled oligonucleotide runs higher on the gel while the displaced FAM-labeled oligo runs lower on the gel. This protocol is written using a 5′ overhang DNA-DNA duplex, but other duplex substrates (e.g. 3′ overhang, RNA-DNA duplexes) can be substituted (Domgaard et al., 2023).

7.1 Materials

- NEB 2.1 buffer
- DNA Oligo 2 (Table 3)
- DNA Oligo 3 (Table 3)
- DNA Oligo 4 (Table 3)
- 50 % glycerol
- Purified CasDinG
- 5x helicase buffer (250 mM Tris, pH 7.5, 5 mM TCEP, 5 mM $MgCl_2$, 0.5 mg/mL rBSA)
- 15 % polyacrylamide 0.5x TBE Native Gel
- 10x TBE buffer (108 g Tris base, 55.2 g boric acid, 7.44 g EDTA, dissolve in 1 L ddH_2O)
- 10 mM Tris buffer, pH 8.0
- 3 M sodium acetate
- DEPC treated water
- 100 mM EDTA solution
- 10 mM ATP solution
- 2x Helicase STOP Buffer (10 mM EDTA, 1 % SDS, 20 % glycerol, 475 nM DNA Oligo 3)
- 1.5 mL microcentrifuge tubes
- 100 % ethanol
- Incubator at 37 °C
- Thermocycler
- −80 °C freezer
- Microcentrifuge
- Nanodrop spectrophotometer

7.2 Duplex substrate annealing

1. Mix complementary oligos (e.g. Oligo 2 and Oligo 4) in an equimolar ratio in the presence of 1X NEB 2.1 buffer
2. Anneal by heating to 95 °C in a thermocycler for 5 min and then lowering to 20 °C at a ramp rate of 1 °C/min
3. Add 50 % glycerol to the annealed reaction at a ratio of 1 μL 50 % glycerol:5 μL sample, to prepare the sample for running on a gel.
4. Run the annealed oligo sample on a 15 % polyacrylamide 0.5x TBE gel at 100 V for 35 min. The gel can be imaged with a filter that excites the FAM label.

5. Extract the oligo by placing the excised gel piece in 0.5 mL of a 10 mM Tris buffer at pH 8.0. Allow it to incubate at 37 °C for at least 3 h but preferably overnight for maximum yield.
6. Ethanol precipitate the oligo by adding 0.3 volumes (0.15 mL) of 3 M sodium acetate and 2.5–3 volumes (1–1.5 mL) of ice-cold 100% ethanol to the 0.5 mL of the extracted duplex. Incubate at −80 °C for at least 1 h.
7. Centrifuge at 4 °C at maximum speed (>13,000 × *g*) for 30 min.
8. Siphon off the unprecipitated solution with a pipette. There should be a small yellow-green pellet at the bottom. Allow this pellet to dry.

   Note: Keep this oligo in foil or in a dark place to prevent photobleaching of the FAM label.
9. Resuspend the dried pellet in 30 μL molecular water.
10. Quantify the oligo spectrophotometrically using the extinction coefficient and molecular weight (e.g. 460,900 mol^{-1} cm^{-1}, 15,184.9 Da) using absorbance at 260 nm.
11. Dilute the extracted duplex to a 300 nM stock solution

Note: We have observed that over time, frozen duplexes will dissociate, resulting in double bands appearing in the controls. To prevent this, make sure that the stock has been annealed no more than 1 month prior. Otherwise, perform the substrate annealing protocol again to get a uniformly annealed substrate.

7.3 Prepare controls

1. Prepare a 10 μL stock of CasDinG at 250 nM
2. Assemble 10 μL controls as described in Table 6, always adding ATP last when required for the 95 °C, no CasDinG, EDTA, and no ATP conditions.
3. Coincubate controls at 37 °C for the duration of the experimental reactions.

7.4 Time course protocol

1. Prepare a diluted stock of CasDinG at 2 μM
2. Prepare a 90 μL reaction solution that lacks ATP with recommended volumes and concentrations listed in Table 7.
3. Equilibrate this mixture in a thermocycler set to 37 °C for at least 1 h.
4. While equilibrating the CasDinG and nucleotide substrate, prepare 9 PCR tubes with 10 μL of 2x STOP buffer in each (see Section 7.1).
5. Once equilibrated, add 10 μL of 10 mM ATP to the 90 μL solution to bring it to 100 μL and to start the helicase reaction (Start a timer at this step). Mix by pipetting up and down. Immediately take a 10 μL sample and quench it

Table 6 Recommended volumes for CasDinG helicase assay controls.

Components	95 °C Control	No CasDinG	EDTA Control	No ATP Control
5x Helicase buffer	2 μL	2 μL	2 μL	2 μL
DEPC treated H_2O	7.5 μL	6.5 μL	4.5 μL	6.5 μL
300 nM Duplexed DNA Oligos 2 and 4	0.5 μL	0.5 μL	0.5 μL	0.5 μL
CasDinG (250 nM)	0 μL	0 μL	1 μL	1 μL
EDTA (100 mM)	0 μL	0 μL	1 μL	0 μL
ATP (10 mM)	0 μL	1 μL	1 μL	0 μL

Note: All reaction volumes should total to 10 μL.

Table 7 CasDinG helicase assay master mix.

Components	[Initial]	[Final]	Volume
CasDinG	2 μM	25 nM	1.25 μL
Duplexed DNA Oligos 2 and 4	300 nM	15 nM	5 μL
Helicase buffer	5x	1x	20 μL
DEPC treated H_2O			63.75 μL
Total			90 μL

Note: This is for one replicate of 10 time points. This can be adapted for multiple replicates by multiplying these values and using a multichannel pipette for the assay.

in the first prepared PCR tube with STOP buffer. Mix thoroughly with the STOP buffer. This will be your first time point (just a few seconds after adding ATP).

6. Take 10 μL samples from the activated helicase reaction in the same manner at the following time points post ATP addition (in seconds): 15, 30, 45, 60, 120, 180, 300, 600.
7. At the end of all the time points, except for the 95 °C control, quench all the controls with 10 μL 2x STOP buffer. Place the 95 °C control on a heat block or thermocycler at 95 °C for 5 min, then quench with 10 μL 2x STOP buffer. Use a benchtop centrifuge to spin down any condensation in the tube.
8. Run 10 μL samples on a 15 % polyacrylamide 0.5x TBE gel at 100 V for ~45 min in a 4 °C environment.
9. Visualize the location of the FAM label using a gel imager with a 530/28 filter with blue epi-illumination, automatically setting exposure time for intense bands.

Note: To make sure the oligos have run far enough down the gel, it can be imaged without removing it from the gel casing to get a rough image. From there, it can be taken out to get a better image or put back in the PAGE apparatus to run longer for better band separation.

8. Conclusion

The study of type IV systems is evolving, with applications derived from these systems already on the horizon (Benz et al., 2024; Čepaitė et al., 2024; Guo et al., 2022; Sanchez-Londono et al., 2024). To improve existing type

IV-based CRISPR applications and develop additional tools, it is vital to understand type IV system functions in nature and their biochemical mechanisms. In these protocols, we have provided ways to assay type IV system activities and mechanistic steps. The protocols summarized here will be vital tools for future work aimed at determining type IV function and mechanisms.

References

Altae-Tran, H., Kannan, S., Suberski, A. J., Mears, K. S., Demircioglu, F. E., Moeller, L., ... Zhang, F. (2023). Uncovering the functional diversity of rare CRISPR-Cas systems with deep terascale clustering. *Science (New York, N. Y.), 382*(6673), eadi1910. https://doi.org/10.1126/science.adi1910.

Benz, F., Camara-Wilpert, S., Russel, J., Wandera, K. G., Čepaitė, R., Ares-Arroyo, M., ... Beisel, C. L., ... Pinilla-Redondo, R. (2024). Type IV-A3 CRISPR-Cas systems drive inter-plasmid conflicts by acquiring spacers in *trans*. *Cell Host & Microbe*. https://doi.org/10.1016/j.chom.2024.04.016.

Čepaitė, R., Klein, N., Mikšys, A., Camar-Wilpert, S., Benz, F., Skorupskaitė, A., ... Pausch, P. (2024). Structural variation of types IV-A1- and IV-A3-mediated CRISPR interference. *Nature Communications, 15*, 9306. https://doi.org/10.1038/s41467-024-53778-1.

Crowley, V. M., Catching, A., Taylor, H. N., Borges, A. L., Metcalf, J., Bondy-Denomy, J., & Jackson, R. N. (2019). A type IV-A CRISPR-cas system in Pseudomonas aeruginosa mediates RNA-guided plasmid interference in vivo. *The CRISPR Journal, 2*(6), 434–440. https://doi.org/10.1089/crispr.2019.0048.

Cui, N., Zhang, J.-T., Liu, Y., Liu, Y., Liu, X.-Y., Wang, C., ... Jia, N. (2023). Type IV-A CRISPR-Csf complex: Assembly, dsDNA targeting, and CasDinG recruitment. *Molecular Cell, 83*(14), 2493–2508.e5. https://doi.org/10.1016/j.molcel.2023.05.036.

Domgaard, H., Cahoon, C., Armbrust, M. J., Redman, O., Jolley, A., Thomas, A., & Jackson, R. N. (2023). CasDinG is a 5′-3′ dsDNA and RNA/DNA helicase with three accessory domains essential for type IV CRISPR immunity. *Nucleic Acids Research, 51*(15), 8115–8132. https://doi.org/10.1093/nar/gkad546.

Domgaard, H., Cahoon, C., Armbrust, M. J., Redman, O., Thomas, A., & Jackson, R. N. (2022). *CasDinG is an ATP-dependent 5'-3' DNA helicase with accessory domains essential for type IV CRISPR immunity* (p. 2022.08.23.504870). bioRxiv. https://doi.org/10.1101/2022.08.23.504870.

Guo, X., Sanchez-Londono, M., Gomes-Filho, J. V., Hernandez-Tamayo, R., Rust, S., Immelmann, L. M., ... Randau, L. (2022). Characterization of the self-targeting Type IV CRISPR interference system in Pseudomonas oleovorans. *Nature Microbiology,* 7(11), https://doi.org/10.1038/s41564-022-01229-2.

Hallmark, T., Williams, A. A., Redman, O., Guinn, B., Judd, C., & Jackson, R. N. (2024). The N-terminal domain of type IV-A1 CRISPR-associated DinG is vulnerable to proteolysis. *microPublication Biology*. https://doi.org/10.17912/micropub.biology.001226.

Liao, C., Ttofali, F., Slotkowski, R. A., Denny, S. R., Cecil, T. D., Leenay, R. T., ... Beisel, C. L. (2019). Modular one-pot assembly of CRISPR arrays enables library generation and reveals factors influencing crRNA biogenesis. *Nature Communications, 10*(1), 2948. https://doi.org/10.1038/s41467-019-10747-3.

Makarova, K. S., Wolf, Y. I., Iranzo, J., Shmakov, S. A., Alkhnbashi, O. S., Brouns, S. J. J., ... Koonin, E. V. (2020). Evolutionary classification of CRISPR–Cas systems: A burst of class 2 and derived variants. *Nature Reviews Microbiology, 18*(2), https://doi.org/10.1038/s41579-019-0299-x.

Moya-Beltrán, A., Makarova, K. S., Acuña, L. G., Wolf, Y. I., Covarrubias, P. C., Shmakov, S. A., ... Quatrini, R. (2021). Evolution of type IV CRISPR-Cas systems: Insights from CRISPR loci in integrative conjugative elements of acidithiobacillia. *The CRISPR Journal, 4*(5), 656–672. https://doi.org/10.1089/crispr.2021.0051.

Mulepati, S., & Bailey, S. (2013). In vitro reconstitution of an Escherichia coli RNA-guided immune system reveals unidirectional, ATP-dependent degradation of DNA target. *The Journal of Biological Chemistry, 288*(31), 22184–22192. https://doi.org/10.1074/jbc.M113.472233.

Newire, E., Aydin, A., Juma, S., Enne, V. I., & Roberts, A. P. (2020). Identification of a type IV-A CRISPR-cas system located exclusively on IncHI1B/IncFIB plasmids in Enterobacteriaceae. *Frontiers in Microbiology, 11*. https://doi.org/10.3389/fmicb.2020.01937.

Özcan, A., Pausch, P., Linden, A., Wulf, A., Schühle, K., Heider, J., ... Randau, L. (2019). Type IV CRISPR RNA processing and effector complex formation in Aromatoleum aromaticum. *Nature Microbiology, 4*(1), https://doi.org/10.1038/s41564-018-0274-8.

Pinilla-Redondo, R., Mayo-Muñoz, D., Russel, J., Garrett, R. A., Randau, L., Sørensen, S. J., & Shah, S. A. (2020). Type IV CRISPR–Cas systems are highly diverse and involved in competition between plasmids. *Nucleic Acids Research, 48*(4), 2000–2012. https://doi.org/10.1093/nar/gkz1197.

Sanchez-Londono, M., Rust, S., Hernandez-Tamayo, R., Gomes-Filho, J. V., Thanbichler, M., & Randau, L. (2024). Visualization of Type IV-A1 CRISPR-mediated repression of gene expression and plasmid replication. *Nucleic Acids Research, 52*(20), 12592–12603. https://doi.org/10.1093/nar/gkae879.

Sinkunas, T., Gasiunas, G., Fremaux, C., Barrangou, R., Horvath, P., & Siksnys, V. (2011). Cas3 is a single-stranded DNA nuclease and ATP-dependent helicase in the CRISPR/Cas immune system. *The EMBO Journal, 30*(7), 1335–1342. https://doi.org/10.1038/emboj.2011.41.

Studier, F. W. (2005). Protein production by auto-induction in high density shaking cultures. *Protein Expression and Purification, 41*(1), 207–234. https://doi.org/10.1016/j.pep.2005.01.016.

Taylor, H. N., Laderman, E., Armbrust, M., Hallmark, T., Keiser, D., Bondy-Denomy, J., & Jackson, R. N. (2021). Positioning diverse type IV structures and functions within class 1 CRISPR-Cas systems. *Frontiers in Microbiology, 12*, 1236. https://doi.org/10.3389/fmicb.2021.671522.

Taylor, H. N., Warner, E. E., Armbrust, M. J., Crowley, V. M., Olsen, K. J., & Jackson, R. N. (2019). Structural basis of type IV CRISPR RNA biogenesis by a Cas6 endoribonuclease. *RNA Biology, 16*(10), 1438–1447. https://doi.org/10.1080/15476286.2019.1634965.

Van Erp, P. B. G., Jackson, R. N., Carter, J., Golden, S. M., Bailey, S., & Wiedenheft, B. (2015). Mechanism of CRISPR-RNA guided recognition of DNA targets in Escherichia coli. *Nucleic Acids Research, 43*(17), 8381–8391. https://doi.org/10.1093/nar/gkv793.

Zhou, Y., Bravo, J. P. K., Taylor, H. N., Steens, J. A., Jackson, R. N., Staals, R. H. J., & Taylor, D. W. (2021). Structure of a type IV CRISPR-Cas ribonucleoprotein complex. *iScience, 24*(3), 102201. https://doi.org/10.1016/j.isci.2021.102201.

CHAPTER SIX

Analysis of metal-dependent DNA nicking activities by Cas endonucleases

Giang T. Nguyen[a], Akshara Raju[b], and Dipali G. Sashital[b,*]
[a]Department of Hematology, St. Jude Children's Research Hospital, Memphis, TN, United States
[b]Roy J. Carver Department of Biochemistry, Biophysics and Molecular Biology, Iowa State University, Ames, IA, United States
*Corresponding author. e-mail address: sashital@iastate.edu

Contents

1. Introduction 118
2. Cleavage assays using negatively supercoiled plasmid DNA substrates 120
   2.1 Plasmid substrate preparation 121
   2.2 Initiate cleavage by mixing Cas RNP and DNA in the presence of Mg^{2+} 122
   2.3 Initiate cleavage through the addition of Mg^{2+} ion to pre-bound RNP-DNA complex 129
   2.4 DNA binding time-course analyzed by determining fraction cleaved 131
3. Cleavage assays using linear DNA substrates 134
   3.1 Linear substrate preparation 135
   3.2 Procedure 137
   3.3 Cleavage assays and separation of nicked products for linear substrates 137
4. Summary and conclusion 139
Acknowledgments 139
References 139

Abstract

CRISPR-Cas systems use RNA-guided CRISPR-associated (Cas) effectors to neutralize infections in bacteria and archaea. In class 2 CRISPR-Cas systems, Cas9 and Cas12 are single-protein Cas effectors that target double-stranded DNA based on complementarity to the guide RNA before cleaving the target DNA using metal-dependent endonuclease domains. Cas9 and Cas12 proteins can be readily programmed to target any DNA of interest by changing the guiding RNA sequence and have been co-opted for genome editing and other biotechnology purposes. The effect of metal ion concentration is an essential consideration in the physiological role of Cas immunity effectors as well as the biotechnological applications of Cas endonucleases. In this chapter, we describe methods for studying the effect of variable divalent metal ion conditions on the DNA binding and cleavage activities of well-studied Cas9 and Cas12a proteins.

Methods in Enzymology, Volume 712
ISSN 0076-6879, https://doi.org/10.1016/bs.mie.2025.01.034

1. Introduction

The effectors of CRISPR-Cas immune systems are Cas proteins guided by CRISPR RNAs (crRNAs) to bind to complementary nucleic acids (Makarova et al., 2020). Cas9 and Cas12a are popular effector proteins that target DNA (Gasiunas, Barrangou, Horvath, & Siksnys, 2012; Jinek et al., 2012; Zetsche et al., 2015). Many orthologs of each protein have been extensively researched and several are effective tools for genome editing (Chen et al., 2017; Cong et al., 2013; Hou et al., 2013; Kim et al., 2017; Mali, Esvelt, & Church, 2013; Ran et al., 2015; Zetsche et al., 2015; Zetsche et al., 2017). DNA binding by Cas9 and Cas12a is initiated by searching for protospacer adjacent motif (PAM) sequences (Anders, Niewoehner, Duerst, & Jinek, 2014; Cofsky, Soczek, Knott, Nogales, & Doudna, 2022; Singh, Sternberg, Fei, Doudna, & Ha, 2016; Singh et al. 2018; Sternberg, Redding, Jinek, Greene, & Doudna, 2014). PAMs are short and conserved motifs located next to target "protospacers" (Mojica, Díez-Villaseñor, García-Martínez, & Almendros, 2009), and are crucial for target location and DNA unwinding (Cofsky et al., 2022; Sashital, Wiedenheft, & Doudna, 2012; Semenova et al., 2011). At PAM sites, Cas9 and Cas12a endonucleases momentarily pause to assess if the adjacent sequence matches the guiding crRNA "spacer" sequence (Fig. 1) (Singh et al., 2018; Sternberg et al., 2014). If a complementary sequence is found, the Cas endonuclease triggers DNA unwinding by forming crRNA-DNA base pairs, starting at the PAM-proximal region of the protospacer. This unwinding progresses directionally away from the PAM to form an R-loop (Szczelkun et al., 2014). Consequently, the initial base pairs formed in the PAM-proximal, or "seed", region play a more significant role in target binding than those in the PAM-distal region (Jinek et al., 2012; Zetsche et al., 2015).

Successful PAM recognition and target binding results in DNA cleavage by Cas9 and Cas12a and formation of a double strand break (DSB) (Fig. 1) (Gasiunas et al., 2012; Jinek et al., 2012; Zetsche et al., 2015). Cas9 generates DSBs in the PAM-proximal region using two metal-dependent nuclease domains, HNH and RuvC (Fig. 1A) (Gasiunas et al., 2012; Jinek et al., 2012). The HNH domain employs a one-metal-ion mechanism to cleave the target strand, while the RuvC domain uses a two-metal-ion mechanism to cleave the non-target strand (Gong, Yu, Johnson, & Taylor, 2018; Nierzwicki et al., 2022). Cas12a uses a single RuvC nuclease domain to cleave both strands of the target DNA in the PAM-distal region (Fig. 1B) (Stella, Alcón, & Montoya, 2017; Stella et al., 2018; Swarts, van der Oost, & Jinek, 2017). Both Cas9 and Cas12a must undergo large

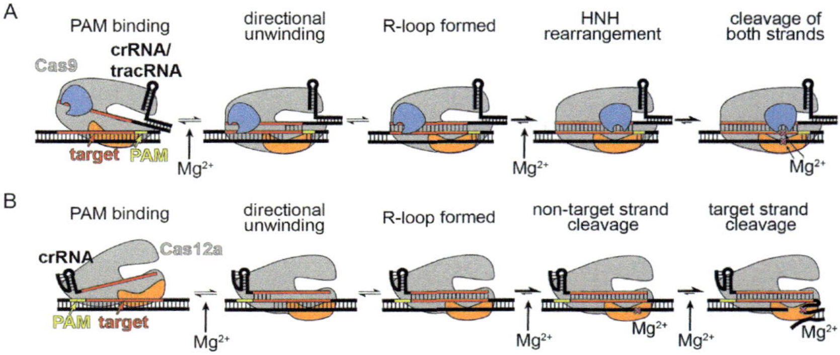

Fig. 1 DNA binding and cleavage mechanisms for (A) Cas9 and (B) Cas12a effectors. Both Cas effectors search for targets and initiate DNA unwinding based on the presence of a PAM sequence adjacent to the target. DNA unwinding occurs directionally away from the PAM. The overall process of DNA binding is dependent on Mg^{2+}. (A) Following complete formation of the R-loop, the Cas9 HNH domain undergoes a Mg^{2+}-dependent conformational rearrangement that is necessary for cleavage of the target strand. Both the HNH and the RuvC domain require divalent metal ion cofactors. (B) Following R-loop formation, Cas12a undergoes two Mg^{2+}-dependent conformational changes, first to position the non-target strand and then the target strand in the RuvC active site. The two successive cleavage events are both carried out by the RuvC domain.

conformational changes to enable cleavage of both strands of DNA (Fig. 1). The Cas9 HNH domain only adopts a catalytic conformation upon formation of a full R-loop (Fig. 1A) (Dagdas, Chen, Sternberg, Doudna, & Yildiz, 2017; Guo et al., 2017; Nierzwicki et al., 2022; Sternberg, Lafrance, Kaplan, & Doudna, 2015). Cas12a must rearrange the DNA substrate following non-target strand cleavage to allow access of the target strand within the RuvC active site (Fig. 1B) (Cofsky et al., 2020; Son et al., 2021; Stella et al., 2018; Strohkendl et al., 2024; Swarts & Jinek, 2019; Swarts et al., 2017). The conformational changes of both Cas9 and Cas12a are metal-ion dependent, underscoring the essential role of metal ions for Cas endonuclease function (Dagdas et al., 2017; Son et al., 2021).

Recent studies have highlighted the impact of physiological metal ion conditions on Cas9 and Cas12a function. Some Cas9 orthologs are unable to bind DNA at the lower Mg^{2+} concentrations present in human cells (Eggers et al., 2024). Engineered Cas9 variants can overcome this limitation, allowing for improved genome editing efficiency. The RuvC domain of Cas9 is also sensitive to lower Mg^{2+} concentration that can be impacted by the presence of biomolecules that coordinate metal ions (Nguyen, Schelling, & Sashital, 2024a). Inhibition of RuvC activity due to low Mg^{2+}

availability results in the formation of a nick through cleavage by only the HNH domain. Cas12a is also prone to nicking at lower Mg^{2+} concentration (Son et al., 2021), especially in the presence of target mismatches in the PAM-distal region (Fu et al., 2019; Murugan, Seetharam, Severin, & Sashital, 2020; Nguyen et al., 2024b). At physiological Mg^{2+} concentrations, Cas12a cleavage of targets containing PAM-distal mismatches is significantly impaired (Schelling, Nguyen, & Sashital, 2023). Conversely, cleavage of targets containing seed mismatches or PAM mutations is relatively fast at lower Mg^{2+} concentrations (Nguyen et al., 2024b).

Analyzing Cas9 and Cas12a kinetics under various metal ion conditions is essential for understanding their biochemical mechanisms and predicting off-target effects, leading to more accurate and safer gene editing technologies. In this chapter, we provide detailed protocols for performing in vitro cleavage assays that allow investigation of the impact of metal ions on Cas effector function, including both DNA binding and cleavage. The protocols provided in this chapter allow for rate measurements for the formation of both products containing cuts on one or both DNA strands using either negatively supercoiled or linear DNA substrates. Both sets of methods allow for detection of cleavage products without the need for expensive labeling techniques like radioactive ^{32}P or fluorescent dyes. Thus, these methods can be readily adapted for a wide variety of Cas effectors and targets.

2. Cleavage assays using negatively supercoiled plasmid DNA substrates

Negatively supercoiled plasmids are commonly used as substrates to measure DNA cleavage by Cas effectors. Cas9 and Cas12a generate double strand breaks, but may initially nick only one strand in the presence of target mutations or low metal ion conditions (Fu et al., 2019; Murugan et al., 2020; Murugan, Suresh, Seetharam, Severin, & Sashital, 2021; Nguyen et al., 2024a, 2024b; Schelling et al., 2023; Son et al., 2021). First-strand cleavage results in relaxation of negatively supercoiled substrates, while second-strand cleavage results in linearization of the plasmid (Fig. 2A). These distinct products have different mobilities during agarose gel electrophoresis, allowing for determination of the rates of both cleavage events. Here we describe standard protocols for cleavage assays using negatively supercoiled plasmid DNA as targets for Cas effector cleavage. We do not include methods for the in vitro synthesis of crRNA and trans activating CRISPR RNA (tracrRNA) or Cas9

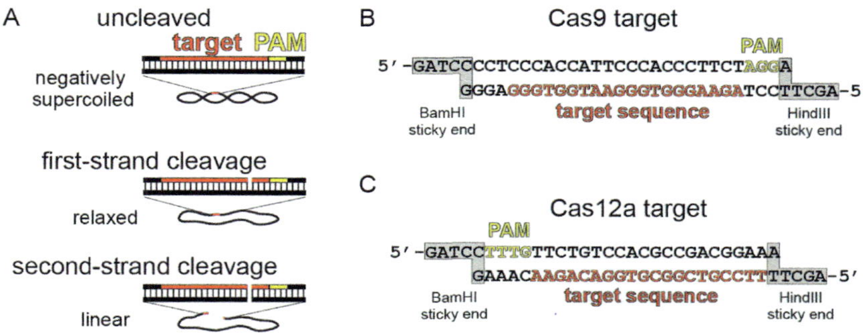

Fig. 2 Negatively supercoiled target DNA substrates allow observation of first- and second-strand cleavage by Cas endonucleases. (A) For a Cas9 target, cleavage of the target strand by the HNH domain results in relaxation of supercoils, which results in slower migration on an agarose gel. Cleavage of the non-target strand by the RuvC domain results in a double-strand break, converting the plasmid to a linear form that migrates between the relaxed and negatively supercoiled bands on the agarose gel. (B–C) Design of oligonucleotides used for (B) a Cas9 target, derived from gene B of the λ phage genome or (C) a Cas12a target, derived from gene L of the λ phage genome. The oligonucleotides contain sticky ends when annealed together, allowing for ligation with doubly-digested plasmid to create the target plasmid.

purification, which have been described in detail in previous method papers (Anders & Jinek, 2014; Mohanraju, Van Der Oost, Jinek, & Swarts, 2018). We describe two distinct protocols for initiating DNA cleavage, either through the mixing of Cas9 RNA-protein complex (RNP) with the DNA substrate in the presence of Mg^{2+}, or through the addition of Mg^{2+} to initiate cleavage following incubation of the Cas9 RNP and DNA. These two methods allow for differentiation between cleavage defects that may occur during or following DNA binding by the Cas effector, respectively. We also provide a third method for determining the rate of Cas effector binding to a DNA substrate in the absence of metal ions.

2.1 Plasmid substrate preparation

Substrate plasmids can be generated using a restriction enzyme cloning protocol that we previously described (Phan, Schelling, Xue, & Sashital, 2019). Oligonucleotides containing a target sequence and PAM flanked by sticky ends for two restriction enzymes can be cloned into a plasmid digested with the corresponding restriction enzymes (Fig. 2B and C). For plasmid cleavage assays, we typically use pUC19 digested with *Bam*HI and *Hin*dIII. After verifying the correct target plasmid sequence by Sanger or whole plasmid sequencing, larger quantities of the substrate plasmid are prepared. We typically use a Midi Kit and follow manufacturer's directions outside of a few optimizations described below.

2.1.1 Equipment

- Incubator at 37 °C.
- Shaker incubator at 37 °C.
- Spectrophotometer (e.g. NanoDrop from Thermo Fisher Scientific).

2.1.2 Buffers and reagents

- *E. coli recA*- competent cells (e.g. NEB 5-alpha from New England Biolabs).
- HiSpeed Plasmid Midi Kit (QIAGEN).

2.1.3 Procedure

1. Transform 1 μL of the verified substrate plasmid into a *recA*- competent *E. coli* strain using standard protocols. Plate 100 μL of the recovery (or a dilution if necessary) on LB-agar plates containing the appropriate antibiotic. Incubate the plate overnight at 37 °C.
2. Using a single colony, inoculate 150 mL LB media supplemented with 100 μg/mL ampicillin. Grow the culture overnight at 37 °C with shaking (between 180−220 rpm).
3. Purify the plasmid DNA from the culture using a HiSpeed Plasmid Midi Kit (QIAGEN), following the manufacturer's protocol with the following modifications:
   - Incubate the overnight culture on ice for at least 30 min prior to harvesting the cells. We find that this incubation ensures that all DNA remains negatively supercoiled upon lysis of the cells.
   - Elute the plasmid with 350 μL nuclease free water that was pre-incubated to 37 °C. This modification yields plasmid DNA with a higher concentration.
4. Determine the absorbance of the DNA at 260 nm using a spectrophotometer.
5. Dilute the plasmid stock to 100 ng/μL and aliquot plasmid into microcentrifuge tubes (50−100 μL each). Store at −20 °C. Each aliquot should be freeze-thawed no more than three times to ensure that plasmid remains negatively supercoiled.

2.2 Initiate cleavage by mixing Cas RNP and DNA in the presence of Mg^{2+}

Cas9 must unwind and bind the target region of plasmid DNA prior to cleavage. The overall rate of unwinding, binding and cleavage can be measured in time course cleavage assays in which the DNA substrate and Cas effector-guide RNA complex are mixed together in the presence of metal ions to initiate cleavage (Fig. 3A) (Gong et al., 2018). In these experiments, the Cas endonuclease protein is first incubated with crRNA (and tracrRNA if necessary) in a 1:1.5 molar ratio in cleavage buffer containing Mg^{2+} ion to form a ribonucleoprotein complex (RNP complex). Cas effectors are generally single turnover enzymes (Chen et al., 2018; Singh et al., 2018; Sternberg et al., 2014; Strohkendl, Saifuddin, Rybarski, Finkelstein, & Russell, 2018). Therefore, the RNP complex concentration is used in excess over plasmid substrates to ensure complete cleavage. The reaction is initiated by adding negatively supercoiled plasmid into the reaction tube containing both RNP complex and Mg^{2+} ion. Aliquots of the reaction are removed at various time points and quenched by precipitating the Cas effector protein using phenol. The nucleic acid remains in the aqueous layer and can be extracted prior to analysis. The reaction is performed with different concentrations of Mg^{2+} added to the reaction buffer to determine the effect of divalent metal ions on the overall rate of cleavage (Nguyen et al., 2024b).

To analyze the rate of cleavage, DNA products are separated by agarose gel electrophoresis (Fig. 3B). The gels are stained with SYBR Safe following electrophoresis to allow visualization of negatively supercoiled, relaxed, and linear DNA. Post-staining the gel is essential to ensure proper migration of the three DNA species, whose migration can be affected when DNA intercalators are present in the gel matrix. Following gel visualization, the intensity of each DNA band can be quantified using ImageJ software (Fig. 3C and D) (Rueden et al., 2017; Schneider, Rasband, & Eliceiri, 2012). The relaxed DNA band contains products that have been nicked on a single strand through a first-strand cleavage event, while the linear DNA band contains products that underwent second-strand cleavage. Thus, separation of a relaxed and linear product enables determination of the rate of both a first- and second-strand cleavage event (Fig. 3E and F) (Nguyen et al., 2024b).

Cas9 has previously been observed to display biphasic cleavage kinetics (Gong et al., 2018; Liu, Gong, Yu, Taylor, & Johnson, 2019). The two phases of cleavage have recently been linked to conformational differences that occur upon DNA binding that result in different cleavage timescales for a Cas9d ortholog (Ocampo et al., 2024). If such biphasic kinetics are observed, it is appropriate to fit the kinetic data to a double-exponential rate equation that estimates the rate constants for the fast and the slow cleavage phase (Liu et al., 2019). However, we sometimes observe that the slow phase cannot be accurately determined by the graphing software. In these cases, it is appropriate to fit the data to a single-exponential rate

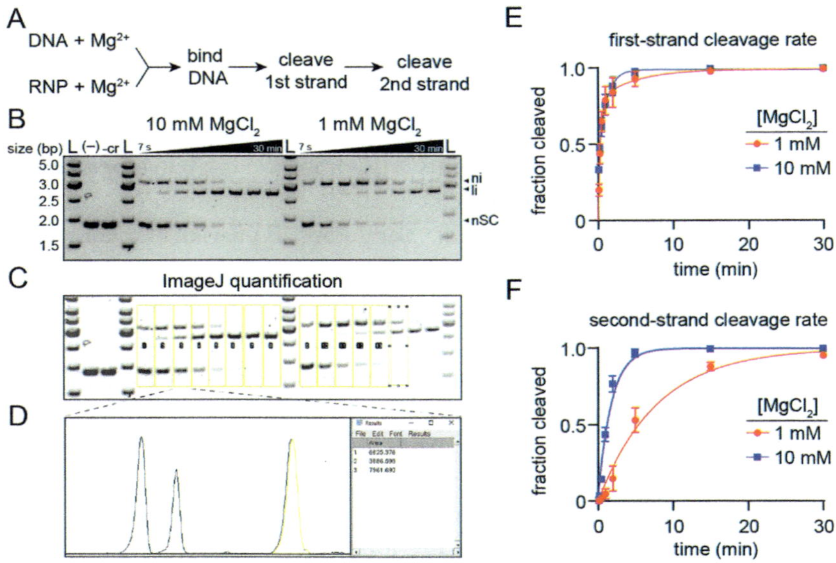

Fig. 3 Cleavage of negatively supercoiled DNA substrate when initiating by mixing DNA and Cas RNP in the presence of metal ions. (A) Schematic of the cleavage assay. The Cas effector and RNA are first incubated in the presence of Mg^{2+} to prepare the Cas RNP. The DNA substrate is also diluted in a buffer containing Mg^{2+}. Both components are mixed to initiate DNA binding and cleavage. The rates determined for the cleavage of each strand encompasses all events that must occur prior to cleavage, including PAM searching, DNA unwinding and R-loop formation, and conformational rearrangements. (B) Agarose gel showing a Cas9 cleavage time course for the gene B target at 10 and 1 mM $MgCl_2$. Nicked DNA (ni), linear DNA (li), and negatively supercoiled DNA (nSC) bands are labeled. The time points were 7 s, 15 s, 30 s, 1 min, 2 min, 5 min, 15 min, and 30 min. The two controls are DNA alone (–) and DNA with Cas9 without crRNA and tracrRNA (-cr). Both controls were performed at the higher $MgCl_2$ concentration for the longest time point. The gel was post-stained with SybrSafe for 1 h before imaging. (C-D) Images showing steps of quantification of the agarose gel in (B) using ImageJ. (C) Following background subtraction, lanes were selected using the rectangle tool. (D) The bands were plotted and integrated using the magic wand tool. The area under the curve reported in the Results window were used to determine the fraction of DNA that had undergone first or second-strand cleavage. (E-F) Rate curves plotting fraction cleaved versus time for (E) first- and (F) second-strand cleavage at the two $MgCl_2$ concentrations. The data in (E) were fit to a double-exponential equation, while the data in (F) fit better to a single-exponential. Curve fitting was performed in GraphPad Prism. The average of three replicates is shown and error bars represent standard deviation. *Data are adapted from Nguyen et al. (2024a).*

equation that models a single cleavage phase. In either case, the rate constants represent the combined rates of all events that lead to cleavage, including PAM searching, target unwinding and accompanying conformational changes, and cleavage catalysis.

We have included exemplar data for a Cas9 target derived from the gene B region of the λ phage genome (Fig. 2B) (Nguyen et al., 2024a). While first-strand cleavage rates were observed to be biphasic, and therefore fit to a double-exponential rate equation (Fig. 3E), second-strand cleavage rates were observed to have a single phase, and were therefore fit to a single-exponential rate equation (Fig. 3F). We observed a similar rate of first-strand cleavage between the two Mg^{2+} conditions tested (Fig. 3E), with some differences between the distribution and rates of the fast and slow phase. In contrast, we observed much slower second-strand cleavage at 1 mM Mg^{2+} in comparison to 10 mM Mg^{2+} (Fig. 3F), suggesting that cleavage by one of the two Cas9 nuclease domains is impaired at lower metal ion conditions. In our previous study, we observed that the slow second-strand cleavage kinetics are due to slower cleavage by the RuvC domain at lower metal ion conditions (Nguyen et al., 2024a).

In this section, we provide detailed methods for performing the cleavage assay initiated by mixing DNA and Cas RNP together in the presence of $MgCl_2$, tips for how best to separate and visualize the products via agarose gel electrophoresis, and a description of how we perform analysis of the data using ImageJ and GraphPad Prism.

2.2.1 Equipment

- Heat block for 37 °C incubation
- Microcentrifuge
- Agarose gel apparatus (e.g. Corning Axygen 15 cm Horizontal Gel Box)
- Agarose gel casting kit with 15 cm gel tray and 30-well 1.5 mm comb
- Power supply
- Orbital shaker
- Agarose gel imaging system (e.g. Bio-Rad Gel Doc EZ)
- Staining tray

2.2.2 Buffers and reagents

- 5X reaction buffer (100 mM HEPES, pH 7.5, 500 mM KCl, 5 mM DTT, 50 mM or 5 mM $MgCl_2$ and 25 % glycerol)
- 25:24:1 phenol-chloroform-isoamyl alcohol (Invitrogen)
- 2X DNA loading buffer (50 mM EDTA, 0.1 % bromophenol blue, 0.1 % xylene cyanol FF, 10 % glycerol).
- TAE buffer (40 mM Tris base, 20 mM acetic acid, and 1 mM EDTA)
- Agarose
- SYBR Safe (Thermo Fisher Scientific)

2.2.3 *Procedure*

2.2.3.1 Perform Cas9 plasmid DNA cleavage time course

1. We typically perform eight time points at 7 s, 15 s, 30 s, 1 min, 2 min, 5 min, 15 min, and 30 min, although these may vary depending on the rate of cleavage for a given target. For each time point, we prepare two tubes, one containing 10 μL of phenol-chloroform-isoamyl alcohol (for time point quenching) and one containing 10 μL of 2X DNA dye (for DNA extraction). Arrange each set of tubes in order of time points to ensure tubes are ready for quenching time points.
2. We typically perform cleavage reactions at two different $MgCl_2$ concentrations, 10 and 1 mM. Reactions can be set up in parallel by adjusting the concentration of $MgCl_2$ in the reaction buffer. Initiation of the reactions should be staggered to ensure that all time points can be taken.
3. Cas9, crRNA and tracrRNA can be prepared as previously described (Schelling et al., 2023). On ice, dilute Cas9 to 1 μM with 1X reaction buffer chilled on ice. On ice, dilute in vitro transcribed crRNA and tracrRNA with 5X reaction buffer and nuclease free water to a final concentration of 1 μM RNA and 1X reaction buffer. Leave the diluted reagents on ice.
4. To form RNP complex, mix 100 nM Cas9 with 150 nM crRNA and 150 nM tracrRNA in pre-chilled 1X reaction buffer in a final volume of 50 μL. Leave the RNP on ice until all reagents are prepared.
5. Dilute the plasmid target with pre-chilled 5X reaction buffer and nuclease-free water to a final volume of 50 μL and a final concentration of 15 ng/mL plasmid and 1X reaction buffer. Leave the DNA on ice until all reagents are prepared.
6. Incubate the diluted RNP complex and plasmid DNA at 37 °C for 10 min
7. Add the contents of the DNA tube to the RNP tube (or vice versa) to initiate the cleavage reaction. Mix well by pipetting up and down 3-4 times. Continue to incubate the reaction at 37 °C.
8. Remove 10 μL of sample at each time-point and rapidly mix with the aliquoted phenol-chloroform-isoamyl alcohol in the appropriate time point tube to quench the reaction. Pipet up and down 5-6 times to ensure mixing.
   - Tip: It is necessary to have a pipet ready with a tip on it to take a time point immediately after initiating cleavage. We typically find that 7 s is the shortest amount of time in which the reaction components can be adequately mixed, and a time point quenched when time points are performed manually.

9. Additional tubes should be prepared as controls. We generally perform one control in which DNA is incubated in reaction buffer without RNP and another in which DNA and Cas9 are incubated without the RNAs. Both controls would be incubated at 37 °C for the longest time point.

2.2.3.2 Prepare samples for running agarose gel

1. Prepare a 1 % agarose gel in TAE buffer using standard protocols. Note that it is important to not add any dye to the gel to ensure the migration of DNA on the gel based on the topology of the DNA. We typically use a 15 cm gel tray and one or two 30-well combs (1.5 mm thickness), depending on how many samples will be run. If using two combs, one is placed at the top of the gel and one in the middle. Different sets of samples can be run in each set of wells. Note that the gel can be prepared prior to performing the time course assay.
2. Centrifuge time point samples in a microcentrifuge at the highest speed for 1 min and then remove 9 μL of DNA from the top aqueous layer into the prepared 2X DNA dye tubes.
    - Tip: Removal of the aqueous layer can be facilitated by adding a small amount (2–5 μL) of 2X DNA dye to the quenched samples prior to centrifugation. The dye will separate into the organic layer and make the aqueous layer more visually distinct. Note that addition of the DNA dye will dilute the aqueous layer.
3. Load 15 μL of each sample into wells and run gel at 110 V for 1 h 30 min to ensure good separation between nicked and linearized products. Run a DNA ladder (e.g. Generuler 1 kb DNA ladder from Thermo Fisher Scientific) in at least one lane adjacent to the time point samples.
4. Dilute 20 μL of SYBR safe into 200 mL of TAE buffer. Stain the gel with SYBR safe for 1 h while shaking gently on an orbital shaker.
5. Image the gel using a gel imager equipped with a UV transilluminator. Export the gel image in a format that is compatible with ImageJ (e.g. TIFF).

2.2.3.3 Quantification of DNA cleavage versus time

1. Gels should have DNA bands corresponding to negatively supercoiled and linear DNA, and may also have relaxed DNA if DNA nicking occurred (Fig. 3B). Controls (e.g. nicked and linearized plasmids prepared using restriction enzymes) can be run to ensure that the migration of DNA bands corresponds to nicked and linearized products.
2. Determine the intensity of each DNA band using Image J (Fig. 3C and D).

- Open the exported gel image in ImageJ.
- Ensure that the gel image is horizontal. Adjust using image rotation tools if necessary.
- Invert the image using Edit > Invert to produce dark DNA bands on a light background.
- Perform background subtraction using the Process > Subtract Background... dialog. We use the "Light background" setting with a rolling ball radius of 25.
- Using the Rectangle tool from the tool bar, draw a rectangle around the first lane containing a time point sample (Fig. 3C). Use the keyboard shortcut control-1 on Windows or command-1 on Mac to select the first lane.
- Drag the rectangle. This should create a new rectangle that can be positioned on the next lane. Use the keyboard shortcut control-2 on Windows or command-2 on Mac to select the next lane. Repeat this for the remaining lanes.
- Once all lanes have been outlined with rectangles, use the keyboard shortcut control-3 on Windows or command-3 on Mac to plot the lanes. This will bring up a new window containing 2D plots of DNA band peaks.
- Use the magic wand tool to determine the area under each peak (Fig. 3D). Lines can be drawn using the line tool from the tool bar to ensure an even baseline and to distinguish between different peaks.
- Copy the peak intensity data to a worksheet in Excel, keeping track of values from the relaxed, linear, and supercoiled peaks (Fig. 3D, inset on right).

3. In Excel, calculate the intensity of total DNA in each sample by adding the intensities of the three bands.
4. First-strand cleavage describes the rate at which the DNA is nicked based on an initial cleavage event. Nicked DNA is eventually linearized, so the fraction of DNA that underwent the first cleavage event can be determined by adding the intensity of the nicked and linear products.

$$\textit{fraction of DNA that underwent first strand cleavage} = \frac{(I_{nicked} + I_{linear})}{I_{all}}$$

where I_{nicked} is intensity of relaxed DNA band, I_{linear} is intensity of linear DNA band, and I_{all} is intensity of all bands

5. DNA that has undergone second strand cleavage should be linearized, and can be determined using

$$\textit{fraction of DNA that underwent second strand cleavage} = \frac{I_{linear}}{I_{all}}$$

where I_{linear} is intensity of linear band and I_{all} is intensity of all bands.

6. Rate constants can be derived from fits of individual replicates to either a single-exponential or a double-exponential rate equation in a graphing software. We typically use GraphPad Prism, using the built-in "One-phase association" non-linear regression for single-exponential rates or the "Two-phase association" for double-exponential rates.
   - The equation for One-phase association is:

$$y = y_0 + (plateau - y_0) \times (1 - e^{(-kx)})$$

   - The equation for Two-phase association is:

$$y = y_0 + SpanFast * (1 - e^{(-k_{fast}x)}) + SpanSlow * (1 - e^{(-k_{slow}x)})$$

   where y_0 is the y value at time 0, *plateau* is the amplitude of the curve, and k is the observed rate constant. We constrain y_0 to 0 and *plateau* to less than 1. SpanFast and SpanSlow estimate the fraction of the signal that is due to the fast or slow phase, respectively.

7. All experiments should be performed at least three times. If rate constants are to be reported, fit each replicate individually to the appropriate rate equation and determine the average and standard deviation of the rate constants determined for each replicate. We depict the rate curves using the average fraction cleaved at each time points, with error bars representing the standard deviation for fraction cleaved at each time point (Fig. 3E and F).

2.3 Initiate cleavage through the addition of Mg^{2+} ion to pre-bound RNP-DNA complex

As described above, cleavage reactions that are initiated by mixing the RNP and DNA in the presence of Mg^{2+} report on the rate of all events leading up to, and including, DNA cleavage. As an alternative, incubation of the RNP and plasmid DNA prior to the addition of Mg^{2+} allows for formation of an R-loop prior to initiation of cleavage (Fig. 4A). Performing the cleavage assay in this manner eliminates the rate of PAM searching, DNA unwinding and target binding from the overall rate of the

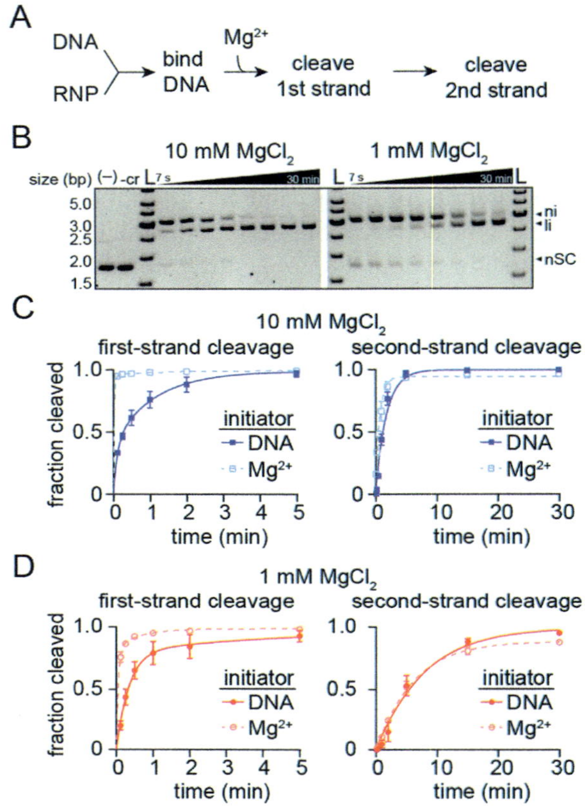

Fig. 4 Cleavage of negatively supercoiled DNA substrate when initiating by adding metal ions. (A) Schematic of the cleavage assay. The Cas RNP and DNA substrate are first incubated in the absence of Mg^{2+}. Cleavage is initiated by adding Mg^{2+}. The rates determined for the cleavage of each strand encompasses Mg^{2+}-dependent events that occur following R-loop formation, including conformational rearrangements and cleavage catalysis. (B) Agarose gel showing a Cas9 cleavage time course for the gene B target initiated with either 10 and 1 mM $MgCl_2$. The time points and labels are as in Fig. 3B. (C–D) Rate curves plotting fraction cleaved versus time for two different initiation methods at (C) 10 mM $MgCl_2$ or (D) 1 mM $MgCl_2$. First-strand cleavage data were fit to double-exponential equations while second-strand cleavage data were fit to single-exponential equations in GraphPad Prism. The average of three replicates is shown and error bars represent standard deviation.

cleavage reaction (Fig. 1) (Liu et al., 2019; Nguyen et al., 2024b). Thus, cleavage rates and rate constants reflect the rate of metal-dependent steps that occur after DNA binding, including conformational changes of endonuclease domains and cleavage catalysis.

In Fig. 4, we have included example data for the λ phage gene B Cas9 target comparing rates when cleavage was initiated with Mg^{2+} with the data shown in Fig. 3E and F. We observed a large increase in rate of first-strand cleavage at both $MgCl_2$ concentrations when cleavage was initiated by adding Mg^{2+} in comparison to when cleavage was initiated by adding DNA (Fig. 4C and D). This difference reflects the slow rate of DNA binding, which limits the rate of cleavage when the reaction is initiated by adding DNA. The difference in rates of first-strand cleavage was larger at 10 mM $MgCl_2$ (Fig. 4C) in comparison to 1 mM $MgCl_2$ (Fig. 4D), consistent with previous observations of slower conformational kinetics of the HNH domain at lower divalent metal ion conditions (Dagdas et al., 2017). Second-strand cleavage rates were similar regardless of initiation methods (Fig. 4C and D), suggesting that pre-binding the Cas9 RNP to DNA has little effect on the rate of cleavage by the RuvC domain (Nguyen et al., 2024a).

In this section, we describe a protocol for the measurement of plasmid DNA cleavage rate by initiating the reaction with Mg^{2+} ions (Fig. 4). All equipment and reagents are the same as for the procedure described in Section 2.2. with the noted exceptions.

2.3.1 Procedure

1. Prepare a 5X reaction buffer without Mg^{2+} containing 100 mM HEPES, pH 7.5, 500 mM KCl, 5 mM DTT, and 25 % glycerol.
2. Prepare 40 μL of 5 mM or 50 mM $MgCl_2$ solution in 1X reaction buffer.
3. Prepare tubes for time points containing phenol solution and 2X DNA dye as described in Section 2.2.2.1.
4. Prepare the RNP and DNA as described in Section 2.2.2.1 using the reaction buffer lacking Mg^{2+}.
5. Incubate the diluted RNP, plasmid and $MgCl_2$ solutions for 10 min at 37 °C.
6. Mix 40 μL of the RNP and the DNA to allow for DNA binding. Mix well by pipetting up and down 3–4 times and incubate for 30 min at 37 °C.
7. Add 20 μL of the $MgCl_2$ solution into the reaction tube and mix well by pipetting up and down 3–4 times.
8. Quench time points, prepare samples for gels, and analyze results as described in Sections 2.2.2.1–2.2.2.3.

2.4 DNA binding time-course analyzed by determining fraction cleaved

We have previously observed that Mg^{2+} may decrease the rate of Cas effector binding to negatively supercoiled plasmids (Nguyen et al., 2024b).

These observations prompted us to develop an assay that enables the measurement of the fraction of DNA that is bound at various time points in the absence of Mg^{2+}. Our assay makes use of the observation that for most DNA targets, Cas effectors cleave the DNA rapidly following target binding in the presence of high Mg^{2+} concentrations (Gong et al., 2018; Nguyen et al., 2024b). Thus, addition of Mg^{2+} for a brief period allows for cleavage of all of the DNA that has been bound by RNP. Measurement of the fraction cleaved provides information on the fraction of DNA that was bound by RNP at a given time point.

To optimize this assay, we first determined how long it was necessary to mix an RNP-DNA complex with Mg^{2+} to see complete cleavage. We found that a Cas12a RNP-DNA complex formed using a reaction buffer that lacked Mg^{2+} for 30 min at 37 °C, could fully cleave the DNA upon mixing with 10 mM $MgCl_2$ for 5 s. This result indicates that any RNP that is bound to DNA would be able to fully cleave within 5 s upon addition of Mg^{2+}. Importantly, the amount of time needed to fully cleave the DNA upon addition of Mg^{2+} may vary between targets, so it is important to perform this control for every DNA target tested.

Using this observation, we developed a protocol for monitoring the fraction of Cas RNP that is bound to DNA in the absence of Mg^{2+} over a time course (Fig. 5A) (Nguyen et al., 2024b). In this assay, the RNP and DNA are mixed in the absence of Mg^{2+} and incubated at 37 °C. At a given time point, an aliquot is removed from this binding reaction and mixed with Mg^{2+} at a final concentration of 10 mM. Cleavage is allowed to proceed for 5 s, and then the RNP-DNA is transferred to a phenol solution for quenching. The fraction cleaved is determined and used to estimate the fraction of DNA that was bound by the RNP at a given time point.

We previously used this technique to measure the rate of cleavage of a Cas12a DNA target derived from gene L of the λ phage genome (Figs. 2C and 5B and C) (Nguyen et al., 2024b). This target contains an A2T mutation within the seed region. The seed mismatch slows the rate of Cas12a DNA binding, especially at higher Mg^{2+} concentrations, resulting in DNA binding being rate-limiting. Thus, measurement of the rate of cleavage using the procedures described in Section 2.2 allowed us to estimate the rate of binding at variable Mg^{2+} concentrations. By using the procedure described below, we were also able to estimate a rate of Cas12a binding to the gene L A2T target in the absence of metal ions. We observed the fastest binding of Cas12a to the target when metal ions were absent during binding and the slowest rate of binding at the highest Mg^{2+} concentration tested (Fig. 5B and C).

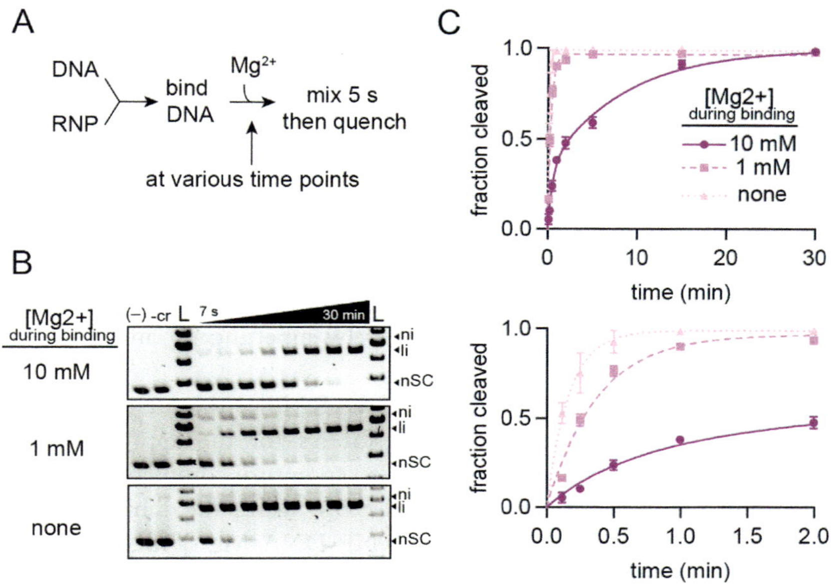

Fig. 5 Measuring the rate of DNA binding in the absence of metal ions. (A) Schematic of the binding assay. Binding is initiated by mixing the Cas RNP and DNA substrate in the absence of Mg^{2+}. At various time points, aliquots are removed from the binding reaction and mixed for 5 s with 10 mM $MgCl_2$ and then immediately quenched. All DNA that is bound by the Cas RNP at that time point will be cleaved, while DNA that remains unbound will remain negatively supercoiled. (B) Agarose gel showing a Cas12a cleavage time course for the gene L A2T target. The first two cleavage reactions were performed by initiating with DNA in the presence of 10 or 1 mM $MgCl_2$. The third reaction is the binding reaction in the absence of $MgCl_2$. The time points and labels are as in Fig. 3B. (C) Rate curves plotting fraction cleaved versus time for the three different $MgCl_2$ conditions that were present during DNA binding. The first-strand cleavage rate is plotted, as binding is rate limiting for first-strand cleavage for the gene L A2T target (Nguyen et al., 2024b). The top graph shows all time points, while the bottom graph shows only the early time points. The average of three replicates is shown and error bars represent standard deviation. *Data are adapted from Nguyen et al. (2024b).*

Below we describe the in-depth procedure for measuring the rate of target binding in the absence of metal ions (Fig. 5). All equipment and reagents are the same as for the procedure described in Section 2.2. with the noted exceptions.

2.4.1 Procedure

1. Prepare a 5X reaction buffer without Mg^{2+} containing 100 mM HEPES, pH 7.5, 500 mM KCl, 5 mM DTT, and 25 % glycerol.

2. Prepare a 50 mM $MgCl_2$ solution in 1X reaction buffer. Prepare tubes containing 2.5 µL of this solution, one for each time point to be tested.
3. Prepare additional tubes for time points containing phenol solution and 2X DNA dye as described in Section 2.2.2.1.
4. Prepare the RNP and DNA as described in Section 2.2.2.1 using the reaction buffer lacking Mg^{2+}.
5. Incubate the diluted RNP and plasmid solutions for 10 min at 37 °C.
6. Mix the contents of the RNP tube with the contents of the DNA tube to allow for DNA binding. Mix well by pipetting up and down 3-4 times. Continue incubating at 37 °C.
7. At each time point, remove 10 µL of the RNP-DNA mixture and mix rapidly by pipetting up and down for 5 s (or the appropriate time determined for individual DNA targets) to ensure complete cleavage of all bound DNA. At the end of the 5 s, using the same pipet tip, immediately move 10 µL of the RNP-DNA-Mg^{2+} mixture to the phenol-containing tube to quench both DNA binding and cleavage. Repeat this process for the remaining time points.
8. Prepare samples for gels, and analyze results as described in Sections 2.2.2.1–2.2.2.3.

3. Cleavage assays using linear DNA substrates

While negatively supercoiled DNA provides a convenient means to monitor nicking events on both the strands of the DNA target, the topology of supercoiled DNA can have significant effects on the ability of Cas effectors to unwind and bind DNA. Linear DNA targets are relevant to the native biology of CRISPR-Cas effectors, as the genomic DNA of bacteriophages is linear at various stages of the viral life cycle (Weigel & Seitz, 2006). Thus, it is also useful to be able to compare cleavage between supercoiled and linear DNA targets.

Short DNA oligonucleotides are commonly used as linear substrates for Cas effector cleavage (Jinek et al., 2012; Liu et al., 2019). For these substrates, individual strands of a double-stranded target can be labeled differentially, either by the addition of different fluorophores to each strand or by labeling each strand radioactively with ^{32}P and performing separate cleavage assays to monitor cleavage on each strand. Although very useful, these labeling techniques can become cost prohibitive, especially when testing a large number of different target sequences and with the rising costs of radioactive ATP. Additionally, short DNA oligonucleotides do not correspond well to the longer linear DNA that Cas effectors would encounter during their native function.

Linearized plasmids or PCR products containing target sequences can serve as alternative substrates that better represent target sequences within the context of longer linear DNA (Zetsche et al., 2015). However, when using a longer linear DNA substrate, it is difficult to observe all cleavage products by gel electrophoresis. Although products that have undergone cleavage events on both strands are readily separated due to the smaller size of the resulting fragments, uncleaved linear DNA and nicked fragments have similar migration patterns and are difficult to separate. This prevents quantification of first-strand DNA cleavage when using linear DNA in the absence of labels on each DNA strand.

To observe all the products from the first and second strand cleavage events in a cleavage assay with linear DNA substrate, we adapted a method using denaturing polyacrylamide gel electrophoresis (Kuhn, Protozanova, & Demidov, 2002). This method relies on the formation of a nick toward the center of a linear DNA fragment (Fig. 6A). When the linear DNA undergoes a first-strand cleavage event, a hinge is formed at the site of nick. Urea disrupts base stacking interactions between base pairs on either side of the nick, allowing bending of the nicked DNA and slower migration on the gel. DNA can be visualized using SYBR Gold stain to avoid the need for labels within the DNA prior to cleavage.

3.1 Linear substrate preparation

To allow for a direct comparison of cleavage between negatively supercoiled and linear substrates, we recommend PCR amplifying Cas9 or Cas12a targets using the target plasmids described in Section 2.1.3 as templates (Fig. 6A). In the development of this method, we PCR amplified DNA targets of various lengths and with targets located in the center of the DNA or toward one of the ends. We found that having the cleavage site at the center of a 600–700 bp fragment provided the best separation between a nicked DNA product and the uncleaved linear DNA. In addition, placing the target at the center of the PCR product results in two shorter fragments of the same length upon second strand cleavage, which improves the visualization of the shorter fragments when staining with SYBR Gold. The procedure below outlines a general framework for the preparation of these substrates.

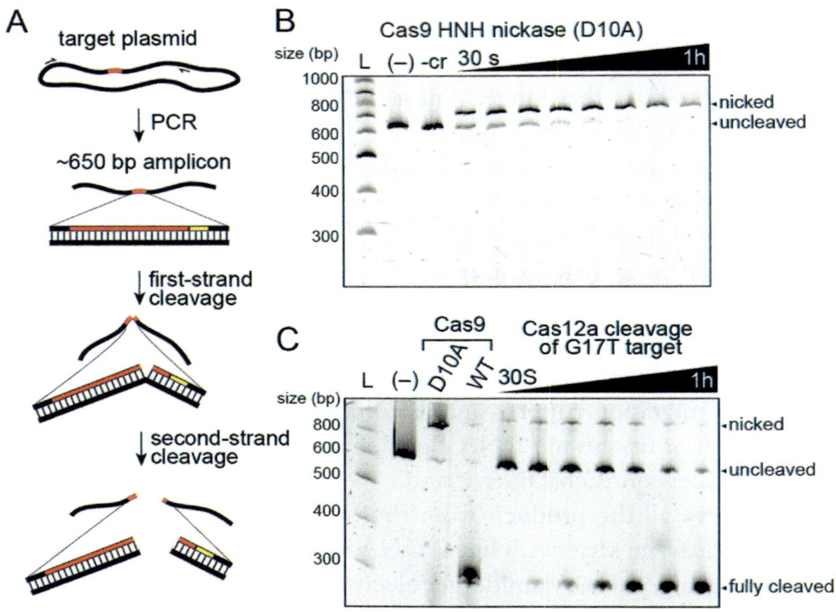

Fig. 6 Cleavage of long linear target DNA substrates. (A) Schematic showing strategy for separation of linear target DNA substrates following first- and second-strand cleavage events. A 600–700 bp fragment can be PCR amplified from a target plasmid. The target should be at the center of the PCR product. First-strand cleavage allows for DNA bending under denaturing conditions, which results in slower migration on a polyacrylamide gel. Second-strand cleavage results in two DNA products of the same length, which co-migrate on the gel and can be visualized readily by SYBR Gold stain. (B) Cleavage of a ~650 bp PCR product using the Cas9 D10A variant, in which only the HNH domain is able to cleave the target, producing a nicked product. A denaturing Urea-PAGE gel enabled separation of the nicked and uncleaved DNA. (C) Cleavage of a ~650 bp PCR product using Cas12a. The target contained a PAM-distal mutation, which results in slow second-strand cleavage and accumulation of a nicked product. A similar length PCR product cleaved by D10A and WT Cas9 were included as controls to ensure that the slowly migrating species is consistent with the migration of a nicked DNA. Gels were visualized using SYBR Gold staining.

3.1.1 Equipment

- Thermal cycler
- Agarose gel apparatus
- Spectrophotemeter (e.g. Thermo Fisher NanoDrop)

3.1.2 Buffers and reagents

- DNA polymerase for PCR (e.g. New England Biolabs Q5 High-Fidelity DNA polymerase)

- Target plasmid (prepared as described in Section 2.1.3)
- PCR primers (designed as described below)
- dNTPs
- Agarose gel with dye for detecting DNA
- PCR cleanup kit (e.g. Promega Wizard Kit or Qiagen QIAquick PCR Purification kit)
- Nuclease-free water

3.2 Procedure

3.2.1 Generation of PCR amplified DNA substrate

1. Prepare a DNA substrate in the range of 600–700 bp by PCR amplifying the target pUC19 plasmids prepared as described in Section 2.1.3. Primers for PCR should be designed in such a way that the target site of the Cas proteins is in the middle of the substrate (Fig. 6A). We typically perform a 200 μL reaction using 30 cycles of a standard PCR protocol.
2. Confirm the length of the PCR product by running a small amount on an agarose gel.
3. Purify the PCR products using a standard PCR cleanup kit following the manufacturer's instructions. Elute the DNA using 30 μL of nuclease-free water.
4. Determine the concentration of the DNA using a spectrophotometer. Dilute the DNA to 100 nM final concentration using nuclease-free water.
5. Aliquot the DNA and store at −20 °C.

3.3 Cleavage assays and separation of nicked products for linear substrates

To verify that we could separate nicked products from uncleaved linear DNA, we used a Cas9 HNH nickase in which the RuvC active site is mutated (D10A) such that only the HNH domain is active (Jinek et al., 2012). Cleavage by the HNH domain results in nicking of only the target strand of the DNA and ensures a nicked product. We performed cleavage of a Cas9 target using the HNH nickase and were able to separate nicked DNA products from the uncleaved DNA by denaturing polyacrylamide gel electrophoresis (Fig. 6B). Having confirmed that nicked products can be separated through this method, we next tested cleavage of a Cas12a target containing a PAM-distal mismatch. We have previously observed that PAM-distal mismatches cause slower second-strand cleavage by Cas12a, resulting in a nicked product when using negatively supercoiled DNA substrates (Murugan et al., 2020; Nguyen et al., 2024b; Schelling et al.,

2023). Similarly, using a linear PCR product as substrate, we observed a DNA product with a similar migration to the Cas9 HNH nickase product (Fig. 6C), indicating that Cas12a also nicks linear DNA containing PAM-distal mismatches.

The cleavage reactions for linear substrates can be performed similarly to procedures described in Sections 2.2.2.1, 2.3.1 or 2.4.1, with a few modifications noted below. The major differences in the below protocols are in the methods by which the DNA is separated and analyzed.

3.3.1 Equipment

- Polyacrylamide gel electrophoresis equipment
- High voltage power supply (e.g. Bio-Rad PowerPac HC)
- Gel imager (e.g. Cytiva Typhoon imager)

3.3.2 Buffers and reagents

- 10X TBE buffer (pH 8.3): 1 M Tris Base, 1 M Boric acid, 0.02 M EDTA
- SYBR Gold Nucleic Acid Gel Stain (Thermo Fisher Scientific)

3.3.3 Procedure

1. For cleavage assays, we typically use a final DNA concentration of 5 nM in the reaction, resulting in a final DNA concentration of 2.5 nM following quenching and dilution with 2X DNA dye. This concentration is sufficient to allow for visualization of the DNA using SYBR Gold stain.
2. With the exception of DNA concentration, cleavage reactions can be performed as described in sections above. For the accompanying data (Fig. 6B and C), we initiated cleavage by mixing RNP and DNA together in the presence of $MgCl_2$ as described in Section 2.2.2.1. Time points, quenching, DNA extraction, and preparation of samples for gel electrophoresis can be performed as described in Section 2.2.2.1.
3. Prepare an 8 % denaturing PAGE gel in 1X TBE with 7 M urea. To ensure visualization of all DNA fragments on the gel, we recommend using an apparatus with gel lengths of 28 cm or longer and spacers of 0.75 mm or thinner.
4. Load 10uL from each reaction to the wells of the TBE PAGE gel.
   - o Tip: Do not heat the sample before loading on the gel. The DNA should remain double-stranded while running on the gel, but the denaturant will serve to disrupt stacking interactions at the nick site and alter the migration of nicked DNA.
5. Separate the cleavage reaction products at 18 W constant current at ambient temperature.

- o Tip: Optimal separation of the nicked and uncleaved DNA requires higher resolution afforded by longer and thinner gels. We run our 28 cm gels for 1 h and 45 min until the bromophenol blue dye front has migrated ~3/4 of the gel length.

6. Dilute the SYBR Gold (10,000X) by adding 2 μL of the dye to 100 mL of distilled water. Remove the gel from the apparatus and plates and carefully transfer to a staining tray containing the diluted SYBR Gold for 10 min.
7. Image the gel using a gel imager. Gels can be quantified as described above in Section 2.2.2.3.

4. Summary and conclusion

In this chapter, we described methods for the analysis of the effect of divalent metal ions on the DNA binding and cleavage activities of Cas effectors. The methods described above allow for detection of the two separate cleavage events by the HNH and RuvC domains of Cas9, or the two successive cleavage events by the single RuvC domain of Cas12 proteins. We demonstrated that nicked fragments can be successfully separated and quantified when using either negatively supercoiled or linear DNA substrates. These methods allow for the determination of rates of the first cleavage event by various Cas effectors. First-strand cleavage rates are especially important when using lower Mg^{2+} concentrations, which have been shown to cause defects in RuvC cleavage by Cas9 and in second-strand cleavage by Cas12a. Initiating cleavage through three separate methods allows for differentiation between binding and cleavage defects, and enables the monitoring of the rate of DNA binding in the absence or presence of Mg^{2+}. Overall, these methods allow for detailed examination of the impact of divalent metal ions on distinct steps of Cas effector activity.

Acknowledgments

We thank members of the Sashital lab for helpful discussions. This work was supported by grants from the NIH (GM140876) and NSF (1652661).

References

Anders, C., & Jinek, M. (2014). In vitro enzymology of Cas9. *Methods in Enzymology, 546*, 1–20. https://doi.org/10.1016/B978-0-12-801185-0.00001-5.

Anders, C., Niewoehner, O., Duerst, A., & Jinek, M. (2014). Structural basis of PAM-dependent target DNA recognition by the Cas9 endonuclease. *Nature, 513*, 569–573. https://doi.org/10.1038/nature13579.

Chen, F., Ding, X., Feng, Y., Seebeck, T., Jiang, Y., & Davis, G. D. (2017). Targeted activation of diverse CRISPR-Cas systems for mammalian genome editing via proximal CRISPR targeting. *Nature Communications, 8*, 14958.

Chen, J. S., Ma, E., Harrington, L. B., Da Costa, M., Tian, X., Palefsky, J. M., & Doudna, J. A. (2018). CRISPR-Cas12a target binding unleashes indiscriminate single-stranded DNase activity. *Science (New York, N. Y.), 360*, 436–439. https://doi.org/10.1126/science.aar6245.

Cofsky, J. C., Karandur, D., Huang, C. J., Witte, I. P., Kuriyan, J., & Doudna, J. A. (2020). CRISPR-Cas12a exploits R-loop asymmetry to form double-strand breaks. *Elife, 9*, e55143. https://doi.org/10.7554/eLife.55143.

Cofsky, J. C., Soczek, K. M., Knott, G. J., Nogales, E., & Doudna, J. A. (2022). CRISPR-Cas9 bends and twists DNA to read its sequence. *Nature Structural & Molecular Biology, 29*, 395–402. https://doi.org/10.1038/s41594-022-00756-0.

Cong, L., Ran, F. A., Cox, D., Lin, S., Barretto, R., Habib, N., ... Zhang, F. (2013). Multiplex genome engineering using CRISPR/Cas systems. *Science (New York, N. Y.), 339*, 819–823. https://doi.org/10.1126/science.1231143.

Dagdas, Y. S., Chen, J. S., Sternberg, S. H., Doudna, J. A., & Yildiz, A. (2017). A conformational checkpoint between DNA binding and cleavage by CRISPR-Cas9. *Science Advances, 3*. https://doi.org/10.1126/sciadv.aao0027.

Eggers, A. R., Chen, K., Soczek, K. M., Tuck, O. T., Doherty, E. E., Xu, B., ... Doudna, J. A. (2024). Rapid DNA unwinding accelerates genome editing by engineered CRISPR-Cas9. *Cell*. https://doi.org/10.1016/j.cell.2024.04.031.

Fu, B. X. H., Smith, J. D., Fuchs, R. T., Mabuchi, M., Curcuru, J., Robb, G. B., & Fire, A. Z. (2019). Target-dependent nickase activities of the CRISPR-Cas nucleases Cpf1 and Cas9. *Nature Microbiology, 4*, 888–897. https://doi.org/10.1038/s41564-019-0382-0.

Gasiunas, G., Barrangou, R., Horvath, P., & Siksnys, V. (2012). Cas9-crRNA ribonucleoprotein complex mediates specific DNA cleavage for adaptive immunity in bacteria. *Proceedings of the National Academy of Sciences of the United States of America, 109*, E2579–E2586. https://doi.org/10.1073/pnas.1208507109.

Gong, S., Yu, H. H., Johnson, K. A., & Taylor, D. W. (2018). DNA unwinding is the primary determinant of CRISPR-Cas9 activity. *Cell Reports, 22*, 359–371. https://doi.org/10.1016/j.celrep.2017.12.041.

Guo, T. W., Bartesaghi, A., Yang, H., Falconieri, V., Rao, P., Merk, A., ... Subramaniam, S. (2017). Cryo-EM structures reveal mechanism and inhibition of DNA targeting by a CRISPR-cas surveillance complex. *Cell, 171*, 414–426.e12. https://doi.org/10.1016/j.cell.2017.09.006.

Hou, Z., Zhang, Y., Propson, N. E., Howden, S. E., Chu, L. F., Sontheimer, E. J., & Thomson, J. A. (2013). Efficient genome engineering in human pluripotent stem cells using Cas9 from Neisseria meningitidis. *Proceedings of the National Academy of Sciences of the United States of America, 110*, 15644–15649. https://doi.org/10.1073/pnas.1313587110.

Jinek, M., Chylinski, K., Fonfara, I., Hauer, M., Doudna, J. A., & Charpentier, E. (2012). A programmable dual-RNA-guided DNA endonuclease in adaptive bacterial immunity. *Science (New York, N. Y.), 337*, 816–821. https://doi.org/10.1126/science.1225829.

Kim, E., Koo, T., Park, S. W., Kim, D., Kim, K., Cho, H.-Y., ... Kim, S. (2017). In vivo genome editing with a small Cas9 orthologue derived from Campylobacter jejuni. *Nature Communications, 8*, 14500.

Kuhn, H., Protozanova, E., & Demidov, V. V. (2002). Monitoring of single nicks in duplex DNA by gel electrophoretic mobility-shift assay. *Electrophoresis, 23*, 2384–2387. https://doi.org/https://doi.org/10.1002/1522-2683(200208)23:15 < 2384::AID-ELPS2384 > 3.0.CO;2-O.

Liu, M.-S., Gong, S., Yu, H.-H., Taylor, D. W., & Johnson, K. A. (2019). Kinetic characterization of Cas9 enzymes. *Methods in Enzymology, 616*, 289–311. https://doi.org/10.1016/bs.mie.2018.10.022.

Makarova, K. S., Wolf, Y. I., Iranzo, J., Shmakov, S. A., Alkhnbashi, O. S., Brouns, S. J. J., ... Koonin, E. V. (2020). Evolutionary classification of CRISPR-Cas systems: A burst of class 2 and derived variants. *Nature Reviews. Microbiology, 18*, 67–83. https://doi.org/10.1038/s41579-019-0299-x.

Mali, P., Esvelt, K. M., & Church, G. M. (2013). Cas9 as a versatile tool for engineering biology. *Nature Methods, 10*, 957–963. https://doi.org/10.1038/nmeth.2649.

Mohanraju, P., Van Der Oost, J., Jinek, M., & Swarts, D. C. (2018). Heterologous expression and purification of CRISPR-Cas12a/Cpf1. *Bio-Protocol, 8*, e2842. https://doi.org/10.21769/BioProtoc.2842.

Mojica, F. J. M., Díez-Villaseñor, C., García-Martínez, J., & Almendros, C. (2009). Short motif sequences determine the targets of the prokaryotic CRISPR defence system. *Microbiology (Reading, England), 155*, 733–740. https://doi.org/10.1099/mic.0.023960-0.

Murugan, K., Seetharam, A. S., Severin, A. J., & Sashital, D. G. (2020). CRISPR-Cas12a has widespread off-target and dsDNA-nicking effects. *The Journal of Biological Chemistry, 295*, 5538–5553. https://doi.org/10.1074/jbc.RA120.012933.

Murugan, K., Suresh, S. K., Seetharam, A. S., Severin, A. J., & Sashital, D. G. (2021). Systematic in vitro specificity profiling reveals nicking defects in natural and engineered CRISPR-Cas9 variants. *Nucleic Acids Research, 49*, 4037–4053. https://doi.org/10.1093/nar/gkab163.

Nguyen, G. T., Schelling, M. A., Raju, A., Buscher, K. A., Sritharan, A., & Sashital, D. G. (2024b). CRISPR-Cas12a exhibits metal-dependent specificity switching. *Nucleic Acids Research*. https://doi.org/10.1093/nar/gkae613.

Nguyen, G. T., Schelling, M. A., & Sashital, D. G. (2024a). CRISPR-Cas9 target-strand nicking provides phage resistance by inhibiting replication. *BioRxiv*. https://doi.org/10.1101/2024.09.05.611540.

Nierzwicki, Ł., East, K. W., Binz, J. M., Hsu, R. V., Ahsan, M., Arantes, P. R., ... Palermo, E. (2022). Principles of target DNA cleavage and the role of Mg2+ in the catalysis of CRISPR–Cas9. *Nature Catalysis, 5*, 912–922. https://doi.org/10.1038/s41929-022-00848-6.

Ocampo, R. F., Bravo, J. P. K., Dangerfield, T., Nocedal, I., Jirde, S., Alexander, L. M., ... Taylor, D. W. (2024). DNA targeting by compact Cas9d and its resurrected ancestor. *BioRxiv*(08), 588528. https://doi.org/10.1101/2024.04.08.588528.

Phan, P. T., Schelling, M., Xue, C., & Sashital, D. G. (2019). Fluorescence-based methods for measuring target interference by CRISPR-Cas systems. *Methods in Enzymology, 616*, 61–85. https://doi.org/10.1016/bs.mie.2018.10.027.

Ran, F. A., Cong, L., Yan, W. X., Scott, D. A., Gootenberg, J. S., Kriz, A. J., ... Zhang, F. (2015). In vivo genome editing using Staphylococcus aureus Cas9. *Nature, 520*, 186–191. https://doi.org/10.1038/nature14299.

Rueden, C. T., Schindelin, J., Hiner, M. C., DeZonia, B. E., Walter, A. E., Arena, E. T., & Eliceiri, K. W. (2017). ImageJ2: ImageJ for the next generation of scientific image data. *BMC Bioinformatics, 18*, 529. https://doi.org/10.1186/s12859-017-1934-z.

Sashital, D. G., Wiedenheft, B., & Doudna, J. A. (2012). Mechanism of foreign DNA selection in a bacterial adaptive immune system. *Molecular Cell, 46*, 606–615. https://doi.org/10.1016/j.molcel.2012.03.020.

Schelling, M. A., Nguyen, G. T., & Sashital, D. G. (2023). CRISPR-Cas effector specificity and cleavage site determine phage escape outcomes. *PLoS Biology, 21*, e3002065. https://doi.org/10.1371/journal.pbio.3002065.

Schneider, C. A., Rasband, W. S., & Eliceiri, K. W. (2012). NIH Image to ImageJ: 25 years of image analysis. *Nature Methods, 9*, 671–675. https://doi.org/10.1038/nmeth.2089.

Semenova, E., Jore, M. M., Datsenko, K. A., Semenova, A., Westra, E. R., Wanner, B., ... Severinov, K. (2011). Interference by clustered regularly interspaced short palindromic repeat (CRISPR) RNA is governed by a seed sequence. *Proceedings of the National Academy of Sciences of the United States of America, 108*, 10098–10103. https://doi.org/10.1073/pnas.1104144108.

Singh, D., Mallon, J., Poddar, A., Wang, Y., Tippana, R., Yang, O., ... Ha, T. (2018). Real-time observation of DNA target interrogation and product release by the RNA-guided endonuclease CRISPR Cpf1 (Cas12a). *Proceedings of the National Academy of Sciences of the United States of America, 115*, 5444–5449. https://doi.org/10.1073/pnas.1718686115.

Singh, D., Sternberg, S. H., Fei, J., Doudna, J. A., & Ha, T. (2016). Real-time observation of DNA recognition and rejection by the RNA-guided endonuclease Cas9. *Nature Communications, 7*, 1277. https://doi.org/10.1038/ncomms12778.

Son, H., Park, J., Hwang, I., Jung, Y., Bae, S., & Lee, S. (2021). Mg(2+)-dependent conformational rearrangements of CRISPR-Cas12a R-loop complex are mandatory for complete double-stranded DNA cleavage. *Proceedings of the National Academy of Sciences of the United States of America, 118*. https://doi.org/10.1073/pnas.2113747118.

Stella, S., Alcón, P., & Montoya, G. (2017). Structure of the Cpf1 endonuclease R-loop complex after target DNA cleavage. *Nature, 546*, 559–563. https://doi.org/10.1038/nature22398.

Stella, S., Mesa, P., Thomsen, J., Paul, B., Alcón, P., Jensen, S. B., ... Montoya, G. (2018). Conformational activation promotes CRISPR-Cas12a catalysis and resetting of the endonuclease activity. *Cell, 175*, 1856–1871.e21. https://doi.org/10.1016/j.cell.2018.10.045.

Sternberg, S. H., Lafrance, B., Kaplan, M., & Doudna, J. A. (2015). Conformational control of DNA target cleavage by CRISPR-Cas9. *Nature, 527*, 110–113. https://doi.org/10.1038/nature15544.

Sternberg, S. H., Redding, S., Jinek, M., Greene, E. C., & Doudna, J. A. (2014). DNA interrogation by the CRISPR RNA-guided endonuclease Cas9. *Nature, 507*, 62–67. https://doi.org/10.1038/nature13011.

Strohkendl, I., Saha, A., Moy, C., Nguyen, A.-H., Ahsan, M., Russell, R., ... Taylor, D. W. (2024). Cas12a domain flexibility guides R-loop formation and forces RuvC resetting. *Molecular Cell, 84*, 2717–2731.e6. https://doi.org/10.1016/j.molcel.2024.06.007.

Strohkendl, I., Saifuddin, F. A., Rybarski, J. R., Finkelstein, I. J., & Russell, R. (2018). Kinetic basis for DNA target specificity of CRISPR-Cas12a. *Molecular Cell, 71*, 816–824.e3. https://doi.org/10.1016/j.molcel.2018.06.043.

Swarts, D. C., & Jinek, M. (2019). Mechanistic insights into the cis- and trans-acting DNase activities of Cas12a. *Molecular Cell, 73*, 589–600.e4. https://doi.org/10.1016/j.molcel.2018.11.021.

Swarts, D. C., van der Oost, J., & Jinek, M. (2017). Structural basis for guide RNA processing and seed-dependent DNA targeting by CRISPR-Cas12a. *Molecular Cell, 66*, 221–233.e4. https://doi.org/10.1016/j.molcel.2017.03.016.

Szczelkun, M. D., Tikhomirova, M. S., Sinkunas, T., Gasiunas, G., Karvelis, T., Pschera, P., ... Seidel, R. (2014). Direct observation of R-loop formation by single RNA-guided Cas9 and Cascade effector complexes. *Proceedings of the National Academy of Sciences of the United States of America, 111*, 9798–9803. https://doi.org/10.1073/pnas.1402597111.

Weigel, C., & Seitz, H. (2006). Bacteriophage replication modules. *FEMS Microbiology Reviews, 30*, 321–381. https://doi.org/10.1111/j.1574-6976.2006.00015.x.

Zetsche, B., Gootenberg, J. S., Abudayyeh, O. O., Slaymaker, I. M., Makarova, K. S., Essletzbichler, P., ... Zhang, F. (2015). Cpf1 is a single RNA-guided endonuclease of a class 2 CRISPR-Cas system. *Cell, 163*, 759–771. https://doi.org/10.1016/j.cell.2015.09.038.

Zetsche, B., Heidenreich, M., Mohanraju, P., Fedorova, I., Kneppers, J., Degennaro, E. M., ... Zhang, F. (2017). Multiplex gene editing by CRISPR-Cpf1 using a single crRNA array. *Nature Biotechnology, 35*, 31–34. https://doi.org/10.1038/nbt.3737.

CHAPTER SEVEN

Purification and in vivo, cell-free, and in vitro characterization of CRISPR-Cas12a2

Friso T. Schut[a,1], Thomson Hallmark[b,1], Oleg Dmytrenko[a], Ryan N. Jackson[b,*], and Chase L. Beisel[a,c,*]

[a]Helmholtz Institute for RNA-based Infection Research (HIRI), Helmholtz Centre for Infection Research (HZI), Würzburg, Germany
[b]Department of Chemistry and Biochemistry, Utah State University, Logan, UT, United States
[c]Medical Faculty, University of Würzburg, Würzburg, Germany
*Corresponding authors. e-mail address: ryan.jackson@usu.edu; chase.beisel@helmholtz-hiri.de

Contents

1. Introduction 144
2. gRNA design for Cas12a2 145
3. Cas12a2 activity in *E. coli* 147
   3.1 Plasmid interference assay 147
   3.2 Growth-based assays 151
4. Cas12a2 activity in cell-free transcription-translation (TXTL) 155
   4.1 Design of DNA expression constructs for TXTL reactions 156
   4.2 DNA preparation for TXTL reactions 157
   4.3 TXTL pre-expression of SuCas12a2 with a gRNA 158
   4.4 TXTL collateral cleavage assay 164
5. Expression and purification of Cas12a2 165
   5.1 Cas12a2 expression 165
   5.2 Cas12a2 purification 168
6. Cas12a2 *in vitro* cleavage assays 172
   6.1 Cas12a2 cleavage visualization 173
   6.2 Sequence-specific target RNA design 174
   6.3 Sequence non-specific, collateral substrate design 176
References 180

Abstract

The CRISPR-associated (Cas) nuclease Cas12a2 from *Sulfuricurvum* sp. PC08–66 (SuCas12a2) binds RNA targets with a complementary guide (g)RNA. Target RNA binding causes a major conformational rearrangement in Cas12a2 that activates a RuvC nuclease domain to collaterally cleave RNA, ssDNA and dsDNA, arresting growth and providing population-level immunity. Here, we report in vivo, cell-free, and in

[1] Equal contributions.

Methods in Enzymology, Volume 712
ISSN 0076-6879, https://doi.org/10.1016/bs.mie.2025.01.032

vitro methods to characterize the collateral cleavage activity of SuCas12a2 as well as a procedure for gRNA design. As part of the in vivo methods, we describe how to capture growth arrest through plasmid interference and induction of an SOS DNA damage response in the bacterium *Escherichia coli*. We further apply cell-free transcription-translation to affirm collateral cleavage activity triggered by an expressed RNA target. Finally, as part of the in vitro methods, we describe how to purify active nuclease and subsequently conduct biochemical cleavage assays. In total, the outlined methods should accelerate the exploration of SuCas12a2 and other related Cas nucleases, revealing new features of CRISPR biology and helping develop new CRISPR technologies for molecular diagnostics and other applications.

1. Introduction

CRISPR-associated (Cas)12a2 is a newly discovered single-effector nuclease distinct from the more well known Cas nucleases used in gene editing. Unlike Cas9 and Cas12a, Cas12a2 recognizes target RNA instead of DNA, and RNA target binding activates the collateral degradation of RNA, ssDNA and dsDNA (Dmytrenko et al., 2023). Furthermore, triggered Cas12a2 indiscriminately cleaves invasive and genomic DNA, activating the SOS response and arresting cell growth. Additionally, the distinct Cas12a2 mechanism of RNA-triggered DNA shredding represents a new programmable tool that could be developed for use in molecular diagnostics and programmable cell killing.

The CRISPR-Cas12a2 immune system is composed of the Cas12a2 RNA-guided nuclease and a CRISPR array composed of unique guide sequences called spacers interspersed by short repeat regions. Cas12a2 binds and cleaves RNA transcribed from the CRISPR array through the 5′ pseudoknot secondary structure formed by the repeat, cleaving the CRISPR transcript into individual guide (g)RNAs. Once Cas12a2 is loaded with a gRNA, it surveils the cell for RNA targets containing two features: (i) sequence complementarity to the 24-nt gRNA guide sequence and (ii) the presence of a protospacer-flanking sequence (PFS) on the 3′ flank of the target RNA. Cas12a2 can be activated by a range of PFS sequences (Dmytrenko et al., 2023), but the common profile is 5′-BAAAN-3′ (where B indicates not adenine) positioned at the 3′ end of the target sequence.

Similar to other class 2 CRISPR nucleases, Cas12a2 has a bilobed architecture comprising a recognition (Rec) lobe and a nuclease (Nuc) lobe (Bravo et al., 2023; Liu et al., 2017; Swarts & Jinek, 2018; Yamano et al., 2016). The Rec lobe contains the Rec1 and Rec2 domains, while the Nuc lobe contains the RuvC nuclease, PFS-interacting (PI) and wedge (WED) domains along with a Zinc Ribbon (ZR) and novel insertion domain (Bravo et al., 2023). Recognition of an appropriate RNA target by Cas12a2 activates sequence non-specific collateral cleavage of RNA, ssDNA and dsDNA through large-scale conformational changes in the Rec lobe that open a large, positively charged binding pocket around the RuvC active site.

Here, we describe methods with supporting experimental data to design gRNAs and characterize collateral cleavage activity of Cas12a2 from *Sulfuricurvum* sp. PC08–66 (SuCas12a2) using various in vitro and in vivo methods. The growth-arrest phenotype caused by the collateral cleavage activity of SuCas12a2 is confirmed in *E. coli* through plasmid interference assays and growth-based assays, which involve detecting an SOS response using a GFP reporter plasmid. Furthermore, the prerequisites for activating the collateral cleavage activity of SuCas12a2, namely the presence of a PFS flanking the target and the target being in RNA form rather than DNA, are confirmed using a cell-free transcription-translation (TXTL) system. We also outline the necessary steps for conducting in vitro cleavage assays to probe collateral cleavage activities on RNA, ssDNA and dsDNA substrates. This includes the purification of the nucleases and the design of synthetic gRNAs. Collectively, the methods outlined in this chapter offer an approach for characterizing collateral cleavage by SuCas12a2, paving the way for the exploration of other CRISPR-Cas subtypes and variants with collateral cleavage activity that could advance CRISPR technologies, including diagnostics and the molecular biology toolkit.

2. gRNA design for Cas12a2

The activity of Cas12a2 depends on target recognition by a gRNA guide with a 24-nt long guide sequence and a 5′ pseudoknot (Dmytrenko et al., 2023). Cas12a2 processes the gRNA upstream of the 5′ pseudoknot. The 3′ end of the gRNA is processed to different lengths, presumably by RNases in the cell, to a minimum activating length of 24 nts.

The processing activities of Cas12a2 allows gRNAs to be co-expressed in vivo from DNA constructs, chemically synthesized or transcribed in vitro. When designing the gRNA, first identify the potential PFS sequences (e.g., conservatively 5′-BAAAN-3′ based on our characterization efforts to-date) within your target gene (Fig. 1). Candidate target

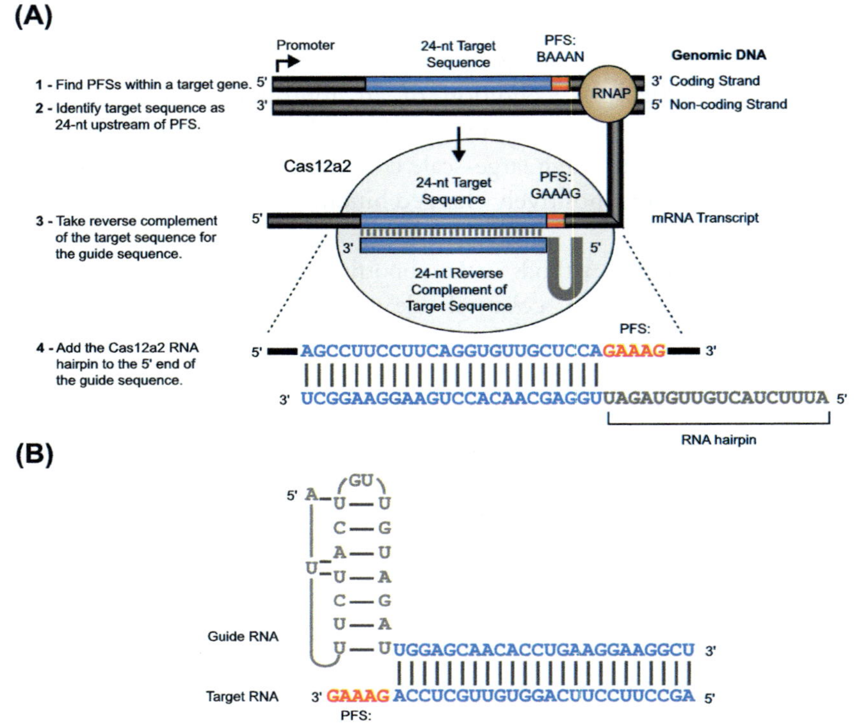

Fig. 1 gRNA design strategy. (A) Schematic of the gRNA design strategy. First, identify the potential PFSs within a target gene (e.g., BAAAN) then identify the preceding 24 nucleotides as the target sequence. Take the reverse complement of the target RNA sequence as the sequence of the guide region of the gRNA. Finally add the Cas12a2 gRNA hairpin sequence to the 5′ end of the guide sequence. (B) Overview of the gRNA - target RNA pair with the gRNA oriented 5′-to-3′ from left to right.

sequences are then the 24-nt immediately upstream on the 5′ side of each PFS. The guide sequence of the gRNA is then the reverse complement of the target sequence converted to RNA (i.e. change T to U). Finally, include the sequence of the 5′ pseudoknot (5′-AUUUCUACUGUUG UAGAU-3′) on the 5′ side of the 24-nt guide sequence. When designing CRISPR arrays, the spacer sequence should be 25- to 30-nt long, identified in the same way as the gRNA guide sequence described above, and the full repeat sequence (5′-GTCAATACGACTTTAAAAATTTCTAC TATTGTAGAT-3′) should flank both sides of a spacer. The last repeat can be excluded to prevent extraneous gRNA production (Liao et al., 2019). Flanking sequences added to the 5′ and 3′ ends of the guide will respectively be cleaved by Cas12a2 processing or not impact targeting activity (Dmytrenko et al., 2023). For in vitro-transcribed gRNAs, DNA templates can be designed to include extra sequence on the 5′ sequence (e.g., a T7 promoter), upstream of the processing site, with the 3′ end designed for run-off transcription as has been described for other type V nucleases (Martin et al., 2023).

3. Cas12a2 activity in *E. coli*

Previous work has demonstrated that SuCas12a2 drives a growth-arrest phenotype upon gRNA-directed recognition of a RNA target flanked by a PFS (Dmytrenko et al., 2023). Hybridization of a gRNA to the target RNA activates the collateral *trans*-cleavage activity of Cas12a2, leading to the non-specific degradation of RNA, ssDNA and dsDNA (Bravo et al., 2023; Dmytrenko et al., 2023). Cas12a achieves antiphage immunity through a distinct mechanism, where hybridization of its gRNA with a double-stranded DNA (dsDNA) target flanked by a protospacer adjacent motif (PAM) leads to on-target *cis*-cleavage of dsDNA (Zetsche et al., 2015).

Similar to Cas12a2, Cas12a also shows promiscuous collateral *trans*-cleavage of ssDNAs (Li et al., 2018). We selected Cas12a as a control to observe growth arrest via plasmid interference by SuCas12a2, because the reported collateral *trans*-cleavage activity of Cas12a does not play a role in bacterial immunity by inducing a growth-arrest or abortive-infection phenotype (Marino, Pinilla-Redondo, & Bondy-Denomy, 2022). Specifically, we picked the Cas12a gene from the microbial community of *Microcerotermes parvus* (MpCas12a), which can use the same gRNA as SuCas12a2 and recognize the PAM present in the target sequence.

In the plasmid interference assay, nuclease plasmids with SuCas12a2 or MpCas12a are transformed into *E. coli* carrying a target plasmid from which a DNA target flanked by a PFS/PAM sequence and a complementary gRNA are expressed. Subsequently, a reduction in colony counts with or without selection for the target plasmid is compared between SuCas12a2 and MpCas12a. In a similar setup, we show that the collateral activity of SuCas12a2 impairs cell growth and activates the SOS DNA damage response in *E. coli*.

3.1 Plasmid interference assay

In this assay, a nuclease-expressing plasmid is introduced into *E. coli* cells already containing a target- and gRNA-expressing plasmid. To measure

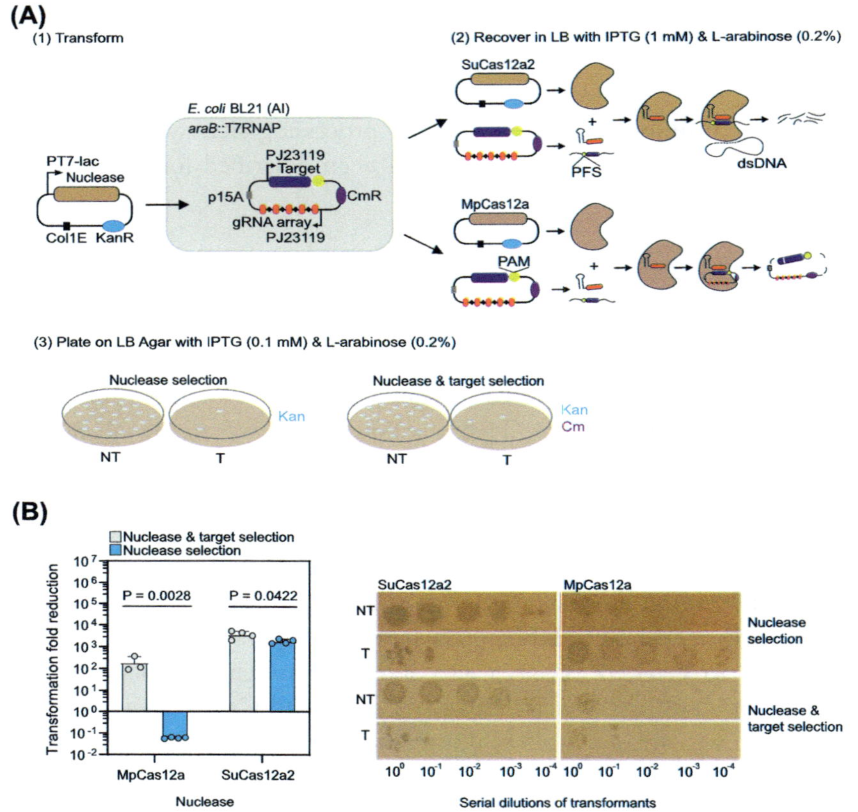

Fig. 2 A plasmid interference assay captures the growth-arrest phenotype induced by SuCas12a2 in E. coli. (A) Diagram of the plasmid interference assay. After electroporation of the nuclease plasmid in target plasmid-carrying E. coli BL21 (AI) cells (1), transformants are recovered in inducing LB medium with IPTG and L-arabinose (2). Selection of the transformants is performed on LB agar plates with IPTG and L-arabinose supplemented with either kanamycin (Kan) and chloramphenicol (Cm) or kanamycin only (3). (B) The reduction in plasmid transformation efficiency for SuCas12a2 and MpCas12a when selecting for the target plasmid and the nuclease plasmid. The dots correspond to replicates. For (A) and (B), the data represent mean and standard deviation each initiated from at least three different colonies. P values were determined using a two-tailed Welch's *t*-test assuming unequal variances.

the reduction in transformation efficiency, a non-targeting gRNA without complementarity to any cellular RNAs is used as a benchmark. After transformation, selection is performed for both plasmids (nuclease and target selection) or solely for the nuclease-expressing plasmid (nuclease selection) (Fig. 2A). We anticipate a reduction in colony count for the dsDNA-cleaving MpCas12a under nuclease and target selection, as clearance of the target plasmid by the nuclease induces antibiotic susceptibility, leading to cell death. However, no colony count reduction is expected for MpCas12a under nuclease selection, since we do not select for the presence of the cleared target plasmid. In contrast, the collateral activity of SuCas12a2 should elicit a growth-arrest phenotype that is expected to result in a reduction of colony count, even when not selecting for the target plasmid.

Supporting this expectation, SuCas12a2 reduced colony counts under nuclease and target selection as well as under nuclease selection, while MpCas12a reduced colony counts only during nuclease and target selection (Fig. 2B). Together, these in vivo results show that this method can be used to assess the growth-arrest phenotype by SuCas12a2 in *E. coli*, at least under antibiotic selection for the nuclease.

3.1.1 Design of the DNA constructs

SuCas12a2 and MpCas12a were both placed under the expression of an IPTG-inducible T7-lac promoter on a plasmid backbone with a Col1E origin of replication and a kanamycin (Kan) resistance marker. The *E. coli* BL21 (AI) strain used in this method has a genomically-integrated T7 RNA polymerase (T7RNAP) gene under the expression of a L-arabinose-inducible araBAD promoter, which is required for nuclease expression. The target plasmids contain the target sequence and an array of five gRNAs, each controlled by its own J23119 constitutive promoter. Furthermore, the target plasmids harbor a chloramphenicol (Cm) resistance marker and a p15A origin of replication. Similar target plasmids are used for SuCas12a2 and MpCas12a since the nucleases can utilize the same gRNA sequence and both recognize the TTTG PAM present in the target sequence (Dmytrenko et al., 2023). As a control, we use a strain with a target plasmid containing a non-targeting gRNA that has no complementary to the target sequence.

3.1.2 Materials for the plasmid interference assay

- *E. coli* Top10 strains with nuclease-expressing plasmids (CBS-5043, CBS-6648)
- *E. coli* BL21 (AI) strains with target- and gRNA-expressing plasmids (CBS-6117, CBS-6118)
- MiniPrep kit (e.g., ZymoPURE Plasmid Miniprep Kit, #D4212 Zymo Research)

- Nanodrop spectrophotometer for DNA concentration measurements
- Spectrophotometer for OD measurements at 600 nm
- 10-mL culture vials
- LB medium (10 g/L tryptone, 5 g/L yeast extract, 5 g/L of NaCl) with Kan (50 μg/mL) or Cm (50 μg/mL)
- Incubator at 37 °C
- 50-mL tubes (e.g., Falcon tubes)
- Centrifuge for 50-mL tubes
- Sterile nuclease-free water (distilled and deionized H_2O) at 4 °C
- Sterile glycerol (10% v/v) at 4 °C
- Pre-cooled electroporation cuvettes (at 4 °C or preferably −20 °C)
- Electroporation system (e.g., Bio-rad Gene Pulser Xcell)
- IPTG stock solution (1 M)
- L-arabinose stock solution (20%)
- 96-well plate (e.g. Corning® 96-well Clear Flat Bottom Polystyrene TC-treated Microplates Cat #3598)
- 1x filter-sterilized PBS solution (80 g/L NaCl, 2 g/L KCl, 17.7 g/L Na_2HPO_4 *2 H_2O, 2.27 g/L KH_2PO_4)
- LB agar plates (10 g/L tryptone, 5 g/L yeast extract, 5 g/L of NaCl, 18 g/L agar) with Cm (34 μg/mL) and Kan (50 μg/mL) or only Kan (50 μg/mL).

3.1.3 Protocol for the plasmid interference assay

1. Purify the plasmid DNA from *E. coli* Top10 overnight cultures with nuclease-expressing plasmids (CBS-5043, CBS-6648) by following the protocol of a MiniPrep kit. Measure the DNA concentration using a standard nanodrop spectrophotometer and dilute the plasmid DNA to a concentration of 50 ng/μL.
2. Inoculate a single colony from the *E. coli* BL21 (AI) strains with target- and gRNA-expressing plasmids (CBS-6117, CBS-6118) in 2 mL of LB medium with Cm (50 μg/mL). Inoculate a total of four separate colonies into individual cultures to serve as biological replicates for each strain. Incubate the cultures overnight at 37 °C with shaking at 220 rpm.
3. Back-dilute 500 μL of the overnight culture into 50 mL of LB medium with Cm (34 μg/mL).
4. Incubate at 37 °C with shaking at 220 rpm until the optical density (OD) at 600 nm reaches ~0.8. This is approximately after 2.5 h.
5. Transfer each culture to a 50-mL conical tube kept on ice, then centrifuge the cells at 4,000 x g at 4 °C for 15 min. After decanting the

supernatant, resuspend the pellet in 50 mL of sterile ice-cold H_2O by gently pipetting up and down.

6. Centrifuge the cells at 4000 x g at 4 °C for 15 min before carefully removing the supernatant with a serological pipette without disturbing the pellet, and resuspend the pellet in 1 mL of sterile ice-cold glycerol.
7. Transfer the cells to a 1.5 mL tube and centrifuge them at 14,000 x g at 4 °C for 1 min in a tabletop centrifuge. Then resuspend the pellet in 240 μL of ice-cold 10% glycerol.
8. For each reaction, electroporate 1 μL of the 50 ng/μL nuclease plasmids in 40 μL of cells. Recover the transformed cells in 460 μL of LB with 1 mM IPTG and 0.2% L-arabinose at 37 °C for approximately 1 h.
9. Create serial dilutions from 10^{-1} to 10^{-5} in 1x PBS using a 96-well plate. Plate 5 μL of each dilution on selective LB agar plates containing:
   a. 1 mM IPTG, 0.2% L-arabinose, 50 μg/mL Kan.
   b. 1 mM IPTG, 0.2% L-arabinose, 50 μg/mL Kan and 34 μg/mL Cm.

3.2 Growth-based assays

According to the plasmid interference assay, SuCas12a2 reduces *E. coli* colony counts under nuclease selection. Here, we show a method demonstrating that the collateral activity of SuCas12a2 impairs cell growth even in the absence of any antibiotic selection. Fig. 3A presents a comprehensive overview of the steps involved in the growth-based assays in *E. coli*, which can be used to measure the growth effect of SuCas12a2 under different antibiotic regimes. As can be seen in Fig. 3B, gRNA-directed SuCas12a2 hampers cell growth of *E. coli* upon recognition of a target during the first 8 h compared with a nuclease guided by a non-targeting gRNA.

3.2.1 Measuring the SOS response

The collateral activity of activated Cas12a2 against dsDNA was tied to the growth-arrest phenotype through damage to the host's chromosomal DNA (Dmytrenko et al., 2023). Here, we present a method to measure the corresponding SOS DNA damage response in *E. coli*. In this assay, we co-transform a plasmid expressing SuCas12a2 with an *E. coli* SOS-biosensor plasmid expressing GFP as a reporter protein under the control of the *recA* promoter (Chen, Lu, Zou, & Wang, 2012). As can be seen in Fig. 3C, SuCas12a2 significantly induces GFP expression from the SOS-construct compared with a nuclease guided by a non-targeting gRNA. This method

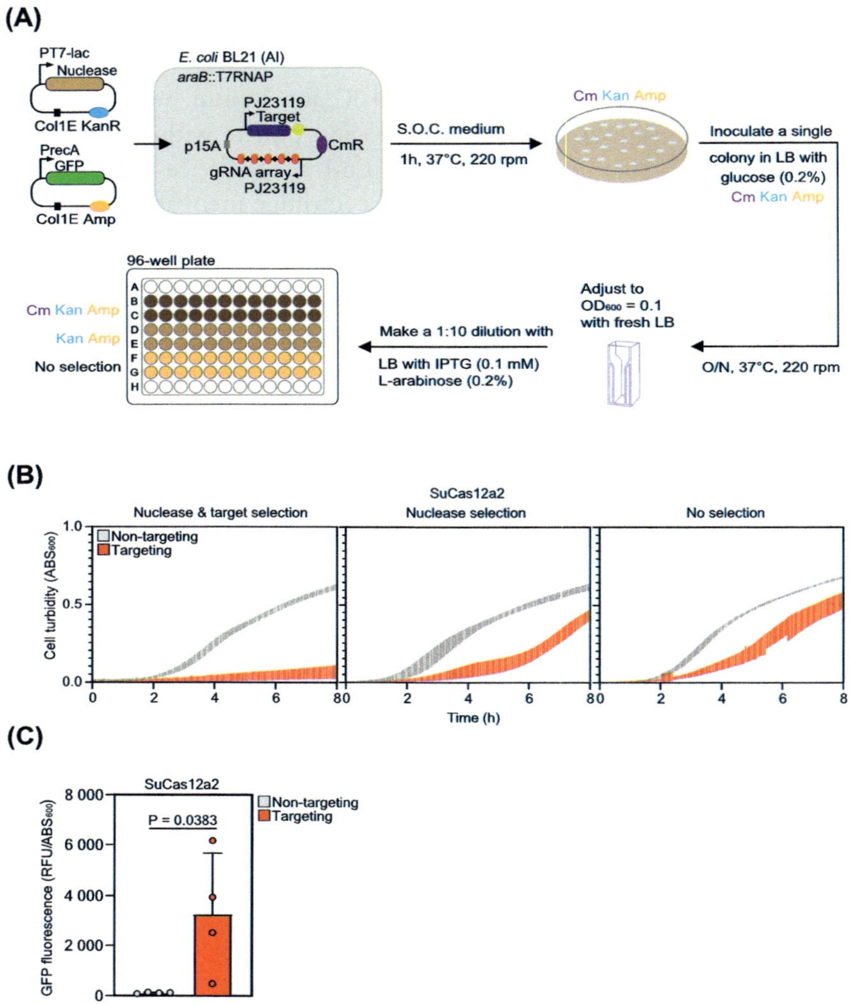

Fig. 3 Growth-based assays with SuCas12a2 show an SOS DNA damage response in E. coli. (A) Diagram of the growth-based assays that can be used to show SuCas12a2 impairs cell growth and elicits a SOS response in E. coli. (B) Growth inhibition of E. coli when SuCas12a2 plasmids are present across various targeting conditions and antibiotic treatments. Cell turbidity was measured by absorbance at 600 nm (ABS_{600}). NT, non-target plasmid; T, target plasmid. Data are mean ± s.d. of at least three independent experiments starting from separate colonies. (C) SOS-responsive expression of GFP in E. coli after 4 h of plasmid targeting by SuCas12a2 without antibiotic selection. Fluorescence was measured at wavelength 485 nm and 528 nm for excitation and emission respectively. RFU, relative fluorescence units. ABS600, absorbance at 600 nm. Data are mean ± s.d. of at least four independent experiments started from separate colonies. The dots correspond to replicates. The statistical analysis was performed using a one-tailed *t*-test assuming unequal variances.

can effectively be used to measure a SOS-response in *E. coli* resulting from the collateral activity of SuCas12a2.

3.2.2 Design of DNA expression constructs

In this assay, SOS-biosensor plasmids and nuclease plasmids are introduced in *E. coli* BL21 (AI) strains previously used in the plasmid clearance assay. The aforementioned strains each contain a target- and gRNA-expressing plasmid. Three antibiotics are used for selection since the nuclease-expressing plasmid contains a kanamycin resistance marker; the target plasmid contains a chloramphenicol resistance marker and the SOS-biosensor plasmid contains an ampicillin (Amp) resistance marker. As controls, we use strains with target plasmids containing non-targeting gRNAs and introduce SOS-biosensor plasmids without a *gfp* gene. 4.3.3. Materials.

- *E. coli* Top10 strains with nuclease plasmids (CBS-5043, CBS-6648)
- *E. coli* Top10 strains with SOS-biosensor plasmids (CBS-3611, CBS-3616)
- *E. coli* BL21 (AI) strains with target- and gRNA-expressing plasmids (CBS-6177, CBS-6178)
- MiniPrep kit (e.g., ZymoPURE Plasmid Miniprep Kit, #D4212 Zymo Research)
- Pre-cooled electroporation cuvettes (at 4 °C)
- Electroporation system (e.g., Bio-rad Gene Pulser Xcell)
- S.O.C. recovery medium (20 g/L tryptone, 5 g/L yeast extract, 0.5 g/L of NaCl, 2.50 mM KCl, 10 mM $MgCl_2.6H_2O$, 20 mM glucose)
- Selective LB agar plates with Cm (34 μg/mL), Kan (50 μg/mL), Amp (100 μg/mL) and 0.2% glucose
- 10-mL culture vials
- LB medium (10 g/L tryptone, 5 g/L yeast extract, 5 g/L of NaCl) with Cm (34 μg/mL) or Kan (50 μg/mL) or Amp (100 μg/mL)
- LB medium (10 g/L tryptone, 5 g/L yeast extract, 5 g/L of NaCl) with Cm (34 μg/mL), Kan (50 μg/mL), Amp (100 μg/mL) and 0.2% glucose
- Nanodrop spectrophotometer for DNA concentration measurements
- Spectrophotometer for OD measurements at 600 nm
- 1.5 mL tubes
- Nuclease and target selection medium: LB (10 g/L tryptone, 5 g/L yeast extract, 5 g/L of NaCl) with Cm (34 μg/mL), Kan (50 μg/mL), Amp (100 μg/mL) and 0.1 mM IPTG and 0.2% L-arabinose

- Target selection medium: LB (10 g/L tryptone, 5 g/L yeast extract, 5 g/L of NaCl) with Kan (50 μg/mL), Amp (100 μg/mL) and 0.1 mM IPTG and 0.2% L-arabinose.
- No selection medium: LB (10 g/L tryptone, 5 g/L yeast extract, 5 g/L of NaCl) with 0.1 mM IPTG and 0.2% L-arabinose
- 96-well plate (e.g., Corning® 96-well Clear Flat Bottom Polystyrene TC-treated Microplates Cat #3598)
- Fluorescence plate reader (e.g., Biotek Synergy Neo2 plate reader)

3.2.3 Protocol

1. Purify the plasmid DNA from *E. coli* Top10 overnight cultures with nuclease plasmids (CBS-5043, CBS-6648) and SOS biosensor plasmids (CBS-3611, CBS-3616) by following the protocol of a MiniPrep kit. Measure the DNA concentration using a standard nanodrop spectrophotometer and dilute the clean and concentrated plasmid DNA to reach a concentration of 10 ng/μL.
2. Electroporate 0.5 μL of both the nuclease plasmids (CBS-5043 or CBS-6648) and SOS-biosensor plasmids (CBS-3611 or CBS-3616) in 40 μL electrocompetent *E. coli* BL21 (AI) strains with target- and gRNA-expressing plasmids (CBS-6177, CBS-6178).
3. Recover the transformed cells in 460 μL of S.O.C. medium for approximately 1 h at 37 °C with shaking at 220 rpm. S.O.C. medium is used because the glucose represses the T7 promoter of the nuclease.
4. Plate 100 μL of each reaction on selective LB agar plates with Cm (34 μg/mL), Kan (50 μg/mL), Amp (100 μg/mL) and 0.2% glucose.
5. The next day, select four colonies from each plate to serve as biological replicates and inoculate them into separate 10-mL culture vials. Each vial should contain 3 mL of selective LB medium supplemented with chloramphenicol (Cm, 34 μg/mL), kanamycin (Kan, 50 μg/mL), ampicillin (Amp, 100 μg/mL), and 0.2% glucose. Incubate the cultures overnight at 37 °C with shaking at 220 rpm.
6. Measure the optical density (OD) of all cultures at 600 nm the next morning using a spectrophotometer with according cuvettes.
7. Transfer 1 mL from each culture to a 1.5 mL tube and spin the tubes down at 5000 x g for 3 min.
8. Remove the supernatant and resuspend the cells in 1 mL of fresh LB medium.
9. Dilute the suspension with fresh LB medium to reach an OD of 0.1.

10. Distribute 20 μL of the diluted LB-only cultures into a 96-well plate with 180 μL of:
    a. Nuclease and target selection medium.
    b. Target selection medium.
    c. No selection medium.
11. Grow at 37 °C with continuous shaking for 15 h. Measure the OD at 600 nm and fluorescence (Ex 485 nm, Em 528 nm) with 5-minute intervals with vigorous shaking in-between.
12. To calculate the change in fluorescence, first subtract the background fluorescence originating from the medium from the fluorescence values of all cultures. Next, subtract the fluorescence values of cultures lacking gfp from the fluorescence values of cultures with gfp at the time point where the fluorescence values peak. We chose the 4.5 h time point because gfp fluorescence was approaching its peak; however, another time point can also be selected. Next, divide the difference in fluorescence by the OD value of the corresponding culture (with a recA promoter with a *gfp* gene) at time point 4.5 h.

4. Cas12a2 activity in cell-free transcription-translation (TXTL)

Here, we demonstrate how a Cell-Free TXTL system can be used for in vitro characterization of SuCas12a2. TXTL is a reaction mixture comprising an *E. coli* lysate, salts, and buffers that supply all 20 amino acids and an ATP regeneration system (Garamella, Marshall, Rustad, & Noireaux, 2016; Shin & Noireaux, 2012). TXTL facilitates rapid functional expression of RNA and protein, circumventing the requirements for cell culturing or protein purification.

In this assay, collateral cleavage activity of SuCas12a2 is measured using a reporter plasmid bearing a *gfp* gene (Fig. 4A). After hybridization of a PFS-flanked RNA target molecule to a gRNA-directed SuCas12a2, non-specific degradation of the dsDNA GFP reporter plasmid by the nuclease will follow. By measuring the fluorescence of the TXTL reaction over time using a plate reader, we can detect the repression of GFP production as a result of collateral activity. As a control for this assay, we use a RNA target without a recognized PFS (non-PFS: 5′-TAGAT-3′), which is not expected to unleash collateral activity. To show that a DNA target will not activate collateral cleavage activation of SuCas12a2, a terminator is placed

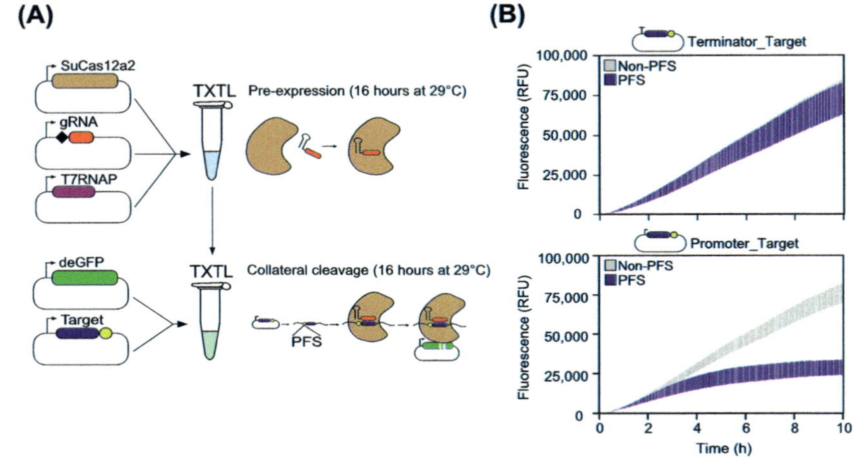

Fig. 4 A cell-free TXTL system validates collateral cleavage activity of SuCas12a2 triggered by an expressed PFS-flanked RNA target. (A) Diagram of the cell-free TXTL reactions used to measure GFP reporter silencing by the collateral trans-cleavage activity of SuCas12a2. A pre-expression step of the nuclease and gRNA in TXTL precedes a GFP reporter-silencing assay in TXTL. (B) Collateral silencing by SuCas12a2 is evaluated by the presence of either a promoter or terminator upstream of the target expression site. Non-PFS, a dysfunctional protospacer flanking sequence. PFS, a functional protospacer flanking sequence. Scatter plots represent averages and standard deviations of at least three independent experiments started from separate colonies.

upstream of the DNA target, since the terminator will prevent transcription of the RNA target. Consistent with our expectation, the repression of GFP production in TXTL by SuCas12a2 is observed only when the target has a PFS and is expressed (Fig. 4B). This in vitro method can thus be used to characterize the collateral cleavage activity of SuCas12a2 in a swift and straightforward manner using TXTL.

4.1 Design of DNA expression constructs for TXTL reactions

The DNA constructs used in TXTL to express the SuCas12a2 nuclease, gRNA, target DNA and GFP reporter can be in linear or plasmid form as long as a strong *E. coli* promoter is used (Marshall, Maxwell, Collins, Beisel, & Noireaux, 2017). An example is the Sigma 70 consensus promoter P70a used to express an adapted form of eGFP specifically designed for TXTL systems (Shin & Noireaux, 2010). It is recommended to use *E. coli* strain KL740 cl857 + for maintaining plasmids with a P70a promoter, because cI represses expression of the strong P70a promoter at temperatures below

30 °C (Shin & Noireaux, 2010). Alternatively, the inducible T7 promoter can be utilized to achieve adequate gRNA expression (Wandera et al., 2020). When using the T7 promoter, however, an additional DNA construct encoding for the otherwise absent T7 RNA polymerase must be added (Wandera et al., 2020). Because plasmid maintenance does not have to be accounted for in TXTL, the same plasmid backbone can be used for different constructs (Wandera et al., 2020). Using linear DNA to avoid time-consuming cloning steps is possible only when GamS or DNA encoding for Chi sites is added to inhibit linear DNA degradation by the RecBCD system present in the *E. coli* lysate-derived TXTL reactions (Marshall et al., 2017; Sitaraman et al., 2004). The DNA sequence of deGFP has been modified through truncation to augment translation and enhance *gfp* expression (Shin & Noireaux, 2012).

4.2 DNA preparation for TXTL reactions

We have learned that highly pure DNA must be provided to the TXTL mix to avoid interference of other components with the transcription-translation machinery. Suitable plasmid DNA is obtained by first using a MidiPrep kit for purification, followed by a separate DNA cleanup protocol. This approach is necessary because MiniPrep kits do not provide DNA of comparable cleanliness and yield more variable results in TXTL experiments. When using linear DNA from a PCR reaction, the DNA is purified using a standard DNA cleanup protocol.

4.2.1 Materials for DNA preparation for TXTL reactions

- *E. coli* strains with nuclease plasmid (CBS-5043), gRNA plasmid (CBS-5044) and T7RNAP plasmid (CBS-338), target plasmids (CBS-3619, CBS-3621, CBS-6650, CBS-6651) and GFP reporter plasmid (CBS-6649)
- 50-mL shake flasks
- LB medium with the appropriate antibiotic (10 g/L tryptone, 5 g/L yeast extract, 5 g/L of NaCl)
- 50-mL tubes (e.g., Falcon tubes)
- Midiprep kit (e.g., ZymoPURE II Plasmid Midiprep Kit, #D4201 Zymo Research)
- DNA cleanup kit (e.g., DNA Clean & Concentrator-25, #D4033 Zymo Research)
- Incubator at 37 °C
- Centrifuge for 50-mL tubes

4.2.2 Protocol for DNA preparation for TXTL reactions

1. Inoculate a single colony from *E. coli* strains with nuclease plasmid (CBS-5043), gRNA plasmid (CBS-5044) and T7RNAP plasmid (CBS-6048), target plasmids (CBS-3619, CBS-3621, CBS-6650, CBS-6651) and GFP reporter plasmid (CBS-6649) in a shake flask with selective LB (~50 mL) with the appropriate antibiotic. Place the culture shaking in an incubator overnight at 37 °C and 220 rpm (Tables 1–4)
2. Transfer the cultures to 50-mL tubes the next day and pellet the cells using centrifugation at 3200 x g for 30 min.
3. Purify the plasmid DNA from the cells by following the protocol of a MidiPrep kit followed by an additional purification step using a DNA cleanup kit. Measure the DNA concentration and dilute the clean and concentrated plasmid DNA to meet the desired stock concentration mentioned in Tables 5 and 6.

4.3 TXTL pre-expression of SuCas12a2 with a gRNA

To ensure the complete formation of active SuCas12a2 RNPs, we pre-express the SuCas12a2 nuclease and gRNA in a commercially available TXTL mix (myTXTL® Sigma 70 Master Mix Kit, Daicel Arbor Biosciences). However, it is also possible to prepare an efficient endogenous *E. coli*-based TXTL system in-house following an established protocol (Sun et al., 2013). A plasmid encoding T7 RNA polymerase is added to ensure gRNA transcription, as this enzyme is not present in the myTXTL® Sigma 70 Master Mix Kit.

4.3.1 Materials for TXTL pre-expression of SuCas12a2 with a gRNA

- myTXTL® Sigma 70 Master Mix Kit (Daicel Arbor Biosciences)
- Clean and concentrated nuclease, gRNA and T7RNAP plasmids from the previous step
- IPTG (1 M)
- PCR tubes
- Bucket with ice
- Incubator or thermocycler at 29 °C

4.3.2 Protocol for TXTL pre-expression of SuCas12a2 with a gRNA

- Calculate the volume of pre-expression TXTL mix that is required for the subsequent DNA cleavage assay. We usually use 2 μL of the pre-expression mixture in a 12-μL GFP reporter cleavage assay.

Table 1 DNA sequences of gRNAs and corresponding target sequences used for in vivo experiments and TXTL.

Description	DNA sequence from 5′ to 3′
CAO1 target flanked by PFS (underlined)	CATCAAGCCTTCCTTCAGGTGTTGCTCCAGAAAG
CAO1 target flanked by non-PFS (underlined)	CATCAAGCCTTCCTTCAGGTGTTGCTCCAATCTA
CAO1-targeting gRNA (Direct repeat is bold)	**GTCAATACGACTTTAAAAATTTCTACTATTGTAGA T**TGGAGCAACACCTGAAGGAAGGCTTG
Non-CAO1 targeting gRNA (Direct repeat is bold)	**GTCAATACGACTTTAAAAATTTCTACTATTGTAGAT** GAGGGATCTCGCTCCTGGAAGATGGTGAT

Table 2 Strains used in the plasmid interference assay.

Strain ID	*E. coli* strain	Plasmid Description	Link
CBS-5043	Top10	SuCas12a2, T7 promoter, Kan	https://benchling.com/s/seq-O5MD7dRC9e3Ev4Z3KeSY?m=slm-tVcy34Iwv2QWhtB7WZzd
CBS-6648	Top10	MpCas12a, T7 promoter, Kan	https://benchling.com/dmytrenko/f/lib_E3sOPw9k-plasmids/seq_1sesOHHO-pod1597-pod987-771775/edit
CBS-6177	BL21 (AI)	CAO1 Target + CAO1 5x array (Targeting), Cm	https://benchling.com/m-krebel/f/lib_oxiZPobk-plasmids/seq_zKO1Sf2V-pod2110/edit
CBS-6178	BL21 (AI)	CAO1 Target + GAPDH T17 5x array (Non-targeting), Cm	https://benchling.com/m-krebel/f/lib_oxiZPobk-plasmids/seq_8h4ED0T3-pod2111/edit

Table 3 Strains used in the growth-based assays.

Strain ID	E.coli strain	Plasmid description	Link
CBS-6649	KL740	GFP expressing plasmid under *recA* promoter, sc101 ori, Amp	https://benchling.com/s/seq-NiQkIzIGb2BwghqjyUKo?m=slm-gwuhBiHAcXtos5SNCU1D
CBS-3616	Top10	Plasmid with *recA* promoter without *gfp* gene, sc101 ori, Amp	https://benchling.com/s/seq-rJXYotGTKZivfgIYCY4E
CBS-5043	Top10	SuCas12a2, T7 promoter, Kan	https://benchling.com/s/seq-O5MD7dRC9e3Ev4Z3KeSY?m=slm-tVcy34Iwv2QWhtB7WZzd
CBS-6648	Top10	MpCas12a, T7 promoter, Kan	https://benchling.com/dmytrenko/f/lib_E3sOPw9k-plasmids/seq_1sesOHHO-pod1597-pod987-771775/edit
CBS-6177	BL21 (AI)	CAO1 Target + CAO1 5x array (Targeting), Cm	https://benchling.com/m-krebel/f/lib_oxiZPobk-plasmids/seq_zKO1Sf2V-pod2110/edit
CBS-6178	BL21 (AI)	CAO1 Target + GAPDH T17 5x array (Non-targeting), Cm	https://benchling.com/m-krebel/f/lib_oxiZPobk-plasmids/seq_8h4ED0T3-pod2111/edit

Table 4 Strains used in the Cell-Free Transcription/Translation (TXTL) assay.

Strain ID	*E. coli* strain	Description	Link
CBS-6649	KL740	Expression Vector for T7 RNA Polymerase for TXTL, Amp	https://benchling.com/s/seq-kowzukmKhKmA8w1ACt2b?m=slm-bh2zFBLFVsnzknVghsfQ
CBS-338	KL740 cI857 +	deGFP under P70a promoter, Amp	https://benchling.com/s/seq-EezkMaIVjmNgFxmLrDKd?m=slm-x3SrqrkOUBLsyJAfemfm
CBS-3619	Top10	CAO1 target under PJ23119 promoter, AUCUA PFS	https://benchling.com/dmytrenko/f/lib_lTlqBtqV-cas-manuscript/seq_eI6f0lAW-pod988/edit
CBS-3621	Top10	CAO1 target under PJ23119 promoter, GAAAG PFS, Kan	https://benchling.com/dmytrenko/f/lib_lTlqBtqV-cas-manuscript/seq_P9BG93fJ-pod1006/edit
CBS-6650	Top10	CAO1 target under terminator, GAAAG PFS, Kan	https://benchling.com/schutfriso/f/lib_JcQTShos-plasmids/seq_9wVdevcb-pod990/edit
CBS-6651	Top10	CAO1 target under terminator, AUCUA PFS, Kan	https://benchling.com/schutfriso/f/lib_JcQTShos-plasmids/seq_1AH7RWWv-pod1007/edit
CBS-5043	Top10	SuCas12a2, T7 promoter, Kan	https://benchling.com/s/seq-O5MD7dRC9e3Ev4Z3KeSY?m=slm-tVcy34Iwv2QWhtB7WZzd
CBS-5044	Top10	CAO1 guide, Amp	https://benchling.com/dmytrenko/f/lib_E3sOPw9k-plasmids/seq_tBdz9uBI-pod1591-pod1323-cao1/edit

Table 5 Components for the TXTL pre-expression reaction. The table states the reaction volume needed for one experiment. We recommend the preparation of a master mix to account for multiple reactions.

Component	Volume (μL)	Stock concentration (nM)	Final concentration (nM)
myTXTL®	9	–	–
Cas12a2 plasmid	0.8	30	2
gRNA plasmid	1.6	30	4
T7RNAP plasmid	0.12	20	0.2
IPTG	0.12	20	0.2
Water	0.36	–	–

Table 6 Components for TXTL collateral cleavage assay. The table indicates the reaction volume necessary for a single experiment. We advise preparing a master mix with myTXTL® and water first to cover a multitude of reactions. Typically, the deGFP plasmid is added as the final step.

Component	Volume (μL)	Stock concentration (nM)	Final concentration (nM)
myTXTL®	8	–	–
Pre-expressed Cas12a2 & gRNA	2	–	–
Target plasmid	0.6	40	2
deGFP plasmid	1.2	12	1.2
Water	0.20	–	–

- Mix the reaction carefully in a PCR tube by slow pipetting and place the reaction in a thermocycler or incubator for 16 h at 29 °C. We recommend making a master mix with myTXTL® mix, IPTG and T7 RNAP plasmid, before aliquoting it to a PCR tube with Cas12a2 plasmid and gRNA plasmid.
- The final pre-expression reaction can be used directly after or kept frozen at −20 °C for future use.

4.4 TXTL collateral cleavage assay

When the pre-expression reaction has been incubated for 16 h, an aliquot of the reaction can be mixed with fresh myTXTL® mix, a target plasmid, a GFP reporter plasmid and water. After mixing all of the components, we measure the production of GFP or the repression thereof over time using a plate reader.

4.4.1 Materials for TXTL collateral cleavage assay

- myTXTL® Sigma 70 Master Mix Kit (Arbor Biosciences myTXTL®)
- Pre-expression TXTL mix from the previous step
- Clean and concentrated target and deGFP reporter plasmids from the previous step
- 96-well plate (e.g., Corning® 96-well Clear V-Bottom, Cat #3363)
- Cover mat (e.g., Fisher Scientific, Corning™ Storage Mat III, Cat #10428571)
- Fluorescence plate reader (e.g., Biotek Synergy Neo2 plate reader)
- 1.5-mL tube
- PCR tubes
- PCR plate cooling-block
- Bucket with ice

4.4.2 Protocol for TXTL collateral cleavage assay

1. To prepare for the assay, turn on the plate reader and allow it to warm up until the internal temperature reaches 29 °C.
2. Add pre-expression TXTL mix to a PCR tube for every reaction and keep the tube on ice. Repeat this three times per reaction to include biological replicates.
3. Prepare a master mix with myTXTL® and water in a 1.5-mL tube and keep it on ice.
4. Transfer an aliquot of the master mix to the PCR tubes with the pre-expression mixture. Make sure to keep the reaction on ice as much as possible.
5. Add the target plasmid to the PCR tubes first, followed by the addition of the deGFP reporter plasmid.Gently mix the mixture by pipetting slowly up and down.
6. Transfer two times 5 μL of each reaction into two separate wells of a 96-well V-bottom plate to include technical replicates. Ensure no bubbles are introduced by pressing the pipet to the first stop only. To keep the

reaction from starting, we recommend placing the 96-well V-bottom plate in or on a cooling block (pre-cooled at −20 °C).

7. Cover the plate with a sealing mat to prevent evaporation of the reaction mix in the plate reader. When the 96-well V-bottom plate touches the ice, make sure to wipe it dry before placing it in the plate reader.
8. Incubate the reactions for 10-16 h at 29 °C and monitor GFP fluorescence at three-minute intervals (Ex 485 nm, Em 528 nm).

5. Expression and purification of Cas12a2

While in vivo assays and TXTL can provide important insights into the function of immunity, in vitro cleavage assays with recombinantly expressed protein allows for the direct characterization of each of the biochemical activities of Cas12a2. These assays can be performed with apo Cas12a2, or, because Cas12a2 processes its own crRNA, with Cas12a2 co-expressed with a CRISPR array. Expression and purification of SuCas12a2 co-expressing the gRNA is a time- and cost-effective alternative to purchasing or transcribing large quantities of synthetic gRNA. However, we have not had success using co-expressed Cas12a2 RNPs in structural studies and co-expressed RNPs at times can be contaminated with activating RNAs.

Purification of apo protein has been essential to determining high-resolution structures and is critically important in leveraging the programmability of SuCas12a2 as a CRISPR–Cas system. For these reasons, we have found it useful to express and purify both apo Cas12a2 and Cas12a2 expressing a CRISPR array. Various methods describing the expression and purification of SuCas12a2 have previously been reported (Bravo et al., 2023; Dmytrenko et al., 2023; Domgaard, 2022; Keiser, 2022). Here we provide a detailed and optimized expression and purification protocol for both apo and co-expressed RNA, along with methods used to reconstitute RNPs using apo Cas12a2 and synthetic guide RNA for structure determination. A schematic overview of the entire method with supporting results can be found in Fig. 5.

5.1 Cas12a2 expression

5.1.1 Materials

- *E. coli* strain NiCo21 (DE3) (NEB) or *E. coli* strain HMS174 (DE3) (Novagen)
- Cas12a2 expression plasmid

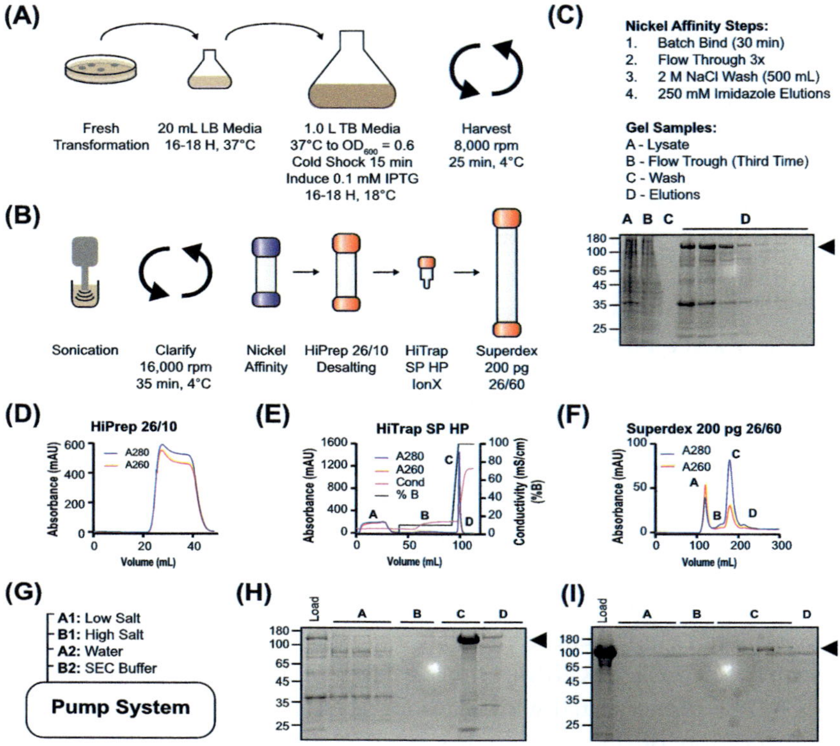

Fig. 5 Expression and purification of Cas12a2. (A) Outline of Cas12a2 expression protocol. (B) Outline of Cas12a2 purification protocol. (C) The individual steps of Cas12a2 purification over nickel affinity chromatography, including a representative gel of the resulting protein. Samples from different stages of the purification are indicated by letters A–D. The Cas12a2 band is indicated with the black arrow. (D) representative chromatogram of Cas12a2 desalting over a HiPrep 26/10 column after nickel chromatography. (E) Representative chromatogram of Cas12a2 ion exchange purification over HiTrap SP HP column. Peaks of the chromatogram are labeled with letters A–D. (F) Representative chromatogram of Cas12a2 purification by size exclusion over a Superdex 200 pg 26/60 column. Peaks of the chromatogram are labeled with letters A–D. (G) Diagram of the usual buffer setup used for purification steps after nickel chromatography. (H) Representative SDS-PAGE of ion exchange chromatography over HiTrap SP HP column. Samples from different peaks are labeled as in (E). Cas12a2 is indicated with the black arrow. (G) Representative SDS-PAGE of size exclusion chromatography over Superdex 200 pg 26/60. Samples from different peaks are labeled as in (F). Cas12a2 is indicated with the black arrow.

- LB medium, autoclaved (GoldBio)
- LB agar plates (Fisher)
- TB media, autoclaved (24 g yeast extract (Fischer), 20 g tryptone (Fischer), 5 g glycerol (Pharmco), 900 mL H_2O)
- Sterile phosphate buffer (0.17 M KH_2PO_4, 0.72 M K_2HPO_4), filtered 0.22 μm.
- 1.0 M Isopropyl ß-D-1-thiogalactopyranoside (IPTG) (Fisher)
- 2–3 gallons of ice

5.1.2 Protocol

1. Thaw 50 μL aliquot of chemically competent *E. coli* NiCo21 (DE3) or HMS173 (DE3) cells on ice for 30 min.
2. Add ~20 ng of purified plasmid DNA of Cas12a2 expression vector. Mix DNA with the cells by flicking the end of the tube three times and incubate on ice for 30 min.
3. Heat-shock cells in 42°C water bath for 45 s then return to ice for 2 min. The tube containing cells should be buried in the ice after heat-shock.
4. Add 250 μL of LB media to the transformations and allow the cells to recover at 37°C with 200 rpm shaking for 45 min. SOC media can also be used at this step, but it has not been shown to make a difference.
5. Plate 60 μL of the recovered cells on agar plates with the appropriate antibiotics.
6. Leave the plates to incubate at 37°C for 16–18 h (overnight).
7. After plates have incubated overnight, remove them from the incubator and place them at room temperature. The protocol may also be paused at this point by placing the plates at 4°C until ready to continue. Plates may be stored up to 3 weeks before a new transformation needs to be made.
8. At the end of the day, use a single colony from the transformation plate to inoculate 20 mL of LB media treated with the appropriate antibiotic. This growth is best performed in a 50 mL conical tube (VWR) or small 250 mL culture flask. For multiple liters of growth, add an additional 20 mL of LB media for each additional liter of growth (e.g., 60 mL overnight for 3 L of growth). The screw lids of 50-mL conical tubes should be left a quarter turn loose, then secured with a piece of lab tape.
9. Incubate the culture at 37°C with 200 rpm shaking overnight.

10. Prepare for expression on Day 3 by autoclaving 900 mL of fresh TB media (omitting phosphate buffer) for each liter of growth. Prepare the phosphate buffer for TB media separately and sterilize by vacuum filtration through a 0.22 μm filter into a sterile bottle.
11. Aseptically combine 100 mL of sterile phosphate buffer with 900 mL of TB media then supplement with the appropriate antibiotic.
12. Inoculate each liter of TB media with 20 mL of overnight culture and allow cells to grow at 37°C with 200 rpm shaking until an OD_{600} of 0.6–0.8. Usually, this takes about 6–8 h, but some mutant constructs can reach OD_{600} in less than 4 h.
13. After cells have reached the appropriate optical density, cold shock the growths in ice for 15 min. The flasks are placed on top of ice then buried by shoveling additional ice over and around the sides of the flask. We recommend cold shocking in a large sink to contain the mess of melting ice.
14. While cells are cold-shocking, cool the incubator to 18°C.
15. After cells have sat on ice for 15 min, induce expression by adding 100 μL of 1 M Isopropyl ß-D-1-thiogalactopyranoside (IPTG) for a final concentration of 0.1 mM and grow cells for 16–18 h (overnight) at 18°C.
16. To harvest cells, centrifuge at 8000 rpm for 25 min at 4°C.
17. While the centrifuge runs, add bleach to each growth flask to begin sterilizing. Cleaning of any other dirty equipment that may also begin at this time.
18. After cells spin down, decant the media supernatant from the centrifuge tube into the bleached growth flasks. Leave the centrifuge tube inverted on a paper towel for ~2 min to allow excess media to drain away.
19. Using a spatula transfer the cell pellet from the bottom of the centrifuge tube into a 50 mL conical tube.
20. Weigh the mass of the cell pellet and record on the outside of the tube along with the plasmid id number and your initials. Cells may be stored in the −80°C freezer for 6 months.

5.2 Cas12a2 purification

5.2.1 Materials

- Buffers prepared according to Table 7, filtered 0.22 μm and chilled to 4°C.

Table 7 Buffers used in SuCas12a2 purification.

Reagents	Lysis	Nickel Wash	Nickel Elution	Low Salt	High Salt	SEC	Cryo-EM
Tris pH 7.2 (mM)	25	25	25	25	25	0	0
HEPES pH 7.2 (mM)	0	0	0	0	0	25	12.5
NaCl (mM)	500	500	500	50	150	0	0
KCl (mM)	0	0	0	0	0	150	150
$MgCl_2$ (mM)	2	2	2	2	2	2	2
Imidazole (mM)	50	50	250	0	0	0	0
Glycerol (%)	10	10	10	10	10	5	0

- Protease inhibitors (0.5 μg/mL aprotinin (GoldBio, A-655-25), 0.7 μg/mL pepstatin A (Fisher, BP26715), 0.5 μg/mL leupeptin (Alfa Aesae, AAJ61188MB), 0.2 mM AEBSF (GoldBio, A-540-500))
- Nickel resin (ThermoFisher, PI88221)
- CrystalCruz Chromatography Column (Kimble, sc-205558)
- HiPrep 26/10 desalting column (Cytiva, 17508701)
- HiTrap SP HP ion exchange column (Cytiva, 17115201)
- HiTrap Q HP column (Cytiva, 17115401).
- Superdex 200 pg 26/60 sizing column (Cytiva, 28989336)
- Superdex 200 10/300 increase GL sizing column (Cytiva, 28990944)
- 100 kDa MWCO spin concentrators (Corning)
- 250 mL Metal beaker chilled on ice.
- Sonifier
- AKTA pure pump system (Cytiva)

5.2.2 Protocol

1. Resuspend cell pellets in 50 mL of lysis buffer treated with 10uL aprotinin, 50uL pepstatin, 50uL leupeptin, 500uL AEBSF, 125uL lysozyme. Allow the pellet to thaw on ice for 30 min.
2. After thawing, mix the cell pellet with the lysis buffer until smooth, trying to avoid bubbles. Transfer slurry to a metal beaker on ice.
3. Sonicate cell mixture with sonifier at 5/50 in 10 second intervals followed by 10 ss rest until the cell mixture color darkens and the viscosity becomes like water. Sonication power settings may need to be adjusted to prevent foaming or overheating of the lysate. Periodic adjustments will be required to ensure that the sonicator probe does not leave the lysate as the ice melts and the metal beaker sinks.
4. Clarify lysate by centrifugation of 16,000 rpm for 35 min at 4 °C.
5. While lysate clarifies, transfer 2–3 mL of Ni-NTA resin into a flex column and equilibrate by first rinsing with filtered water, followed by 10 mL of lysis buffer.
6. Incubate clarified lysate with Ni-NTA resin in flex column for 30 min at 4 °C with rocking.
7. Flow lysate over the column, collecting the flowthrough, then re-flow the lysate over the column two more times.
8. Wash resin with 500 mL of Nickel Wash Buffer. Washes can be expedited by pushing the buffer through the flex column with an air filled syringe or by siphoning the buffer through a silicone tube.

   a. **Note:** Cas12a2 with co-expressed RNA should be washed with 100 mL of Lysis Buffer instead of Nickel Wash Buffer. Washing with Nickel Wash Buffer will wash away the bound gRNA.
9. After washing the Ni-NTA resin, elute Cas12a2 from the column with ten, 4 mL fractions of nickel elution buffer. Incubate the first three elutions for 30 min before collecting each elution. Usually most of the protein comes off in the first three elutions. The remaining seven elutions can be collected with or without incubation.
10. Collect samples of each stage of the nickel purification (eg. lysis, flowthrough, washes, etc.) and analyze the protein content by SDS-PAGE.
11. Prepare the pump system with Low Salt Buffer on line A1 and High Salt buffer on line B1. Perform a pump wash for both buffers and then equilibrate the whole system in Low Salt Buffer.
12. Load fractions containing Cas12a2 from nickel affinity chromatography over a Hiprep 26/10 desalting column. The 26/10 column has a capacity of 16 mL per run, so larger purifications may require desalting of the nickel elutions in multiple batches.
13. Collect fractions of desalted protein and pool to load over ion exchange.
   a. **Note:** Apo Cas12a2 is purified with a HiTrap SP HP ion exchange column (Cytiva) while Cas12a2 co-expressed with a gRNA guide is purified with the HiTrap Q HP column (Cytiva).
14. Flush the sample pump with a low salt buffer and then load the desalted protein sample over the ion exchange column. Collect the flow-through.
15. After the sample has been loaded over the ion exchange column, continue to wash with low salt until the UV absorbance returns to baseline.
16. Wash the column with 5 CV of 10% B and collect fractions.
17. Elute protein from column with a 5 CV gradient to 100% B (High Salt Buffer) with the system in up-flow. Collect the elution peaks in 2 mL fractions.
18. Collect and analyze samples from desalting and ion exchange by SDS-PAGE to identify fractions containing Cas12a2.
19. Pool fractions containing clean Cas12a2 as determined by SDS-PAGE.
20. Concentrate Cas12a2 in a 100 kDa MWCO spin concentrator by centrifugation between 1200 and 2500 rpm in 9 min intervals at 4 C. Speed may need to be adjusted during the course of concentration as

fast speeds with highly concentrated proteins can lead to precipitation. Concentrate Cas12a2 down to 1 mL.

21. While concentrating, start the Size Exclusion Chromatography program on the AKTA. It will take several hours for the column to be equilibrated before the sample is injected from the sample loop. While the column is equilibrating, attach the 1 mL sample loop to the column and flush with 5 mL of filtered water followed by 5 mL of Low Salt Buffer.
22. Load concentrated Cas12a2 into the sample loop through the injection port. Important! One end of the sample line is open to the waste position to allow injection of a sample. Make sure to leave the syringe used to load Cas12a2 in the injection port even after injection to prevent siphoning of the sample into the waste.
23. Analyze peak fraction from SEC by SDS-PAGE.
24. Cryo-EM structures are prepared with apo Cas12a2 complexed with synthetic gRNA.
25. Pool and concentrate SEC fractions that contain clean apo Cas12a2 in a 100 kDa MWCO spin concentrator. Quantify by nano-drop when the protein sample reaches ~500 μL.
26. Prepare gRNA for complexation by incubating a 500 μM aliquot of synthetic gRNA at 65 °C for 3 min followed by cooling 1 °C min^{-1} to room temperature.
27. Combine heated gRNA with apo Cas12a2 in a 1:1.2 molar ratio to a final volume of 500 μL.
28. While forming the binary complex, start the Size Exclusion Chromatography program on the AKTA to equilibrate a Superdex 200 10/300 increase GL sizing column (Cytiva) into Cryo-EM Buffer. While the column is equilibrating, attach the 0.5 mL sample loop to the column and flush with 5 mL of filtered water followed by 5 mL of Size Exclusion Buffer.
29. Load Cas12a2 binary complex into the sample loop through the injection port using a microtip injection needle (Hampton).

6. Cas12a2 *in vitro* cleavage assays

Characterizing the in vitro activity of Cas12a2 requires purified protein (as described above) along with gRNA, target RNA and/or sequence non-specific substrates. In the following sections we provide a description of how to design sequence specific and non-specific cleavage

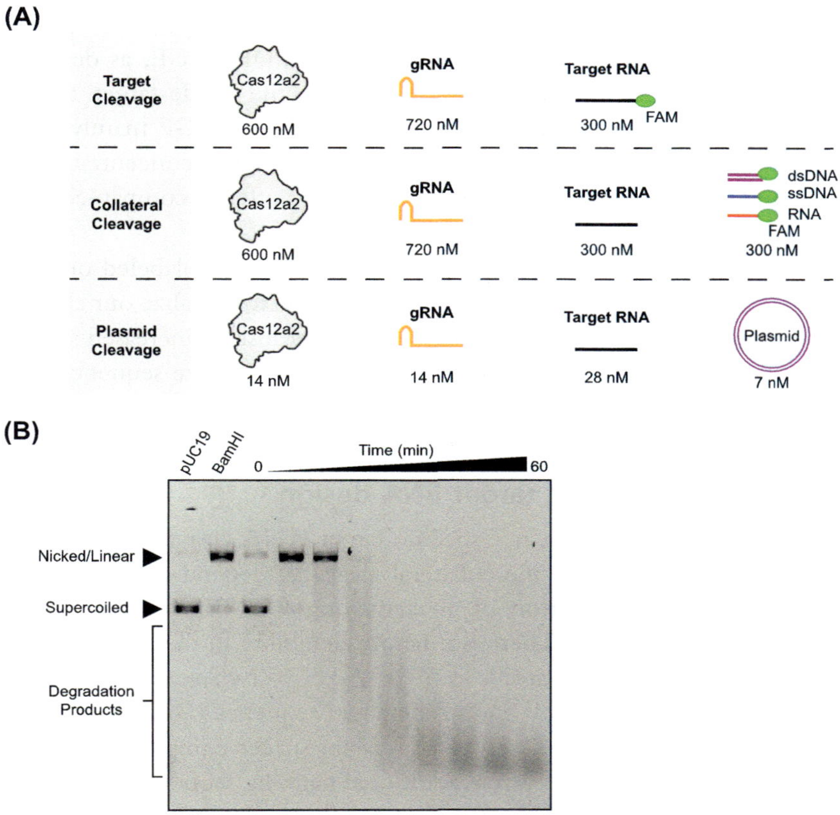

Fig. 6 Cas12a2 in vitro cleavage assays. (A) graphical representation of the components needed for Cas12a2 target (top), collateral (middle) and plasmid (bottom) cleavage assays including final concentrations for each reaction component. (B) Representative time course for supercoiled plasmid cleavage by Cas12a2. Time points were collected at 0, 1, 2, 5, 10, 20, 30, 45 min and 60 min.

substrates, along with providing detailed methods for testing the biochemical activity of SuCas12a2. A schematic overview of the different in vitro cleavage assay with supporting results can be found in Fig. 6.

6.1 Cas12a2 cleavage visualization

Denaturing PAGE analysis is the preferred method to observe nucleic acid cleavage by Cas12a2. Denaturing PAGE conditions like 7 M urea PAGE is a common method used to demonstrate nucleic acid cleavage and is used below to visualize sequence non-specific collateral cleavage. However, we found that 7 M urea-PAGE was often insufficient for consistent denaturation of the guide-target duplex leading to multiple, shifted bands making interpretation difficult. We found that FDF-PAGE, as described previously, was more effective at denaturing the guide-target duplex. (Harris, Molnar, Müller, & Baulcombe, 2015). FDF-PAGE mainly differs from traditional urea-PAGE by the inclusion of higher concentrations of formamide and formaldehyde in the loading buffer accompanied by a 15-minute denaturing step at 55°C.

Cleavage products could be visualized by either radiolabeled or fluorescein (FAM) labeled cleavage substrate. We chose to visualize our cleavage experiments using FAM labeled target RNA because of increased stability of the FAM label. An additional method used to visualize sequence non-specific cleavage activity is running cleaved plasmid DNA on an agarose gel treated with ethidium bromide (EtBr).

6.2 Sequence-specific target RNA design

As mentioned above, sequence specific target RNA cleavage, and activation of sequence non-specific collateral cleavage, depend on guide-target binding and the recognition of an activating PFS (Bravo et al., 2023; Dmytrenko et al., 2023). There is inherent flexibility in the activating PFS profile, but the common profile is 5′-BAAAN-3′ (where B indicates not adenine) positioned at the 3′ end of the target sequence (Dmytrenko et al., 2023). The minimal target consists of a 24-nt spacer complement flanked on the 3′ side by the PFS. While additional flanking sequence at the 5′ or 3′ ends is not strictly required for Cas12a2 activation, substrates between 50–75-nt result in dramatic shifts during polyacrylamide gel electrophoresis and are thus preferred when analyzing sequence specific target cleavage.

6.2.1 Materials for target cleavage assay

- NEB 3.1 Buffer
- Purified Cas12a2
- gRNA
- Target RNA (5′ FAM labeled)
- Nuclease Free Water (DEPC treated)
- Phenol-chloroform (pH 6.0)
- Denaturing Buffer (for 990 μL, 267.3 μL 37% formaldehyde, 655.2 μL formamide, 67.5 μL 10x MOPS buffer)
- FDF-PAGE (4.2 g urea, 0.5 mL 10 x MOPS buffer, 3 mL 40% acrylamide, fill to 10 mL with ddH2O, per gel)

- 10x MOPS buffer (200 mM MOPS, 50 mM sodium acetate, 10 mM EDTA)
- 10% ammonium persulfate
- Tetramethylethylenediamine (TEMED)

6.2.2 Protocol for target cleavage assay

1. Prepare FDF-PAGE gels as described in (Harris et al., 2015). Polymerize FDF-PAGE mix (4.2 g urea, 0.5 mL 10 x MOPS buffer, 3 mL 40% acrylamide, fill to 10 mL with ddH2O, per gel) with 80 μL 10% APS and 10 μL TEMED. Allow gels to polymerize for at least 30 min.
2. After gels are polymerized, flush the wells of the gel with a p1000 pipet and pre-run the gel in 0.5 × MOPS buffer (10 mM MOPS, 2.5 mM sodium acetate, 0.5 mM EDTA) for at least 30 min.
3. Prepare individual reactions as described in Table 8.
4. Combine all reaction components in a PCR tube and incubate for 1.0 h at 37°C. Prepare a no cleavage control by omitting the gRNA from the mix and replacing its volume with water.
5. Quench reactions by adding 10 μL of acidic (pH 6.0) phenol-chloroform. Securely close the tube and mix by flicking the end of the tube 3 times. Separate aqueous and organic layers of the quenched reactions by briefly centrifuging the tubes for 1 min.
6. Extract 4 μL from the aqueous layer of the quenched reactions and combine with 13 μL of formaldehyde denaturing buffer. Incubate extracted samples at 55°C for 15 min in a thermocycler.

Table 8 Target cleavage reaction condition.

Components	[Initial]	[Final]	Volume (μL)
NEB 3.1 Buffer (X)	10	1	1.0
Cas12a2 (μM)	5	0.6	1.2
gRNA (μM)	2	0.72	3.6
FAM Target RNA (μM)	2	0.3	1.5
H_2O			2.7
Total			10.0

Note: For Cas12a2 co-expressing gRNA, replace the volume of gRNA with H_2O.

7. Re-flush the wells of the FDF-PAGE gel with the p1000 pipet. Then load 15 μL of sample into FDF - PAGE wells. If not all the wells are being used, load 15 μL of formaldehyde denaturing buffer to prevent "smiling" in the gel.
8. Run the gel at 50 V for 15 min to allow samples to enter the gel, then adjust the voltage to 150 V for 1 h. Cover the gel box with a cardboard box to protect FAM labeled substrates from photobleaching. Usually, 1 h is sufficient to fully resolve the cleaved and uncleaved target RNA.

6.3 Sequence non-specific, collateral substrate design

Sequence non-specific collateral substrates may be RNA, ssDNA, dsDNA and cleavage has been observed using either synthetic, linear substrates or supercoiled (SC) plasmid DNA. Synthetic substrates are designed to contain no sequence complementarity with the gRNA guide and can be any other sequence, usually 50–75-nt to maximize the shift observed by PAGE analysis. Example sequences can be found in Table 9 Similarly, a SC plasmid is selected for sequence non-specific cleavage assays by identifying a plasmid with no complementarity to the gRNA guide that is large enough to produce significant shifts when run on an agarose gel.

Table 9 Representative collateral cleavage sequences.

Name	Sequence
Collateral RNA	5′-CAGAGAUAAGUGACGCGCGGCGAGUGGCG CGCCACGUCGGAAAUCUAGAGGCG-3′
Collateral ssDNA	5′-AACTGATATGACAATTGCGCGTAGCACGA CGACGATATGACACTTGCGCATAACGACGA CGATACAATGAT-3′
Collateral dsDNA	5′-AACTGATATGACAATTGCGCGTAGCACGA CGACGATATGACACTTGCGC 3′-TTGACTATACTGTTAACGCGCATCGTGCT GCTGCTATACTGTGAACGCG ATAACGACGACGATACAATGAT-3′ TATTGCTGCTGCTATGTTACTA-5′

Note: All substrates are FAM labeled on the 5′ end. Labeled dsDNA can be made by annealing the unlabeled complement to the Collateral ssDNA substrate.

6.3.1 Materials for sequence non-specific, collateral cleavage assay

- NEB 3.1 Buffer
- Purified Cas12a2
- gRNA
- Target RNA
- Collateral Substrate - RNA, ssDNA, or dsDNA (5′ FAM labeled)
- Nuclease Free Water (DEPC treated)
- Phenol-chloroform (pH 6.0)
- 50% glycerol solution
- 12% urea-PAGE (4.8 g urea, 2 mL 5x TBE, 3 mL 40% acrylamide, fill to 10 mL with ddH2O, per gel)
- 5 x TBE buffer (450 mM Tris-borate, 10 mM EDTA)
- 10% ammonium persulfate (APS)
- Tetramethylethylenediamine (TEMED)

6.3.2 Protocol for sequence non-specific, collateral cleavage assay

1. Prepare 12% urea-PAGE gels by polymerize urea-PAGE mix (4.8 g urea, 2 mL 5x TBE, 3 mL 40% acrylamide, fill to 10 mL with ddH2O, per gel) with 80 μL 10% APS and 10 μL TEMED. Allow gels to polymerize for at least 30 min.
2. After gels are polymerized, flush the wells of the gel with a p1000 pipet and pre-run the gel in 1 x TBE buffer (90 mM Tris-borate, 2 mM EDTA) for at least 30 min.
3. Prepare individual reactions as described in Table 10.
4. Combine all reaction components in a PCR tube and incubate for 1.0 H at 37°C. Prepare a no cleavage control by omitting the gRNA from the mix and replacing its volume with water.
5. Quench reactions by adding 10 μL of acidic (pH 6.0) phenol-chloroform. Securely close the tube and mix by flicking the end of the tube 3 times. Separate aqueous and organic layers of the quenched reactions by briefly centrifuging the tubes for 1 min.
6. Extract 5 μL from the top aqueous layer and combine with 5 μL of 50% glycerol.
7. Re-flush the wells of the urea-PAGE gel with the p1000 pipet. Then load 9 μL of sample into urea - PAGE wells. If not all the wells are being used, load 9 μL of 25% glycerol to prevent "smiling" in the gel.
8. Run the gel at 50 V for 15 min to allow samples to enter the gel, then adjust the voltage to 150 V for 1 h. Cover the gel box with a cardboard

Table 10 Sequence non-specific, collateral cleavage reaction conditions.

Components	[Initial]	[Final]	Volume (μL)
NEB 3.1 Buffer (X)	10	1	1.0
Cas12a2 (μM)	5	0.6	1.2
gRNA (μM)	2	0.72	3.6
Target RNA (μM)	2	0.3	1.5
FAM non-target RNA, ssDNA or dsDNA (μM)	2	0.3	1.5
H_2O			1.2
Total			10.0

Note: For Cas12a2 co-expressing gRNA, replace the volume of gRNA with H_2O.

box to protect FAM labeled substrates from photobleaching. Usually, 1 h is sufficient to fully resolve the cleaved and uncleaved target RNA.

6.3.3 Materials for plasmid cleavage assay

- NEB 3.1 Buffer
- Purified Cas12a2
- Target RNA
- Supercoiled pUC19 plasmid (or similar), isolated by Miniprep (e.g., E.Z.N.A. Plasmid DNA mini kit, Omega Bio-tek, D6942-00), then supercoiled fraction gel extracted from a 1% agarose and purified (e.g., E.Z.N.A. gel extraction kit. Omega Bio-tek, D2500-02)
- Nuclease Free Water (DEPC treated)
- Phenol-chloroform (pH 7–8)
- 50% glycerol solution
- Agarose (GoldBio)
- TAE buffer (40 mM Tris pH 8.5, 20 mM glacial acetic acid, 1 mM EDTA)
- Ethidium bromide (Apex)

6.3.4 Protocol for plasmid cleavage assay

1. Prepare a 1% agarose gel by combining 0.5 g agarose (GoldBio) with 50 mL of 1 x TAE buffer (40 mM Tris pH 8.5, 20 mM glacial acetic

Table 11 Plasmid cleavage assay reaction conditions.

Reagent	[Initial]	[Initial]	Volume (μL)
NEB 3.1 (X)	10	1	10
Cas12a2 (nM)	500	14	2.8
gRNA (nM)	500	14	2.8
SC-pUC19 (ng/μL)	25	12.2	48.8
H_2O			30.6
Target RNA (nM)	500	25	4.5
Total			100

Note: The table above is for a time course assay. A 10 μL total volume can be used for single time points.

acid, 1 mM EDTA). Microwave in 30 second increments until all agarose is dissolved.

2. Before agarose gel polymerizes, spike the solution with one drop of ethidium bromide solution (Apex) and mix in. Then pour the agarose gel mixture into a mold and let it cool.
3. While agarose gel cools, prepare reaction mixtures as described in Table 11.
4. Combine all the reagents except target RNA in a PCR tube and incubate at 37°C for 30 min. For assays with a single time point combine all the reagents and incubate for 1.0 H.
5. For time course assays, prepare 9 PCR tubes, each with 10 μL of pH 8.0 phenol-chloroform. For end-point assays, reactions will be quenched directly in the reaction tube.
6. Before initiating the time course, prepare a zero time point by adding 0.5 μL of ddH_2O to the first tube (T = 0), then add 9.5 uL of the incubated reaction mixture and mix by flicking the tube 3 times.
7. Add 4.5 μL target RNA to initiate the reaction. Quench each time point by transferring 10 μL of the reaction mixture to the prepared tubes of phenol-chloroform and mix by flicking 3 times. Time points are collected at T = 1, 2, 5, 10, 20, 30, 45, and 60 min.
8. After collecting all the time points, ensure that all the tubes are securely closed before mixing again, then spin down for at least 30 ss.

9. Extract 5 μL from the aqueous layer of the phenol-chloroform and add to 5 μL of 50% glycerol.
10. Prepare markers for SC, OC and LN plasmid in the gel by diluting each sample to 12.2 ng/μL then diluting again 1:1 in 50% glycerol.
11. Load 9 μL of topological markers and time course samples into the agarose gel in chronological order. Load 2 μL of 1 kbp + DNA ladder (GoldBio).
12. Run the gel at 100 V until the blue dye front in the DNA ladder lane reaches about halfway down the gel.

References

Bravo, J. P. K., Hallmark, T., Naegle, B., Beisel, C. L., Jackson, R. N., & Taylor, D. W. (2023). RNA targeting unleashes indiscriminate nuclease activity of CRISPR-Cas12a2. *Nature, 613*(7944), 582–587.

Chen, Z., Lu, M., Zou, D., & Wang, H. (2012). An E. coli SOS-EGFP biosensor for fast and sensitive detection of DNA damaging agents. *Journal of Environmental Sciences, 24*(3), 541–549.

Dmytrenko, O., Neumann, G. C., Hallmark, T., Keiser, D. J., Crowley, V. M., Vialetto, E., Mougiakos, I., et al. (2023). Cas12a2 elicits abortive infection through RNA-triggered destruction of dsDNA. *Nature, 613*(7944), 588–594.

Domgaard, H. (2022). *Investigations of prokaryotic defense systems.* Utah State University. https://doi.org/10.26076/cebb-fa0a.

Garamella, J., Marshall, R., Rustad, M., & Noireaux, V. (2016). The all E. coli TX-TL toolbox 2.0: A platform for cell-free synthetic biology. *ACS Synthetic Biology, 5*(4), 344–355.

Harris, C. J., Molnar, A., Müller, S. Y., & Baulcombe, D. C. (2015). FDF-PAGE: A powerful technique revealing previously undetected small RNAs sequestered by complementary transcripts. *Nucleic Acids Research, 43*(15), 7590–7599.

Keiser, D. J. (2022). Functional characterization of the newly discovered type V CRISPR-Cas protein Cas12a2. *Utah State University.* https://doi.org/10.26076/7ac8-f685.

Li, S.-Y., Cheng, Q.-X., Liu, J.-K., Nie, X.-Q., Zhao, G.-P., & Wang, J. (2018). CRISPR-Cas12a has both Cis- and Trans-cleavage activities on single-stranded DNA. *Cell Research, 28*(4), 491–493.

Liao, C., Ttofali, F., Slotkowski, R. A., Denny, S. R., Cecil, T. D., Leenay, R. T., ... Beisel, C. L. (2019). Modular one-pot assembly of CRISPR arrays enables library generation and reveals factors influencing crRNA biogenesis. *Nature Communications, 10*(1), 2948.

Liu, L., Li, X., Ma, J., Li, Z., You, L., Wang, J., ... Wang, Y. (2017). The molecular architecture for RNA-guided RNA cleavage by Cas13a. *Cell, 170*(4), 714–726 e10.

Marino, N. D., Pinilla-Redondo, R., & Bondy-Denomy, J. (2022). CRISPR-Cas12a targeting of ssDNA plays no detectable role in immunity. *Nucleic Acids Research, 50*(11), 6414–6422.

Marshall, R., Maxwell, C. S., Collins, S. P., Beisel, C. L., & Noireaux, V. (2017). Short DNA containing χ sites enhances DNA stability and gene expression in E. coli cell-free transcription-translation systems. *Biotechnology and Bioengineering, 114*(9), 2137–2141.

Martin, L., Rostami, S., & Rajan, R. (2023). Optimized protocols for the characterization of Cas12a activities. *Methods in Enzymology, 679*, 97–129.

Shin, J., & Noireaux, V. (2010). Efficient cell-free expression with the endogenous E. coli RNA polymerase and sigma factor 70. *Journal of Biological Engineering, 4*(June), 8.

Shin, J., & Noireaux, V. (2012). An E. coli cell-free expression toolbox: Application to synthetic gene circuits and artificial cells. *ACS Synthetic Biology, 1*(1), 29–41.

Sitaraman, K., Esposito, D., Klarmann, G., Le Grice, S. F., Hartley, J. L., & Chatterjee, D. K. (2004). A novel cell-free protein synthesis system. *Journal of Biotechnology, 110*(3), 257–263.

Sun, Z. Z., Clarmyra, A. H., Shin, J., Caschera, F., Murray, R. M., & Noireaux, V. (2013). Protocols for implementing an Escherichia coli based TX-TL cell-free expression system for synthetic biology. *Journal of Visualized Experiments: JoVE, 79*, e50762.

Swarts, D. C., & Jinek, M. (2018). Cas9 versus Cas12a/Cpf1: Structure-function comparisons and implications for genome editing. *Wiley Interdisciplinary Reviews. RNA, 9*(5), e1481.

Wandera, K. G., Collins, S. P., Wimmer, F., Marshall, R., Noireaux, V., & Beisel, C. L. (2020). An enhanced assay to characterize anti-CRISPR proteins using a cell-free transcription-translation system. *Methods (San Diego, Calif.), 172*, 42–50.

Yamano, T., Nishimasu, H., Zetsche, B., Hirano, H., Slaymaker, I. M., Li, Y., Fedorova, I., et al. (2016). Crystal structure of Cpf1 in complex with guide RNA and target DNA. *Cell, 165*(4), 949–962.

Zetsche, B., Gootenberg, J. S., Abudayyeh, O. O., Slaymaker, I. M., Makarova, K. S., Essletzbichler, P., Volz, S. E., et al. (2015). Cpf1 is a single RNA-guided endonuclease of a class 2 CRISPR-Cas system. *Cell, 163*(3), 759–771.

CHAPTER EIGHT

Experimental strategy for characterization of novel TnpB orthologs

Gytis Druteika, Tautvydas Karvelis*, and Virginijus Šikšnys*
Institute of Biotechnology, Life Sciences Center, Vilnius University, Vilnius, Lithuania
*Corresponding authors. e-mail address: tautvydas.karvelis@bti.vu.lt; siksnys@ibt.lt

Contents

1. Introduction 184
2. Methods 186
   2.1 Bioinformatic curation of TnpB and guide RNA 186
   2.2 Construction of randomized TAM (7 N) plasmid library 186
   2.3 TnpB and reRNA expression and complex assembly 187
   2.4 Plasmid library cleavage 188
   2.5 Capture of cleavage products *via* adapter ligation 188
   2.6 Illumina library generation and deep sequencing 190
   2.7 Cleavage position and TAM identification 191
3. Concluding remarks 193
Acknowledgments 193
References 193

Abstract

TnpB proteins encoded in IS200/IS605 and IS607 mobile genetic elements are among the most widespread proteins in the microbial world. They function as RNA-guided DNA nucleases that play a critical role in transposon proliferation and are the predecessors of CRISPR-Cas12 effector proteins of the type V CRISPR-Cas family. Small size of TnpB nucleases makes them an attractive alternative for larger Cas9 and Cas12 proteins in genome editing applications. However, only a small fraction of TnpB nucleases characterized to date are active in human cells, highlighting the need to identify new TnpB variants that can function as genome editors. Here, we present an experimental pipeline for the characterization of TnpB proteins by combining *in silico* analysis with *in vitro* assays. To validate it we determined guide RNA and identified TAM for a set of TnpB orthologs. The proposed workflow can be employed for rapid screening and characterization of the huge TnpB protein family to identify novel TnpB variants that might expand the genome editing toolbox.

Methods in Enzymology, Volume 712
ISSN 0076-6879, https://doi.org/10.1016/bs.mie.2025.01.056

1. Introduction

Mobile genetic elements (MGEs) such as plasmids, bacteriophages, transposons, and insertion sequences (IS) are known to be the driving force of evolution that shaped prokaryotic genomes. Insertion sequences (IS) are the simplest MGEs and only encode the genes necessary for transposition (Frost, Leplae, Summers, & Toussaint, 2005; Koonin, Makarova, Wolf, & Krupovic, 2020). The IS200/IS605 family insertion sequences are among the most widespread groups of IS. The transposition of these elements occurs through a circular single-stranded (ss) DNA intermediate and is mediated by TnpA transposase from the tyrosine recombinase family (He et al. 2015). Some IS200/IS605 family transposons also carry a gene encoding TnpB protein (Fig. 1A), which is among the most widespread

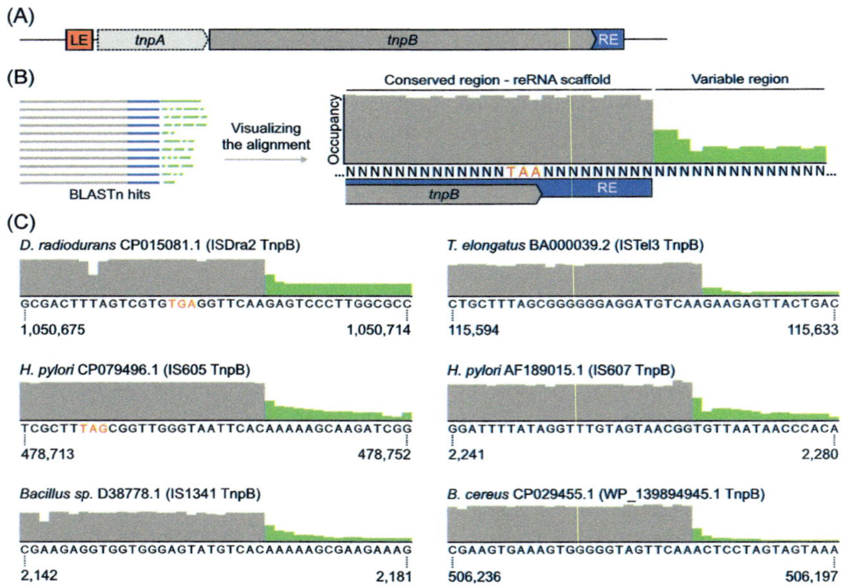

Fig. 1 Bioinformatic determination of the 3′-end of reRNA. (A) Organization of IS200/IS605 and IS607 family transposons. (B) Schematic representation of the identification of the 3′-end of reRNA. Retrieved BLASTn hits (as described in Section 2.1) are visualized to identify conserved and variable sequences within the alignment. The 3′-end of the reRNA/transposon is identified as the point where sequence conservation significantly decreases. (C) Determination of the 3′-end of the reRNA for selected TnpB variants. The green color indicates a region of low conservation, corresponding to the variable, target-complementary region of the reRNA.

proteins in prokaryotes. However, TnpB role in transposition remained unknown until recently (Altae-Tran et al. 2021; Karvelis et al. 2021).

Comparative sequence analysis indicated that the RuvC-like nuclease domain of compact (~400 amino acids) TnpB proteins is related to the RuvC domain of much larger Cas12 nucleases (>900 amino acids) from CRISPR-Cas systems (Shmakov et al. 2015, 2017). Biochemical characterization of TnpB revealed that it associates with RNA and acts as an RNA-guided DNA nuclease to cleave DNA, confirming that TnpB is an evolutionary predecessor of Cas12 nuclease family, and that its association with CRISPR arrays gave rise to CRISPR-Cas12 systems (Altae-Tran et al. 2021; Karvelis et al. 2021).

In contrast to Cas12 nucleases that are guided by crRNAs transcribed from CRISPR arrays, TnpB forms the complex with transposon right end (RE) (Fig. 1A) derived non-coding RNA (named reRNA or ωRNA) which overlaps with the coding sequence of TnpB and includes the variable guide sequence of reRNA derived from a sequence encoded next to transposon RE (Altae-Tran et al. 2021; Karvelis et al. 2021). A short sequence called the transposon adjacent motif (TAM) 5′ to the DNA target complimentary to the RNA guide is required for DNA recognition and cleavage which is analogous to the protospacer adjacent motif (PAM) recognized by Cas12. Interestingly, the TAM sequence recognized by TnpB coincides with the sequence adjacent to the left end (LE) element, which is required for TnpA-mediated transposon excision and integration (Karvelis et al. 2021). TAM requirement and complementarity to guide RNA ensure that TnpB nuclease can cleave donor joint formed after transposon excision next to the TAM sequence to initiate homologous recombination from sister chromatid resulting in TnpB-mediated transposon retention (Karvelis et al. 2021; Meers et al. 2023).

The discovery that compact TnpB proteins act as RNA-guided nucleases has significant implications for molecular biology and biotechnology. The demonstrated programmable DNA cleavage activity in eukaryotic cells makes them attractive alternatives to larger Cas9 and Cas12 nucleases for unconstrained delivery into the cells using adeno-associated viruses (AAVs) (Karvelis et al. 2021; Li et al. 2024; Marquart et al. 2024; Ren et al. 2024; Xiang et al. 2024). Furthermore, the prevalence of TnpB in prokaryotes as well as the discovery of TnpB homologs (Fanzors) in eukaryotes might provide a vast reservoir of untapped TnpB variants with diverse targeting capabilities (Altae-Tran et al. 2023; Jiang et al. 2023; Saito et al. 2023; Yoon et al. 2023).

Here, we describe a pipeline for the establishment of DNA cleavage requirements (reRNA sequence and TAM sequence) of novel TnpB proteins. The proposed approach combines bioinformatic determination for reRNA 3′-end with experimental characterization of TAM sequences that is a prerequisite for further testing of TnpB nucleases for their genome-editing potential.

2. Methods

2.1 Bioinformatic curation of TnpB and guide RNA

TnpB WP_010889991.1 from *Helicobacter pylori* HpGP-ITA-012 (accession no. CP079496.1), BAA07634.1 from *Bacillus* sp. PS3 (D38778.1), BAC07687.1 from *Thermosynechococcus elongatus* BP-1 (BA000039.2), WP_001274345.1 from *Helicobacter pylori* CPY0041 (AF189015.1), and WP_139894945.1 from *Bacillus cereus* strain FORC087 (CP029455.1), along with the previously characterized TnpB WP_010887311.1 from *Deinococcus radiodurans* ISDra2 transposon (CP015081.1) were selected for characterization.

To identify 3′-end of reRNA scaffold, retrieve DNA sequence encoding TnpB ORF with 200 nt length region downstream of the stop codon from NCBI (Sayers et al. 2022). Run the search using NCBI BLASTn in the nr/nt database (Camacho et al. 2009). If there are few sequences (hits) for the query, we recommend switching the parameters from 'megablast' to 'discontiguous megablast'.

Note. Filtering out the sequences with <20 % query coverage might reduce the background noise, interfering with the analysis.

Download the alignment and visualize using Jalview (Waterhouse, Procter, Martin, Clamp, & Barton, 2009) (or alternative tool for multiple sequence alignment visualization and analysis) and identify conserved region downstream of the TnpB stop codon. The reduced DNA sequence conservation at the 3′-end represents the intersection between the reRNA scaffold and the guiding sequence (Fig. 1B–C).

2.2 Construction of randomized TAM (7 N) plasmid library

Various strategies can be utilized for plasmid library construction to obtain degenerate 7–8 N sequence 5′ to the target sequence (Altae-Tran et al. 2021; Karvelis et al. 2015; Walton, Hsu, Joung, & Kleinstiver, 2021). We adopted protocol for TAM plasmid library construction described in detail elsewhere (Karvelis, Young, & Siksnys, 2019a, 2019b) as it allows to produce the plasmid library with uniform representation at the 7 N region. The sequence map of the 7 N plasmid library construct can be found at (https://benchling.com/s/seq-J6bTZYOzH9WEBpDS18iF?m=slm-L6V8z5qX7XHKCZVoO9f8).

1. Incubate 4 pools of dsDNA fragments (200 ng of each) containing SdaI restriction site and 6 bp degenerate sequence (NNNNZNN), where Z is A, T, G or C in each pool, with 1 μl of FastDigest SdaI (Thermo Fisher Scientific, FD1194) in 50 μl 1 × FD buffer for 1 h at 37 °C. Purify the reaction products using GeneJET PCR Purification Kit (Thermo Fisher Scientific, K0701). The prepared DNA fragment can be stored at −20 °C.
2. Cleave linear DNA vector pTZ57R/T (5 μg) from InsTAclone PCR Cloning Kit (Thermo Fisher Scientific, K1214, discontinued) with FastDigest *Pst*I restriction endonuclease (Thermo Fisher Scientific, FD0614) in 100 μl 1 × FD buffer for 1 h at 37 °C. Purify the DNA fragments using a GeneJET PCR Purification Kit (Thermo Fisher Scientific, K0701). The prepared vector DNA can be stored at −20 °C.
3. Ligate 1 μg of precleaved DNA vector with 80 ng of each library (molar ratio 1:3) in a 100 μl reaction volume for 30–60 min at room temperature using a Rapid DNA Ligation Kit (Thermo Fisher Scientific, K1422).
4. Transform *E. coli* DH5α Ca^{2+} competent cells with the ligation mixtures. Recover the cells for 1 h at 37 °C in LB media and transfer to 250 ml of LB media supplemented with carbenicillin (Carl Roth, 6344.1). Incubate overnight at 37 °C with vigorous shaking.
5. Purify the plasmid DNA from the overnight cultures using a GeneJET Plasmid Maxiprep Kit (Thermo Fisher Scientific, K0491).
6. Combine equimolar amounts of 4 libraries to obtain 7 N plasmid library. The obtained plasmid library can be stored at −20 °C.

2.3 TnpB and reRNA expression and complex assembly

For rapid TnpB complex characterization *in vitro*, TnpB and guide RNA complex can be obtained using the cell-free *in vitro* transcription translation (IVTT) reaction to co-express complex components which allows to form the TnpB and guide RNA RNP complex directly in the reaction mixture. Alternatively, lysates from *E. coli* or HEK293T cells expressing TnpB complex components, or purified TnpB complex can also be used (Karvelis, Young, & Siksnys, 2019a, 2019b; Walton et al. 2021). The TnpB PCR template corresponds to the protein's ORF, while a 250-nucleotide sequence upstream of the 3′-end of the reRNA is used as the template for the TnpB guide RNA.

1. The sequences encoding TnpB protein and guide RNA are amplified by PCR using Phusion Hot Start II DNA Polymerase (Thermo Fisher Scientific, F579L) and an individual set of primers to add additional sequences. *tnpB* gene is PCR amplified using pair of oligos 5′-GCGAAT TAATACGACTCACTATAGGGCTTAAGTATAAGGAGGAAAA AAT… bases matching *tnpb* 5′-end starting with ATG…−3′ and 5′-AAACCCCTCCGTTTAGAGAGGGGTTATGCTAG…bases matching *tnpb* 3′-end including stop codon-3′ to add T7 promoter, RBS and T7-like terminator, respectively. The reRNA encoding fragment is PCR amplified using 5′-GCGAATTAATACGACTCACTATAGGG… bases matching guide RNA 5′-end…−3′ and 5′-CCGGCGACGTTGG GTCAACT…bases matching guide RNA 3′-end…−3′ to add T7 promoter and 20 nt length plasmid library targeting sequence, respectively.
2. Purify the obtained PCR fragments using GeneJET PCR Purification Kit (Thermo Fisher Scientific, K0701). The samples can be stored at −20 °C.
3. For cell-free synthesis of TnpB and reRNA the PURExpress *In Vitro* Protein Synthesis Kit (New England Biolabs, E6800L) is used. Initiate the *in vitro* protein synthesis reactions by mixing TnpB and reRNA encoding DNA templates (125 ng of each) with 5 μl of Solution A, 3.75 μl Solution B, 1 U/μl RiboLock RNase Inhibitor (Thermo Fisher Scientific, EO0381) in the final volume of 12.5 μl.
4. Incubate IVTT reaction for 2 h at 37 °C. After incubation transfer the tubes on ice for short-term storage until use (2.4. paragraph) (Fig. 2A).

2.4 Plasmid library cleavage

1. Mix 5 μl of TnpB RNPs containing solution from IVTT (2.3 paragraph) with 500 ng of the 7 N randomized plasmid library (2.2. paragraph) in 50 μl of reaction buffer, containing 10 mM Tris-HCl (pH 7.5 at 37 °C), 100 mM NaCl, 10 mM $MgCl_2$, 1 mM DTT. For the negative control sample, the plasmid library is incubated with IVTT mix without added DNA templates.
2. Incubate the reaction mixture for 1 h at 37 °C. We typically use PCR tubes and incubate in PCR cycler with heated lid to prevent evaporation.
3. Inactivate the cleavage reaction by incubating the samples at 75 °C for 10 min

2.5 Capture of cleavage products *via* adapter ligation

1. Add 0.5 μl of T4 DNA Polymerase (5 U/μl) (Thermo Fisher Scientific, EP0061), 0.5 μl of 10 mM dNTP mix (Thermo Fisher Scientific, R0191) directly into the previous reaction and incubate at 11 °C for

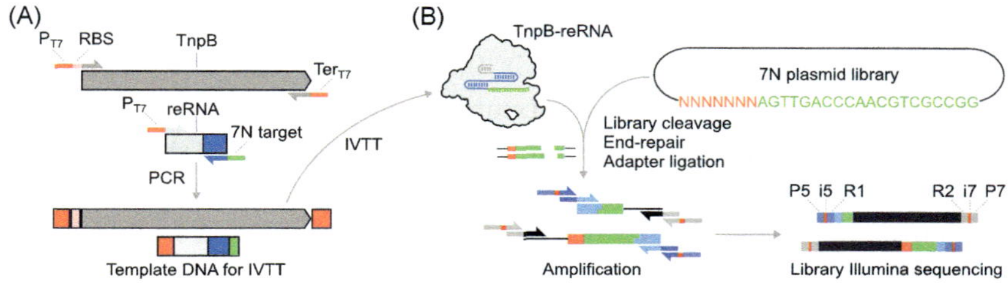

Fig. 2 Workflow for experimental TAM determination of TnpB variants. (A) Cell-free synthesis of the TnpB-reRNA RNP complex. TnpB and reRNA templates for *in vitro* transcription-translation (IVTT) are generated by PCR using oligonucleotides designed to introduce the following elements: a T7 promoter, an RBS, and a T7 terminator for TnpB; and a T7 promoter and a 7 N library complementary target for reRNA. These templates are then mixed with IVTT components to produce the TnpB-reRNA RNP complexes. (B) 7 N library cleavage and NGS library preparation. The IVTT samples containing TnpB-reRNA RNP are combined with a 7 N plasmid library in a cleavage reaction mix. Cleavage of the 7 N plasmid results in a variety of products with various lengths and overhangs. The sticky ends of the cleavage products are processed to add 3′-dA, allowing ligation to a pre-annealed 3′-dT overhang-containing adapter. The cleavage products are then amplified upstream and downstream of the cleavage site. Two rounds of PCR amplification are carried out to introduce the Illumina Read1 and Read2 sequences followed by addition of unique Illumina i5 and i7 indexes to each sample. The PCR products are pooled to create a library for next generation sequencing (NGS), which is performed on Illumina MiniSeq platform. A detailed description of the method is provided in Sections 2.3–2.6.

20 min followed by heat inactivation for 10 min at 75 °C to repair the DNA ends after the cleavage.

2. Add 0.5 μl of DreamTaq polymerase (5 U/μl) (Thermo Fisher Scientific, EP0705) and 0.5 μl of 10 mM dATP (Thermo Fisher Scientific) to the reaction mixture and incubate for 30 min at 72 °C to add 3′-dA overhangs.
3. To remove the remaining RNA, add 1 μl of RNase A/T1 Mix (Thermo Fisher Scientific, EN0551) and incubate for 15 min at 37 °C.
4. Purify the reaction products using GeneJET PCR Purification Kit (Thermo Fisher Scientific, K0701).
5. To prepare the adapter incubate 2.5 μM of oligonucleotide 5′-GATC GGAAGAGCGGTTCAGCAGGAATGCCG-3′ with 1 μl of T4 Polynucleotide Kinase (10 U/μl) (Thermo Fisher Scientific, EK0031), 10 mM ATP (Thermo Fisher Scientific, R0441), and 1 × Reaction Buffer A (Thermo Fisher Scientific) in the total volume of 100 μl. Incubate the mixture at 37 °C for 30 min followed by inactivation at 75 °C for 10 min. Mix 0.5 μM of phosphorylated oligonucleotide with 0.5 μM of reverse oligonucleotide 5′-CGGCATTCCTGCTGAACC

GCTCTTCCGATCT-3′ in 500 μl of the annealing solution containing 5 mM Tris-HCl (pH 7.5 at 37 °C), 0.5 mM EDTA, and 50 mM NaCl, and anneal by incubating the mixture in the thermocycler at 95 °C for 5 min and cooling down to room temperature (1 °C/10 s).

6. Ligate 100 ng 3′-dT overhang containing dsDNA adapters to the 100 ng of purified 3′-dA-tailed cleavage products using 1 μl of T4 DNA Ligase (Thermo Fisher Scientific, EL0011) in 20 μl 1 × T4 DNA ligase buffer (Thermo Fisher Scientific) supplemented with 5% of PEG4000 (Thermo Fisher Scientific). Incubate ligation mixture for 1 h at 22 °C. Inactivate the ligase by incubating the samples for 10 min at 75 °C. The adapter ligated cleavage products can be stored at −20 °C.

2.6 Illumina library generation and deep sequencing

1. Amplify the adapter containing DNA fragments by performing 12 cycles of PCR using Phusion Hot Start II DNA Polymerase (Thermo Fisher Scientific, F579L), and 2 μl of the adapter ligation reaction mixture as a template. For amplification of the cleavage product libraries we use 5′-ACACTCTTTCCCTACACGACGCTCTTCCGATCTCGGCA TTCCTGCTGAAC-3′ primer specific to the ligated adapter in combination with 5′-TTCCTTGGCACCCGAGAATTCCAACTATAG GGAAAGCTTGCATG-3′ for determination of the TAM sequence and target strand (TS) cleavage position and 5′-TTCCTTGGCACCC GAGAATTCCATGCTGCAAGGCGATTAAG-3′ for determination of the non-target strand (NTS) cleavage site, which is complementary to the 7 N plasmid library backbone (the introduced Read1 (R1) and Read2 (R2) sequences are underlined). As the negative control, the 7 N plasmid library incubated with only the IVTT mix components (as described in paragraph 2.4) and subjected to PCR amplification as well. The initial constructed 7 N library is PCR amplified using 5′-ACAC TCTTTCCCTACACGACGCTCTTCCGATCTTCGGTACCTCG CGAATG-3′ and 5′-TTCCTTGGCACCCGAGAATTCCAACTAT AGGGAAAGCTTGCATG-3′ primers which are complementary to plasmid backbone upstream and downstream the 7 N region (R1 and R2 sequences are underlined). This control reaction allows the assessment of nucleotide biases in TAM region that may occur during library preparation and is subtracted from individual samples during the analysis of NGS data. Note. To avoid library homogeneity, we suggest using a mixture of eight equimolarly mixed R1 oligonucleotides, each containing an insertion of 0 to 7 nucleotides between the Read1

(5′-ACACTCTTTCCCTACACGACGCTCTTCCGATCT-3′) and the adapter complementary (5′-CGGCATTCCTGCTGAAC-3′) sequences. A similar strategy could be used for the amplification of the 7 N library control sample where the mixture of oligonucleotides containing an 0–7 nt insertion between the Read1 and the vector backbone complementary sequence. To incorporate Illumina P5 and P7 adapter sequences, and i5 and i7 indexes, perform the 2nd PCR amplification of 20 cycles, using a different combination of custom 5′-AATGATACGGCGACCACCGAGATCTACAC-i5-ACACTCTTTCCCTACACGACGCTCTTCCGATCT-3′ (complementary to R1, P5 sequence underlined) and 5′-CAAGCAGAAGACGGCATACGAGAT-i7-GTGACTGGAGTTCCTTGGCACCCGAGAATTCCA-3′ (complementary to R2, P7 sequence underlined) primers for each sample (i5 and i7 represent unique 8 nt length Illumina sequencing indexes). The PCR is performed using Phusion Hot Start II DNA Polymerase (Thermo Fisher Scientific, F579L) and 1/20 of reaction volume from 1st PCR as a template.

2. Run samples on 1 % agarose gel. In the case of multiple samples, quantify the intensity of the fragments using ImageJ (Schneider, Rasband, & Eliceiri, 2012), pool accordingly in an equimolar ratio, and re-run the electrophoresis. Extract fragments in the size range of interest from agarose gel and purify using GeneJET Gel Extraction Kit (Thermo Fisher Scientific, K0691). Quantify with Qubit 4 Fluorimeter using Qubit 1X dsDNA HS Assay Kit (Thermo Fisher Scientific, Q33230). Additionally, final quality check can be performed by running purified DNA fragments on Agilent 2100 Bioanalyzer (Agilent) using Agilent High Sensitivity DNA Kit (5067-4626).

Run pair-end sequencing (2 × 75 bp) of the library using MiniSeq High Output Reagent Kit (150-cycles) (Illumina, FC-420–1002) on a MiniSeq System (Illumina) according to manufacturer's recommended instructions.

2.7 Cleavage position and TAM identification

The custom Python scripts for cleavage position and TAM identification are based on analyses described elsewhere (Bigelyte et al. 2024). Briefly, the obtained sequences are analyzed to detect, and count reads with adapter ligated at positions 1 to 30 bp from the 5′ TAM sequence. The significant enrichment of such reads (in comparison with negative control) indicates the cleavage by TnpB as well as preferred cleavage position at targeted and non-targeted DNA strands (Fig. 3).

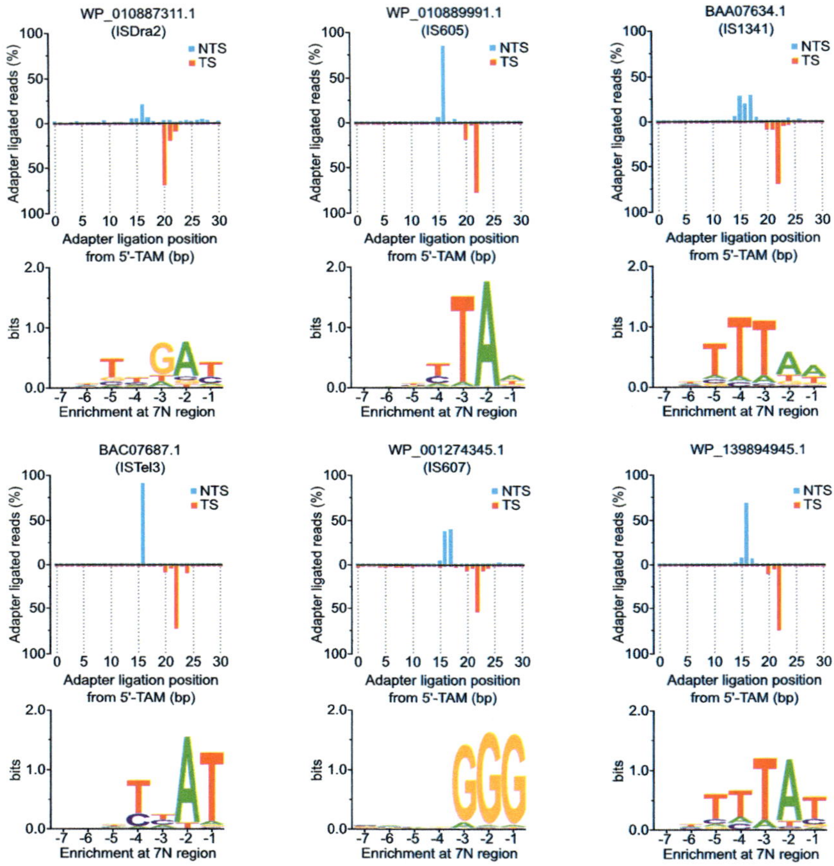

Fig. 3 Cleavage position and TAM sequences of TnpB orthologs. For each TnpB variant, the cleavage site (shown above) and the TAM sequence (shown below) were determined by analyzing NGS data from 7 N library cleavage samples. The cleavage of all selected TnpB homologs resulted in 5′- sticky ends located 15–22 nucleotides away from the TAM sequence. TS – target strand, NTS – non-target strand.

Next, the TAM sequences are extracted from sequenced reads. To eliminate the inherent biases in the constructed TAM plasmid library, normalization against the original 7 N PAM library is performed. To minimize the impact of background noise only TAM sequences with over 5-fold enrichment compared to the initial library are selected. Finally, the extracted TAM sequences are presented in WebLogo format (Fig. 3) (Crooks, 2004).

3. Concluding remarks

The CRISPR-Cas RNA-guided DNA endonucleases revolutionized the genome editing field. However, their large size and PAM sequence requirements constrain their delivery using AAVs and targetable sequences, respectively (Doudna, 2020). Recently, transposon-encoded RNA-guided nucleases have emerged as a new class of compact RNA-guided genome editors with the potential to address these challenges. Nonetheless, only a limited number of TnpB proteins with restricted TAM sequence variations showed activity in human cells (Karvelis et al. 2021; Li et al. 2024; Marquart et al. 2024; Ren et al. 2024; Xiang et al. 2024). Therefore, the characterization of novel TnpB variants could expand the TnpB-based toolbox for genome editing.

In this chapter we provide a user-friendly protocol for the characterization of TnpB orthologs. As proof of principle, we experimentally characterized five novel TnpB variants along with previously characterized TnpB from *D. radiodurans* (ISDra2) identifying their reRNA sequence, cleavage positions and diverse TAM requirements licensing the cleavage. The described workflow could also be adopted for characterization of TnpB homologs found in eukaryotes (Fanzors) as well as IscB proteins. Therefore, we believe that this chapter provides detailed guidelines for the researchers interested in characterization of novel RNA-guided TnpB/Fanzor nucleases paving the way for development of novel genome editing tools.

Acknowledgments

T.K. is supported by Central Project Management Agency, Lithuania [02002-P-0001].

References

Altae-Tran, H., Kannan, S., Demircioglu, F. E., Oshiro, R., Nety, S. P., McKay, L. J., Dlakić, M., et al. (2021). The widespread IS200/IS605 transposon family encodes diverse programmable RNA-guided endonucleases. *Science (New York, N. Y.), 374*(6563), 57–65. https://doi.org/10.1126/science.abj6856.

Altae-Tran, H., Shmakov, S. A., Makarova, K. S., Wolf, Y. I., Kannan, S., Zhang, F., & Koonin, E. V. (2023). Diversity, evolution, and classification of the RNA-guided nucleases TnpB and Cas12. *Proceedings of the National Academy of Sciences, 120*(48), e2308224120. https://doi.org/10.1073/pnas.2308224120.

Bigelyte, G., Duchovska, B., Zedaveinyte, R., Sasnauskas, G., Sinkunas, T., Dalgediene, I., Tamulaitiene, G., et al. (2024). Innate programmable DNA binding by CRISPR-Cas12m effectors enable efficient base editing. *Nucleic Acids Research, 52*(6), 3234–3248. https://doi.org/10.1093/nar/gkae016.

Camacho, C., Coulouris, G., Avagyan, V., Ma, N., Papadopoulos, J., Bealer, K., & Madden, T. L. (2009). BLAST+: Architecture and applications. *BMC Bioinformatics, 10*(1), 421. https://doi.org/10.1186/1471-2105-10-421.

Crooks, G. E. (2004). WebLogo: A sequence logo generator. *Genome Research, 14*(6), 1188–1190. https://doi.org/10.1101/gr.849004.

Doudna, J. A. (2020). The promise and challenge of therapeutic genome editing. *Nature, 578*(7794), 229–236. https://doi.org/10.1038/s41586-020-1978-5.

Frost, L. S., Leplae, R., Summers, A. O., & Toussaint, A. (2005). Mobile genetic elements: The agents of open source evolution. *Nature Reviews. Microbiology, 3*(9), 722–732. https://doi.org/10.1038/nrmicro1235.

He, S., Corneloup, A., Guynet, C., Lavatine, L., Caumont-Sarcos, A., Siguier, P., Marty, B., et al. (2015). The IS200/IS605 family and 'Peel and Paste' single-strand transposition mechanism. *Microbiology Spectrum, 3*(4), 3.4.02. https://doi.org/10.1128/microbiolspec.MDNA3-0039-2014.

Jiang, K., Lim, J., Sgrizzi, S., Trinh, M., Kayabolen, A., Yutin, N., Bao, W., et al. (2023). Programmable RNA-guided DNA endonucleases are widespread in eukaryotes and their viruses. *Science Advances, 9*(39), eadk0171. https://doi.org/10.1126/sciadv.adk0171.

Karvelis, T., Druteika, G., Bigelyte, G., Budre, K., Zedaveinyte, R., Silanskas, A., ... Siksnys, V. (2021). Transposon-associated TnpB Is a programmable RNA-guided DNA endonuclease. *Nature, 599*(7886), 692–696. https://doi.org/10.1038/s41586-021-04058-1.

Karvelis, T., Gasiunas, G., Young, J., Bigelyte, G., Silanskas, A., Cigan, M., & Siksnys, V. (2015). Rapid characterization of CRISPR-Cas9 protospacer adjacent motif sequence elements. *Genome Biology, 16*(1), 253. https://doi.org/10.1186/s13059-015-0818-7.

Karvelis, T., Young, J. K., & Siksnys, V. (2019a). *A pipeline for characterization of novel Cas9 orthologs. Methods in Enzymology*. Elsevier219–240. https://doi.org/10.1016/bs.mie.2018.10.021.

Karvelis, T., Young, J. K., & Siksnys, V. (2019b). *Chapter ten—A pipeline for characterization of novel Cas9 orthologs. Methods in enzymology, CRISPR-Cas enzymes*. Scott Bailey: Academic Press, 219–240. https://doi.org/10.1016/bs.mie.2018.10.021.

Koonin, E. V., Makarova, K. S., Wolf, Y. I., & Krupovic, M. (2020). Evolutionary entanglement of mobile genetic elements and host defence systems: Guns for hire. *Nature Reviews. Genetics, 21*(2), 119–131. https://doi.org/10.1038/s41576-019-0172-9.

Li, Z., Guo, R., Sun, X., Li, G., Shao, Z., Huo, X., Yang, R., et al. (2024). Engineering a transposon-associated TnpB-ωRNA system for efficient gene editing and phenotypic correction of a tyrosinaemia mouse model. *Nature Communications, 15*(1), 831. https://doi.org/10.1038/s41467-024-45197-z.

Marquart, K. F., Mathis, N., Mollaysa, A., Müller, S., Kissling, L., Rothgangl, T., Schmidheini, L., et al. (2024). Effective genome editing with an enhanced ISDra2 TnpB system and deep learning-predicted ωRNAs. *Nature Methods*. https://doi.org/10.1038/s41592-024-02418-z.

Meers, C., Le, H. C., Pesari, S. R., Hoffmann, F. T., Walker, M. W. G., Gezelle, J., ... Sternberg, S. H. (2023). Transposon-encoded nucleases use guide RNAs to promote their selfish spread. *Nature, 622*(7984), 863–871. https://doi.org/10.1038/s41586-023-06597-1.

Ren, K., Zhou, F., Zhang, F., Yin, M., Zhu, Y., Wang, S., Chen, Y., et al. (2024). Discovery and structural mechanism of DNA endonucleases guided by RAGATH-18-derived RNAs. *Cell Research: 1–16*. https://doi.org/10.1038/s41422-024-00952-1.

Saito, M., Xu, P., Faure, G., Maguire, S., Kannan, S., Altae-Tran, H., Vo, S., et al. (2023). Fanzor Is a eukaryotic programmable RNA-guided endonuclease. *Nature, 620*(7974), 660–668. https://doi.org/10.1038/s41586-023-06356-2.

Sayers, E. W., Bolton, E. E., Brister, J. R., Canese, K., Chan, J., Comeau, D. C., Connor, R., et al. (2022). Database resources of the national center for biotechnology information. *Nucleic Acids Research, 50*(D1), D20–D26. https://doi.org/10.1093/nar/gkab1112.

Schneider, C. A., Rasband, W. S., & Eliceiri, K. W. (2012). NIH image to ImageJ: 25 Years of image analysis. *Nature Methods, 9*(7), 671–675. https://doi.org/10.1038/nmeth.2089.

Shmakov, S., Abudayyeh, O. O., Makarova, K. S., Wolf, Y. I., Gootenberg, J. S., Semenova, E., Minakhin, L., et al. (2015). Discovery and functional characterization of diverse class 2 CRISPR-Cas systems. *Molecular Cell, 60*(3), 385–397. https://doi.org/10.1016/j.molcel.2015.10.008.

Shmakov, S., Smargon, A., Scott, D., Cox, D., Pyzocha, N., Yan, W., Abudayyeh, O. O., et al. (2017). Diversity and evolution of class 2 CRISPR–Cas systems. *Nature Reviews. Microbiology, 15*(3), 169–182. https://doi.org/10.1038/nrmicro.2016.184.

Walton, R. T., Hsu, J. Y., Joung, J. K., & Kleinstiver, B. P. (2021). Scalable characterization of the PAM requirements of CRISPR-Cas enzymes using HT-PAMDA. *Nature Protocols, 16*(3), 1511–1547. https://doi.org/10.1038/s41596-020-00465-2.

Waterhouse, A. M., Procter, J. B., Martin, D. M. A., Clamp, M., & Barton, G. J. (2009). Jalview version 2—A multiple sequence alignment editor and analysis workbench. *Bioinformatics, 25*(9), 1189–1191. https://doi.org/10.1093/bioinformatics/btp033.

Xiang, G., Li, Y., Sun, J., Huo, Y., Cao, S., Cao, Y., Guo, Y., et al. (2024). Evolutionary mining and functional characterization of TnpB nucleases identify efficient miniature genome editors. *Nature Biotechnology, 42*(5), 745–757. https://doi.org/10.1038/s41587-023-01857-x.

Yoon, P. H., Skopintsev, P., Shi, H., Chen, L. X., Adler, B. A., Al-Shimary, M., Craig, R. J., et al. (2023). Eukaryotic RNA-guided endonucleases evolved from a unique clade of bacterial enzymes. *Nucleic Acids Research, 51*(22), 12414–12427. https://doi.org/10.1093/nar/gkad1053.

CHAPTER NINE

Hydrolytic endonucleolytic ribozyme (HYER): Systematic identification, characterization and potential application in nucleic acid manipulation

Zi-Xian Liu* **and Jun-Jie Gogo Liu***

Beijing Frontier Research Center for Biological Structure, Center of Synthetic and Systems Biology, State Key Laboratory of Membrane Biology, Tsinghua-Peking Center for Life Sciences, School of Life Sciences, Tsinghua University, Beijing, P.R. China

*Corresponding authors. e-mail address: liuzx21@mails.tsinghua.edu.cn; junjiegogoliu@tsinghua.edu.cn

Contents

1. Introduction 198
2. Identification of GII introns 200
3. The structural studies and the mechanisms of GII intron splicing and propagation 201
4. GII intron-derived tools for DNA manipulation 203
5. Protocol 204
   5.1 Materials 204
   5.2 Bioinformatics pipeline 207
   5.3 *In vitro* DNA cleavage activity characterization 208
   5.4 *In vivo* activity characterization 212
   5.5 Structure determination of HYERs 217
   5.6 Engineering and optimization of HYER1 218
6. Perspectives on the application of HYERs 219

Acknowledgments 220

References 220

Abstract

Group II introns are transposable elements that can propagate in host genomes through the "copy and paste" mechanism. They usually comprise RNA and protein components for effective propagation. Recently, we found that some bacterial GII-C introns without protein components had multiple copies in their resident genomes, implicating their potential transposition activity. We demonstrated that some of these systems are active for hydrolytic DNA cleavage and proved their DNA manipulation capability in bacterial or mammalian cells. These introns are therefore named HYdrolytic Endonucleolytic Ribozymes (HYERs). Here, we provide a detailed protocol

Methods in Enzymology, Volume 712

ISSN 0076-6879, https://doi.org/10.1016/bs.mie.2025.01.033

for the systematic identification and characterization of HYERs and present our perspectives on its potential application in nucleic acid manipulation.

1. Introduction

GII intron is a retroelement comprising an RNA and a protein component. The RNA component consists of six domains and is responsible for multiple essential functions: catalyzing the RNA splicing reaction, encoding the intron-encoded protein (IEP) with an open reading frame (ORF), and cooperating with the IEP to propagate in the host genome (Haack et al., 2019; Marcia & Pyle, 2012; Toor, Keating, Taylor, & Pyle, 2008). Among all the functional domains, domain I (DI) is the largest and has multiple stem-loops, serving as a structural scaffold. Notably, the exon binding site (EBS) crucial for substrate recognition *via* base pairings resides in DI. Domain II and III (DII and DIII) are also important structural elements with many long-range interaction motifs involved in RNA folding. Domain IV (DIV) often contains an ORF for IEP encoding. The nucleotides responsible for coordinating Mg^{2+} ions to form the catalytic core reside in domain V (DV). Domain VI (DVI) contains the branch-point adenosine (A) for the transesterification reaction in splicing (Fig. 1A and B) (Chan et al., 2018; Chung et al., 2022; Haack et al., 2019; Haack, Rudolfs, Zhang, Lyumkis, & Toor, 2024; Manigrasso et al., 2020; Marcia & Pyle, 2012; Toor, Keating, et al., 2008).

The protein component of the GII intron often contains a DNA endonuclease domain (EN), a DNA-binding domain (D), a maturase domain (X), and a reverse transcriptase domain (RT) (Fig. 1A). The splicing and propagation of GII introns rely on the intricate coordination of these two components (Lambowitz & Zimmerly, 2004, 2011; Pyle, 2016; Yang, Zimmerly, Perlman, & Lambowitz, 1996; Zimmerly & Semper, 2015).

Various GII introns have been identified in bacteria, archaea, mitochondria, and chloroplasts (Candales et al., 2012; Dai & Zimmerly, 2003; Dai, Toor, Olson, Keeping, & Zimmerly, 2003; Simon et al., 2008). Based on the RNA components, GII introns can be divided into A, B, and C classes. These classes of GII introns have distinct RNA sequences, structural characteristics and substrate recognition mechanisms. Specifically, GII-A introns recognize exon sequences using EBS1, EBS2 and δ motif, while GII-B and GII-C introns use EBS1-EBS2-EBS3 combination and EBS1-EBS3

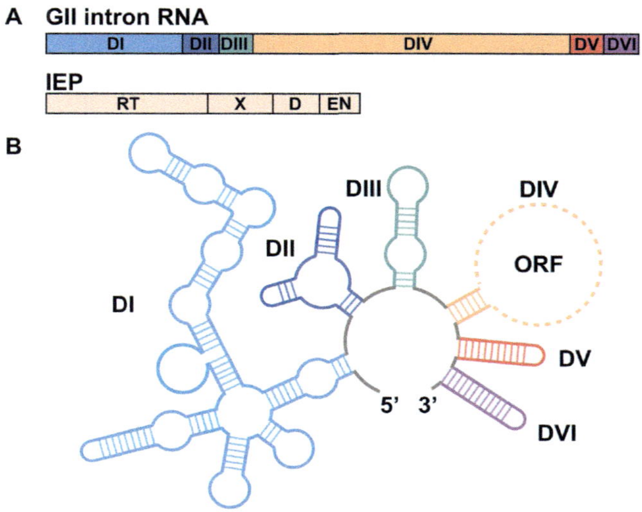

Fig. 1 Schematics of a canonical GII intron. (A) The RNA and IEP domain organization of a canonical GII intron. DI is colored in blue, DII in dark blue, DIII in cyan, DIV in light orange, DV in red, and DVI in purple. The IEP domains are labeled in the panel. (B) The secondary structure model of a canonical GII-C intron. Domains labeled in the panel and colored according to (A).

combination respectively (Chan et al., 2018; Lambowitz & Zimmerly, 2011; Marcia & Pyle, 2012; Pyle, 2016; Qu et al., 2016; Toor, Rajashankar, Keating, & Pyle, 2008). Within these three classes of GII introns, bacterial GII-C introns are considered more ancestral and have around half the size of GII-A and GII-B introns (Lambowitz & Zimmerly, 2004; Rest & Mindell, 2003; Simon et al., 2008).

Recently, we discovered a group of uncharacterized GII-C introns with no ORF embedded in their RNA sequence and, therefore, do not encode IEP in the genome database from the Earth's Microbiomes catalog (Nayfach et al., 2021). Interestingly, we noticed that some ORF-less GII-C introns have more than one copy in the genome, presumably indicating their capability to cleave the target DNA to facilitate intron mobility. Further analysis showed that some of these multicopy ORF-less GII-C introns are active for site-specific DNA and RNA cleavage *in vitro* through the hydrolysis mechanism. Therefore, they are named hydrolytic endonucleolytic ribozymes (HYERs). We also demonstrated HYERs' capability to deplete plasmids in the bacterial cells and introduce the genomic DNA breaks in the mammalian cells, thus shedding light on a potential RNA-based platform for nucleic acid manipulation (Liu et al., 2024). Here, we briefly summarize the previous study on GII introns and present the methods we have employed in our research.

2. Identification of GII introns

Previously, profound work has been carried out to systematically identify GII introns in bacteria and eukaryotic organelles, leading to comprehensive studies on their mechanisms, structures, and evolutionary trajectories (Candales et al., 2012; Dai & Zimmerly, 2003; Dai et al., 2003; Simon et al., 2008). Within the GII introns cataloged in databases, many are canonical systems with ORF in DIV, while systems that lack ORF and do not encode IEP have also been discovered. Notably, all 42 ORF-less introns belong exclusively to the GII-A and GII-B classes (Candales et al., 2012; Dai & Zimmerly, 2003; Dai et al., 2003; Simon et al., 2008). The existence of ORF-less GII-C introns remains elusive in the previous studies.

In our study, driven by the hypothesis that ORF-less GII-C introns with multiple copies in the genome are presumably capable of cleaving the target DNA to facilitate the intron propagation, we established a bioinformatics pipeline to systematically discover multicopy ORF-less GII-C introns in the bacterial genome database and further characterized their biochemical activities to obtain ribozymes for DNA cleavage. First, we collected the 92 reported ORF-containing GII-C introns from the GII intron databases (Candales et al., 2012; Dai et al., 2003; Nawrocki & Eddy, 2013; Simon et al., 2008). These sequences are divided into three parts for further analysis: DI-DIII, ORF region, and DV-DVI. Two covariance models were constructed respectively for DI-DIII and DV-DVI based on RNA sequence and secondary structure. Then, we searched in the genomes from Earth's Microbiomes catalog and selected candidates matching both covariance models with high confidence (Nayfach et al., 2021). Next, we clustered highly similar copies in the same species-level operational taxonomic units (OTUs) to find candidates with more than one copy and obtained 469 multicopy GII-C introns in the genomes. Furthermore, we utilized the reported ORF-containing GII-C systems to construct the hidden Markov model to identify the candidates without ORFs (Finn, Clements, & Eddy, 2011). Finally, we choose candidates with recognizable target recognition sites (TRSs) and presumably stable structure for further experimental characterization (Liu et al., 2024).

Our efforts identified 93 multicopy ORF-less GII-C introns in total, supplementing the GII intron families with novel subgroups. Furthermore, among the nine candidates selectively characterized through wet-lab experiments, a significant proportion of six demonstrated DNA cleavage activity *in vitro*, underscoring the paramount importance of comprehensively elucidating the activities and functions of GII introns (Liu et al., 2024).

3. The structural studies and the mechanisms of GII intron splicing and propagation

During the past a few decades, efforts to determine the RNA structure have yielded many high-resolution RNA structures, boosting the progress on unveiling the folding, dynamic patterns, and mechanisms of RNA machineries and expanding the field's understanding of RNA functions in multiple biological processes. The very first RNA three-dimensional (3D) structure is the phenylalanine tRNA structure determined in the early 1970s using X-ray crystallography (Kim et al., 1973; Robertus et al., 1974). In the following half of the century, many methods have been used to study the fascinating structures of RNA, such as nuclear magnetic resonance (NMR) (Barnwal, Yang, & Varani, 2017; Wacker et al., 2020), cryogenic electron microscopy (cryo-EM) (Chung et al., 2022; Haack et al., 2019; Liu, Thelot, Piccirilli, Liao, & Yin, 2022; Liu et al., 2024; Luo et al., 2023), and small angle X-ray scattering (SAXS) (Bhandari et al., 2017; Funari et al., 2000).

The milestone 3D structure of the GII intron is characterized by the Pyle group in 2008 (Toor, Keating, et al., 2008). In this work, the structure of a GII-C intron in the linear form generated through hydrolytic self-splicing *in vitro* is solved by X-ray crystallography. Although this work employed a truncated construct of the GII-C intron for crystallization, the high-resolution overall structure and detailed interactions within and between domains provided the field with novel insights into the intricate folding and mechanisms of GII introns. Following this work, more structural studies have been carried out. Combining the structural insights, biochemical and *in vivo* assays, researchers made rapid progress in elucidating the splicing and propagation processes.

The self-splicing process of GII introns is an important example of RNA catalysis, involving several distinct steps orchestrated by multiple structural elements within the intron RNA. GII introns typically fold into a complex structure and bind the 5' exon through base pairings using the

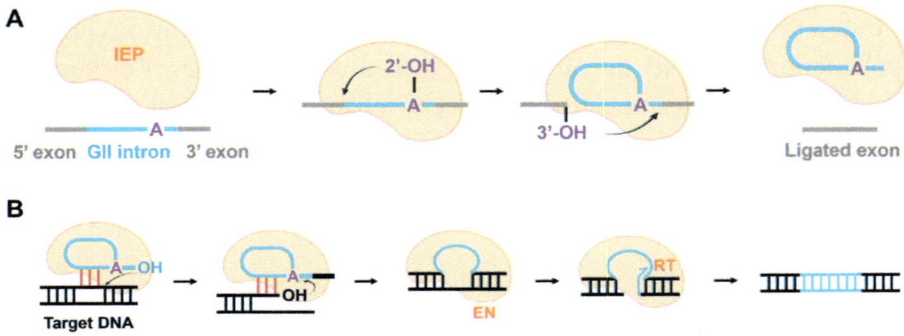

Fig. 2 Canonical mechanisms of a GII intron's self-splicing and retrotransposition processes. (A) The self-splicing process of a canonical GII intron. The IEP is colored in light orange. The GII intron is colored in blue, and its flanking 5′ and 3′ exons are colored in gray. The branch-point A is marked. (B) The retrotransposition process of a canonical GII intron. The RNA, IEP, and branch-point A are colored or labeled according to (A). The DNA for intron integration is colored in black. Base pairs between the DNA target site in the genome and the GII intron are colored in yellow. The EN and RT domains are labeled separately.

EBS in DI (for GII-A introns, the δ motif is also involved in substrate recognition), which brings the 5' exon and the branch-point A to the catalytic core. Then, the pre-RNA is spliced *via* the two-step splicing reaction. In the first step, the 2'-OH of the branch-point A attacks the phosphodiester bond at the 5' splice site, resulting in the formation of a lariat intermediate. In the second step, the 3' -OH of 5' exon attacks the phosphodiester bond at the 3' splice site, leading to the ligation of flanking exons and the release of the intron RNA in the lariat form (Fig. 2A) (Chan et al., 2018; Haack et al., 2024; Lambowitz & Zimmerly, 2004, 2011; Manigrasso et al., 2020; Marcia & Pyle, 2012; Peebles et al., 1986; Pyle, 2010, 2016; Toor, Robart, Christianson, & Zimmerly, 2006; Toor, Keating, et al., 2008; Zhao & Pyle, 2017).

The retrotransposition or propagation process of GII introns involves the cooperation of RNA and protein components, enabling them to insert into new genomic loci. After the self-splicing of intron RNA, the lariat product can form ribonucleoprotein (RNP) complex with IEP and recognize specific DNA sequences at target sites. The intron RNA is integrated into the sense strand of DNA through two sequential transesterification reactions with the assistance of IEP. Then, the IEP cleaves the antisense strand with EN domain and reverse transcribes the intron RNA into complementary DNA (cDNA) with RT domain through target-primed reverse transcription (TPRT). Finally, *via* the host's DNA repair pathways, the DNA copy of the intron is

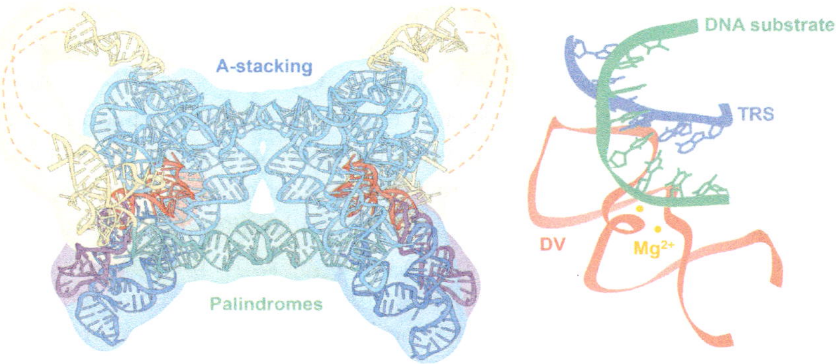

Fig. 3 The overall structure and catalytic core of HYER1. Left, the atomic model of HYER1 fitted into the transparent map. Domains labeled in the panel and colored according to Fig. 1. The homodimerization of HYER1 is mediated by the A-stacking interactions in DI and palindromes in DIII, which are labeled in the figure. Right, the catalytic core of HYER1 for hydrolytic DNA cleavage. The DNA substrate (green) is recognized and captured by the TRS (blue) in DI. Two Mg^{2+} ions in the catalytic core are colored in orange.

inserted into the new DNA target site (Fig. 2B) (Chung et al., 2022; Cousineau et al., 1998; Haack et al., 2019; Lambowitz & Zimmerly, 2004, 2011; Marcia & Pyle, 2012; Yang et al., 1996; Zimmerly, Guo, Eskes, et al., 1995; Zimmerly, Guo, Perlman, & Lambowitz, 1995).

Interestingly, aside from their well-documented splicing and propagation through transesterification mechanisms, prior research has demonstrated their capacity to achieve self-splicing *via* hydrolysis rather than transesterification, illuminating another dimension of GII introns' versatility. This alternative pathway leads to the generation of linear intron RNA (Granlund, Michel, & Norgren, 2001; Marcia & Pyle, 2012; Podar, Chu, Pyle, & Perlman, 1998; Toor et al., 2006; Toor, Keating, et al., 2008). In our study, we uncovered GII intron's DNA cleavage capability *via* the hydrolytic pathway. Structural analysis revealed the homodimerized structure of HYER1, in which the DNA substrate is recognized and captured by the TRS within the catalytic core involving two Mg^{2+} ions for DNA hydrolysis (Fig. 3) (Liu et al., 2024).

4. GII intron-derived tools for DNA manipulation

Leveraging the retrotransposition capability of GII introns, researchers have developed a site-specific DNA integration tool, Targetron, for gene manipulation (Enyeart, Mohr, Ellington, & Lambowitz, 2014; Guo et al., 2000; Karberg et al., 2001; Yao, Zhong, & Lambowitz, 2005; Zhong, Karberg, & Lambowitz, 2003). A synthetic ORF-deleted GII intron gene and an IEP gene are placed at separate expression sites in the plasmid vector. After the delivery of this vector into the target host cells, the GII intron RNA and IEP protein are expressed and subsequently assemble into a functional RNP complex. This complex then identifies the DNA target site and integrates the intron sequence into the target gene, thereby disrupting its function. However, the integration of exogenous sequences into the genome becomes an unavoidable byproduct, posing a challenge to the broad application of this tool.

Distinct from the Targetron, which involves protein components of GII introns to achieve DNA manipulation, our work focused solely on the RNA components. We selected the ORF-less GII intron candidates with multiple copies in the bacterial genomes and found that within the six candidates capable of DNA hydrolytic cleavage *in vitro*, two depleted plasmids in *E. coli*. Most strikingly, one candidate, HYER1, introduced DNA breaks without intron-insertion byproduct in the HEK293T genome, highlighting the potential of establishing RNA-based platforms for DNA manipulation.

5. Protocol

5.1 Materials

5.1.1 Equipment

1. Nanodrop (ThermoFisher)
2. PCR Thermocycler (LongGene)
3. Centrifuges (Eppendorf)
4. Electrophoresis apparatus (BioRad)
5. GelDoc Go Imaging System (BioRad)
6. 100 kDa molecular weight cut-off (MWCO) centrifugal filters (Merck Millipore)
7. Amersham Typhoon 5 (GE Healthcare)
8. Biosafety cabinet (Labconco)
9. Heracell Vios 160i CR CO_2 incubator (ThermoFisher)
10. Nikon ECLIPSE Ts2-FL
11. Cell electroporation apparatus (Celetrix biotechnologies)
12. Superose 6 Increase 10/300 (GE Healthcare)
13. Superose 6 Increase 5/150 GL (GE Healthcare)

14. Graphene oxide grid (Quantifoil Au 1.2/1.3, 300 mesh)
15. Plasma cleaner (HARRICK PLASMA).
16. FEI Vitrobot Marke IV
17. Titan Krios electron microscope operated at 300 kV (ThermoFisher)
18. Filter papers (Ted Pella)
19. CFX96 real-time system (BIO-RAD)

5.1.2 Reagents and buffers

1. DEPC-treated water
2. Q5 High-Fidelity 2 × Master Mix (New England Biolabs, NEB)
3. *In vitro* transcription (IVT) buffer (10×): 300 mM Tris (pH 8.1), 250 mM $MgCl_2$, 0.1 % (v/v) Triton X-100, and 20 mM spermidine
4. NTP mixture (5×): 25 mM ATP, 25 mM GTP, 25 mM UTP, and 25 mM CTP, pH7.5
5. 1 M Dithiothreitol (DTT)
6. RNase inhibitor (Promega)
7. RNase-Free DNase I (Promega)
8. DNase Buffer (10×): 400 mM Tris-HCl (pH 8.0), 100 mM $MgSO_4$, and 10 mM $CaCl_2$ (Promega)
9. Proteinase K (Solarbio, 20 mg/mL)
10. Purification buffer: 40 mM MOPS (pH 7.5), 10 mM $MgCl_2$, and 15 % (v/v) glycerol
11. Formamide
12. 0.5 M Ethylenediaminetetraacetic acid (EDTA) (pH 8.0)
13. 8–20 % urea polyacrylamide gel
14. TBE buffer (10×): 890 mM Tris base, 890 mM boric acid, and 20 mM EDTA
15. SuperStain (10,000× in water, CWBio)
16. 2× formamide loading buffer: 95 % (v/v) formamide, 0.02 % (w/v) SDS, 1 mM EDTA, 0.01 % (w/v) xylene cyanol, and 0.02 % (w/v) bromphenol blue
17. 0.1 M $NaHCO_3$
18. 20 mM Cy5-NHS (Lumiprobe) in DMSO
19. 3 M NaOAc (pH 5.2)
20. Absolute Ethanol (Sangon)
21. 1 M $MgCl_2$
22. 2.5 M NH_4Cl
23. 400 mM MOPS (pH 7.5)
24. 1 μg/μL BSA (Sigma)

25. 3 M NaCl
26. 1 M and 3 M KCl
27. 1 M Tris (pH 8.0)
28. RNase A (Beyotime, 10 mg/mL)
29. Nt.BspQI (NEB)
30. *Eco*RI-HF (NEB)
31. Competent Cell Preparation Kit (TaKaRa)
32. *ccdB* plasmid (p11-LacY-wtx1, addgene: 69056)
33. Streptomycin (Sangon)
34. Ampicillin (Sangon)
35. 1 M Arabinose (Sangon)
36. 20 % (w/v) Glucose (Sangon)
37. Liquid LB medium
38. LB plates
39. TIANScript II RT Kit (Tiangen Biotech).
40. Talent qPCR PreMix (SYBR Green) reagent (Tiangen Biotech)
41. PiggyBac integration plasmid and transposase plasmid (System Biosciences)
42. Opti-MEM (Gibco)
43. Blasticidin (MedChemExpress)
44. DMEM (Hyclone)
45. FBS (Biological Industries)
46. Penicillin-Streptomycin (Hyclone)
47. Hygromycin B (MedChemExpress)
48. Puromycin (MedChemExpress)
49. Universal Genomic DNA Kit (CWBIO)
50. 2 × Phanta Max Master Mix (Vazyme)
51. VAHTS DNA Clean Beads (Vazyme)
52. T7 Endonuclease I (Vazyme)
53. SEC buffer:10 mM $MgCl_2$, 50 mM KCl, and 20 mM HEPES (pH 7.5)

5.1.3 Bacterial strain and cell line

1. *E. coli* BW25141 (CGSC: 7635)
2. HEK293T (ATCC: CRL-11268)

5.1.4 Software for HYER identification and structure determination

1. Clustal Omega (Sievers et al., 2011)
2. INFERNAL (Nawrocki & Eddy, 2013).
3. HMMER (Finn et al., 2011)
4. Prodigal (Hyatt et al., 2010)

5. RNAfold (Lorenz et al., 2011)
6. CryoSPARC (Punjani, Rubinstein, Fleet, & Brubaker, 2017)
7. Coot (Emsley & Cowtan, 2004)
8. UCSF Chimera (Pettersen et al., 2021)
9. PHENIX (Liebschner et al., 2019)

5.2 Bioinformatics pipeline

5.2.1 Construction of covariance models

1. Collect the 92 reported GII-C intron sequences and divide the introns into two parts, DI-DIII and DV-DVI, for building multi-covariance models (Candales et al., 2012; Dai et al., 2003; Simon et al., 2008).
2. Remove the duplicated sequences and perform sequence alignments using Clustal Omega.
3. Construct the covariance models for DI-DIII and DV-DVI separately based on the sequence alignment results using INFERNAL with default parameters.

5.2.2 Identification of ORF-less candidates with multiple copies

1. Within the genomes of 45,599 species-level operational taxonomic units (OTUs) in Earth's Microbiomes catalog (Nayfach et al., 2021), select the candidates aligning with the two following criteria for further analysis: (1). The candidate must match both DI-DIII and DV-DVI models with an E-value lower than 10^{-5}; (2). The RNA sequence length must be shorter than 4000 nt.
2. Cluster and find highly identical copies in each OTU using BLASTN. Note that the candidates with query coverage per subject and percentage of identical matches both higher than 90% are considered as highly identical introns.
3. Build the hidden Markov model of IEPs using HMMER based on the IEP sequences encoded by the ORF in DIV of the reported GII-C introns.
4. Predict the proteins encoded in the candidate sequences using Prodigal to obtain ORF-less systems.

5.2.3 Ranking of candidates

To further identify suitable candidates for subsequent experimental characterization, the following three criteria are implemented:

1. The DIV should be less than 200 nt to obtain more compact introns.
2. Predict the secondary structures with RNAfold and manually screen for candidates with recognizable TRS.

3. To obtain the highly stable candidates, rank the thermal stabilities calculated from RNAfold output (ensemble diversity normalized by length).

5.3 *In vitro* DNA cleavage activity characterization

Here presents the pipeline we employed during the identification and characterization of the active HYERs for DNA cleavage *in vitro* for demonstration. The procedures can be modified by researchers for specific experimental designs and purposes. To identify HYERs with DNA cleavage capability, we designed DNA substrates with five continuous targets for initial screening, maximizing the possibility of DNA cleavage by HYERs, while for further comprehensive characterization, we only placed one target on the DNA substrates to simplify the experiment design and analysis (Fig. 4).

5.3.1 *Preparation of HYERs*

1. Templates for IVT are prepared through PCR. 50 μL PCR systems are set as follows: add 1 μL of template plasmid encoding HYER (0.1 ng/μL), 2.5 μL of forward primer (10 uM), 2.5 μL of reverse primer (10 uM) and 25 μL of Q5 High-Fidelity 2 × Master Mix to 19 μL of DEPC-treated

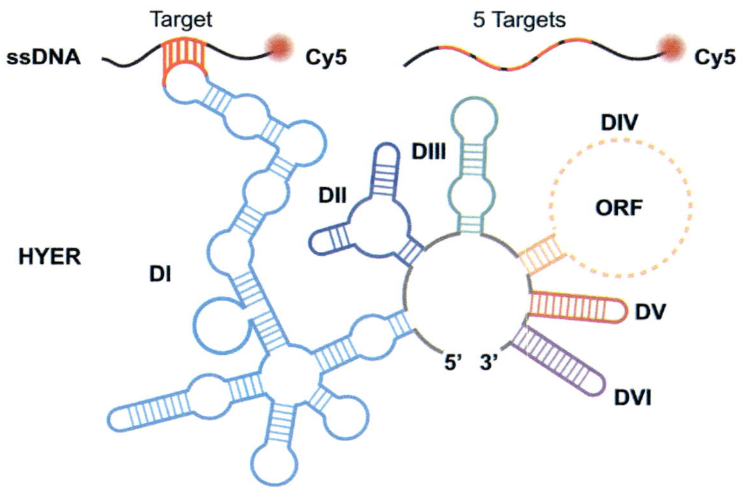

Fig. 4 Schematics of the *in vitro* ssDNA cleavage assay. ssDNA with one or five targets are labeled in the figure. The base pairs between TRS and ssDNA are labeled in red. Cy5 fluorophore on the 3′ end is labeled with a red dot. HYER domains are labeled and colored according to Fig. 1.

water. Thermocycling conditions are set according to the manufacturer's instructions.

2. 500 μL IVT of each HYER is carried out as follows: add 45 μL of PCR product, 50 μL of IVT buffer (10×), 100 μL of NTP mixture (5×), 5 μL of 1 M DTT, 20 μL of RNase inhibitor and 10 μL T7 RNA polymerase (2 mg/mL) to 270 μL of DEPC-treated water. Incubate at 37 °C for 4 h.
3. Add 50 μL of DNase Buffer (10×) and 25 μL of RNase-Free DNase I. Continue to incubate at 37 °C for 1 h.
4. Add 6.25 μL of proteinase K and incubate at 37 °C for 1 h.
5. Centrifuge at 15,000 g, 4 °C for 8 min to pellet the magnesium pyrophosphate precipitation.
6. Transfer the supernatant to 100 kDa MWCO centrifugal filters and centrifuge at 13,000 g, 4 °C until around 100 μL liquid is left in the filter.
7. Add 400 μL of purification buffer and centrifuge at 13,000 g, 4 °C until around 100 μL liquid is left in the filter. Perform this step for eight times in total.
8. Determine the RNA concentration with Nanodrop.
9. Mix around 250 ng of RNA products with the same volumes of 2 × formamide loading buffer. Heat the samples at 95 °C for 3 min and put on ice immediately. Use 8 % Urea-PAGE to examine the RNA quality by electrophoresis. Saint the gel with SuperStain and visualize the gel using the GelDoc Go Imaging System.
10. Flash-freeze the RNA products in liquid nitrogen and stored at −80 °C.

5.3.2 Labeling of ssDNA or ssRNA substrate

1. Order 3'-NH_2 labeled ssDNA from Sangon Biotech and dissolve with 0.1 M $NaHCO_3$ to 1 mM.
2. Mix 3 μL of 1 mM 3'-NH_2 labeled DNA and 3 μL of 1 mM Cy5-NHS. Incubate at 25 °C for 4 h.
3. Add 0.66 μL 3 M NaOAc (pH 5.2) and 24 μL absolute ethanol. Store at −80 °C overnight.
4. Centrifuge at 15,000 g, 4 °C for 1 h and remove the supernatant.
5. Add 200 μL of 75 % (v/v) cold ethanol (−20 °C) to wash the pellet and centrifuge at 15,000 g, 4 °C for 10 min. Repeat this step.
6. Resuspend the pellet with 30 μL DEPC-treated water.
7. Precipitate and wash the ssDNA with ethanol again. Generally, performing ethanol precipitation for 2–3 times in total is enough for purification. The supernatant should be colorless, and the pellet should be blue.

8. Resuspend the 3'-Cy5-labeled ssDNA with 50 μL DEPC-treated water.
9. Dilute 1 μL of 3'-Cy5-labeled ssDNA with 9 μL DEPC-treated water. Determine the ssDNA concentration and absorbance at 646 nm with Nanodrop. The molar extinction coefficient for Cy5 is 250,000 $cm^{-1}M^{-1}$. The concentration of 3'-Cy5-labeled ssDNA is calculated by dividing the absorbance at 646 nm with the molar extinction coefficient for Cy5. Labeling efficacy is calculated by dividing the concentration of 3'-Cy5-labeled ssDNA with the total concentration of ssDNA, which should be around 50 % in general.
10. ssRNA can also be labeled with Cy5 following the same protocol.

5.3.3 ssDNA or ssRNA cleavage assay of HYERs

1. For initial ssDNA cleavage activity screening, five copies of target site are placed continuously on the ssDNA (Fig. 4). To prepare a 34 μL cleavage system, mix 4.25 μL of 1 M $MgCl_2$, 6.8 μL of 2.5 M NH_4Cl, 3.4 μL of 400 mM MOPS (pH 7.5), 3.4 μL of 1 μg/μL BSA, ssDNA substrate (final concentration: 0.1 μM) and HYER (final concentration: 1.5 μM). Split the system into 7 tubes (4.4 μL/tube) for monitoring the cleavage process at 7 different time points: 0 min, 5 min, 10 min, 20 min, 40 min, 1 h, and 2 h.
2. Incubate at 50 °C and terminate the reaction at each time point by adding 1.376 μL of 0.5 M EDTA.
3. Add RNase A to a final concertation of 0.1 mg/mL and incubate at 37 °C for 1 h.
4. Add proteinase K to a final concertation of 0.1 mg/mL and incubate at 50 °C for 1 h.
5. Add formamide to a final concertation of 50 % (v/v). Heat at 95 °C for 3 min and put on ice immediately.
6. The reaction samples are analyzed in 15 % Urea-PAGE. 100 fmol of 3'-Cy5-labeled ssDNA is loaded into each lane.
7. Visualize the gel using Amersham Typhoon 5.
8. Use ImageJ to quantify the percentage cleaved and plot the curves using the One-Phase-Association model in GraphPad Prism 7. All experiments should be performed with at least three replicates.
9. The ssRNA cleavage assay is carried out likewise with a few alterations: formamide is directly added to a final concentration of 50 % (v/v) after the addition of EDTA; RNase A and proteinase K treatment are not involved; 20 % Urea-PAGE is used to analyze the ssRNA cleavage products.

5.3.4 Determination of cleavage sites of HYERs on ssDNA substrates

1. For simplicity, one target site for each HYER is placed on the ssDNA substrate to determine the cleavage sites (Fig. 4).
2. To generate a DNA ladder through random hydrolysis, mix the ssDNA substrate (final concentration: 0.18 μM), 1 μL of DNase Buffer (10 ×), and 0.5 U RNase-Free DNase I in a 10 μL reaction system. Incubate at 37 °C for 15 min. Add formamide to a final concentration of 50 % (v/v). Heat at 95 °C for 3 min and put on ice immediately.
3. To prepare a 10 μL cleavage system, mix 1.25 μL of 1 M $MgCl_2$, 2 μL of 2.5 M NH_4Cl, 1 μL of 400 mM MOPS (pH 7.5), 1 μL of 1 μg/μL BSA, ssDNA substrate (final concentration: 0.1 μM) and HYER (final concentration: 1.5 μM). Incubate at 50 °C for 24 h. Terminate the reaction, perform RNase A and proteinase K treatment, and analyze the cleavage products using 20 % Urea-PAGE following the same protocol in 5.3.3.

5.3.5 In-depth biochemical characterization of HYERs

1. To investigate the cleavage activity of HYERs at different concentrations of Mg^{2+} (10 mM, 50 mM or 125 mM), mix 0.54, 2.7 or 6.75 μL of 1 M $MgCl_2$, 2.7 μL of 3 M KCl, 5.4 μL of 400 mM MOPS (pH 7.5), 5.4 μL of 1 μg/μL BSA, ssDNA with one target site (final concentration: 0.1 μM) and HYER (final concentration: 1.5 μM) in a 54 μL cleavage system. The incubation temperature is 37 °C.
2. To investigate the effects of monovalent ions on HYERs' DNA cleavage activity, mix 0.18, 2.7, or 9 μL of 3 M KCl or 3 M NaCl (final concentrations are 10 mM, 150 mM or 500 mM), 2.7 μL of 1 M $MgCl_2$, 5.4 μL of 400 mM MOPS (pH 7.5), 5.4 μL of 1 μg/μL BSA, ssDNA with one target site (final concentration: 0.1 μM) and HYER (final concentration: 1.5 μM) in a 54 μL cleavage system. The incubation temperature is 37 °C.
3. To investigate the cleavage activity of HYERs at different temperatures, mix 2.7 μL of 1 M $MgCl_2$, 0.18 μL of 3 M KCl, 5.4 μL of 400 mM MOPS (pH 7.5), 5.4 μL of 1 μg/μL BSA, ssDNA with one target site (final concentration: 0.1 μM) and HYER (final concentration: 1.5 μM) in a 54 μL cleavage system. Incubate at 25 °C, 37 °C, 42 °C or 50 °C.
4. Split the systems above into 11 tubes (4.4 μL/tube) to monitor the cleavage process at the following time points: 0 min, 5 min, 10 min, 20 min, 40 min, 1 h, 2 h, 4 h, 8 h, 12 h, and 24 h. Terminate the reactions above with EDTA, and perform RNase A and proteinase K treatment, and analyze the cleavage products using 20 % Urea-PAGE following the same protocol in 5.3.3.

5.3.6 Cleavage of multiple forms of DNA by HYERs

5.3.6.1 Double-stranded DNA (dsDNA) cleavage assay

1. Dissolve the non-labeled ssDNA fully or partially complementary to the 3'-Cy5-labeled ssDNA substrate with DEPC-treated water and diluted to 10 μM.
2. To obtain fully annealed or bubbled dsDNA, mix 0.25 μL of 1 M KCl, 0.25 μL of 1 M Tris (pH 8.0), 1.2 μL of 10 μM non-labeled ssDNA and 3'-Cy5-labeled ssDNA substrate with one target site (final concentration: 1 μM) in a 10 μL system. Heat at 95 °C for 3 min and allow the system to cool down to 4 °C gradually.
3. Mix 1 μL of 1 M $MgCl_2$, 0.2 μL of 1 M KCl, 2 μL of 400 mM MOPS (pH 7.5), 2 μL of 1 μg/μL BSA and 2 μL of 1 μM dsDNA and HYER (final concentration: 1.5 μM) in a 20 μL cleavage system. Split the system into 4 tubes (4.4 μL/tube) for monitoring the cleavage process at following time points: 0 h, 3 h, 8 h, and 24 h.
4. Incubate at 37 °C. Terminate the reaction, perform RNase A and proteinase K treatment, and analyze the cleavage products using 15 % Urea-PAGE following the same protocol in 5.3.3.

5.3.6.2 Plasmid cleavage assay

1. Mix 0.2 μL of 1 M $MgCl_2$, 3 μL of 1 M KCl, 2 μL of 400 mM MOPS (pH 7.5), 2 μL of 1 μg/μL BSA and plasmid with target sites (final concentration: 0.03 μM) and HYER (final concentration: 1.5 μM) in a 20 μL cleavage system. Split the system into 4 tubes (4.4 μL/tube) for monitoring the cleavage process at following time points: 0 h, 3 h, 8 h, and 24 h.
2. Terminate the reaction with EDTA and perform RNase A and proteinase K treatment following the same protocol in 5.3.3.
3. Add glycerol to a final concentration of 15 % (v/v). Analyze the reaction products in 0.9 % agarose gels and visualize the gels using the GelDoc Go Imaging System. Open-circular and linearized plasmid can be prepared using Nt.BspQI and *Eco*RI-HF according to the manufacturer's protocols.

5.4 *In vivo* activity characterization

5.4.1 Plasmid interference assay in bacteria

To evaluate HYERs' DNA cleavage activity in bacterial cells, we established the *ccdB* reporter system for characterization. ccdB toxin inhibits

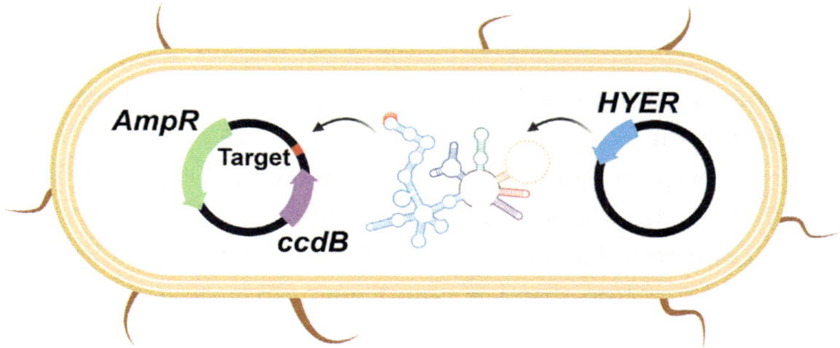

Fig. 5 Schematics of the plasmid depletion assay in bacteria. The HYER-expressing plasmid, target plasmid, *ccdB* gene, and ampicillin-resistance gene (*AmpR*) are labeled in the figure. HYER domains are labeled and colored according to Fig. 1.

DNA replication in *E. coli* and can result in bacterial death (Bahassi et al., 1999; Chen & Zhao, 2005). After the delivery of the HYER-expressing plasmid, *E. coli* will survive only if the *ccdB* plasmid is depleted by HYER (Fig. 5).

5.4.1.1 Preparation of competent bacterial cells containing the *ccdB* plasmid target

1. Take *E. coli* BW25141 strain and thaw on ice. Streak for single colonies on an LB plate without antibiotics. Culture overnight at 37 °C.
2. Inoculate a single colony into 3 mL of LB liquid medium without antibiotics. Culture at 37 °C until OD_{600} is 0.4–0.5 (OD_{600} should not exceed 0.5).
3. Transfer 0.5 mL of the culture into 50 mL of LB liquid medium without antibiotics. Culture at 37 °C until OD_{600} is 0.4–0.5 (OD_{600} should not exceed 0.5).
4. Centrifuge at 1700 g, 4 °C for 10 min. Discard the supernatant and resuspend cells with 5 mL of solution A in the competent cell preparation kit (TaKaRa).
5. Centrifuge at 1700 g, 4 °C for 5 min. Discard the supernatant and resuspend cells with 5 mL of solution B in the competent cell preparation kit (TaKaRa).
6. Split the competent cells to sterile tubes (50 μL/tube). Freeze the bacteria in liquid nitrogen and store at −80 °C.
7. Take *E. coli* BW25141 competent bacterial cells and thaw on ice. Add 50 ng *ccdB* plasmid and incubate for 30 min on ice.

8. Heat shock at 42 °C for 60 s in a water bath, and then incubate on ice for 2 min
9. Add 1 mL of LB liquid medium with 0.2 % (w/v) glucose and recover at 37 °C for 1 h.
10. Streak for single colonies on an LB plate with 100 μg/mL ampicillin and 0.2 % (w/v) glucose. Culture overnight at 37 °C.
11. Prepare competent bacteria containing the *ccdB* plasmid following steps 2-6 in 5.4.1.1. Note that LB liquid medium with 100 μg/mL ampicillin and 0.2 % (w/v) glucose should be used to culture the bacteria here.

5.4.1.2 Plasmid interference activity validation

1. Take *E. coli* BW25141 competent bacterial cells containing the *ccdB* plasmid and thaw on ice. Transform the bacteria with 100 ng HYER-expressing plasmid following steps 8-10 in 5.4.1.1. Note that the LB plate used here contains 50 μg/mL streptomycin and 0.2 % (w/v) glucose.
2. Inoculate a single colony into 0.5 mL LB liquid medium with 50 μg/mL streptomycin and 0.2 % (w/v) glucose. Culture at 37 °C for 12 h.
3. Transfer 5 μL bacterial culture into 0.5 mL fresh LB liquid medium with 50 μg/mL streptomycin and 0.2 % (w/v) glucose. Culture at 37 °C for 12 h. Repeat this step until continuously culture the bacteria for 72 h in total.
4. Spot 5 μL of bacterial culture on LB plates with gradient dilutions from 10^0 to 10^{-7} and incubate overnight at 37 °C. Bacterial culture should be spotted on three plates for plasmid interference activity validation: (1). LB plate with 50 μg/mL streptomycin and 0.2 % (w/v) glucose to confirm the existence of the HYER-expressing plasmid; (2). LB plate with 50 μg/mL streptomycin and 10 mM arabinose to characterize the depletion of *ccdB* plasmid; (3). LB plate with 100 μg/mL ampicillin and 0.2 % (w/v) glucose to confirm the *ccdB* plasmid depletion results on the second plate.
5. Extract bacterial RNA with 1 mL culture and reverse transcribe the RNA into cDNA using TIANScript II RT Kit according to the manufacturer's instructions.
6. Perform quantitative PCR (qPCR) using Talent qPCR PreMix (SYBR Green) on the CFX96 real-time system. The thermocycling conditions are set as follows: 95 °C for 3 min; 40 cycles of 95 °C for 5 s, 55 °C for 10 s and 72 °C for 15 s

7. Normalize the relative mRNA level of *ccdB* gene to the 16S rRNA level of the same sample and analyze the significance of relative gene expression by one-way analysis of variance (ANOVA).

5.4.2 Genomic DNA cleavage capability validation in HEK293T cells

To evaluate HYERs' capability to cleave genomic DNA, a 3 n + 1 frame-shifted puromycin resistance gene (*puro**) driven by a CMV promoter is inserted into the genome of HEK293T cells as a reporter for DNA cleavage events. A target cassette comprising three target sites with 30-bp intervals is positioned between the *puro** gene and the upstream CMV promoter. All the three target sites have the same TRS complementary sequence on both top and bottom DNA strands (Fig. 6). Double-stranded DNA breaks (DSBs) generated by HYER can be repaired by endogenous pathways and may result in insertions and deletions (INDELs), restoring the functional ORF for puromycin resistance gene (*puro*) expression, which then confers puromycin resistance to the cells (Fig. 6).

5.4.2.1 Construction of the reporter cell line

1. Electroporate 1×10^6 HEK293T cells with 2 μg PiggyBac transposase plasmid and 2 μg integration plasmid in 20 μL of Opti-MEM at 450 V. The PiggyBac integration plasmid carries the target cassette and *puro** gene driven by the CMV promoter.
2. One day later, add blasticidin to a final concentration of 10 μg/mL. Select the cells with successful insertions for 10 days. Passage the cells to maintain sub-confluent conditions.
3. Culture the HEK293T-*puro** cells in DMEM with 10 % (v/v) FBS and 1 % (v/v) penicillin-streptomycin at 37 °C in 5 % CO_2.

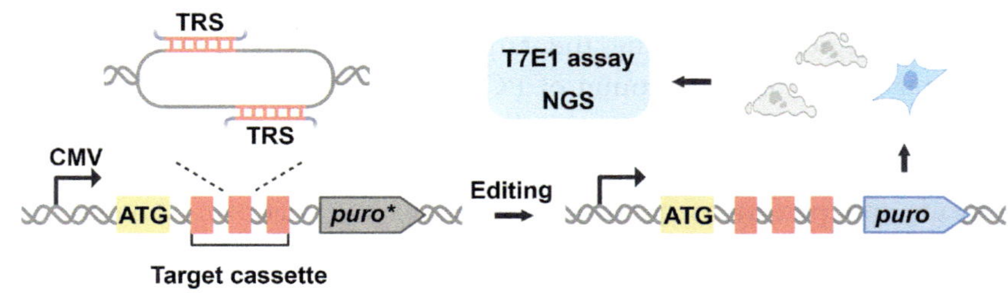

Fig. 6 Schematics of the "out-of-frame" *puro reporter system for genome cleavage capability validation.** The target recognition details within the target cassette are zoomed in on the top. The target cassette contains three identical target sites. The 3 n + 1 frame-shifted *puro** is colored in gray. After HYER editing, the frame-restored *puro* is colored in blue.

5.4.2.2 Validation of genomic DNA cleavage capability of HYERs

1. Electroporate 1×10^6 HEK293T cells with 2 μg PiggyBac transposase plasmid and 2 μg integration plasmid in 20 μL of Opti-MEM at 450 V. The PiggyBac integration plasmid carries the native *HYERs* and non-targeting *HYERs* with mutated TRS driven by a U6 promoter. The selection marker in the PiggyBac integration plasmid is replaced by *hygBR* (hygromycin B resistance gene).
2. One day later, add hygromycin B to a final concentration of 50 μg/mL. Select the cells with integrated native or non-targeting *HYERs* for two weeks. Passage the cells to maintain sub-confluent conditions.
3. Add puromycin to a final concentration of 5 μg/mL for two weeks to enrich the edited cells. Passage the cells to maintain sub-confluent conditions.
4. Ten days after the administration of puromycin, survival cells can be observed in HYER1 group and are photographed with Nikon ECLIPSE Ts2-FL.
5. Collect the cells before and after puromycin selection and extract the genome using the Universal Genomic DNA Kit according to the manufacturer's instructions.
6. The DNA cleavage events can be validated through the T7E1 assay (Mashal, Koontz, & Sklar, 1995). PCR with the forward primer on the CMV promoter and the reverse primer on the *puro** to obtain the regions covering the target cassette using 2 × Phanta Max Master Mix. Then, purify the PCR product with VAHTS DNA Clean Beads. Around 200 ng of the purified DNA is used for the T7E1 assay following the manufacturer's protocol. Analyze the reaction products using 2 % agarose gels.
7. To characterize the DNA cleavage events through NGS, amplify the 254-bp regions containing the target cassette using 2 × Phanta Max Master Mix in the first round of PCR. Then, add sequencing adaptors through PCR with primers containing MiSeq adaptor sequences in the second round of PCR. Purify the amplicons with VAHTS DNA Clean Beads and perform sequencing on the MiSeq platform. Analyze the sequencing reads with CRISPResso2 for editing events (parameters used here: quantification window sizes are 20 bp for each single target site and 60 bp for the whole target cassette; plot window sizes are 20 bp for each single target site and 60 bp for the whole target cassette). Calculate the INDEL percentage of each target site according to the CRISPResso2 output.

5.5 Structure determination of HYERs

5.5.1 Purification of HYERs through size exclusion chromatography

1. Prepare the RNA samples through IVT and digest the DNA templates with RNase-Free DNase I as described in steps 1-3 in 5.3.1.
2. Centrifuge at 15,000 g, 4 °C for 8 min to remove the precipitation.
3. To obtain homogeneous HYER samples for structure determination, purify the RNA with size exclusion chromatography (SEC) column Superose 6 Increase 10/300 GL or Superose 6 Increase 5/150 GL in the SEC buffer at 4 °C: equilibrate the column with 30 mL or 5 mL SEC buffer at a flow rate of 0.5 or 0.1 mL/min, and then load the supernatant at a flow rate of 0.5 or 0.1 mL/min for the two types of columns separately.
4. Collect the RNA fractions corresponding to the chromatogram peaks and analyze the RNA quality by electrophoresis using 8 % Urea-PAGE.
5. Concentrate the homogeneous RNA fractions using 100 kDa MWCO centrifugal filters.
6. Flash-freeze the purified RNA in liquid nitrogen and stored at −80 °C.

5.5.2 Cryo-EM sample preparation and data collection

1. Pretreat the graphene oxide grid (Quantifoil Au 1.2/1.3, 300 mesh) by glow discharging in the plasma cleaner for 10-15 s at low or middle level after 2-min evacuation.
2. Dilute the purified RNA samples to around 1.5 μM with SEC buffer.
3. Load a pair of 55-mm filter paper onto FEI Vitrobot Marke IV arms and preincubate the chamber to 8 °C with 100 % humidity.
4. Load the pretreated graphene oxide grid with a tweezer into the FEI Vitrobot Marke IV and apply 4 μL of diluted RNA to the grid.
5. Set the blot force to −2 and blot time to 2.5 s to blot the grid with the filter paper and flash-freeze the grid in liquid ethane using FEI Vitrobot Marke IV.
6. Note that the glow discharging conditions, RNA concentration, blot time, and blot force can be optimized according to the equipment and environment conditions for better performance.
7. Transfer the grids to a Titan Krios electron microscope operated at 300-kV acceleration voltage with a Gatan K3 direct electron detector to collect the cryo-EM data. Micrographs are recorded in counting mode at a nominal magnification of 81,000×, corresponding to a physical pixel size of 1.0825 Å per pixel. The defocus value is set between −1.0 to −2.5 μm. The total exposure time of each movie stack leads to a total accumulated dose of 50 electrons per $Å^2$, which is fractionated into 32 frames.

5.5.3 Image processing and model building

1. The raw dose-fractionated image stacks are 2 × Fourier binned, aligned, dose-weighted, and summed using MotionCor2 within cryoSPARC.
2. Perform CTF estimation, blob particle picking, 2D reference-free classification, initial model generation, 3D reference-based classification, final 3D refinement, and local resolution estimation in cryoSPARC.
3. Manually build the models in Coot according to the refined cryo-EM maps.
4. Combine the secondary structure and geometry restraints to refine the models based on EM maps using PHENIX in real space.
5. Validate the final models using PHENIX and visualize in UCSF Chimera.

5.6 Engineering and optimization of HYER1

Various mutations and engineering strategies can be applied to HYER molecules according to the researchers' designs and experiment purposes. Here provide some engineering examples we have conducted for demonstration. To improve HYER1's DNA targeting and cleavage performance, we replaced the A223 (spatially adjacent to the TRS) with a 14-nt recruiting sequence (RS), which increased both the DNA targeting specificity and cleavage activity of HYER1. We also engineered the palindromes into non-palindromic complementary sequences and redesigned the TRS sequences to target the top and bottom DNA strands at different positions, resulting in customizable cleavage patterns (Fig. 7).

5.6.1 In vitro cleavage assays for HYERs with engineered TRS and RS insertion

1. Construct HYERs with TRS of different sequences or lengths and RS-inserted HYERs through site-directed mutagenesis by PCR.

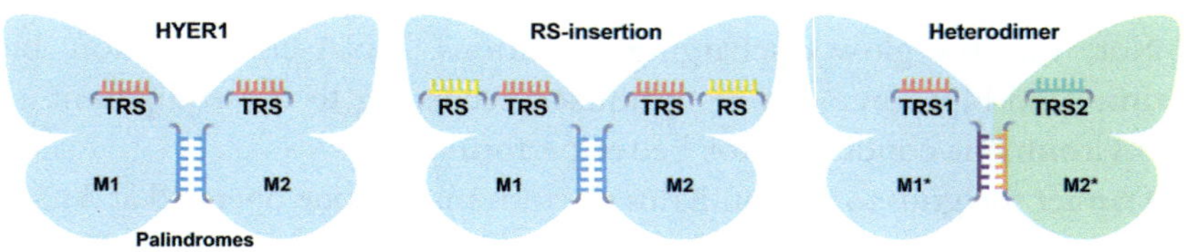

Fig. 7 The cartoon model of native HYER1, engineered HYER1 with RS insertion, and heterodimerized HYER1. The model of HYER1 comprising two monomers, monomer 1 (M1) and monomer 2 (M2), is simplified as a butterfly, with TRS colored in red and palindromes colored in blue. The RS is colored in yellow. In the heterodimerization design, two monomers carrying different TRS sequences and modified palindromes are denoted as M1 * (blue) and M2 * (green).

2. Prepare the RNA through IVT and conduct ssDNA cleavage assay in a buffer containing 50 mM $MgCl_2$, 10 mM KCl, 40 mM MOPS (pH 7.5), and 100 ng/μL BSA following the procedures presented in 5.3.

5.6.2 Cleavage of bubbled dsDNA by heterodimerized HYER1

1. Mutations are introduced to the palindromes and the TRS region in HYER1 through site-directed mutagenesis by PCR.
2. Order the 3'-NH_2 and 3'-FAM labeled ssDNA (corresponding to top and bottom strands, respectively) from Sangon Biotech and label the 3'-NH_2 ssDNA with NHS-Cy5 following the descriptions in 5.3.2.
3. Anneal the top and bottom strands of DNA and perform cleavage assay on the bubbled dsDNA following the descriptions in 5.3.6.1 with a few alterations: both strands of DNA are annealed at 1:1 molar ratio; terminate the reactions at 0 h, 3 h, 6 h, and 18 h; analyze the cleavage products in 10 % Urea-PAGE.
4. Merge Cy5 and FAM fluorescence channels with different pseudo colors with ImageJ to evaluate the cleavage of both strands of DNA.

6. Perspectives on the application of HYERs

Our work systemically discovered ORF-less GII-C introns with multiple copies in the genomes, which are capable of cleaving multiple forms of DNA and RNA *in vitro* and generating DNA breaks in bacterial and mammalian cells. To improve the specificity and efficacy of HYER1, we employed structure-based engineering strategies, including TRS extension, RS insertion, and heterodimerization, which led to the enhanced DNA targeting and cleavage capability. Despite the relatively low gene editing efficacy in mammalian cells, HYER1 represents a pioneering example of utilizing RNA's catalytic proprieties for *in vivo* DNA cleavage (Liu et al., 2024). The major hindrance to developing HYERs into efficient gene editing tools is their limited ability to melt the dsDNA and form base pairings with the target sites. Several strategies could address this issue. First, incorporating DNA helicase factors into the HYER systems may assist to melt the DNA strands for HYER targeting and cleavage. It has been shown that fusing single strand targeting deaminase to DNA unwinding factors can improve the mutagenesis rate throughout the chromosome (Wang et al., 2021). Second, while our study focused solely on GII-C introns, GII-A and GII-B introns with more complex structures and functional domains may interact with substrates *via* more stable interactions and offer new opportunities for DNA editing. Third, *in vitro* selection methods could be utilized to identify aptamers that can provide dsDNA invasion capability or increase the affinity of HYERs to DNA targets. In addition to DNA editing, it is noteworthy that HYERs also exhibit RNA cleavage activity *in vitro*, thus posing the possibility for applications in RNA editing scenarios.

Acknowledgments

This work is supported by the National Natural Science Foundation of China No.324B2051 (ZXL), National Natural Science Foundation of China No.32150018 (JJGL), the STI2030-Major Projects 2022ZD0213900 (JJGL), the Ministry of Agriculture and Rural Affairs of China (JJGL), and the start-up funds from Tsinghua University, Beijing (JJGL). JJGL is an awardee of Xplorer Prize by New Cornerstone Science Foundation.

References

Bahassi, E. M., O'Dea, M. H., Allali, N., Messens, J., Gellert, M., & Couturier, M. (1999). Interactions of CcdB with DNA gyrase. Inactivation of Gyra, poisoning of the gyrase-DNA complex, and the antidote action of CcdA. *The Journal of Biological Chemistry, 274*(16), 10936–10944. https://doi.org/10.1074/jbc.274.16.10936.

Barnwal, R. P., Yang, F., & Varani, G. (2017). Applications of NMR to structure determination of RNAs large and small. *Archives of Biochemistry and Biophysics, 628*, 42–56. https://doi.org/10.1016/j.abb.2017.06.003.

Bhandari, Y. R., Fan, L., Fang, X., Zaki, G. F., Stahlberg, E. A., Jiang, W., ... Wang, Y. X. (2017). Topological structure determination of RNA using small-angle X-ray scattering. *Journal of Molecular Biology, 429*(23), 3635–3649. https://doi.org/10.1016/j.jmb.2017.09.006.

Candales, M. A., Duong, A., Hood, K. S., Li, T., Neufeld, R. A., Sun, R., ... Zimmerly, S. (2012). Database for bacterial group II introns (Database issue) *Nucleic Acids Research, 40*, D187–D190. https://doi.org/10.1093/nar/gkr1043.

Chan, R. T., Peters, J. K., Robart, A. R., Wiryaman, T., Rajashankar, K. R., & Toor, N. (2018). Structural basis for the second step of group II intron splicing. *Nature Communications, 9*(1), 4676. https://doi.org/10.1038/s41467-018-06678-0.

Chen, Z., & Zhao, H. (2005). A highly sensitive selection method for directed evolution of homing endonucleases. *Nucleic Acids Research, 33*(18), e154. https://doi.org/10.1093/nar/gni148.

Chung, K., Xu, L., Chai, P., Peng, J., Devarkar, S. C., & Pyle, A. M. (2022). Structures of a mobile intron retroelement poised to attack its structured DNA target. *Science (New York, N. Y.), 378*(6620), 627–634. https://doi.org/10.1126/science.abq2844.

Cousineau, B., Smith, D., Lawrence-Cavanagh, S., Mueller, J. E., Yang, J., Mills, D., ... Belfort, M. (1998). Retrohoming of a bacterial group II intron: Mobility via complete reverse splicing, independent of homologous DNA recombination. *Cell, 94*(4), 451–462. https://doi.org/10.1016/s0092-8674(00)81586-x.

Dai, L., Toor, N., Olson, R., Keeping, A., & Zimmerly, S. (2003). Database for mobile group II introns. *Nucleic Acids Research, 31*(1), 424–426. https://doi.org/10.1093/nar/gkg049.

Dai, L., & Zimmerly, S. (2003). ORF-less and reverse-transcriptase-encoding group II introns in archaebacteria, with a pattern of homing into related group II intron ORFs. *RNA (New York, N. Y.), 9*(1), 14–19. https://doi.org/10.1261/rna.2126203.

Emsley, P., & Cowtan, K. (2004). Coot: model-building tools for molecular graphics. *Acta Crystallographica. Section D, Biological Crystallography, 60*(Pt 12 Pt 1), 2126–2132. https://doi.org/10.1107/S0907444904019158.

Enyeart, P. J., Mohr, G., Ellington, A. D., & Lambowitz, A. M. (2014). Biotechnological applications of mobile group II introns and their reverse transcriptases: Gene targeting, RNA-seq, and non-coding RNA analysis. *Mobile DNA, 5*(1), 2. https://doi.org/10.1186/1759-8753-5-2.

Finn, R. D., Clements, J., & Eddy, S. R. (2011). HMMER web server: Interactive sequence similarity searching (Web Server issue) *Nucleic Acids Research, 39*, W29–W37. https://doi.org/10.1093/nar/gkr367.

Funari, S. S., Rapp, G., Perbandt, M., Dierks, K., Vallazza, M., Betzel, C., ... Svergun, D. I. (2000). Structure of free Thermus flavus 5 S rRNA at 1.3 nm resolution from synchrotron X-ray solution scattering. *The Journal of Biological Chemistry, 275*(40), 31283–31288. https://doi.org/10.1074/jbc.M004974200.

Granlund, M., Michel, F., & Norgren, M. (2001). Mutually exclusive distribution of IS1548 and GBSi1, an active group II intron identified in human isolates of group B streptococci. *Journal of Bacteriology, 183*(8), 2560–2569. https://doi.org/10.1128/JB.183.8.2560-2569.2001.

Guo, H., Karberg, M., Long, M., Jones, J. P., 3rd, Sullenger, B., & Lambowitz, A. M. (2000). Group II introns designed to insert into therapeutically relevant DNA target sites in human cells. *Science (New York, N. Y.), 289*(5478), 452–457. https://doi.org/10.1126/science.289.5478.452.

Haack, D. B., Rudolfs, B., Zhang, C., Lyumkis, D., & Toor, N. (2024). Structural basis of branching during RNA splicing. *Nature Structural & Molecular Biology, 31*(1), 179–189. https://doi.org/10.1038/s41594-023-01150-0.

Haack, D. B., Yan, X., Zhang, C., Hingey, J., Lyumkis, D., Baker, T. S., & Toor, N. (2019). Cryo-EM structures of a group II intron reverse splicing into DNA. *Cell, 178*(3), 612–623.e612. https://doi.org/10.1016/j.cell.2019.06.035.

Hyatt, D., Chen, G. L., Locascio, P. F., Land, M. L., Larimer, F. W., & Hauser, L. J. (2010). Prodigal: Prokaryotic gene recognition and translation initiation site identification. *BMC Bioinformatics, 11*, 119. https://doi.org/10.1186/1471-2105-11-119.

Karberg, M., Guo, H., Zhong, J., Coon, R., Perutka, J., & Lambowitz, A. M. (2001). Group II introns as controllable gene targeting vectors for genetic manipulation of bacteria. *Nature Biotechnology, 19*(12), 1162–1167. https://doi.org/10.1038/nbt1201-1162.

Kim, S. H., Quigley, G. J., Suddath, F. L., McPherson, A., Sneden, D., Kim, J. J., ... Rich, A. (1973). Three-dimensional structure of yeast phenylalanine transfer RNA: Folding of the polynucleotide chain. *Science (New York, N. Y.), 179*(4070), 285–288. https://doi.org/10.1126/science.179.4070.285.

Lambowitz, A. M., & Zimmerly, S. (2004). Mobile group II introns. *Annual Review of Genetics, 38*, 1–35. https://doi.org/10.1146/annurev.genet.38.072902.091600.

Lambowitz, A. M., & Zimmerly, S. (2011). Group II introns: Mobile ribozymes that invade DNA. *Cold Spring Harbor Perspectives in Biology, 3*(8), a003616. https://doi.org/10.1101/cshperspect.a003616.

Liebschner, D., Afonine, P. V., Baker, M. L., Bunkoczi, G., Chen, V. B., Croll, T. I., ... Adams, P. D. (2019). Macromolecular structure determination using X-rays, neutrons and electrons: Recent developments in Phenix. *Acta Crystallographica Section D: Structural Biology, 75*(Pt 10), 861–877. https://doi.org/10.1107/S2059798319011471.

Liu, D., Thelot, F. A., Piccirilli, J. A., Liao, M., & Yin, P. (2022). Sub-3-A cryo-EM structure of RNA enabled by engineered homomeric self-assembly. *Nature Methods, 19*(5), 576–585. https://doi.org/10.1038/s41592-022-01455-w.

Liu, Z. X., Zhang, S., Zhu, H. Z., Chen, Z. H., Yang, Y., Li, L. Q., ... Liu, J. G. (2024). Hydrolytic endonucleolytic ribozyme (HYER) is programmable for sequence-specific DNA cleavage. *Science (New York, N. Y.), 383*(6682), eadh4859. https://doi.org/10.1126/science.adh4859.

Lorenz, R., Bernhart, S. H., Honer Zu Siederdissen, C., Tafer, H., Flamm, C., Stadler, P. F., & Hofacker, I. L. (2011). ViennaRNA Package 2.0. *Algorithms for Molecular Biology: AMB, 6*, 26. https://doi.org/10.1186/1748-7188-6-26.

Luo, B., Zhang, C., Ling, X., Mukherjee, S., Jia, G., Xie, J., ... Su, Z. (2023). Cryo-EM reveals dynamics of Tetrahymena group I intron self-splicing. *Nature Catalysis, 6*(4), 298–309. https://doi.org/10.1038/s41929-023-00934-3.

Manigrasso, J., Chillon, I., Genna, V., Vidossich, P., Somarowthu, S., Pyle, A. M., ... Marcia, M. (2020). Visualizing group II intron dynamics between the first and second steps of splicing. *Nature Communications, 11*(1), 2837. https://doi.org/10.1038/s41467-020-16741-4.

Marcia, M., & Pyle, A. M. (2012). Visualizing group II intron catalysis through the stages of splicing. *Cell, 151*(3), 497–507. https://doi.org/10.1016/j.cell.2012.09.033.

Mashal, R. D., Koontz, J., & Sklar, J. (1995). Detection of mutations by cleavage of DNA heteroduplexes with bacteriophage resolvases. *Nature Genetics, 9*(2), 177–183. https://doi.org/10.1038/ng0295-177.

Nawrocki, E. P., & Eddy, S. R. (2013). Infernal 1.1: 100-fold faster RNA homology searches. *Bioinformatics (Oxford, England), 29*(22), 2933–2935. https://doi.org/10.1093/bioinformatics/btt509.

Nayfach, S., Roux, S., Seshadri, R., Udwary, D., Varghese, N., Schulz, F., ... Eloe-Fadrosh, E. A. (2021). A genomic catalog of Earth's microbiomes. *Nature Biotechnology, 39*(4), 499–509. https://doi.org/10.1038/s41587-020-0718-6.

Peebles, C. L., Perlman, P. S., Mecklenburg, K. L., Petrillo, M. L., Tabor, J. H., Jarrell, K. A., & Cheng, H. L. (1986). A self-splicing RNA excises an intron lariat. *Cell, 44*(2), 213–223. https://doi.org/10.1016/0092-8674(86)90755-5.

Pettersen, E. F., Goddard, T. D., Huang, C. C., Meng, E. C., Couch, G. S., Croll, T. I., ... Ferrin, T. E. (2021). UCSF ChimeraX: Structure visualization for researchers, educators, and developers. *Protein Science: A Publication of the Protein Society, 30*(1), 70–82. https://doi.org/10.1002/pro.3943.

Podar, M., Chu, V. T., Pyle, A. M., & Perlman, P. S. (1998). Group II intron splicing in vivo by first-step hydrolysis. *Nature, 391*(6670), 915–918. https://doi.org/10.1038/36142.

Punjani, A., Rubinstein, J. L., Fleet, D. J., & Brubaker, M. A. (2017). cryoSPARC: Algorithms for rapid unsupervised cryo-EM structure determination. *Nature Methods, 14*(3), 290–296. https://doi.org/10.1038/nmeth.4169.

Pyle, A. M. (2010). The tertiary structure of group II introns: Implications for biological function and evolution. *Critical Reviews in Biochemistry and Molecular Biology, 45*(3), 215–232. https://doi.org/10.3109/10409231003796523.

Pyle, A. M. (2016). Group II intron self-splicing. *Annual Review of Biophysics, 45*, 183–205. https://doi.org/10.1146/annurev-biophys-062215-011149.

Qu, G., Kaushal, P. S., Wang, J., Shigematsu, H., Piazza, C. L., Agrawal, R. K., ... Wang, H. W. (2016). Structure of a group II intron in complex with its reverse transcriptase. *Nature Structural & Molecular Biology, 23*(6), 549–557. https://doi.org/10.1038/nsmb.3220.

Rest, J. S., & Mindell, D. P. (2003). Retroids in archaea: Phylogeny and lateral origins. *Molecular Biology and Evolution, 20*(7), 1134–1142. https://doi.org/10.1093/molbev/msg135.

Robertus, J. D., Ladner, J. E., Finch, J. T., Rhodes, D., Brown, R. S., Clark, B. F., & Klug, A. (1974). Structure of yeast phenylalanine tRNA at 3 A resolution. *Nature, 250*(467), 546–551. https://doi.org/10.1038/250546a0.

Sievers, F., Wilm, A., Dineen, D., Gibson, T. J., Karplus, K., Li, W., ... Higgins, D. G. (2011). Fast, scalable generation of high-quality protein multiple sequence alignments using Clustal Omega. *Molecular Systems Biology, 7*, 539. https://doi.org/10.1038/msb.2011.75.

Simon, D. M., Clarke, N. A., McNeil, B. A., Johnson, I., Pantuso, D., Dai, L., ... Zimmerly, S. (2008). Group II introns in eubacteria and archaea: ORF-less introns and new varieties. *RNA (New York, N. Y.), 14*(9), 1704–1713. https://doi.org/10.1261/rna.1056108.

Toor, N., Keating, K. S., Taylor, S. D., & Pyle, A. M. (2008). Crystal structure of a self-spliced group II intron. *Science (New York, N. Y.), 320*(5872), 77–82. https://doi.org/10.1126/science.1153803.

Toor, N., Rajashankar, K., Keating, K. S., & Pyle, A. M. (2008). Structural basis for exon recognition by a group II intron. *Nature Structural & Molecular Biology, 15*(11), 1221–1222. https://doi.org/10.1038/nsmb.1509.

Toor, N., Robart, A. R., Christianson, J., & Zimmerly, S. (2006). Self-splicing of a group IIC intron: 5' exon recognition and alternative 5' splicing events implicate the stem-loop motif of a transcriptional terminator. *Nucleic Acids Research, 34*(22), 6461–6471. https://doi.org/10.1093/nar/gkl820.

Wacker, A., Weigand, J. E., Akabayov, S. R., Altincekic, N., Bains, J. K., Banijamali, E., ... Zetzsche, H. (2020). Secondary structure determination of conserved SARS-CoV-2 RNA elements by NMR spectroscopy. *Nucleic Acids Research, 48*(22), 12415–12435. https://doi.org/10.1093/nar/gkaa1013.

Wang, J., Zhao, D., Li, J., Hu, M., Xin, X., Price, M. A., ... Zhang, X. (2021). Helicase-AID: A novel molecular device for base editing at random genomic loci. *Metabolic Engineering, 67*, 396–402. https://doi.org/10.1016/j.ymben.2021.08.005.

Yang, J., Zimmerly, S., Perlman, P. S., & Lambowitz, A. M. (1996). Efficient integration of an intron RNA into double-stranded DNA by reverse splicing. *Nature, 381*(6580), 332–335. https://doi.org/10.1038/381332a0.

Yao, J., Zhong, J., & Lambowitz, A. M. (2005). Gene targeting using randomly inserted group II introns (targetrons) recovered from an Escherichia coli gene disruption library. *Nucleic Acids Research, 33*(10), 3351–3362. https://doi.org/10.1093/nar/gki649.

Zhao, C., & Pyle, A. M. (2017). Structural Insights into the Mechanism of Group II Intron Splicing. *Trends in Biochemical Sciences, 42*(6), 470–482. https://doi.org/10.1016/j.tibs.2017.03.007.

Zhong, J., Karberg, M., & Lambowitz, A. M. (2003). Targeted and random bacterial gene disruption using a group II intron (targetron) vector containing a retrotransposition-activated selectable marker. *Nucleic Acids Research, 31*(6), 1656–1664. https://doi.org/10.1093/nar/gkg248.

Zimmerly, S., Guo, H., Eskes, R., Yang, J., Perlman, P. S., & Lambowitz, A. M. (1995). A group II intron RNA is a catalytic component of a DNA endonuclease involved in intron mobility. *Cell, 83*(4), 529–538. https://doi.org/10.1016/0092-8674(95)90092-6.

Zimmerly, S., Guo, H., Perlman, P. S., & Lambowitz, A. M. (1995). Group II intron mobility occurs by target DNA-primed reverse transcription. *Cell, 82*(4), 545–554. https://doi.org/10.1016/0092-8674(95)90027-6.

Zimmerly, S., & Semper, C. (2015). Evolution of group II introns. *Mobile DNA, 6*, 7. https://doi.org/10.1186/s13100-015-0037-5.

CHAPTER TEN

Methods for Cas13a expression and purification for use in CRISPR diagnostics

Her Xiang Chai, Rebecca S. Bamert, and Gavin J. Knott*
Department of Biochemistry and Molecular Biology, Monash Biomedicine Discovery Institute, Monash University, Clayton, VIC, Australia
*Corresponding author. e-mail address: gavin.knott@monash.edu

Contents

1. Introduction 226
2. Before you begin 228
   2.1 Preparation of purification buffers 228
   2.2 Preparing working solutions for RNA detection assay (Table 1) 230
3. Key materials and equipment tables 230
   3.1 Protein purification (Table 2) 230
   3.2 Nucleic acid detection buffer components (Table 3) 232
   3.3 Oligonucleotides (Table 4) 232
   3.4 Materials and equipment 232
   3.5 Stock solution 233
4. Expression of LbuCas13a 234
   4.1 TB media preparation 234
   4.2 Transformation 234
   4.3 Overnight culture 234
   4.4 Expression (Day 1) 234
   4.5 Expression (Day 2) 235
5. Purification of LbuCas13a 235
   5.1 Purification (Day 1) 235
   5.2 Purification (Day 2) 236
6. Benchmarking LbuCas13a in an RNA detection assay 239
   6.1 Preparation of LbuCas13a RNP (Ribonucleic acid protein) complex 239
   6.2 Preparing target RNA and substrate reporter mix 239
   6.3 Conducting assay 240
7. Expected outcomes from purification 240
8. Troubleshooting and optimisation 241
   8.1 Overgrown cell pellet after overnight expression 241
   8.2 Degradation of protein during purification 242
   8.3 Incomplete cleavage of MBP tag 242

Methods in Enzymology, Volume 712
ISSN 0076-6879, https://doi.org/10.1016/bs.mie.2025.01.030

8.4 RNase contamination 242
8.5 Poor limit of detection 242
9. Concluding remarks 243
References 243

Abstract

The threat of emerging infectious diseases (e.g., SARS-CoV-2 the RNA virus responsible for the COVID-19 pandemic) has highlighted the importance of accurate and rapid testing for screening, patient diagnosis, and effective treatment of infectious disease. Nucleic acid diagnostic tools such as qPCR are considered the gold standard, providing a sensitive, accurate, and robust method of detection. However, these conventional diagnostic platforms are resource intensive, limited in some applications, and are almost always confined to laboratory settings. With the increasing demand for low-cost, rapid, and accurate point-of-care diagnostics, CRISPR-based systems have emerged as powerful tools to augment detection capabilities. Of note is the potent RNA detection enzyme, *Leptotrichia buccalis* (Lbu) Cas13a, which is capable of rapid RNA detection in complex mixtures with or without pre-amplification. To support its wide-spread use, we describe a detailed method for the expression, purification, and validation of LbuCas13a for use in molecular diagnostics.

Highlights

- A comprehensive guide to expression and purification of active LbuCas13a
- A quantitative assay for benchmarking purified LbuCas13a in CRISPR diagnostics
- Troubleshooting suggestions

1. Introduction

Nucleic acid-based diagnostics such as quantitative Polymerase Chain Reaction (qPCR) play an essential role in the diagnosis of various viral/bacterial pathogens. The technology can detect trace amounts of DNA/RNA via amplification of the target nucleic acid. The versatility, robustness and sensitivity of PCR has made it the gold standard for detection of pathogenic DNA/RNA in patient samples (de Vries et al., 2012; Oliveira, Veigas, & Baptista, 2021; Waller et al., 2020). However, the cost of reagents for this technology can be particularly expensive, especially in countries where sophisticated instruments and trained personnel are harder to source. Although there have been various advances such as isothermal amplification, which bypass the requirement of thermal cyclers – non-specific amplification can result in lower detection specificity (Kaminski, Abudayyeh, Gootenberg, Zhang, & Collins, 2021; Mahony et al., 2009).

There is a growing demand for new technologies that amalgamate the sensitivity of PCR and rapid detection of antigen test kits into a cost effective, quick, and sensitive point-of-care diagnostic.

Clustered regularly interspaced short palindromic repeat (CRISPR) and their CRISPR associated (Cas) enzymes have the potential to address areas of unmet need. The CRISPR Cas system is an adaptive immune system that can be found in prokaryotes and archaea (Ahmed, Hafeez, Ahmad, & Mahmood, 2018). This system functions in microbial adaptive immunity and grants the host defence against mobile genetic elements such as bacteriophages or foreign plasmids. The CRISPR-Cas system is currently categorised into two classes – Class I & Class II (Makarova & Koonin, 2015). The Class I CRISPR-Cas systems employ multiple Cas effector proteins as an effecter module including the Type I, III, IV (Adler et al., 2024). In contrast, Class II CRISPR-Cas systems are large single multi-domain Cas effectors including the Type II, V, VI. Of relevance for molecular diagnostics are the Type VI system which consist of four well studied subtypes VI A-D, also known as Cas13a, Cas13b, Cas13c, Cas13d, respectively (Adler et al., 2024).

The Type VI CRISPR-Cas13 effectors are ribonucleases, that upon specific hybridisation of their crRNA to target RNA activate nuclease (HEPN) domains resulting in collateral degradation of single stranded RNA (ssRNA) (Abudayyeh et al., 2016; East-Seletsky et al., 2016). This collateral nuclease activity has found use as a diagnostic tool, allowing measurement of fluorescent probe ssRNA cleavage in response to the presence of the activating target RNA (East-Seletsky et al., 2016). Cas13a (formerly known as C2c2) has been studied extensively and deployed in various platforms, such as SHERLOCK, to detect viral RNA pathogens as a point-of-care diagnostic (Gootenberg et al., 2017). More recently, researchers have developed CRISPR-based RNA detection methods that are so called 'amplification free'. For example, *Leptotrichia buccalis* (LbuCas13a) has been shown to detect SARS-CoV2 RNA in patient samples with a Limit of Detection (LoD) of 100 copies/μL without the need for target RNA amplification through crRNA multiplexing (Fozouni et al., 2021). LbuCas13a coupled with Csm6 (Class I Type III CRISPR-Cas system) enhances the LoD to 30 copies/μL of target RNA under 20 min (Liu et al., 2021). The application of these methods in droplets further improves the sensitivity to bring it on par with the gold standard qPCR (Son et al., 2021). With the immense potential of Cas13a based point-of-care diagnostics, we prepared the following detailed protocol to express, purify and benchmark LbuCas13a for applications in RNA detection and molecular diagnostics.

2. Before you begin

This protocol describes a method for the expression and purification of LbuCas13a from bacteria (*Escherichia coli* BL21 DE3). It is important to note that the Cas13a protein is prone to protease degradation, oxidation, and temperature-dependent denaturation. Therefore, prepare and chill the materials associated with protein purification (e.g., buffers, reducing agent stocks, equilibrated columns etc.) beforehand to increase efficiency. Note that buffers containing reducing agents will have a limited shelf-life when stored at 4 °C.

This protocol also describes a method for testing the expressed and purified LbuCas13a in an RNA detection assay. Note that Cas13 RNA detection assays are sensitive to RNase contamination so appropriate care should be taken to minimise the risk of RNase contamination.

2.1 Preparation of purification buffers

Ensure all buffer solutions are filtered (0.22 μm) once made.

Prepare the purification buffers without adding any reducing agents and protease inhibitors. When purifying protein, decant the desired volume of buffers and add the appropriate volume of reducing agents/protease inhibitors from stock solutions. Once a reducing agent or protease inhibitor is added, the buffers have a shelf life of about 1 week at 4 °C.

The pH for all buffers are adjusted at room temperature.

a. Resuspension buffer (2 L)
50 mM Tris (pH 7.0 with HCl)
1 M NaCl
5 % (v/v) Glycerol

b. Lysis buffer (2 L)
50 mM Tris (pH 7.0 with HCl)
500 mM NaCl
10 mM Imidazole
5 % (v/v) Glycerol
1 mM TCEP
0.25 mM PMSF
cOmplete, EDTA-free protease inhibitor cocktail

c. Wash I buffer (2 L)
50 mM Tris (pH 7.0 with HCl)
500 mM NaCl
10 mM Imidazole
5 % (v/v) Glycerol
1 mM TCEP

d. Wash II buffer (2 L)
50 mM Tris (pH 7.0 with HCl)
500 mM NaCl
20 mM Imidazole
5 % (v/v) Glycerol
1 mM TCEP

e. Elution buffer (2 L)
50 mM Tris (pH 7.0 with HCl)
500 mM NaCl
300 mM Imidazole
5 % (v/v) Glycerol
1 mM TCEP

f. Dialysis buffer (1 L)
50 mM Tris (pH 7.0 with HCl)
150 mM NaCl
5 % (v/v) Glycerol
1 mM TCEP

g. IEX A buffer (2 L)
50 mM HEPES (pH 7.0 with KOH)
200 mM KCl
5 % (v/v) Glycerol
1 mM TCEP

h. IEX B buffer (1 L)
50 mM HEPES (pH 7.0 with KOH)
1 M KCl
5 % (v/v) Glycerol
1 mM TCEP

i. SEC buffer (2 L)
20 mM HEPES (pH 7.0 with KOH)
200 mM KCl
10 % (v/v) Glycerol
1 mM TCEP

j. 5X Assay buffer (50 mL)
100 mM HEPES (pH 6.8 with KOH)
250 mM KCl
25 mM MgCl2
0.05 % IGEPAL (CA-630)
10 % (v/v) Glycerol

2.2 Preparing working solutions for RNA detection assay (Table 1)

Table 1 Components and reagents to prepare oligonucleotides to working concentration.

Components	Working concentration (μM)	Diluent
Purified LbuCas13a	1	Cas13a SEC buffer
crRNA	1	THE RNA storage solution
target RNA	1	THE RNA storage solution
Substrate reporter	10	THE RNA storage solution

3. Key materials and equipment tables

3.1 Protein purification (Table 2)

Table 2 Key information on the reagents/components used to purify LbuCas13a.

Reagents or components	Manufacturer/source	Identifier
Expression/purification buffers	Laboratory developed	
Escherichia coli BL21 DE3	Thermo Fisher Scientific	EC0114
LbuCas13a plasmid	Addgene	Plasmid#83482
Yeast Extract	Millipore	1.03753
Tryptone (Vegetable)	Millipore	16922
Potassium phosphate monobasic	Sigma Aldrich	P5655 -1KG
Potassium phosphate dibasic	Supelco	1.05104

Glycerol	Ajax Finechem	AJA242-2.5PL
Tris	MP Biomedical	01203133-CF
HEPES	Sigma Aldrich	H40341KG
NaCl	Sigma Aldrich	S3014-1KG
KCl	Sigma Aldrich	P9541-1KG
Imidazole	Sigma Aldrich	I2399-500G
TCEP-HCl	GoldBio	TCEP100
Ampicillin	GoldBio	A-301-500
UltraYield® 2.5 L Flask	Thomson Instrument Company	931136-B
cOmplete, EDTA-free Protease Inhibitor Cocktail	Roche	5056489001
Ni-NTA superflow (500 mL)	Qiagen	#30450
Econo-Column® Chromatography columns	BioRad	#7372522
Dialysis tubing cellulose membrane	Sigma Aldrich	D9652
HiTrap SP HP	Cytiva	17115201
Superdex™ 200 Increase 10x300mm[a]	Cytiva	28990944
384-well dark low volume plate	Corning	3820
THE RNA storage solution	ThermoScientific	AM7001
Plate adhesive seals	ThermoScientific	AB0558
TEV protease	Laboratory developed	

[a]Depending on the scale of the purification - larger purification batches we recommend HiLoad Superdex 200 pg preparative SEC columns (#28989335).

3.2 Nucleic acid detection buffer components (Table 3)

Table 3 Regents/components used to make a 5X assay buffer.

Reagents or components	Manufacturer/ source	Identifier
1 M HEPES	Sigma Aldrich	H3537-mL
1 M $MgCl_2$	Sigma Aldrich	63069-100 ML
IGEPAL® CA-630	Sigma Aldrich	I8896-mL
Glycerol	MP Biomedical	0219399680

3.3 Oligonucleotides (Table 4)

Table 4 The nucleic acid sequences of RNA used in RNA detection assay.

Oligo ID	Sequence (5′–3′)
crRNA[a]	*UAGACCAGCCCAAAAAUGAAGGGCACUA AAAC***GCAGCGCCUCUUGCAACGAUUAAA**
Target RNA	UUUAAUCGUUGCAAGAGGCGCUGCUC
Substrate reporter[b]	6-FAM 5'-UUUUU-3' Iowa Black

[a]crRNA sequence – Direct repeat Spacer sequence (complementary to target RNA sequence).
[b]Synthesised by IDT (Integrated DNA Technologies) at 250 nmol scale with RNase-free HPLC purification.

3.4 Materials and equipment

- pH metre
- Stir plate and stir bar
- Rotary instruments for rotating 50 mL conical tubes and Econo columns
- Hamilton syringes
- A set of single channel* and multichannel pipette**
  - o 0.5 – 2 uL*
  - o 1 – 10 uL*
  - o 2 – 20 uL* and 2 – 20 uL**
  - o 20 – 200 uL*
  - o 200 – 1000 uL*
- Pipettor with 5 mL, 25 mL, 50 mL pipettes
- 1.5 mL microcentrifuge tubes
- 15- and 50-mL conical tubes

- Amicon Ultra Centrifugal filters (4 mL and 15 mL capacity with 30 kDa MWCO)
- 40 % acrylamide (37.5: 1 acrylamide: bis-acrylamide; BioRad)
- N,N,N',N' – tetramethylethylenediamine (TEMED; BioRad)
- Mini-PROTEAN Tetra Vertical Electrophoresis cell (BioRad)
- Homogeniser (Wiggens)
- Electrophoresis power supply
- Benchtop centrifuge (Eppendorf 5804 R)
- Heat block with digital thermostat
- Petri dishes
- 50 mL conical flask
- Aluminium foil
- Autoclave tape
- 1 L and 2 L measuring cylinder
- 0.25 L, 0.5 L, 1 L and 2 L Schott Bottle
- Plate reader
- AKTA FPLC
- Autoclave
- High speed centrifuge
- Refrigerated incubator
- Spectrophotometer
- Sonicator
- Protein ladder (10–250 kDa)

3.5 Stock solution

- 5 M KOH
- 5 M NaOH
- 0.5 M TCEP pH 7.0 (store at −20 °C)
- Ampicillin 100 mg/mL (store at −20 °C)
- 0.25 M PMSF (store at −20 °C)
- 50 % (v/v) Glycerol, autoclaved
- 2 M Imidzaole
- 2 M KCl
- 2 M NaCl
- 1 M HEPES
- 1 M Tris
- 10 % (w/v) APS (store at −20 °C)
- 10 % (w/v) SDS
- 10X SDS running buffer

- Agar
- Coomassie blue stain
- 1.5 M Tris-Cl (pH 8.8)
- 0.5 M Tris-Cl (pH 6.8)

4. Expression of LbuCas13a

4.1 TB media preparation

1. Dissolve 115.2 g Yeast extract, 57.6 Tryptone (vegetable), 11 g Potassium phosphate monobasic, 60.2 g Potassium phosphate dibasic into 4.8 L of MilliQ H_2O
2. Transfer 800 mL of TB media into 2 L UltraYield Flask.
3. Cover the flask with foil and autoclave for 20 min at 120 °C.
4. This procedure yields 6 flask of TB media. Media can be stored at room temperature for months prior to use.
5. 50 % (v/v) Glycerol is added before seeding the flask with overnight culture.

4.2 Transformation

1. Transform p2CT-His-MBP-Lbu-C2c2-WT (https://www.addgene.org/83482/) into BL21 DE3 chemically competent cells.
2. Plate the cells onto LB agar containing Ampicillin (100 μg/mL).
3. Incubate the plates overnight (16 h) at 37 °C.

4.3 Overnight culture

1. Store plates with colonies at 4 C until the afternoon.
2. Inoculate a single colony into 30 mL of fresh LB media with Ampicillin (100 μg/mL)
3. Incubate the culture overnight (16 h) at 200 RPM at 37 °C.

4.4 Expression (Day 1)

1. Store the overnight culture at 4 °C until large scale expression commences later in the afternoon. It is important that the induced cultures do not grow for more than 16 h overnight, hence delaying the start of expression ensures that harvest times can be within usual working hours.
2. Allow the overnight culture to sit at room temperature for 10 min before seeding the flask

3. To each 800 mL flask of pre-prepared TB media, add 6.4 mL 50 % (v/v) Glycerol and 800 μL Ampicillin (100 mg/mL) and mix thoroughly by swirling
4. To each flask seed 4 mL overnight culture and incubate at 37 °C 180 RPM until the culture has reached OD_{600} 0.5. This will take approximately 4-5 h; hence it is suggested to commence the large scale expression at ~1 p.m. and check the OD every 20 min after 3 p.m. The culture should reach 0.5 OD_{600} around 4.30 p.m.
5. Once the desired OD_{600} is achieved, cold shock the cultures in an ice bath for 20–30 min.
6. Whilst cooling cultures, lower the incubator temperature to 16 °C.
7. After 20–30 min of cold shock, induce the cells with 0.5 mM IPTG
8. Return the flasks into the incubator.
9. Incubate the flasks overnight (16 h) at 16 °C 180 RPM.

4.5 Expression (Day 2)

1. Pellet the cells by transferring the culture into a centrifuge bottle and spin down at 4000 x g for 15 min
2. Decant the supernatant. Important note: Examine the pellet. For a successful expression the pellet should weigh around 3–4–g for each flask and it should have a pale yellow to blonde colour.
3. Resuspend the cell pellet with 30 mL of Resuspension buffer by pipetting. Transfer resuspended cells into 50 mL Falcon tubes and pellet the cells by centrifuging at 4000 x g for15 minutes.
4. Decant the supernatant. Snap-freeze the cell pellets using liquid nitrogen or dry-ice. Store the pellets at −80 °C until purification.

5. Purification of LbuCas13a

5.1 Purification (Day 1)

1. Ensure to carry out the purification in cold conditions to minimise the degradation of protein.
2. Thaw the cell pellets on ice for an hour or until the cells have fully thawed.
3. Decant the following volume of buffers into bottles on ice for purification and add reducing agent to all buffers (as instructed in see Section 2.1):
   a. 300 mL Lysis buffer*

b. 250 mL Wash I buffer
c. 250 mL Wash II buffer
d. 100 mL Elution buffer

*note also add PMSF and EDTA-free cOmplete protease inhibitor tablet to lysis buffer just prior to use.

4. Resuspend the Cas13a expression pellets in Lysis buffer (10 mL/g wet cell paste).
5. Create a homogenous suspension using a handheld homogeniser until no clumps are present.
6. Transfer the homogenous cell suspension into a glass beaker in an ice bath
7. Sonicate to lyse the cells at 10 kHz with the beaker in an ice bath. Sonicate for 30 s followed by 30 s of rest. Repeat this step 4 to 6 times.
8. Clarify the cell lysates via centrifugation for 32,000 x g for 45 min at 4 °C.
9. Transfer 10 mL of 50 % Ni-NTA resin into an Econo column and wash with 10X resin volumes of H2O followed by 10X resin volumes of Lysis buffer
10. Transfer the clarified lysate to the Econo column containing the 5 mL pre-equilibrated Ni-NTA resin and incubate on a roller for 1 h at 4 °C.
11. Drain the flow-through and wash the resin with 10X resin volume of Wash buffer I followed by Wash buffer II.
12. Elute bound Cas13a from the column using 1X resin volumes of Elution buffer. Repeat this step 4 times and collect the elutions at each step.
13. Load 10 μL of each elution onto a 10 % SDS-PAGE gel to visualise the quality of purification (Fig. 1).
14. Measure A280 and A260/A280 of each fraction on a spectrophotometer to determine the purity of each fraction (Fig. 1).
15. Pool the Cas13a fractions with high protein concentration (see Elution 1&2 Fig. 1) and measure the pooled protein concentration by spectrophotometer again.
16. Thaw and add TEV protease with a mass ratio of 1:100 (mg of TEV: mg of His-MBP Cas13a)
17. Transfer the solution into a dialysis tubing and dialyse overnight against 50X volumes of dialysis buffer with slow stirring at 4 °C.

5.2 Purification (Day 2)

1. Centrifuge the dialysed solution at 4,000 x g for 30 min to sediment precipitate. Note: As the protein undergoes proteolytic cleavage by TEV protease precipitation can occur.

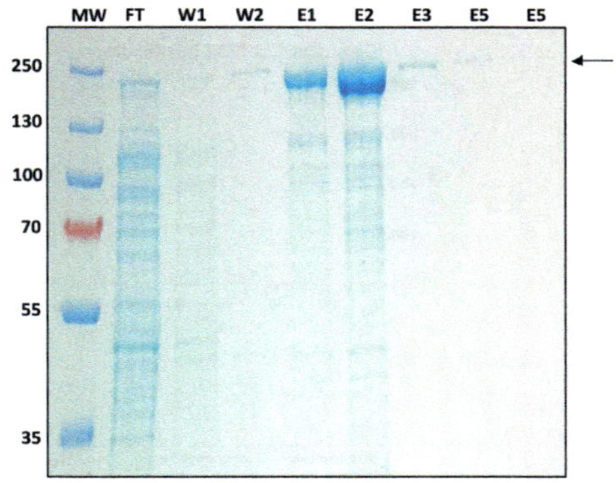

Sample	Concentration (mg/mL)	A_{260}/A_{280}
E1	0.640	1.047
E2	1.840	0.609
E3	0.096	1.156
E4	0.052	1.481
E5	0.084	1.690

Fig. 1 Ni-NTA purification of His-MBP-LbuCas13a visualised by SDS-PAGE (left) and protein concentration, and A_{260}/A_{280} ratios of eluted fraction, as measured by spectrophotometer (right). Lane 1 illustrates the protein ladder of known molecular weight. Adjacent to the molecular weight markers (MW), the gel illustrates the His-MBP-LbuCas13a enrichment and purification from Ni-NTA: FT (Lysate flow through), W1 (10 mM Imidazole wash), W2 (20 mM Imidazole wash), E1-E5 (Elution 1–5). Black arrow indicates the expected protein band that corresponds to His-MBP-LbuCas13a at a molecular weight of 183 kDa. 10 µL of each sample was loaded onto a 10 % Tris-Glycine SDS-PAGE gel and resolved for 1 h at 140 V. The gel was stained with Coomassie blue.

2. Pass the supernatant through a 0.22 µm filter to remove any trace precipitate before loading onto cation exchange.
3. Load the filtered, centrifuged and dialysed protein onto a 5 mL HiTrap SP column equilibrated with IEX Buffer A
4. Elute the protein with a linear gradient (0–100 % over 5X CV) of Cas13a IEX Buffer B (Fig. 2A).
5. Load 10 µL of each elution onto a 10 % SDS-PAGE gel to visualise the quality of purification.
6. Measure A_{280} and A_{260}/A_{280} of each fraction. Pool the desired Cas13a fractions together (single peak at chromatogram in Fig. 2A) and measure the pooled concentration by spectrophotometer.
7. Concentrate Cas13a to the desired injection volume for SEC (no more than 5 % of total column volume) using an Amicon Ultra 30 K MWCO by centrifugation at 4000 x g for 5 min at a time at 4 °C. After each spin, pipette to evenly redistribute before centrifuging again.
8. Load the concentrated fractions onto a pre-equilibrated (SEC buffer) HiLoad Superdex 200 pg preparative SEC column and elute with 1CV of SEC buffer at 0.4 mL/min (Fig. 2B). Anticipate that Cas13a will elute at approximately 65 mL, corresponding to a monomeric MW of 130 kDa.

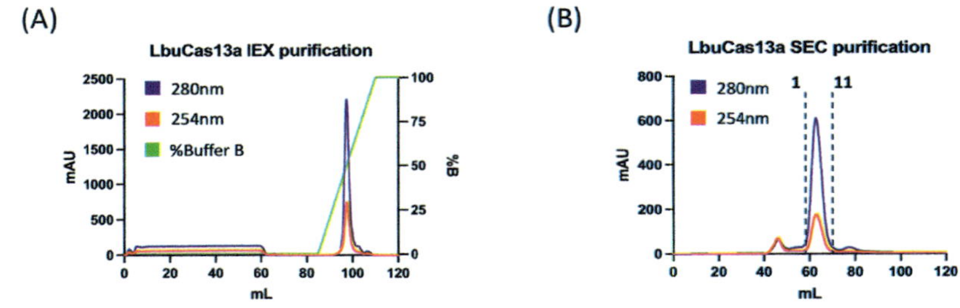

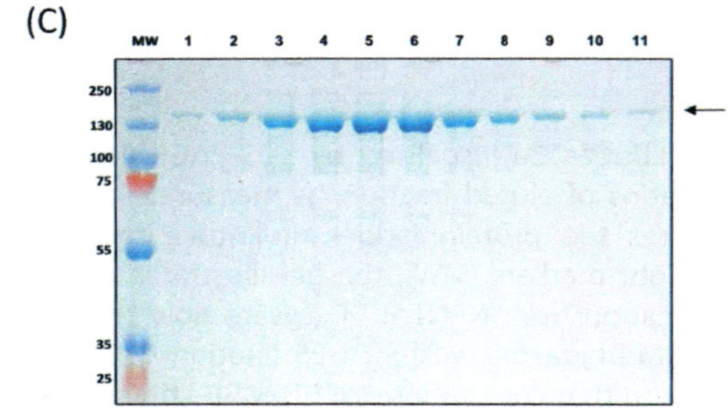

Size exclusion fraction No.	Concentration (mg/mL)	A_{260}/A_{280}
1	0.128	0.563
2	0.337	0.460
3	0.698	0.468
4	1.151	0.425
5	1.324	0.422
6	1.131	0.453
7	0.788	0.468
8	0.472	0.506
9	0.25	0.640

Fig. 2 Chromatograms and SDS-PAGE gel of the second purification process. (A) Chromatogram of LbuCas13a purification over HiTrap SP 5 mL column. The protein was eluted from the column at ~600 mM KCl. (B) Chromatogram of LbuCas13a purification over S200 16/600 column. The protein elutes as a single monomer-sized peak. (C) 10 % SDS-PAGE gel of fraction 1–11 from SEC (Fig. 2B) stained with Coomassie blue, black arrow indicates purified Cas13 (left); protein concentration and A_{260}/A_{280} ratio of size exclusion fractions, as measured by spectrophotometer (right).

9. Load 10 μL of each fraction across peak onto a 10 % SDS-PAGE gel to visualise the quality of purification (Fig. 2C).
10. Measure A_{280} and A_{260}/A_{280} of each fraction. Pool Cas13a fractions with single band on SDS-PAGE gel (size exclusion fractions 4–6, Fig. 2C) and measure the protein concentration by spectrophotometer
11. Concentrate the protein using a 30 K MWCO until 50–60 μM and aliquot 50 μL into PCR tubes. Use the formula below to calculate the molar concentration of purified protein.

$$Protein concentration\,(\mu M) = \left(\frac{A_{280}}{118280 M^{-1}\,cm^{-1}}\right) X\ 10^{6}$$

12. Snap freeze the protein aliquots in liquid nitrogen and store at −80 °C.

6. Benchmarking LbuCas13a in an RNA detection assay

The following assay allows for assessment of the activity of purified LbuCas13a. Upon binding target RNA, the guide(crRNA)-bound LbuCas13a nuclease activity results in cleavage of a fluorescent substrate reporter RNA, and a rapid, quantitative increase in fluorescent signal. The limit of detection in this assay can be used as an indicator of the purified proteins quality. Using this method, we observe reproducible activity down to a target RNA concentration of 10fM (Table 5).

6.1 Preparation of LbuCas13a RNP (Ribonucleic acid protein) complex

1. Rapid thaw concentrated aliquot of LbuCas13a in between the fingers. Once thawed, dilute to 1uM with ice cold SEC buffer and store on ice.
2. Thaw the RNA at room temperature and once thawed immediately return to ice.
3. Prepare a 100 nM target RNA solution and use to generate a 1/10 serial dilution series (10 nM, 1 nM, 0.1 nM, 10 pM, 1 pM, 0.1 pM, 10 fM, 0 fM) in water
4. Prepare RNP mix by mixing the following components (Table 6). Pipette gently until the phases are mixed well together.
5. Incubate the RNP mix at room temperature for at least 15 min and no longer than 1 h.
6. While waiting for RNP complex to incubate prepare the target RNA and reporter substrate mix.

6.2 Preparing target RNA and substrate reporter mix

1. Add the following components in Table 7 in order, making 10ul of each dilution of target RNA reporter mix.

Table 5 Final concentration of protein and RNA components in the 384 well plate.

Components	Concentration
LbuCas13a	100 nM
crRNA	100 nM
Target RNA	1 nM–1 fM
Substrate reporter	400 nM

Table 6 The components and volume required to make Cas13a RNP complex at room temperature.

Components	Volume (μL)
5X Assay buffer	32
H2O	88
1 μM crRNA	20
1 μM LbuCas13a	20

Table 7 The reagents required to make the target RNA substrate mix.

Reagents	1 nM	0.1 nM	10 pM	1 pM	0.1 pM	10 fM	1 fM	0 fM
5X Assay buffer	2	2	2	2	2	2	2	2
H_2O	2.4	2.4	2.4	2.4	2.4	2.4	2.4	2.4
Target RNA (from serial dilution)	4	4	4	4	4	4	4	4
Reporter (10 μM)	1.6	1.6	1.6	1.6	1.6	1.6	1.6	1.6
Final volume	10	10	10	10	10	10	10	10

6.3 Conducting assay

1. 1. Pipette 5 μL of target RNA reporter substrate into a black 384- well plate.
2. Next pipette 15 μL of RNP mix (Table 6, above) into each well and mix by pipette gently.
3. Cover the plate with a clear adhesive seal and use a KimWipe to press firmly to ensure the plate is sealed evenly.
4. Load the plate into a plate reader prewarmed to 37 °C and measure fluorescence output every 5 min, for 120 min total assay duration (Fig. 3). Excitation and emission frequencies will depend upon substrate reporter fluorophore selected (6-FAM excitation and emission frequencies are 495 nm and 520 nm, respectively).

7. Expected outcomes from purification

Using this purification method should result in 600 μL of 60 μM (~7 mg/mL) pure Cas13a from 4.8 L of TB media. Purified LbuCas13a

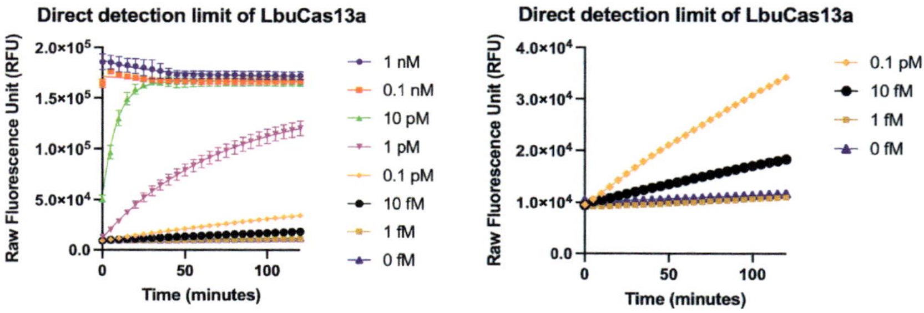

Fig. 3 The limit of detection of Target RNA using LbuCas13a without pre-amplification from 1 nM, 0.1 nM, 10 pM, 1 pM, 0.1 pM, 10 fM, 1 fM, 0 fM. Right most panel is a detail of the lower target RNA concentrations traces seen in the left panel. Excitation and emission frequencies of 495 nm and 520 nm.

protein should show a limit of detection of 10 fM and little to no evident background RNase activity in the absence of target RNA (0 fM condition).

8. Troubleshooting and optimisation

8.1 Overgrown cell pellet after overnight expression

Should your pellets be excessively heavier (10–12 g) than the recommended pellet it might be an indication that the protein was not expressed. LbuCas13a overexpression can be toxic to bacteria, hampering cell growth. Mutation of the plasmid and selection pressure during expression will allow cells to recover and grow at a quicker pace, leading to larger pellets with less protein expressed. There are several measures you can take to solve this and its relatively easy to troubleshoot:

1. Adding glucose to your overnight culture.
   - p2CT plasmids are prone to leaky expression. Supplementing the overnight culture with a small percentage of glucose (0.5-1.0 %, w/v) can reduce the basal expression of LbuCas13a (Tahara et al., 2021)
2. Freshly transformed BL21 (DE3) and competent cells
   - From previous experience, using BL21 (DE3) cells that have been transformed with p2CT-His-MBP-LbuCas13a plasmids more than 1 week prior results in increased cell mass and decreased protein yield. Therefore, it is best to use freshly transformed cells (less than 1 week old), in addition to that it is to always use freshly made competent cells (less than 2 months old). As such we recommend that when expressing

LbuCas13a is to inoculate from a freshly transformed BL21 on an Agar plate and not glycerol stock.

8.2 Degradation of protein during purification

LbuCas13a a large multidomain protein that is prone to degradation from proteases during purification. By adding protease inhibitors, working quickly, and at 4 °C as much as possible, this can be minimised. In this protocol the lysis buffer is supplemented with protease inhibitor cocktails and PMSF. Adding protease inhibitors to other purification buffers can be helpful too. Additionally, if degradation is occurring during or after dialysis consider the addition of 1 mM EDTA to chelate divalents.

8.3 Incomplete cleavage of MBP tag

Increasing the ratio of TEV protease to LbuCas13a from 1:100 to 1:50 should help to ensure all the MBP tag has been cleaved off. It is recommended to run a sample from post TEV digestion on an SDS-PAGE gel to confirm all the MBP tag has been cleaved. In the case of incomplete cleavage of MBP-tag during large scale expression, attaching a MBP Trap HP in series, before the HiTrap SP column can help capture undigested proteins. Note that lack of complete digestion can suggest MBP-LbuCas13a aggregation.

8.4 RNase contamination

Due to the reporter in this assay being an RNA, trace amounts of RNase present in the buffers or reagents can generate a signal in the absence of activated target bound Cas13a (0 fM target DNA/negative control). RNase contamination is most likely a result of improper handling of reagents and buffer. Aseptic RNA handling of reagents and buffers can reduce the likelihood of RNase contamination. For example: wearing a mask, use filter pipette tips, use DEPC treated water, wipe down workspace with 80 % (v/v) Ethanol and RNaseZap™ (Nielsen, 2011).

8.5 Poor limit of detection

If the limit of detection is within the nanomolar range (and above the reported femtomolar range described above) there is a possibility that the protein preparation has reduced activity. Load the purified LbuCas13a again on a 10 % SDS-PAGE gel to verify protein has not been degraded. If SDS-PAGE analysis confirms no protein degradation (no additional bands from previous SDS-PAGE gel), supplement the 5X Assay buffer with

1 mM TCEP/DTT. If adding additional reducing agent does not lead to efficient RNA detection, the protein should be re-expressed and purified under ideal conditions to recover full activity.

9. Concluding remarks

The expression, purification, and validation of LbuCas13a requires multiple purification steps and careful attention to detail. The procedures described in this protocol provide a comprehensive guide to allow expression and purification of catalytically active, pure LbuCas13a that displays femtomolar limits of detection. By enabling large scale LbuCas13a production this method will aid in the development of improved point-of-care diagnostics.

References

Abudayyeh, O. O., Gootenberg, J. S., Konermann, S., Joung, J., Slaymaker, I. M., Cox, D. B. T., et al. (2016). C2c2 is a single-component programmable RNA-guided RNA-targeting CRISPR effector. *Science, 353*(6299), aaf5573.

Adler, B. A., Trinidad, M. I., Bellieny-Rabelo, D., Zhang, E., Karp, H. M., Skopintsev, P., et al. (2024). CasPEDIA database: A functional classification system for class 2 CRISPR-Cas enzymes. *Nucleic Acids Research, 52*(D1), D590-d6.

Ahmed, W., Hafeez, M. A., Ahmad, R., & Mahmood, S. (2018). CRISPR-Cas system in regulation of immunity and virulence of bacterial pathogens. *Gene Reports, 13*, 151–157.

de Vries, J. J., van der Eijk, A. A., Wolthers, K. C., Rusman, L. G., Pas, S. D., Molenkamp, R., et al. (2012). Real-time PCR versus viral culture on urine as a gold standard in the diagnosis of congenital cytomegalovirus infection. *Journal of Clinical Virology, 53*(2), 167–170.

East-Seletsky, A., O'Connell, M. R., Knight, S. C., Burstein, D., Cate, J. H. D., Tjian, R., & Doudna, J. A. (2016). Two distinct RNase activities of CRISPR-C2c2 enable guide-RNA processing and RNA detection. *Nature, 538*(7624), 270–273.

Fozouni, P., Son, S., Díaz de León Derby, M., Knott, G. J., Gray, C. N., D'Ambrosio, M. V., et al. (2021). Amplification-free detection of SARS-CoV-2 with CRISPR-Cas13a and mobile phone microscopy. *Cell, 184*(2), 323–333.e9.

Gootenberg, J. S., Abudayyeh, O. O., Lee, J. W., Essletzbichler, P., Dy, A. J., Joung, J., et al. (2017). Nucleic acid detection with CRISPR-Cas13a/C2c2. *Science, 356*(6336), 438–442.

Kaminski, M. M., Abudayyeh, O. O., Gootenberg, J. S., Zhang, F., & Collins, J. J. (2021). CRISPR-based diagnostics. *Nature Biomedical Engineering, 5*(7), 643–656.

Liu, T. Y., Knott, G. J., Smock, D. C. J., Desmarais, J. J., Son, S., Bhuiya, A., et al. (2021). Accelerated RNA detection using tandem CRISPR nucleases. *Nature Chemical Biology, 17*(9), 982–988.

Mahony, J. B., Blackhouse, G., Babwah, J., Smieja, M., Buracond, S., Chong, S., et al. (2009). Cost analysis of multiplex PCR testing for diagnosing respiratory virus infections. *Journal of Clinical Microbiology, 47*(9), 2812–2817.

Makarova, K. S., & Koonin, E. V. (2015). Annotation and classification of CRISPR-Cas systems. *Methods in Molecular Biology, 1311*, 47–75.

Nielsen, H. (2011). Working with RNA. In H. Nielsen (Ed.). *RNA: Methods and protocols* (pp. 15–28). Totowa, NJ: Humana Press.

Oliveira, B. B., Veigas, B., & Baptista, P. V. (2021). Isothermal amplification of nucleic acids: The race for the next "Gold Standard". *Frontiers in Sensors, 2*.

Son, S., Lyden, A., Shu, J., Stephens, S. I., Fozouni, P., Knott, G. J., et al. (2021). Sensitive and multiplexed RNA detection with Cas13 droplets and kinetic barcoding. *medRxiv* 2021.08.02.21261509.

Tahara, N., Tachibana, I., Takeo, K., Yamashita, S., Shimada, A., Hashimoto, M., et al. (2021). Boosting auto-induction of recombinant proteins in Escherichia coli with glucose and lactose additives. *Protein and Peptide Letters, 28*(10), 1180–1190.

Waller, J. V., Kaur, P., Tucker, A., Lin, K. K., Diaz, M. J., Henry, T. S., & Hope, M. (2020). Diagnostic tools for coronavirus disease (COVID-19): Comparing CT and RT-PCR viral nucleic acid testing. *American Journal of Roentgenology, 215*(4), 834–838.

CHAPTER ELEVEN

Using CRISPR for viral nucleic acid detection

Maaike S. Grimm[a] **and Cameron Myhrvold**[a,b,c,d,*]
[a]Department of Molecular Biology, Princeton University, Princeton, NJ, United States
[b]Department of Chemical and Biological Engineering, Princeton University, Princeton, NJ, United States
[c]Omenn-Darling Bioengineering Institute, Princeton University, Princeton, NJ, United States
[d]Department of Chemistry, Princeton University, Princeton, NJ, United States
*Corresponding author. e-mail address: cmyhrvol@princeton.edu

Contents

1. Introduction 246
2. Existing nucleic acid detection methods 247
3. Class 2 CRISPR-Cas systems 249
   3.1 Cas9 249
   3.2 Cas12 249
   3.3 Cas13 251
4. First-generation detection methods 253
5. Single-step detection methods 254
6. Amplification-free methods 257
7. Multiplexed detection methods 259
8. How to perform CRISPR-based viral detection 261
   8.1 Comparison of CRISPR-Cas viral nucleic acid detection methods 263
9. Challenges and future improvements 263
10. Conclusion 268
References 269

Abstract

Pathogenic microorganisms, such as viruses, have threatened human health and will continue to contribute to future epidemics and pandemics, highlighting the importance of developing effective diagnostics. To contain viral outbreaks within populations, fast and early diagnosis of infected individuals is essential. Although current standard methods are highly sensitive and specific, like RT-qPCR, some can have slow turnaround times, which can hinder the prevention of viral transmission.

The discovery of CRISPR-Cas systems in bacteria and archaea initially revolutionized the world of genome editing. Intriguingly, CRISPR-Cas enzymes also have the ability to detect nucleic acids with high sensitivity and specificity, which sparked the interest of researchers to also explore their potential in diagnosis of viral pathogens. In particular, the CRISPR-Cas13 system has been used as a tool for detecting viral nucleic acids. Cas13's capability to detect both target RNA and non-specific RNAs has led to the development of detection methods that leverage these characteristics through

Methods in Enzymology, Volume 712
ISSN 0076-6879, https://doi.org/10.1016/bs.mie.2025.01.031

designing specific detection read-outs. Optimization of viral sample collection, amplification steps and the detection process within the Cas13 detection workflow has resulted in assays with high sensitivity, rapid turnaround times and the capacity for large-scale implementation. This review focuses on the significant innovations of various CRISPR-Cas13-based viral nucleic acid detection methods, comparing their strengths and weaknesses while highlighting Cas13's great potential as a tool for viral diagnostics.

1. Introduction

Viral pathogens like Influenza A viruses and Coronaviruses have caused numerous pandemics and epidemics worldwide and will continue to threaten human health (Buchy et al., 2021). To prevent disease progression and limit spread amongst populations, detecting these pathogens in infected individuals is crucial. Currently, nucleic acid amplification tests (NAATs), such as quantitative polymerase chain reaction (qPCR) and reverse transcription qPCR (RT-qPCR), are the gold standard for detection due to their sensitivity and specificity (Bohn et al., 2020; Maleki & Hojati, 2022). Nevertheless, their sample processing and reliance on specialized equipment and trained personnel slow-down the turnaround time (Maleki & Hojati, 2022), thereby limiting their utility in responding to outbreaks and pandemics, where fast diagnosis is crucial to contain the viral spread. Furthermore, PCR only detects the presence of a virus but often sequencing is needed to determine the specific strain or variant, which is costly and time-consuming. Hence, there is a need for faster, user-friendly, high-throughput, and highly sensitive and specific detection methods for viral nucleic acids. Such methods would be advantageous for point-of-care (POC) and point-of-need (PON) settings, aiding in prevention and containment of future viral outbreaks.

The discovery of Clustered Regularly Interspaced Short Palindromic Repeats (CRISPR), integral to the adaptive immunity of bacteria and archaea, has paved the way for various biotechnological applications, including nucleic acid detection. CRISPR, along with CRISPR-associated (Cas) proteins, form CRISPR-Cas systems, which are used to acquire immune memory against invading genetic elements, such as bacteriophages and plasmids (Barrangou et al., 2007; Brouns et al., 2008; Rajan et al., 2022). The loci of these systems include a CRISPR array with various spacers derived from these genetic elements, flanked by direct repeat sequences and Cas genes encoding the related Cas proteins

(Davidson et al., 2020; Yang & Patel, 2024). The immune memory acquisition involves three steps: adaptation, crRNA maturation and interference. Adaptation includes the incorporation of invading nucleic acids (proto-spacers) into the CRISPR array (Garneau et al., 2010; Heler, Marraffini, & Bikard, 2014). After, the CRISPR array gets transcribed, resulting in pre-crRNA, which subsequently gets processed by Cas proteins to generate mature crRNAs (Gesner, Schellenberg, Garside, George, & Macmillan, 2011; Haurwitz, Jinek, Wiedenheft, Zhou, & Doudna, 2010). Lastly, during interference, mature crRNAs bound to the Cas protein form the effector complex that uses the crRNA as a guide to detect foreign nucleic acids complementary to that crRNA, which leads to cleavage of the target DNA or RNA (Marraffini & Sontheimer, 2010a; Wiedenheft, Sternberg, & Doudna, 2012; Yang & Patel, 2024). CRISPR-Cas systems are categorized into two classes based on their effector proteins and diverse loci arrangements. Class 1 CRISPR-Cas systems include types I, III and IV and use multiple Cas proteins for adaptation and interference. In contrast, class 2 systems include types II, V and VI and use a single multi-domain CRISPR effector protein for degradation of invading nucleic acids (Makarova et al., 2015; Makarova, Zhang, & Koonin, 2017a; Makarova, Zhang, & Koonin, 2017b; Shmakov et al., 2017).

Combining CRISPR-Cas complexes' ability to detect specific nucleotide sequences with various amplification methods has led to the development of CRISPR-based diagnostics (CRISPR-Dx). Some Cas enzymes, upon activation, can cleave non-specific DNA or RNA, a process known as *trans* cleavage. Researchers have utilized this property by designing fluorescent DNA and RNA reporters that emit fluorescence upon cleavage to measure nucleic acid detection (Su et al., 2023). Various CRISPR-Dx methods have been explored, particularly those involving Class 2 Cas enzymes (Hillary & Ceasar, 2023). This review highlights exciting developments in the field of CRISPR-Dx, focusing on the applications of the Class 2 CRISPR-Cas13 system in viral nucleic acid detection.

2. Existing nucleic acid detection methods

Reverse Transcription-PCR (RT-PCR) remains the gold standard for nucleic acid detection due to its high sensitivity and adaptability. RT-PCR combines reverse transcription with PCR, where viral RNA is first reverse transcribed into complementary DNA (cDNA) by a reverse transcriptase, and the cDNA is used to generate dsDNA by a DNA polymerase. The dsDNA is then amplified through PCR (Artika, Wiyatno, & Ma'roef, 2020; Artika, Dewi, Nainggolan, Siregar, & Antonjaya, 2022). Despite its reliability, RT-PCR requires expensive reagents and equipment, including thermocyclers, and trained professionals, which makes it challenging to perform outside laboratory settings (Yan et al., 2020).

To simplify the process, isothermal amplification methods, such as recombinase polymerase amplification (RPA) and loop-mediated isothermal amplification (LAMP), have been developed. RPA utilizes a strand-displacing DNA polymerase combined with a recombinase-primer complex and single-strand DNA binding proteins to mediate amplification (Piepenburg, Williams, Stemple, & Armes, 2006). LAMP uses a DNA polymerase with high replication and strand displacement activity and multiple primer sets to amplify the target sequence (Notomi et al., 2000). RPA and LAMP, which operate at a constant temperature, eliminate the need for thermocyclers, but RPA can be less specific than PCR and LAMP involves complex primer design (Wang, Brewster, Paul, & Tomasula, 2015). Furthermore, it is challenging to multiplex both of these methods due to competition between amplicons during amplification.

During the coronavirus disease 2019 (COVID-19) pandemic, various methods were employed to diagnose infected individuals, with RT-PCR playing a crucial role (Chu et al., 2020; Corman et al., 2020; Vogels et al., 2020). However, due to its dependence on specific equipment and trained personnel and the high demand for diagnostic tests, RT-PCR had slow sample-to-answer times (Arizti-Sanz et al., 2022; Mina & Andersen, 2021). Despite not being nucleic acid detection methods, serological and antigen tests were also used for detecting severe acute respiratory syndrome coronavirus 2 (SARS-CoV-2). Although these tests are faster and do not require specialized equipment, serological tests face cross-reactivity issues, and antigen tests, lacking an amplification step, are less sensitive (Jaaskelainen et al., 2021; Yuce, Filiztekin, & Ozkaya, 2021).

The pandemic underscored the need for next-generation diagnostics that are as easy to use and cost-effective as isothermal amplification methods and as specific as PCR. CRISPR diagnostics present a potential solution to this problem, offering high versatility and single-nucleotide specificity through its crRNA (Kaminski, Abudayyeh, Gootenberg, Zhang, & Collins, 2021). In the next section, we provide an overview of Class 2 CRISPR-Cas systems and discuss how they can each be leveraged for nucleic acid detection.

3. Class 2 CRISPR-Cas systems

3.1 Cas9

The most widely used CRISPR-Cas effector protein is Cas9, derived from the *S. pyogenes* bacteria and is used for genome editing. Cas9 is part of the class 2 type II system and acts as an RNA-guided DNA endonuclease with two nuclease domains, RuvC and HNH, that create double-stranded DNA (dsDNA) breaks in target sequences (Jinek et al., 2012; Tyumentseva, Tyumentsev, & Akimkin, 2023). A transactivating crRNA (tracrRNA), complementary to the crRNA repeat sequences, directs the processing of pre-crRNA into mature crRNA by RNase III, and forms a tracrRNA:crRNA complex that assembles with Cas9 (Deltcheva et al., 2011). This complex can also form a single guide RNA (sgRNA) for targeted DNA cleavage (Jinek et al., 2012; Knott & Doudna, 2018). Protospacer adjacent motifs (PAMs) are short, conserved sequences of 2 to 5 base pairs located in close proximity to the protospacer on the target DNA (Jiang & Doudna, 2017; Marraffini & Sontheimer, 2010b). PAMs are essential for distinguishing target DNA from the CRISPR loci, preventing auto-immunity (Gleditzsch et al., 2019; Marraffini & Sontheimer, 2010b). After PAM recognition, Cas9 searches the adjacent target sequence for complementarity to its sgRNA (Jiang & Doudna, 2017; Jiang, Zhou, Ma, Gressel, & Doudna, 2015). The RuvC and HNH nuclease domains then cleave opposite DNA strands, creating double-stranded breaks (DSBs) with blunt ends (Fig. 1) (Chen, Choi, & Bailey, 2014; Gasiunas, Barrangou, Horvath, & Siksnys, 2012). The DSBs are repaired by non-homologous end joining (NHEJ) or homology-directed repair (HDR), enabling desired genome editing (Ran et al., 2013). The sgRNA can be easily altered to any target region by changing the spacer sequence, making this system highly versatile for genome editing (Cong et al., 2013; Jinek et al., 2013; Knott & Doudna, 2018; Mali et al., 2013). Recently, Cas9 has also been discovered to *trans*-cleave single-stranded DNA (ssDNA) or RNA (ssRNA) when activated by ssDNA, dsDNA, and ssRNA, providing new opportunities to utilize Cas9's *trans* cleavage capability in viral nucleic acid detection assays (Chen et al., 2024).

3.2 Cas12

In addition to Cas9, the Cas12 (formerly known as Cpf1) enzyme of the Class 2 type V CRISPR-Cas systems is used for genome editing and for nucleic acid detection. Cas12 is an RNA-guided DNA endonuclease

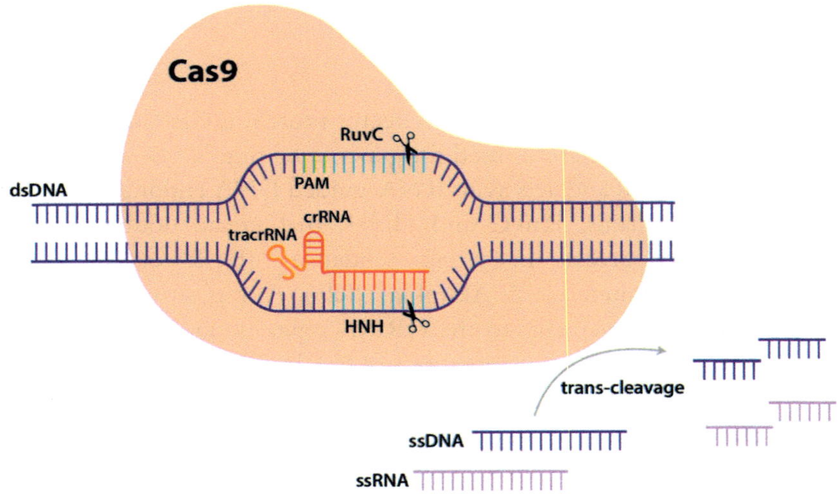

Fig. 1 Cas9 cleavage. Binding of the tracrRNA:crRNA complex to the target sequence (*light blue*) activates Cas9's HNH and RuvC nuclease domains. The HNH domain cleaves the DNA strand complementary to the tracrRNA:crRNA complex and the RuvC domain cleaves the opposite DNA strand. *Cis* cleavage results in double-stranded DNA breaks with blunt ends. Target recognition also activates Cas9's *trans* cleavage activity, followed by cleavage of non-specific ssRNA or ssDNA sequences.

with a RuvC-like nuclease domain that cleaves dsDNA or ssDNA targets (Zetsche et al., 2015). Unlike Cas9, which requires a G-rich PAM, Cas12 targets sequences with a T-rich PAM (Fonfara, Richter, Bratovic, Le Rhun, & Charpentier, 2016; Zetsche et al., 2015). Moreover, Cas12 can process pre-crRNA independently and does not need both tracrRNA and RNAse III (Fonfara et al., 2016; Zetsche et al., 2015). Cas12 not only cleaves its target DNA but also *trans*-cleaves non-target DNA. Upon assembling with a crRNA and binding to a complementary ssDNA or dsDNA target, Cas12's RuvC-like domain is activated and causes specific (*cis*) cleavage of target DNA, creating DSBs with staggered ends, and *trans* cleavage of non-target ssDNA (Fig. 2) (Chen et al., 2018; Hillary & Ceasar, 2023; Zhou, Li, Seun Olajide, & Wang, 2024). Here, the DSBs are also repaired by NHEJ or HDR, facilitating genome editing (Senthilnathan et al., 2023). Cas12's *trans* cleavage capability has been particularly valuable for developing new detection methods using ssDNA reporter systems.

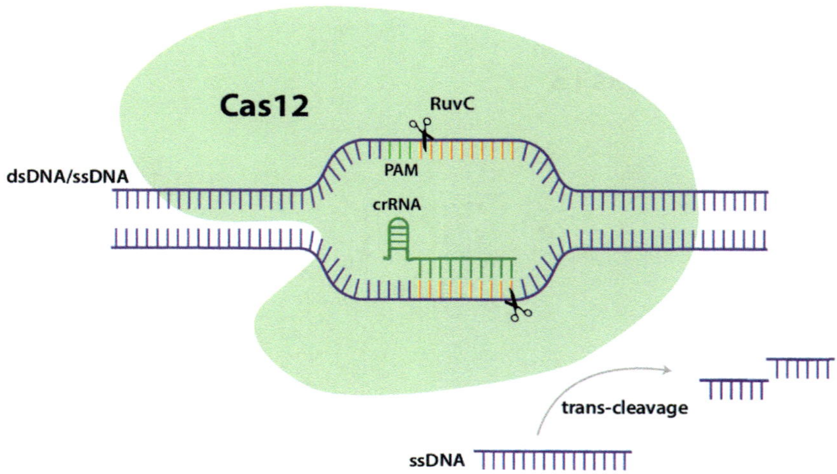

Fig. 2 Cas12 cleavage. Binding of the crRNA (*dark green*) to the target sequence (*orange*) activates Cas12's RuvC nuclease domains resulting in *cis* cleavage of the DNA strands, leaving double-stranded DNA breaks with staggered ends. Target recognition also activates Cas12's *trans* cleavage activity, followed by cleavage of non-specific ssDNA sequences.

3.3 Cas13

In 2015, Shmakov et al. identified the Cas13 (originally called C2c2) effector protein in the bacterium *Leptotrichia shahii* (Lsh). Unlike Cas9 and Cas12, Cas13 is an RNA-guided RNase that is part of the type VI Class 2 CRISPR-Cas systems (Abudayyeh et al., 2016; Shmakov et al., 2015). Cas13 has two distinct RNase activities: one for pre-crRNA processing and one for ssRNA degradation. After processing pre-crRNAs into mature crRNAs, the Cas13-crRNA complex is used to detect ssRNA sequences complementary to the crRNA (East-Seletsky et al., 2016). Upon binding, Cas13 cleaves target RNA (*cis* cleavage) and non-target RNA (*trans* cleavage) through its Higher Eukaryotes and Prokaryotes Nucleotide binding (HEPN) domains (Fig. 3) (Anantharaman, Makarova, Burroughs, Koonin, & Aravind, 2013; East-Seletsky et al., 2016; Liu et al., 2017; Shmakov et al., 2015). Cas13 targets RNA and does not require PAMs but instead uses a protospacer flanking sequence (PFS) on the target RNA, although the dependency on this PFS can differ between Cas13 orthologs and subtypes. For Cas13 orthologs dependent on a PFS, it is necessary that it consists of either an adenine, uracil, or a cytosine nucleotide at the 3' end of the protospacer since a guanine (G)- nucleotide in this region resulted in

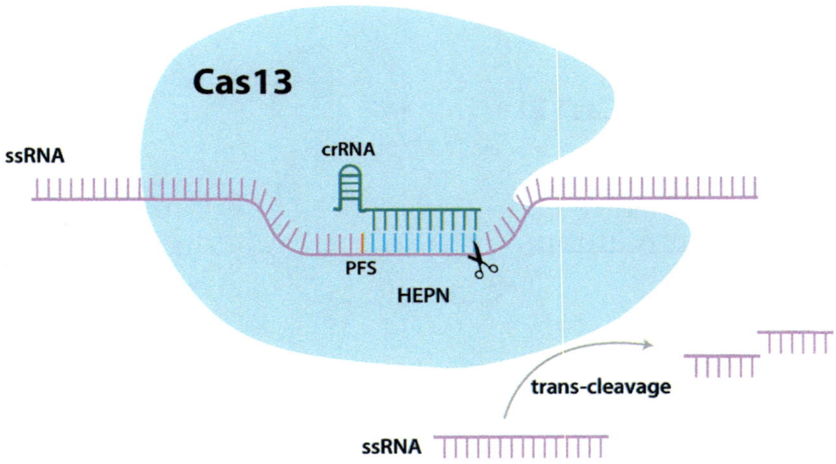

Fig. 3 Cas13 cleavage. Binding of the crRNA (*dark blue*) to the target sequence (*light blue*) activates Cas13's HEPN nuclease domain resulting in *cis* cleavage of target RNA sequence. Target recognition also activates Cas13's *trans* cleavage activity, followed by cleavage of non-specific ssRNA sequences.

reduced RNase activity of Cas13 (Abudayyeh et al., 2016). This is likely caused by the direct binding of the G-containing PFS to the direct repeat of the crRNA, thereby interfering with the Cas13-crRNA complex and preventing its cleavage activities (Meeske & Marraffini, 2018).

There are four Cas13 subtypes: Cas13a, Cas13b, Cas13c and Cas13d, each with unique properties. Cas13a and Cas13b can cleave collateral RNAs, but Cas13b's activity is influenced by cofactors (Csx27 as an inhibitor and Csx28 as an activator) (Smargon et al., 2017). Cas13c is less efficient at RNA targeting and is rarely used in studies (Huynh, Depner, Larson, & King-Jones, 2020; Kavuri, Ramasamy, Qi, & Mandadi, 2022). Cas13a, -b, and -c vary in dependence on the PFS, while Cas13d does not depend on the PFS at all and is the smallest variant with 930 amino acids (Konermann et al., 2018).

Beyond genome editing, researchers have explored the use of Cas enzymes in viral detection due to their ability to recognize and bind specific sequences. Cas9, Cas12a, and Cas13a can all *trans*-cleave DNA and/or RNA, respectively, making them useful for detecting pathogenic microorganisms. Cas9 and Cas12 require a specific PAM, while Cas13a can target any RNA sequence with a PFS other than guanine, making it more versatile (Huang, Fang, Zhou, Gong, & Xiang, 2022). However, Cas13a orthologues can differ in PFS preference and cleavage activity. For instance, Abudayyeh et al. discovered that Cas13a from *Leptotrichia wadei* (LwaCas13a) does not require a PFS at all for *cis* cleavage and is more efficient than LshCas13a, and enables RNA knockdown and tracking in mammalian cells (Abudayyeh et al., 2017). Cas13's ability to detect target RNA and *trans*-cleave non-specific RNAs upon activation has led to its exploration in nucleic acid detection, particularly for detecting RNA viruses.

4. First-generation detection methods

Choosing the appropriate Cas enzyme is crucial for detecting viral nucleic acids: Cas9 and Cas12 are well suited for targeting dsDNA, while Cas13 targets ssRNA. Depending on the Cas enzyme and viral genome, amplification and reverse transcription may be required prior to detection. First generation CRISPR-Dx methods therefore often include a separate target amplification step before the detection reaction.

In 2016, Pardee et al. developed Nucleic Acid Sequence Based Amplification and CRISPR Cleavage (NASBACC), a CRISPR-based assay using Cas9, which involves LAMP amplification to incorporate sequences including a toehold switch RNA sensor that controls the LacZ enzyme translation. Cas9 cleavage of target DNA results in yellow substrate due to truncated DNA lacking the LacZ trigger, while untruncated DNA turns the substrate purple. With NASBACC, clinically relevant concentrations of Zika virus were detected rapidly and at a low-cost, with single-base resolution discrimination between viral strains (Pardee et al., 2016).

Gootenberg et al. developed Specific High-sensitivity Enzymatic Reporter unLOCKing (SHERLOCK), a platform leveraging LwaCas13a's *trans* cleavage activity for high sensitivity detection (Gootenberg et al., 2017). The assay combines ssRNA targets, Cas13a-crRNA, and quenched fluorescent RNA reporters, with RPA and T7 RNA polymerase converting amplified DNA to RNA for detection. SHERLOCK can differentiate between similar strains of Zika (ZIKV) and Dengue (DENV) viruses, with attomolar sensitivity comparable to digital-droplet PCR and qPCR. Its lyophilized reagents simplify storage and enhance field deployment for POC diagnostics.

To minimize the requirement of special equipment and accelerate the workflow, Myhrvold et al. developed Heating Unextracted Diagnostic Samples to Obliterate Nucleases (HUDSON), enabling SHERLOCK to detect viral nucleic acids directly from bodily fluids (Myhrvold et al., 2018).

HUDSON heats and chemically reduces samples to inactivate RNases and lyse viral particles, allowing direct addition to RPA reactions. Combination of HUDSON and SHERLOCK enables detection and differentiation of DENV serotypes and ZIKV strains without special equipment, reducing SHERLOCK's turnaround time from 2.5 to 2 h.

Similar to SHERLOCK is DNA endonuclease-targeted CRISPR *trans* reporter (DETECTR), a DNA-detection method utilizing *Lachnospiraceae bacterium* (Lb)Cas12's *trans* cleavage of ssDNA. DETECTR combines RPA of target dsDNA with Cas12a-crRNA complexes to activate *trans* cleavage of ssDNA reporters, allowing for specific detection of and discrimination between high-risk human papilloma virus (HPV) strains HPV16 and HPV18 (Chen et al., 2018).

Likewise, HOLMES (one-HOur Low-cost Multipurpose highly Efficient System) uses LbCas12a to detect target DNA and RNA which, after PCR amplification, is added to Cas12a-crRNA complexes combined with quenched fluorescent ssDNA reporters. HOLMES requires no special equipment or expensive reagents, providing rapid, specific, and low-cost detection (Li et al., 2018).

In response to the need for POC devices for SARS-CoV-2 detection, Lopez-Valls et al. developed CRISPR/CAS-based Colorimetric nucleic Acid DEtection (CASCADE), that combines Cas13's RNA detection with colloidal aggregation of gold nanoparticles (AuNPs) for a visual read-out (Lopez-Valls et al., 2022). Target-activated Cas13 cleaves ssRNAs that stabilize the AuNPs, causing them to aggregate and change color from red to purple. CASCADE, coupled with isothermal amplification methods, showed high sensitivity, detecting SARS-CoV-2 RNA down to 3 femtomolar with RT-RPA and 40 attomolar with nucleic acid sequence-based amplification (NASBA). Although CASCADE showed high sensitivity and specificity in clinical samples and naked eye detection was possible, the workflow, especially target preparation, is quite complex. This limits its use in POC and PON settings, where simple and rapid on-site detection methods are favorable.

5. Single-step detection methods

Even though the aforementioned methods showed high sensitivity and specificity in detecting viral DNA and RNA, their multi-step workflows increase the chance of cross-contamination, ultimately limiting their applicability at the POC and PON. To combat this problem, second-generation methods have been developed that combine the sample amplification and detection steps.

Li et al. upgraded HOLMES to HOLMESv2 by integrating sample amplification and detection in one step using Cas12b, which can *cis*-cleave target dsDNA and ssDNA and *trans*-cleave non-specific ssDNA. Unlike HOLMES, which uses PCR, HOLMESv2 employs LAMP, reducing the target concentration needed for Cas12b *trans* cleavage activation (Li et al., 2019). HOLMESv2, requiring only simple instruments, enables rapid detection of a wide range of clinical pathogens and is suitable for POC testing (Li et al., 2019).

To streamline viral RNA detection and reduce contamination risks, Arizti-Sanz et al. developed SHINE (Streamlined Highlighting of Infections to Navigate Epidemics), a single-step method combining amplification and Cas13-based detection of SARS-CoV-2 (Arizti-Sanz et al., 2020). SHINE involves a 10-minute HUDSON treatment for viral inactivation and lysis, followed by a single-step SHERLOCK detection assay. This assay allows direct use of patient samples, requiring minimal reagents and equipment. Moreover, lateral flow strips or in-tube fluorescence can be used as read-outs for detection and it has a turnaround time of 50 min, making it an attractive tool for combating the high test-demand during the pandemic (Arizti-Sanz et al., 2020).

To make SHINE less complex and easier to distribute, Arizti-Sanz et al. developed SHINEv.2, which eliminates the need for cold storage of reagents and, by introducing a new lysis reagent, the need for multiple heating steps (Arizti-Sanz et al., 2022). This lysis reagent is active at ambient temperatures and inactivates RNases and SARS-CoV-2 virions without compromising the assay's sensitivity. Additionally, lyophilizing the detection assay reagents simplifies the reaction set-up, distribution, and reduced workflow time. Overall, SHINEv.2 is a user-friendly, equipment-free, easy-to-distribute, sensitive detection assay suitable for community surveillance testing of variants of concern of SARS-CoV-2.

Like SHINE and SHINEv.2, the minimally instrumented SHERLOCK (MiSHERLOCK) method enables the direct use of patient samples for a low-cost, user-friendly and sensitive detection of SARS-CoV-2 variants (de Puig et al., 2021). MiSHERLOCK features a device with a sample preparation chamber that collects and concentrates viral RNA from saliva. The RNA is then transferred to a detection chamber with lyophilized SHERLOCK reagents, which are rehydrated before adding the viral RNA. MiSHERLOCK uses RT-RPA for sample amplification and

Cas12a for target detection, which is visualized with a built-in transilluminator, allowing assessment after 55-minutes. This simple, instrument-free device can be useful for on-site POC testing.

Lu et al. developed the sPAMC (suboptimal PAM Cas12-based) assay to detect SARS-CoV-2 RNA in unextracted clinical samples. By targeting sequences with suboptimal PAMs, Cas12's *trans* cleavage activity is slowed, enhancing target amplification and fluorescence signal, thereby increasing sensitivity. Due to the abundance of suboptimal PAM sequences, this method is highly flexible. sPAMC is a simple, one-step detection reaction that can identify SARS-CoV-2 and human cytomegalovirus in unextracted samples within 15 min, with a sensitivity of 1 cp/μL (Lu et al., 2022).

For diagnosing SARS-CoV-2 and simultaneously determining the causative variant, a single-pot multiplexed RT-RPA Cas13 detection system was developed (Patchsung et al., 2023). This system amplifies two different viral genes, each detected by a different Cas13 ortholog of which cleavage activity is visualized with an unique fluorescence channel. While easily field-deployable through lyophilization of reagents and simple detection visualization, it does not support the use of extraction-free clinical samples.

The CESSAT (Chemical additive-Enhanced Single-Step Accurate CRISPR/Cas13 Testing) method, similar to MiSHERLOCK, offers extraction-free, single-step RT-RPA-Cas13 detection and genotyping of SARS-CoV-2 variants in clinical samples (Wang et al., 2023). Glycine addition enhances CESSAT's sensitivity to a limit of detection (LOD) of 1 cp/μL. It tests a single sample for eight different genotypes within 40 min, with fluorescence imaging results interpreted via a smartphone. Due to using a portable device with lyophilized reagents and minimal equipment, CESSAT is suitable for field deployment.

Another detection method suitable for on-site or in-home testing was developed and named REVERSE-2. Tong et al. discovered that Cas13 degrades viral RNA, thus interfering with target amplification in SHINE's workflow. With REVERSE-2, RNA transcribed from amplified DNA is reverse complementary to viral RNA, enhancing Cas13's *trans* cleavage specificity and the assay's sensitivity (Tong et al., 2024). REVERSE-2 detects extracted and unextracted SARS-CoV-2 at a concentration of 0.5 cp/μL and is visualized through in-tube fluorescence, visible with the naked eye or through the use of a smartphone. The detection reaction only takes 30 min and can be performed at room temperature, making it a field-deployable test.

Recently, a field-deployable POC device called SCOPE (Streamlined CRISPR On Pod Evaluation platform) was developed for rapid, on-site detection of monkeypox virus (MPXV). SCOPE lyses viral DNA directly from patient samples, followed by a single-step RPA-Cas13 detection reaction, with all reagents lyophilized for easier transportation and storage (Wang et al., 2024). A special device integrates SCOPE's entire workflow, minimizing user operation, and connects to a smartphone for assay interpretation. SCOPE detects MPXV at 0.5 cp/μL within 15 min, making it a valuable method for low-resource settings.

During viral outbreaks, fast diagnosis is crucial for controlling transmission amongst the population. To reduce turnaround time of detection assays, researchers have tried simplifying them by, for instance, introducing amplification-free methods.

6. Amplification-free methods

To reduce turnaround time and reagent use in Cas9 detection assays, Hajian et al. developed the CRISPR-Chip. This amplification-free method uses Cas9's DNA sequence detection combined with a graphene-based field-effect transistor (gFET) for high sensitivity (Hajian et al., 2019). Immobilized catalytically deactivated Cas9 (dCas9)-sgRNA complexes on the gFET's graphene surface detect Duchenne muscular dystrophy mutations in whole genomic DNA, generating an electrical signal upon target binding. Results are obtained within 15 min with a sensitivity of 1.7 fM, though amplification may still be needed for viral nucleic acid detection as viral copy numbers are often low.

Moon et al. also used dCas9 to develop a colorimetric detection method for SARS-CoV-2 and drug-resistant influenza. After viral lysate and biotin-labeled PAM-presenting oligonucleotide (PAMmer) are added, the on a microplate immobilized dCas9-gRNA complex binds to the target. Streptavidin-horseradish peroxidase (HRP) binds to the biotin-PAMmer, and tetramethylbenzidine peroxide (TMB) substrate causes a color change upon reaction with the HRP and PAMmer. This method successfully identified SARS-CoV-2 and oseltamivir-resistant H1N1 in clinical samples by leveraging Cas9's sequence-specific binding activity (Moon et al., 2020).

In 2021, Fozouni et al. developed a CRISPR-Cas13 based assay for detecting SARS-CoV-2, which unlike SHERLOCK and DETECTR, does not require pre-amplification of target RNA (Fozouni et al., 2021). They optimized Cas13a's activation with crRNAs targeting the nucleocapsid gene, followed by adding pre-assembled Cas13a-crRNA complexes to a mix of target RNA and quenched fluorescent RNA reporter. Using multiple crRNAs increased Cas13's sensitivity, and although the initial sensitivity was reported at 100 cp/μL, subsequent studies suggest that this is likely inaccurate (Ramachandran & Santiago, 2021). Without amplification, quantitative RNA measurements of viral loads are possible, which help manage viral spread by estimating an individual's infectivity and recovery. The use of the mobile phone to measure the fluorescent signal confirmed the feasibility of direct SARS-CoV-2 detection outside specialized labs.

Another way to increase Cas13's sensitivity in an amplification-free detection method is to combine the enzyme with the RNA endonuclease Csm6, which was done to develop Fast Integrated Nuclease Detection In Tandem (FIND-IT) (Liu et al., 2021). Combined with eight different crRNAs targeting a ssRNA target, the single-pot reaction of Cas13 and Csm6 made detection of SARS-CoV-2 RNA at a LOD of 31 cp/μL within 20 min possible. FIND-IT is an attractive tool for POC diagnostics because of its highly sensitive, rapid, low-cost and simple workflow.

Shinoda et al. developed the amplification-free Cas13-based SATORI assay, integrating their microchamber technology with CRISPR-Cas13 for RNA detection (Shinoda et al., 2021). SATORI uses pre-assembled LwaCas13a-crRNA complexes and fluorophore-quenched RNA reporters placed in a microchamber device, with results visible within 5 min via fluorescence microscopy. This assay detects RNA at single-molecule levels with a LOD of 5.7 fM and avoids amplification errors. Due to SATORI's resistance to contaminants, raw patient samples can be used. Despite its sensitivity compared to other amplification-free methods, SATORI is still less sensitive than SHERLOCK and DETECTR.

Like SATORI, the ultralocalized Cas13a assay achieves single-molecule detection by emulsifying target RNA, Cas13a, crRNAs and quenched fluorescent RNA reporters into picolitre-sized droplets (Tian et al., 2021). The droplets assemble into an array, and fluorescence is visualized when Cas13a *trans*-cleaves the reporter upon target recognition. Absolute quantitation is possible since each droplet contains a single target RNA. This amplification-free system is notable for its high sensitivity as it can detect SARS-CoV-2 RNA at a concentration of 6 cp/μL in clinical samples.

The Autocatalytic Cas12a Circular DNA Amplification Reaction (AutoCAR) system, recently developed by Deng et al., uses a DNA nanostructure to amplify detection signals and stimulate autocatalysis of Cas12 ribonucleoproteins (RNPs) (Deng et al., 2024). The system employs a Cir-mediator, which contains a PAM-containing dsDNA region complementary to the sgRNA and circularizes via a connecting ssDNA region. Target-activated Cas12 cleaves the ssDNA, linearizing the dsDNA and activating an autocatalytic loop for detecting SNP mutations and SARS-CoV-2 RNA with 1 aM sensitivity in 15 min. Detection visualization through fluorescence or colorimetric read-outs, and without the need for temperature cycling or special equipment, make AutoCAR suitable for POC applications.

Even though these detection methods keep improving to become more sensitive, specific, and easier to use, they often have limited use because they detect only one specific target. Furthermore, some methods require extraction steps (Fozouni et al., FIND-IT) or complex instrumentation for the read-out (SATORI, ultralocalized Cas13a assay). This is far from ideal, especially in the context of viral infection surveillance to prevent or contain viral outbreaks.

7. Multiplexed detection methods

Due to the cocirculation of multiple pathogens and the probability of co-infection of individuals, multiplexed assays can be beneficial as they allow for the detection of numerous pathogens in a single sample. With the opportunity to test for different subtypes of these pathogens and to test multiple samples simultaneously, multiplexed assays significantly enhance pathogen diagnostics (Lamb, Kang, & Myhrvold, 2023).

Gootenberg et al. developed the SHERLOCKv2 method, which addresses SHERLOCK's lack of quantitation and dependence on special equipment for fluorescence read-out. SHERLOCKv2 uses the orthogonal base preferences of four Cas enzymes to detect multiple targets within a single sample, with an unique fluorescent reporter for each Cas ortholog (Gootenberg et al., 2018). Multiplexed detection of four viruses in one sample lowers the costs and increases the scale of this assay. Furthermore, the assay provides quantitative measurements at low target concentrations. Addition of Csm6, an auxiliary CRISPR-associated enzyme, further enhanced Cas13's activity and the assay's sensitivity. Finally, a lateral flow read-out can be used for SHERLOCKv2, making it a portable, easy-to-use assay. Even though using multiple Cas12 and Cas13 orthologs made

SHERLOCKv2 more multiplexed, finding more than four orthologs with different cleavage preferences is difficult and limits the multiplexing capacity of the technology.

For large-scale viral nucleic acid detection, Ackerman et al. developed Combinatorial Arrayed Reactions for Multiplexed Evaluation of Nucleic acids (CARMEN), integrating CRISPR-Cas13a into a microwell array system (Ackerman et al., 2020). The targets include viral DNA or RNA that is amplified through PCR or RPA. The detection mixes include T7 for transcription and Cas13, crRNAs and cleavage reporter for the detection reaction. Both target and detection mixes are uniquely barcoded with solution-based fluorescent color codes and emulsified into sample and detection droplets. Once the droplet emulsions are added to the microwell, the droplets organize themselves into different wells, simultaneously forming all pairwise combinations of sample and detection droplets. After the droplets merge through exposure to an electric field, the fluorescence is used as a read-out for Cas13 detection. With a massive capacity chip (mChip), over 4500 reactions can be performed using CARMEN, but its need for custom-made chips and extensive workflow make the assay challenging to use in clinical settings.

To improve practicality, Welch et al. developed microfluidic CARMEN (mCARMEN) to combine the CARMEN technology with commercially available Fluidigm (now known as Standard Biotools) microfluidics, eliminating the need for color-coding or dropletization (Welch et al., 2022). mCARMEN allows testing of 192 samples with 24 detection assays or 96 samples with 96 detection assays on special integrated fluidic circuits (IFCs). A Fluidigm controller moves the sample and detection assays through the channels to their reaction chambers, after which the Fluidigm Biomark measures fluorescence from Cas13's *trans* cleavage of quenched fluorescent reporters. mCARMEN showed high specificity and sensitivity for a respiratory virus panel of 21 viruses and was helpful in SARS-CoV-2 variant detection during the Omicron wave in December 2021. Automating RNA extraction from samples, using one-step amplification and a faster detection read-out all reduced manual labor and processing time to under five hours. However, mCARMEN requires specialized and expensive equipment, making it less suitable for low-resource settings.

In search of a more ideal CRISPR-based multiplexed detection platform, the MiCaR assay was developed using LbCas12a (Xu et al., 2022). It involves direct multiplex RPA of samples, followed by loading onto a microfluidic chip with 30 channels, each containing a unique Cas12-crRNA complex and a fluorescence reporter. MiCaR was validated with nine different HPV subtypes, and with a turnaround time of 40 min and a sensitivity of 0.26 aM, provides a rapid, accurate and easy on-site diagnostic tool.

Another effort to simplify multiplex detection assays was done by coupling color-coded beads to specific crRNAs to enable separate localization and using droplets to separate each Cas13-detection reaction. This enables the bead-based CARMEN (bbCARMEN) assay to perform multiple reactions in parallel, and the use of the beads might offer a sensitive, multiplex detection assay that is low-cost and suitable in resource-limited settings (Siddiqui et al., 2023).

8. How to perform CRISPR-based viral detection

CRISPR-based viral detection methods generally involve target preparation, the detection reaction, and visual read-out. The SHINE assay combines the first two steps and multiple visualization options, making it adaptable to various settings. Notably, SHINE allows for rapid detection of SARS-CoV-2 RNA directly from bodily fluid samples (Arizti-Sanz et al., 2020). A detailed description of the workflow is listed here, as described in the paper.

SHINE workflow (Fig. 4):

1. HUDSON (10-minute viral lysis and inactivation)
   a. Collect clinical samples (saliva or nasopharyngeal swabs);
   b. To each sample, add TCEP, EDTA, and murine RNase inhibitor at a final concentration of 100 millimolar (mM), 1 mM, and 0.8 U/μL, respectively;
   c. Incubate for 5 min at 40 °C;
   d. Final incubation of 5 min at 70 °C for nasopharyngeal samples and at 95 °C for saliva samples.

Note: addition of RNase inhibitor to the viral/universal transport media is important to reduce lysis and inactivation time.

2. Single-step SHERLOCK
   a. First, completely resuspend the RPA pellets with nuclease-free water before adding sample or other detection reagents;
   b. Add resuspended RPA pellets to the HUDSON-treated samples;
   c. Add the detection reagents at the following final concentrations: 1X SHINE buffer (20 mM HEPES pH 8.0 with 60 mM KCl and 5%

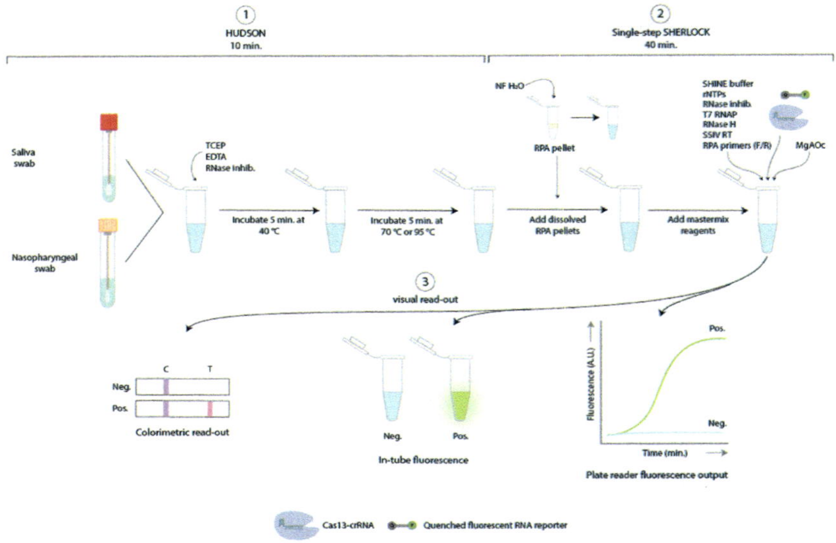

Fig. 4 SHINE workflow. The sample preparation (HUDSON), detection reaction (Single-step SHERLOCK) and visual read-out steps are shown. Tris(2-carboxyethyl)phosphine) (TCEP), Ethylenediaminetetraacetic acid (EDTA) and RNase inhibitors (RNase inhib.) are added to the samples for HUDSON treatment. SHINE mastermix reagents are added to the treated samples, starting with dissolved RPA pellets. Magnesium acetate (MgAOc) is added last, right before visualization. The visual read-out can be performed in three ways. Colorimetric read-out includes lateral flow strips with control (C) and test (T) lines. In-tube fluorescence showing fluorescence (*green*) only for positive (pos.) and not for negative (neg.) samples. The plate reader output will show increasing fluorescence (*green*) for pos. samples and a steady fluorescent signal for neg. samples.

PEG), 2 mM of each rNTP, 1 U/μL murine RNase inhibitor, 1 U/μL T7 RNA polymerase, 0.1 U/μL RNase H, 2 U/μL SuperScript IV reverse transcriptase, 120 nanomolar (nM) forward and reverse RPA primers, 22.5 nM crRNA, 45 nM LwaCas13a, 125 nM polyU FAM quenched reporter ***;

*For colorimetric read-out: use 1 μM biotinylated FAM reporter

** For in-tube fluorescence read-out: use 62.5 nM polyU FAM

Note: Addition of RNase H improves the sensitivity of Cas13

d. Add 14 mM magnesium acetate right before visualizing the detection reaction.

3. Visual read-out

a. Colorimetric read-out:

Incubate samples for 1–3 h at 37 °C, then dilute in Milenia

HybridDetect buffer before adding the Milenia HybriDetect lateral flow strips. Collect images of the strip after 5 min of incubation.

i. Positive samples will show lines at the control and test lines, while negative samples only show the C line.

b. In-tube fluorescence:

Incubate samples at 37 °C and collect images, using a transilluminator, every 30 min until results are clear.

i. Positive samples will show fluorescence visible with the naked eye, while negative samples will not show fluorescence beyond the background signal

c. Plate-reader fluorescence:

Measure fluorescence kinetics at 37 °C using a microplate reader.

i. Positive samples will show increasing fluorescence over time due to *trans* cleavage of the quenched fluorescent RNA reporter. Negative samples will show a steady baseline fluorescence signal over time.

8.1 Comparison of CRISPR-Cas viral nucleic acid detection methods

Over the years, numerous CRISPR-Cas-based detection methods have been developed, each with unique advantages and challenges (Table 1). The discovery of Cas13, which can detect RNA and *trans*-cleave non-specific RNA, significantly advanced the diagnostics field. While many techniques offer high sensitivity for viral RNA detection, some have relatively slow turnaround times (e.g., SHERLOCK and CASCADE), and others are unable to quantify viral loads or perform multiplex assays (e.g., SHINE, SHINEv2). Recent methods like SHERLOCKv2, CARMEN and mCARMEN achieve high sensitivity and reasonable turnaround times on a large scale, but they often require specialized equipment, limiting their applicability in low-resource settings. The "ideal" Cas13-based viral nucleic acid detection method, balancing sensitivity, rapid turnaround time and practicality for on-site diagnostics, has yet to be developed.

9. Challenges and future improvements

Considering the different steps of CRISPR-based detection methods, each of them can pose challenges to the overall applicability of the method. Sample preparation often requires DNA or RNA extraction, needing additional equipment which may not be feasible in low-resource

Table 1 Comparison of the different Cas-based viral nucleic acid detection methods mentioned in this review.

Method	Name of detection assay	Cas orthologue	Target	Amplification method	Read out method	LOD	Time to result (hours)	Multiplex? (Y/N)	Quantitative? (Y/N)	(Viral) targets	References
Multi-step	NASBACC	SpCas9	RNA	LAMP	Colorimetric: LacZ	2.8 fM	3–3.5	N	N	ZIKV	Pardee et al. (2016)
	SHERLOCK	LwaCas13a	DNA, RNA	RPA, RT-RPA	Fluorescence: quenched fluorescent RNA reporter	2.1 aM	2.5	N	N	ZIKV, DENV	Gootenberg et al. (2017)
	HUDSON + SHERLOCK	LwaCas13a	DNA, RNA	RPA, RT-RPA	Fluorescence: quenched fluorescent RNA reporter Colorimetric: lateral flow strips	0.9–90 aM	< 2	N	N	ZIKV, DENV	Myhrvold et al. (2018)
	DETECTR	LbCas12a	DNA	RPA	Fluorescence: ssDNA FQ-reporter	1 aM	1	N	N	HPV16, HPV18	Chen et al. (2018)
	HOLMES	LbCas12a	DNA, RNA	PCR, RT-PCR	Fluorescence: ssDNA FQ-reporter	1–10 aM	1	N	N	PRV, JEV, SNP	Li et al. (2018)
	CASCADE	LwaCas13a	RNA	NASBA, RT-RPA, RPA	Colloidal aggregation of gold nanoparticles (AuNPs)	40 aM (NASBA) 3 fM (RPA)	<2.5	N	N	SARS-CoV-2	Lopez-Valls et al. (2022)
Single-step	HOLMESv2	AacCas12b	DNA, RNA	LAMP, RT-LAMP, PCR	Fluorescence: HEX-N12-BHQ1 ssDNA probe	10 aM	1	N	Y	JEV	Li et al. (2019)
	SHINE	LwaCas13a	RNA	RT-RPA	Fluorescence: quenched fluorescent RNA reporter Colorimetric: lateral flow strips	10–100 cp/μL	0.8	N	N	SARS-CoV-2	Arizti-Sanz et al. (2020)

	SHINEv2	LwaCas13a	RNA	RT-RPA	Fluorescence: quenched fluorescent RNA reporter, Colorimetric: lateral flow strips	200 cp/μL	< 1.5	N	N	SARS-CoV-2	Arizti-Sanz et al. (2022)
	MiSHERLOCK	LbaCas12a	RNA	RT-RPA	Fluorescence: ssDNA FQ-reporter	1 cp/μL	< 1	N	N	SARS-CoV-2	de Puig et al. (2021)
	sPAMC	LbCas12a	DNA, RNA	RPA, RT-RPA	Fluorescence: ssDNA FQ-reporter	1 cp/μL	0.2–0.25	N	N	HCMV, SARS-CoV-2	Lu et al. (2022)
	Multiplexed RT-RPA Cas13-detection	LwaCas13a PsmCas13b	RNA	RT-RPA	Fluorescence: quenched fluorescent RNA reporter	Ct~37	1–1.5	N	N	SARS-CoV-2	Patchsung et al. (2023)
	CESSAT	LwaCas13a	RNA	RT-RPA	Fluorescence: quenched fluorescent RNA reporter	1 cp/μL	0.67	N	N	SARS-CoV-2	Wang et al. (2023)
	REVERSE-2	LwaCas13a	RNA	RT-RPA	Fluorescence: quenched fluorescent RNA reporter	0.5 cp/μL	0.5	N	N	SARS-CoV-2	Tong et al. (2024)
	SCOPE	LwaCas13a	DNA	RPA	Fluorescence: quenched fluorescent RNA reporter	0.5 cp/μL	0.25	N	N	MPXV	Wang et al. (2024)
Amplification-free	CRISPR-Chip	SpdCas9	DNA	—	gFET	1.7 fM	0.25	N	Y	SNP	Hajian et al. (2019)
	CRISPR-dCas9	SpdCas9	RNA	—	Colorimetric: HRP	30 pM	2	N	Y	SARS-CoV-2, H1N1	Moon et al. (2020)
	Amplification-free Cas13 detection assay [a]	LbuCas13a	RNA	—	Fluorescence: quenched fluorescent RNA reporter	100 cp/μL	< 0.5	N	Y	SARS-CoV-2	Fozouni et al. (2021)

(continued)

Table 1 Comparison of the different Cas-based viral nucleic acid detection methods mentioned in this review. (*cont'd*)

Method	Name of detection assay	Cas orthologue	Target	Amplification method	Read out method	LOD	Time to result (hours)	Multiplex? (Y/N)	Quantitative? (Y/N)	(Viral) targets	References
	FIND-IT	LbuCas13a	RNA	—	Fluorescence: fluorescent reporter oligonucleotides	31 cp/μL	0.33	N	N	SARS-CoV-2	Liu et al. (2021)
	SATORI	LwaCas13a	RNA	—	Fluorescence: quenched fluorescent RNA reporter	5–10 fM	0.08	N	N	SARS-CoV-2	Shinoda et al. (2021)
	Ultralocalized Cas13a assay	LbuCas13a	RNA	—	Fluorescence: quenched fluorescent RNA reporter	6 cp/ μL	1–1.2	N	Y	microRNAs, 16s rRNAs, SARS-CoV-2	Tian et al. (2021)
	AutoCAR	LbaCas12a	DNA, RNA	—	Fluorescence: quenched fluorescent RNA reporter Colorimetric: lateral flow	1 aM	0.25	N	Y	SNP, SARS-CoV-2	Deng et al. (2024)
Multiplexed	SHERLOCKv2	LwaCas13a, PsmCas13b, CcaCas13b, AsCas12a	DNA, RNA	RPA, RT-RPA	Fluorescence: quenched fluorescent RNA reporter, homopolymer and di-nucleotide reporter Colorimetric: lateral flow	2 aM	1–3	Y,	Y	ZIKV, DENV, *S. aureus* thermonuclease and *P. aeruoginosa* acyltransferase, cfDNA	Gootenberg et al. (2018)
	CARMEN	LwaCas13a	DNA, RNA	RPA, PCR, RT-RPA, RT-PCR	Fluorescence: quenched fluorescent RNA reporter	0.9 aM	0.5–3	Y	Y	169 human-associated viruses	Ackerman et al. (2020)

	mCARMEN	LwaCas13a	DNA, RNA	RPA, PCR, RT-RPA, RT-PCR	Fluorescence: quenched fluorescent RNA reporter	0.5–20 cp/μL	< 3	Y	Y	HMPV HRSV HPIV-3 FLUBV FLUAV HCoV-NL63 HCoV-OC43 HCoV-HKU1	Welch et al. (2022)
	MiCaR	LbCas12a	DNA	RPA	Fluorescence: ssDNA FQ-reporter	0.26 aM	1	Y	N	HPV, FLUBV, HCoV-NL63, HCoV-OC43, HRSV, HCoV-HKU1, SARS-CoV-2, HPIV-3, HMPV	Xu et al. (2022)

LOD, limit of detection; *fM*, femtomolar; *aM*, attomolar; *cp/ μL*, copies per μL; *Ct*, RT-qPCR Ct-value; *Y*, yes; *N*, no; *ZIKV*, Zika virus; *HPV*, human papillomavirus; *SARS-CoV-2*, severe acute respiratory syndrome coronavirus 2; *H1N1*, subtype of influenza A virus; *PRV*, pseudorabies virus; *JEV*, Japanese encephalitis virus; *HCMV*, human cytomegalovirus; *MPXV*, Monkeypox virus; *SNP*, single nucleotide polymorphism; *DENV*, Dengue virus; *HMPV*, human metapneumovirus; *HRSV*, human respiratory syncytial virus; *HPIV*, human parainfluenza virus; *FLUBV*, influenza B virus; *FLUAV*, influenza A virus; *HCoV*, human coronavirus. [a]No official name of the assay was mentioned in the paper.

settings. While some assays might be robust against contaminants, these often affect the assay's sensitivity. Combining HUDSON and SHERLOCK allows direct virus detection from bodily fluids, making extraction-free detection methods desirable for field deployment (Myhrvold et al., 2018). Additionally, more recent methods like MiSHERLOCK, sPAMC, CESSAT, REVERSE-2 and SCOPE are all compatible with unextracted clinical samples, making the detection reaction faster and more simple.

Another challenge can be the amplification of target DNA and RNA, often requiring thermocyclers. Isothermal amplification methods, like RPA and LAMP, can avoid this, though they often still require a heating block and can introduce amplification errors. To avoid this, amplification-free Cas13 detection methods have been explored with the extra advantage of being able to quantify the viral load in samples (Fozouni et al., 2021; Shinoda et al., 2021). However, a lack of amplification of target DNA/RNA can translate into a decrease in assay sensitivity. Enhancement of the sensitivity through the use of a CRISPR-Cas-associated enzyme, such as Csm6, has shown to be helpful for this (Liu et al., 2021).

Detection, especially the read-out, can be problematic in low-resource settings if it requires a fluorescence microscope or plate reader. However, colorimetric read-outs, such as lateral flow strips, offer a solution by making the assays easy to distribute, intuitive and cost-effective (Arizti-Sanz et al., 2020; Gootenberg et al., 2018).

Improvement of all the steps of a CRISPR-based detection method will ultimately yield a sensitive, specific, user-friendly, low-cost and simple detection method that requires minimal reagents and equipment for sample preparation, amplification and visualization.

10. Conclusion

Viruses will continue to pose significant health threats, making it essential to enhance current diagnostics and develop rapid and accurate methods for detecting viral infections. The discovery of CRISPR-Cas systems, particularly Class 2 Cas enzymes, has revolutionized genome editing and pathogen detection. As highlighted in this review, various detection methods leveraging Cas13's RNA binding and cleavage abilities have shown promise in targeting viral nucleic acids. While each method has its strengths and weaknesses, including suitability for detection, quantification, or distinguishing between variants of viral strains, they have collectively improved viral nucleic acid detection. These improvements include high sensitivity and specificity, use of minimal equipment, fast turnaround times and ease of use. However, an ideal technique that encompasses all these qualities has yet to be developed. Continuous improvement of detection methods and understanding of CRISPR-Cas enzymes could lead to the design of a viral nucleic acid detection tool that is fast, cost-effective, easy-to-use, high-throughput, and highly sensitive and specific.

References

Abudayyeh, O. O., Gootenberg, J. S., Essletzbichler, P., Han, S., Joung, J., Belanto, J. J., ... Zhang, F. (2017). RNA targeting with CRISPR-Cas13. *Nature, 550*(7675), 280–284. https://doi.org/10.1038/nature24049.

Abudayyeh, O. O., Gootenberg, J. S., Konermann, S., Joung, J., Slaymaker, I. M., Cox, D. B., ... Zhang, F. (2016). C2c2 is a single-component programmable RNA-guided RNA-targeting CRISPR effector. *Science, 353*(6299), aaf5573. https://doi.org/10.1126/science.aaf5573.

Ackerman, C. M., Myhrvold, C., Thakku, S. G., Freije, C. A., Metsky, H. C., Yang, D. K., ... Sabeti, P. C. (2020). Massively multiplexed nucleic acid detection with Cas13. *Nature, 582*(7811), 277–282. https://doi.org/10.1038/s41586-020-2279-8.

Anantharaman, V., Makarova, K. S., Burroughs, A. M., Koonin, E. V., & Aravind, L. (2013). Comprehensive analysis of the HEPN superfamily: Identification of novel roles in intra-genomic conflicts, defense, pathogenesis and RNA processing. *Biology Direct, 8*, 15. https://doi.org/10.1186/1745-6150-8-15.

Arizti-Sanz, J., Bradley, A., Zhang, Y. B., Boehm, C. K., Freije, C. A., Grunberg, M. E., ... Myhrvold, C. (2022). Simplified Cas13-based assays for the fast identification of SARS-CoV-2 and its variants. *Nature Biomedical Engineering, 6*(8), 932–943. https://doi.org/10.1038/s41551-022-00889-z.

Arizti-Sanz, J., Freije, C. A., Stanton, A. C., Petros, B. A., Boehm, C. K., Siddiqui, S., ... Myhrvold, C. (2020). Streamlined inactivation, amplification, and Cas13-based detection of SARS-CoV-2. *Nature Communications, 11*(1), 5921. https://doi.org/10.1038/s41467-020-19097-x.

Artika, I. M., Dewi, Y. P., Nainggolan, I. M., Siregar, J. E., & Antonjaya, U. (2022). Real-time polymerase chain reaction: Current techniques, applications, and role in COVID-19 diagnosis. *Genes (Basel), 13*(12), https://doi.org/10.3390/genes13122387.

Artika, I. M., Wiyatno, A., & Ma'roef, C. N. (2020). Pathogenic viruses: Molecular detection and characterization. *Infection, Genetics and Evolution, 81*, 104215. https://doi.org/10.1016/j.meegid.2020.104215.

Barrangou, R., Fremaux, C., Deveau, H., Richards, M., Boyaval, P., Moineau, S., ... Horvath, P. (2007). CRISPR provides acquired resistance against viruses in prokaryotes. *Science, 315*(5819), 1709–1712. https://doi.org/10.1126/science.1138140.

Bohn, M. K., Lippi, G., Horvath, A., Sethi, S., Koch, D., Ferrari, M., ... Adeli, K. (2020). Molecular, serological, and biochemical diagnosis and monitoring of COVID-19: IFCC taskforce evaluation of the latest evidence. *Clinical Chemistry and Laboratory Medicine, 58*(7), 1037–1052. https://doi.org/10.1515/cclm-2020-0722.

Brouns, S. J., Jore, M. M., Lundgren, M., Westra, E. R., Slijkhuis, R. J., Snijders, A. P., ... van der Oost, J. (2008). Small CRISPR RNAs guide antiviral defense in prokaryotes. *Science, 321*(5891), 960–964. https://doi.org/10.1126/science.1159689.

Buchy, P., Buisson, Y., Cintra, O., Dwyer, D. E., Nissen, M., Ortiz de Lejarazu, R., & Petersen, E. (2021). COVID-19 pandemic: Lessons learned from more than a century of pandemics and current vaccine development for pandemic control. *International Journal of Infectious Diseases, 112*, 300–317. https://doi.org/10.1016/j.ijid.2021.09.045.

Chen, H., Choi, J., & Bailey, S. (2014). Cut site selection by the two nuclease domains of the Cas9 RNA-guided endonuclease. *The Journal of Biological Chemistry, 289*(19), 13284–13294. https://doi.org/10.1074/jbc.M113.539726.

Chen, J., Chen, Y., Huang, L., Lin, X., Chen, H., Xiang, W., & Liu, L. (2024). Trans-nuclease activity of Cas9 activated by DNA or RNA target binding. *Nature Biotechnology*. https://doi.org/10.1038/s41587-024-02255-7.

Chen, J. S., Ma, E., Harrington, L. B., Da Costa, M., Tian, X., Palefsky, J. M., & Doudna, J. A. (2018). CRISPR-Cas12a target binding unleashes indiscriminate single-stranded DNase activity. *Science, 360*(6387), 436–439. https://doi.org/10.1126/science.aar6245.

Chu, D. K. W., Pan, Y., Cheng, S. M. S., Hui, K. P. Y., Krishnan, P., Liu, Y., ... Poon, L. L. M. (2020). Molecular diagnosis of a novel coronavirus (2019-nCoV) causing an outbreak of pneumonia. *Clinical Chemistry, 66*(4), 549–555. https://doi.org/10.1093/clinchem/hvaa029.

Cong, L., Ran, F. A., Cox, D., Lin, S., Barretto, R., Habib, N., ... Zhang, F. (2013). Multiplex genome engineering using CRISPR/Cas systems. *Science, 339*(6121), 819–823. https://doi.org/10.1126/science.1231143.

Corman, V. M., Landt, O., Kaiser, M., Molenkamp, R., Meijer, A., Chu, D. K., ... Drosten, C. (2020). Detection of 2019 novel coronavirus (2019-nCoV) by real-time RT-PCR. *Euro Surveillance, 25*(3), https://doi.org/10.2807/1560-7917.ES.2020.25.3.2000045.

Davidson, A. R., Lu, W. T., Stanley, S. Y., Wang, J., Mejdani, M., Trost, C. N., ... Sontheimer, E. J. (2020). Anti-CRISPRs: Protein inhibitors of CRISPR-Cas systems. *Annual Review of Biochemistry, 89*, 309–332. https://doi.org/10.1146/annurev-biochem-011420-111224.

de Puig, H., Lee, R. A., Najjar, D., Tan, X., Soeknsen, L. R., Angenent-Mari, N. M., ... Collins, J. J. (2021). Minimally instrumented SHERLOCK (miSHERLOCK) for CRISPR-based point-of-care diagnosis of SARS-CoV-2 and emerging variants. *Science Advances,* 7(32), https://doi.org/10.1126/sciadv.abh2944.

Deltcheva, E., Chylinski, K., Sharma, C. M., Gonzales, K., Chao, Y., Pirzada, Z. A., ... Charpentier, E. (2011). CRISPR RNA maturation by trans-encoded small RNA and host factor RNase III. *Nature, 471*(7340), 602–607. https://doi.org/10.1038/nature09886.

Deng, F., Li, Y., Yang, B., Sang, R., Deng, W., Kansara, M., ... Goldys, E. M. (2024). Topological barrier to Cas12a activation by circular DNA nanostructures facilitates autocatalysis and transforms DNA/RNA sensing. *Nature Communications, 15*(1), 1818. https://doi.org/10.1038/s41467-024-46001-8.

East-Seletsky, A., O'Connell, M. R., Knight, S. C., Burstein, D., Cate, J. H., Tjian, R., & Doudna, J. A. (2016). Two distinct RNase activities of CRISPR-C2c2 enable guide-RNA processing and RNA detection. *Nature, 538*(7624), 270–273. https://doi.org/10.1038/nature19802.

Fonfara, I., Richter, H., Bratovic, M., Le Rhun, A., & Charpentier, E. (2016). The CRISPR-associated DNA-cleaving enzyme Cpf1 also processes precursor CRISPR RNA. *Nature, 532*(7600), 517–521. https://doi.org/10.1038/nature17945.

Fozouni, P., Son, S., Diaz de Leon Derby, M., Knott, G. J., Gray, C. N., D'Ambrosio, M. V., ... Ott, M. (2021). Amplification-free detection of SARS-CoV-2 with CRISPR-Cas13a and mobile phone microscopy. *Cell, 184*(2), 323–333.e9. https://doi.org/10.1016/j.cell.2020.12.001.

Garneau, J. E., Dupuis, M. E., Villion, M., Romero, D. A., Barrangou, R., Boyaval, P., ... Moineau, S. (2010). The CRISPR/Cas bacterial immune system cleaves bacteriophage and plasmid DNA. *Nature, 468*(7320), 67–71. https://doi.org/10.1038/nature09523.

Gasiunas, G., Barrangou, R., Horvath, P., & Siksnys, V. (2012). Cas9-crRNA ribonucleoprotein complex mediates specific DNA cleavage for adaptive immunity in bacteria. *Proceedings of the National Academy of Sciences of the United States of America, 109*(39), E2579–E2586. https://doi.org/10.1073/pnas.1208507109.

Gesner, E. M., Schellenberg, M. J., Garside, E. L., George, M. M., & Macmillan, A. M. (2011). Recognition and maturation of effector RNAs in a CRISPR interference pathway. *Nature Structural & Molecular Biology, 18*(6), 688–692. https://doi.org/10.1038/nsmb.2042.

Gleditzsch, D., Pausch, P., Muller-Esparza, H., Ozcan, A., Guo, X., Bange, G., & Randau, L. (2019). PAM identification by CRISPR-Cas effector complexes: Diversified mechanisms and structures. *RNA Biology, 16*(4), 504–517. https://doi.org/10.1080/15476286.2018.1504546.

Gootenberg, J. S., Abudayyeh, O. O., Kellner, M. J., Joung, J., Collins, J. J., & Zhang, F. (2018). Multiplexed and portable nucleic acid detection platform with Cas13, Cas12a, and Csm6. *Science, 360*(6387), 439–444. https://doi.org/10.1126/science.aaq0179.

Gootenberg, J. S., Abudayyeh, O. O., Lee, J. W., Essletzbichler, P., Dy, A. J., Joung, J., ... Zhang, F. (2017). Nucleic acid detection with CRISPR-Cas13a/C2c2. *Science, 356*(6336), 438–442. https://doi.org/10.1126/science.aam9321.

Hajian, R., Balderston, S., Tran, T., deBoer, T., Etienne, J., Sandhu, M., ... Aran, K. (2019). Detection of unamplified target genes via CRISPR-Cas9 immobilized on a graphene field-effect transistor. *Nature Biomedical Engineering, 3*(6), 427–437. https://doi.org/10.1038/s41551-019-0371-x.

Haurwitz, R. E., Jinek, M., Wiedenheft, B., Zhou, K., & Doudna, J. A. (2010). Sequence- and structure-specific RNA processing by a CRISPR endonuclease. *Science, 329*(5997), 1355–1358. https://doi.org/10.1126/science.1192272.

Heler, R., Marraffini, L. A., & Bikard, D. (2014). Adapting to new threats: The generation of memory by CRISPR-Cas immune systems. *Molecular Microbiology, 93*(1), 1–9. https://doi.org/10.1111/mmi.12640.

Hillary, V. E., & Ceasar, S. A. (2023). A review on the mechanism and applications of CRISPR/Cas9/Cas12/Cas13/Cas14 proteins utilized for genome engineering. *Molecular Biotechnology, 65*(3), 311–325. https://doi.org/10.1007/s12033-022-00567-0.

Huang, Z., Fang, J., Zhou, M., Gong, Z., & Xiang, T. (2022). CRISPR-Cas13: A new technology for the rapid detection of pathogenic microorganisms. *Frontiers in Microbiology, 13*, 1011399. https://doi.org/10.3389/fmicb.2022.1011399.

Huynh, N., Depner, N., Larson, R., & King-Jones, K. (2020). A versatile toolkit for CRISPR-Cas13-based RNA manipulation in Drosophila. *Genome Biology, 21*(1), 279. https://doi.org/10.1186/s13059-020-02193-y.

Jaaskelainen, A. E., Ahava, M. J., Jokela, P., Szirovicza, L., Pohjala, S., Vapalahti, O., ... Kurkela, S. (2021). Evaluation of three rapid lateral flow antigen detection tests for the diagnosis of SARS-CoV-2 infection. *Journal of Clinical Virology, 137*, 104785. https://doi.org/10.1016/j.jcv.2021.104785.

Jiang, F., & Doudna, J. A. (2017). CRISPR-Cas9 structures and mechanisms. *Annual Review of Biophysics, 46*, 505–529. https://doi.org/10.1146/annurev-biophys-062215-010822.

Jiang, F., Zhou, K., Ma, L., Gressel, S., & Doudna, J. A. (2015). Structural biology. A Cas9-guide RNA complex preorganized for target DNA recognition. *Science, 348*(6242), 1477–1481. https://doi.org/10.1126/science.aab1452.

Jinek, M., Chylinski, K., Fonfara, I., Hauer, M., Doudna, J. A., & Charpentier, E. (2012). A programmable dual-RNA-guided DNA endonuclease in adaptive bacterial immunity. *Science, 337*(6096), 816–821. https://doi.org/10.1126/science.1225829.

Jinek, M., East, A., Cheng, A., Lin, S., Ma, E., & Doudna, J. (2013). RNA-programmed genome editing in human cells. *Elife, 2*, e00471. https://doi.org/10.7554/eLife.00471.

Kaminski, M. M., Abudayyeh, O. O., Gootenberg, J. S., Zhang, F., & Collins, J. J. (2021). CRISPR-based diagnostics. *Nature Biomedical Engineering, 5*(7), 643–656. https://doi.org/10.1038/s41551-021-00760-7.

Kavuri, N. R., Ramasamy, M., Qi, Y., & Mandadi, K. (2022). Applications of CRISPR/Cas13-based RNA editing in plants. *Cells, 11*(17), https://doi.org/10.3390/cells11172665.

Knott, G. J., & Doudna, J. A. (2018). CRISPR-Cas guides the future of genetic engineering. *Science, 361*(6405), 866–869. https://doi.org/10.1126/science.aat5011.

Konermann, S., Lotfy, P., Brideau, N. J., Oki, J., Shokhirev, M. N., & Hsu, P. D. (2018). Transcriptome engineering with RNA-targeting type VI-D CRISPR effectors. *Cell, 173*(3), 665–676.e614. https://doi.org/10.1016/j.cell.2018.02.033.

Lamb, C. H., Kang, B., & Myhrvold, C. (2023). Multiplexed CRISPR-based methods for pathogen nucleic acid detection. *Current Opinion in Biomedical Engineering, 27*. https://doi.org/10.1016/j.cobme.2023.100471.

Li, L., Li, S., Wu, N., Wu, J., Wang, G., Zhao, G., & Wang, J. (2019). HOLMESv2: A CRISPR-Cas12b-assisted platform for nucleic acid detection and DNA methylation quantitation. *ACS Synthetic Biology, 8*(10), 2228–2237. https://doi.org/10.1021/acssynbio.9b00209.

Li, S. Y., Cheng, Q. X., Wang, J. M., Li, X. Y., Zhang, Z. L., Gao, S., ... Wang, J. (2018). CRISPR-Cas12a-assisted nucleic acid detection. *Cell Discovery, 4*, 20. https://doi.org/10.1038/s41421-018-0028-z.

Liu, L., Li, X., Ma, J., Li, Z., You, L., Wang, J., ... Wang, Y. (2017). The molecular architecture for RNA-guided RNA cleavage by Cas13a. *Cell, 170*(4), 714–726.e10. https://doi.org/10.1016/j.cell.2017.06.050.

Liu, T. Y., Knott, G. J., Smock, D. C. J., Desmarais, J. J., Son, S., Bhuiya, A., ... Doudna, J. A. (2021). Accelerated RNA detection using tandem CRISPR nucleases. *Nature Chemical Biology, 17*(9), 982–988. https://doi.org/10.1038/s41589-021-00842-2.

Lopez-Valls, M., Escalona-Noguero, C., Rodriguez-Diaz, C., Pardo, D., Castellanos, M., Milan-Rois, P., ... Sot, B. (2022). CASCADE: Naked eye-detection of SARS-CoV-2 using Cas13a and gold nanoparticles. *Analytica Chimica Acta, 1205*, 339749. https://doi.org/10.1016/j.aca.2022.339749.

Lu, S., Tong, X., Han, Y., Zhang, K., Zhang, Y., Chen, Q., ... Yin, H. (2022). Fast and sensitive detection of SARS-CoV-2 RNA using suboptimal protospacer adjacent motifs for Cas12a. *Nature Biomedical Engineering, 6*(3), 286–297. https://doi.org/10.1038/s41551-022-00861-x.

Makarova, K. S., Wolf, Y. I., Alkhnbashi, O. S., Costa, F., Shah, S. A., Saunders, S. J., ... Koonin, E. V. (2015). An updated evolutionary classification of CRISPR-Cas systems. *Nature Reviews. Microbiology, 13*(11), 722–736. https://doi.org/10.1038/nrmicro3569.

Makarova, K. S., Zhang, F., & Koonin, E. V. (2017a). SnapShot: Class 1 CRISPR-Cas systems. *Cell, 168*(5), 946–946.e1. https://doi.org/10.1016/j.cell.2017.02.018.

Makarova, K. S., Zhang, F., & Koonin, E. V. (2017b). SnapShot: Class 2 CRISPR-Cas systems. *Cell, 168*(1-2), 328–328.e1. https://doi.org/10.1016/j.cell.2016.12.038.

Maleki, B., & Hojati, Z. (2022). A precise review on NAATs-based diagnostic assays for COVID-19: A motion in fast POC molecular tests. *European Journal of Clinical Investigation, 52*(11), e13853. https://doi.org/10.1111/eci.13853.

Mali, P., Yang, L., Esvelt, K. M., Aach, J., Guell, M., DiCarlo, J. E., ... Church, G. M. (2013). RNA-guided human genome engineering via Cas9. *Science, 339*(6121), 823–826. https://doi.org/10.1126/science.1232033.

Marraffini, L. A., & Sontheimer, E. J. (2010a). CRISPR interference: RNA-directed adaptive immunity in bacteria and archaea. *Nature Reviews Genetics, 11*(3), 181–190. https://doi.org/10.1038/nrg2749.

Marraffini, L. A., & Sontheimer, E. J. (2010b). Self versus non-self discrimination during CRISPR RNA-directed immunity. *Nature, 463*(7280), 568–571. https://doi.org/10.1038/nature08703.

Meeske, A. J., & Marraffini, L. A. (2018). RNA guide complementarity prevents self-targeting in type VI CRISPR systems. *Molecular Cell, 71*(5), 791–801.e3. https://doi.org/10.1016/j.molcel.2018.07.013.

Mina, M. J., & Andersen, K. G. (2021). COVID-19 testing: One size does not fit all. *Science, 371*(6525), 126–127. https://doi.org/10.1126/science.abe9187.

Moon, J., Kwon, H. J., Yong, D., Lee, I. C., Kim, H., Kang, H., ... Kang, T. (2020). Colorimetric detection of SARS-CoV-2 and drug-resistant pH1N1 using CRISPR/dCas9. *ACS Sensors, 5*(12), 4017–4026. https://doi.org/10.1021/acssensors.0c01929.

Myhrvold, C., Freije, C. A., Gootenberg, J. S., Abudayyeh, O. O., Metsky, H. C., Durbin, A. F., ... Sabeti, P. C. (2018). Field-deployable viral diagnostics using CRISPR-Cas13. *Science, 360*(6387), 444–448. https://doi.org/10.1126/science.aas8836.

Notomi, T., Okayama, H., Masubuchi, H., Yonekawa, T., Watanabe, K., Amino, N., & Hase, T. (2000). Loop-mediated isothermal amplification of DNA. *Nucleic Acids Research, 28*(12), E63. https://doi.org/10.1093/nar/28.12.e63.

Pardee, K., Green, A. A., Takahashi, M. K., Braff, D., Lambert, G., Lee, J. W., ... Collins, J. J. (2016). Rapid, low-cost detection of Zika virus using programmable biomolecular components. *Cell, 165*(5), 1255–1266. https://doi.org/10.1016/j.cell.2016.04.059.

Patchsung, M., Homchan, A., Aphicho, K., Suraritdechachai, S., Wanitchanon, T., Pattama, A., ... Uttamapinant, C. (2023). A multiplexed Cas13-based assay with point-of-care attributes for simultaneous COVID-19 diagnosis and variant surveillance. *The CRISPR Journal, 6*(2), 99–115. https://doi.org/10.1089/crispr.2022.0048.

Piepenburg, O., Williams, C. H., Stemple, D. L., & Armes, N. A. (2006). DNA detection using recombination proteins. *PLoS Biology, 4*(7), e204. https://doi.org/10.1371/journal.pbio.0040204.

Rajan, A., Shrivastava, S., Janhawi, Kumar, A., Singh, A. K., & Arora, P. K. (2022). CRISPR-Cas system: From diagnostic tool to potential antiviral treatment. *Applied Microbiology and Biotechnology, 106*(18), 5863–5877. https://doi.org/10.1007/s00253-022-12135-2.

Ramachandran, A., & Santiago, J. G. (2021). CRISPR enzyme kinetics for molecular diagnostics. *Analytical Chemistry, 93*(20), 7456–7464. https://doi.org/10.1021/acs.analchem.1c00525.

Ran, F. A., Hsu, P. D., Wright, J., Agarwala, V., Scott, D. A., & Zhang, F. (2013). Genome engineering using the CRISPR-Cas9 system. *Nature Protocols, 8*(11), 2281–2308. https://doi.org/10.1038/nprot.2013.143.

Senthilnathan, R., Ilangovan, I., Kunale, M., Easwaran, N., Ramamoorthy, S., Veeramuthu, A., & Kodiveri Muthukaliannan, G. (2023). An update on CRISPR-Cas12 as a versatile tool in genome editing. *Molecular Biology Reports, 50*(3), 2865–2881. https://doi.org/10.1007/s11033-023-08239-1.

Shinoda, H., Taguchi, Y., Nakagawa, R., Makino, A., Okazaki, S., Nakano, M., ... Watanabe, R. (2021). Amplification-free RNA detection with CRISPR-Cas13. *Communications Biology, 4*(1), 476. https://doi.org/10.1038/s42003-021-02001-8.

Shmakov, S., Abudayyeh, O. O., Makarova, K. S., Wolf, Y. I., Gootenberg, J. S., Semenova, E., ... Koonin, E. V. (2015). Discovery and functional characterization of diverse class 2 CRISPR-cas systems. *Molecular Cell, 60*(3), 385–397. https://doi.org/10.1016/j.molcel.2015.10.008.

Shmakov, S., Smargon, A., Scott, D., Cox, D., Pyzocha, N., Yan, W., ... Koonin, E. V. (2017). Diversity and evolution of class 2 CRISPR-Cas systems. *Nature Reviews Microbiology, 15*(3), 169–182. https://doi.org/10.1038/nrmicro.2016.184.

Siddiqui, S. M., Welch, N. L., Nguyen, T. G., Razmi, A., Chang, T., Senft, R., ... Myhrvold, C. (2023). Bead-based approaches to CRISPR diagnostics. *medRxiv*, 2023.2009.2003.23294926. https://doi.org/10.1101/2023.09.03.23294926.

Smargon, A. A., Cox, D. B. T., Pyzocha, N. K., Zheng, K., Slaymaker, I. M., Gootenberg, J. S., ... Zhang, F. (2017). Cas13b Is a type VI-B CRISPR-associated RNA-guided RNase differentially regulated by accessory proteins Csx27 and Csx28. *Molecular Cell, 65*(4), 618–630.e7. https://doi.org/10.1016/j.molcel.2016.12.023.

Su, W., Li, J., Ji, C., Chen, C., Wang, Y., Dai, H., ... Liu, P. (2023). CRISPR/Cas systems for the detection of nucleic acid and non-nucleic acid targets. *Nano Research,* 1–14. https://doi.org/10.1007/s12274-023-5567-4.

Tian, T., Shu, B., Jiang, Y., Ye, M., Liu, L., Guo, Z., ... Zhou, X. (2021). An ultralocalized Cas13a assay enables universal and nucleic acid amplification-free single-molecule RNA diagnostics. *ACS Nano, 15*(1), 1167–1178. https://doi.org/10.1021/acsnano.0c08165.

Tong, X., Zhang, K., Han, Y., Li, T., Duan, M., Ji, R., ... Yin, H. (2024). Fast and sensitive CRISPR detection by minimized interference of target amplification. *Nature Chemical Biology, 20*(7), 885–893. https://doi.org/10.1038/s41589-023-01534-9.

Tyumentseva, M., Tyumentsev, A., & Akimkin, V. (2023). CRISPR/Cas9 landscape: Current state and future perspectives. *International Journal of Molecular Sciences, 24*(22), https://doi.org/10.3390/ijms242216077.

Vogels, C. B. F., Brito, A. F., Wyllie, A. L., Fauver, J. R., Ott, I. M., Kalinich, C. C., ... Grubaugh, N. D. (2020). Analytical sensitivity and efficiency comparisons of SARS-CoV-2 RT-qPCR primer-probe sets. *Nature Microbiology, 5*(10), 1299–1305. https://doi.org/10.1038/s41564-020-0761-6.

Wang, D. G., Brewster, J. D., Paul, M., & Tomasula, P. M. (2015). Two methods for increased specificity and sensitivity in loop-mediated isothermal amplification. *Molecules, 20*(4), 6048–6059. https://doi.org/10.3390/molecules20046048.

Wang, Y., Chen, H., Gao, H., Wei, H., Wang, Y., Mu, K., ... Wang, S. (2023). CESSAT: A chemical additive-enhanced single-step accurate CRISPR/Cas13 testing system for field-deployable ultrasensitive detection and genotyping of SARS-CoV-2 variants of concern. *Biosensors & Bioelectronics, 229*, 115238. https://doi.org/10.1016/j.bios.2023.115238.

Wang, Y., Chen, H., Lin, K., Han, Y., Gu, Z., Wei, H., & Wang, S. (2024). Ultrasensitive single-step CRISPR detection of monkeypox virus in minutes with a vest-pocket diagnostic device. *Nature Communications, 15*(1), 3279. https://doi.org/10.1038/s41467-024-47518-8.

Welch, N. L., Zhu, M., Hua, C., Weller, J., Mirhashemi, M. E., Nguyen, T. G., ... Myhrvold, C. (2022). Multiplexed CRISPR-based microfluidic platform for clinical testing of respiratory viruses and identification of SARS-CoV-2 variants. *Nature Medicine, 28*(5), 1083–1094. https://doi.org/10.1038/s41591-022-01734-1.

Wiedenheft, B., Sternberg, S. H., & Doudna, J. A. (2012). RNA-guided genetic silencing systems in bacteria and archaea. *Nature, 482*(7385), 331–338. https://doi.org/10.1038/nature10886.

Xu, Z., Chen, D., Li, T., Yan, J., Zhu, J., He, T., ... Liu, M. (2022). Microfluidic space coding for multiplexed nucleic acid detection via CRISPR-Cas12a and recombinase polymerase amplification. *Nature Communications, 13*(1), 6480. https://doi.org/10.1038/s41467-022-34086-y.

Yan, C., Cui, J., Huang, L., Du, B., Chen, L., Xue, G., ... Yuan, J. (2020). Rapid and visual detection of 2019 novel coronavirus (SARS-CoV-2) by a reverse transcription loop-mediated isothermal amplification assay. *Clinical Microbiology and Infection, 26*(6), 773–779. https://doi.org/10.1016/j.cmi.2020.04.001.

Yang, H., & Patel, D. J. (2024). Structures, mechanisms and applications of RNA-centric CRISPR-Cas13. *Nature Chemical Biology, 20*(6), 673–688. https://doi.org/10.1038/s41589-024-01593-6.

Yuce, M., Filiztekin, E., & Ozkaya, K. G. (2021). COVID-19 diagnosis—A review of current methods. *Biosensors & Bioelectronics, 172*, 112752. https://doi.org/10.1016/j.bios.2020.112752.

Zetsche, B., Gootenberg, J. S., Abudayyeh, O. O., Slaymaker, I. M., Makarova, K. S., Essletzbichler, P., ... Zhang, F. (2015). Cpf1 is a single RNA-guided endonuclease of a class 2 CRISPR-Cas system. *Cell, 163*(3), 759–771. https://doi.org/10.1016/j.cell.2015.09.038.

Zhou, J., Li, Z., Seun Olajide, J., & Wang, G. (2024). CRISPR/Cas-based nucleic acid detection strategies: Trends and challenges. *Heliyon, 10*(4), e26179. https://doi.org/10.1016/j.heliyon.2024.e26179.

CHAPTER TWELVE

Preparation of high-purity RNPs of CRISPR-based DNA base editors

Mitchell J. McAndrew[a,b], **Madeleine B. King**[a,b], and **Audrone Lapinaite**[b,c,d,*]
[a]School of Molecular Sciences, Arizona State University, Tempe, AZ, United States
[b]Gavin Herbert Eye Institute - Center for Translational Vision Research, University of California Irvine, School of Medicine, Irvine, CA, United States
[c]Department of Ophthalmology, University of California Irvine, School of Medicine, Irvine, CA, United States
[d]Department of Biomedical Engineering, University of California Irvine, Irvine, CA, United States
*Corresponding author. e-mail address: alapinai@hs.uci.edu

Contents

1. Introduction 278
2. Materials 281
   2.1 Expression of recombinant DNA base editors 281
   2.2 Purification of ABE and CBE enzymes 282
   2.3 In vitro transcription (IVT) of sgRNA 284
   2.4 In vitro DNA deamination assay 286
3. Methods 287
   3.1 Transformation of ABE or CBE plasmid (20 h) 287
   3.2 Expression of recombinant DNA base editors (48 h) 288
   3.3 Purification of ABE and CBE enzymes (14.5 h) 289
   3.4 *In vitro* transcription (IVT) of sgRNA (30 h) 296
   3.5 *In vitro* DNA deamination assay (7-8 h) 301
4. Summary and conclusion 304
Acknowledgement 305
Funding 305
Data availability 305
5. Buffer Appendix 305
   5.1 *Purification buffers* 305
References 313

Abstract

Since their introduction, CRISPR-based DNA base editors (BEs) have become essential in the field of precision genome editing, revolutionizing the correction of pathogenic SNPs for both basic research and therapeutic applications. As this technology advances, more laboratories are implementing these tools into their workflow. The delivery of BEs as BE-guide RNA complexes (RNPs), rather than as mRNA or plasmids, has been shown to exhibit lower off-target effects, establishing it as the preferred

Methods in Enzymology, Volume 712
ISSN 0076-6879, https://doi.org/10.1016/bs.mie.2025.01.019

method of delivery. However, there are no protocols describing in detail how to obtain high-purity and highly active BE RNPs. Here, we offer a comprehensive guide for the expression, purification, RNP reconstitution, and *in vitro* activity assessment of TadA-based BEs. The protocol includes guidance on performing activity assays using commercial denaturing gels, which is convenient and uses standard molecular biology equipment. This allows for rapid quality control testing of reconstituted BE RNPs prior to more expensive and time-consuming in vivo genome editing experiments. Overall, this protocol aims to empower more laboratories to generate tailored BE RNPs for diverse *in vitro* and in vivo applications.

1. Introduction

Current state-of-the-art genome editing approaches are based on RNA-guided DNA endonucleases such as Cas9 and Cas12 originating from Type II and Type V Clustered Regularly Interspaced Short Palindromic Repeats (CRISPR) systems (Adler et al., 2024; Makarova et al., 2019), which function as adaptive immune mechanisms protecting prokaryotic cells from invading phages and mobile genetic elements (Barrangou et al., 2007; Marraffini & Sontheimer, 2008; Mojica, Díez-Villaseñor, García-Martínez, & Soria, 2005). Cas9 and Cas12 enzymes are favored for their efficiency and programmability in introducing double-stranded DNA breaks (DSBs) at precise genomic locations in human cells, making them ideal for engineering various genome editing tools (Cong et al., 2013; Jinek et al., 2012; Jinek et al., 2013; Mali et al., 2013; Zetsche et al., 2015).

For genome editing of eukaryotic DNA, Cas9 is directed to a target site by an engineered single guide RNA (sgRNA), the spacer region of which is designed to match the sequence of the specific target of interest (Jinek et al., 2012). Subsequently, Cas9 introduces a DSB into the genomic DNA, which is then repaired via one of the competing cellular DNA repair pathways: homology-directed repair (HDR) or non-homologous end joining (NHEJ) (Doudna & Charpentier, 2014; Hou et al., 2013; Song, Yang, Hwang, Yu, & Bae, 2021), with only HDR allowing precise changes in the DNA sequence (Xue, 2021). However, HDR is only active in certain cell types and during certain cell cycle phases, which limits the application of CRISPR-based tools for precision genome editing (Cong et al., 2013; Ran et al., 2013). Thus, new precise genome editing tools such as base editors (BEs) (Gaudelli et al., 2017; Komor, Kim, Packer, Zuris, & Liu, 2016) and prime editors (PEs) (Anzalone et al., 2019) have been developed to edit DNA without relying on DSB repair pathways.

BEs are CRISPR-based genome editing tools that introduce single nucleotide mutations in a programmable manner without the introduction of DSBs or the use of donor templates (Gaudelli et al., 2017; Komor et al., 2016). There are two main types of BEs: cytosine base editors (CBEs, convert C-G to T-A base pairs) and adenine base editors (ABEs, convert A-T to G-C base pairs). Both consist of a catalytically impaired nickase, Cas9 (nCas9^{D10A}), fused to a deaminase protein specific for cytosine or adenine deamination, respectively (Fig. 1A). While the Cas9 locates the target DNA sequence by complementarity to the spacer region of the sgRNA and nicks it, the deaminase domain performs the base modification and converts adenine to inosine (ABEs) (Chen, McAndrew, & Lapinaite, 2023; Gaudelli et al., 2017; Lapinaite et al., 2020) or cytosine to uracil (CBEs) on the intact DNA strand (Komor et al., 2016). The nicked strand is then repaired by the cellular DNA repair machinery, using the modified DNA strand as a template to ensure that both strands are accurately modified (Fig. 1B) (reviewed in (Gu et al., 2021)).

Currently, many versions of ABEs and CBEs are used in both basic research and biomedical applications (Alves et al., 2024; Chiesa et al., 2023; Hooper, Tang, & Burnett, 2024; Knipping et al., 2022). ABEs are usually based on *Escherichia coli* tRNA-specific adenine deaminase (EcTadA) (Gaudelli et al., 2017), while CBEs encode either activation-induced cytidine deaminase

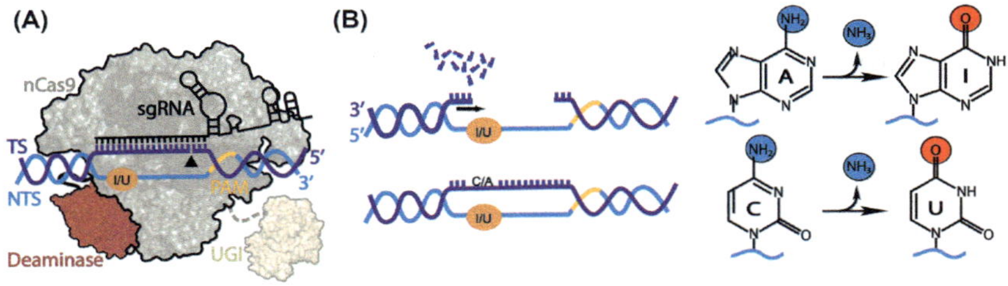

Fig. 1 Mechanism of CRISPR-based ABEs and CBEs. (A) The general architecture of base editors consists of a catalytically impaired Cas9 protein (nCas9, gray) linked to a deaminase domain (red). In the case of CBEs, the construct also includes a uracil DNA glycosylase inhibitor (UGI, pale yellow). The Cas9 protein complexes with an sgRNA (black) and scans the DNA for a protospacer adjacent motif (PAM, orange). Once it recognizes a PAM and the adjacent sequence complementary to the sgRNA, it forms an R-loop. The single-stranded non-target DNA strand (NTS, cyan) then serves as a substrate for the deaminase. (B) The deaminase domain catalyzes the conversion of adenine to inosine (ABEs) or cytosine to uracil (CBEs) in the NTS. The nicking of the target strand (TS, navy) by nCas9 activates the single-stranded break DNA repair machinery, allowing for editing of the TS using the modified NTS as a template.

(AID) (Nishida et al., 2016) or apolipoprotein B mRNA editing enzyme (APOBEC) (Komor et al., 2016). However, it's worth noting that these CBEs tend to exhibit more off-target effects compared to their ABE counterparts (Doman, Raguram, Newby, & Liu, 2020; Zuo et al., 2019). Thus, newer versions of CBEs have been developed, featuring a cytosine deaminase derived from EcTadA that has been evolved to target cytosine instead of adenine (Chen et al., 2023; Lam et al., 2023; Neugebauer et al., 2023; Zhang, Neugebauer, Krasnow, & Liu, 2024). These CBEs were shown to have overall reduced DNA and RNA off-target editing while retaining similar levels of on-target editing (Zhang et al., 2024).

To perform precision genome editing in vivo using BEs, they must first be delivered into the cells and they can be introduced in three distinct formulations: as plasmids, mRNA, or *in vitro* reconstituted BE-sgRNA complexes (BE RNPs). The introduction of BE RNPs offers advantages over the other two methods because RNPs are immediately active (eliminating the need to synthesize proteins or assemble RNPs), and have the shortest lifespan (Kim, Kim, Cho, Kim, & Kim, 2014). These factors collectively contribute to reducing off-target effects compared to plasmid or mRNA delivery approaches (Jang et al., 2021; Rees et al., 2017; Zuris et al., 2015). However, the success of RNP delivery-based genome editing is highly dependent on the quality of the reconstituted BE RNPs.

Currently, there are several protocols established for preparing ABE RNPs, which support both the investigation of their molecular mechanisms through structural and biochemical studies (Lapinaite et al., 2020; Richter et al., 2020) and their delivery into human cell lines (Huang, Newby, & Liu, 2021). However, the preparation of high-quality, active CBE RNPs remains challenging, mostly due to the toxicity of the APOBEC-derived deaminase (Grünewald et al., 2019). A couple of protocols are available for the expression and purification of APOBEC-based CBEs using either human cell expression systems or *E. coli* (Jang et al., 2021; Rees et al., 2017). However, CBEs purified using these protocols demonstrate reduced activity in vivo genome editing applications (Jang et al., 2021; Rees et al., 2017).

Here, we outline the steps for expressing, purifying, and preparing ABE RNPs using an *E. coli* expression system (Fig. 2)—similar to the protocol used for producing ABE8e for structural studies (Lapinaite et al., 2020)—and adapt these methods for the preparation of TadA-based CBE RNPs. The sgRNAs are prepared via *in vitro* transcription, and the activity of reconstituted RNPs is tested *in vitro* to confirm their suitability for various *in vitro* and in vivo DNA editing applications.

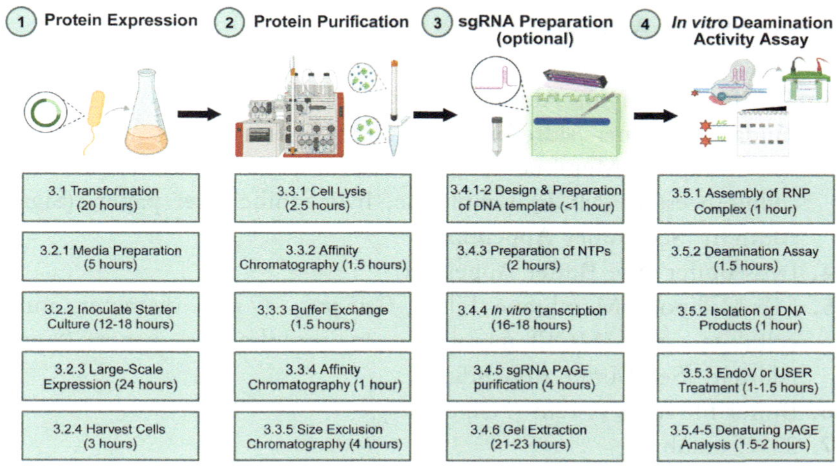

Fig. 2 Overview of BE RNP preparation and characterization. The protocol is sectioned into four main parts: protein expression (1), protein purification (2), sgRNA preparation (3), and *in vitro* deamination activity assay (4).

2. Materials

2.1 Expression of recombinant DNA base editors

1. Bacterial expression plasmid encoding ABE or other TadA-based DNA base editor of interest, e.g., bacterial expression of ABE8e or CBE-T1.52 in a pBR322-based plasmid.
2. *E. coli* expression strain: Rosetta™ 2(DE3)pLysS Competent Cells (Novagen) if gene encoding DNA base editor is not codon optimized for bacterial expression; if a codon-optimized plasmid is used, other strains such as BL21 (DE3)pLysS can be used.
3. LB Broth (MP Biomedicals)
4. Terrific Broth Modified Powder (Research Products International)
   a. Casein Digest Peptone: 12 g/L, Yeast Extract: 24 g/L, Dipotassium phosphate: 9.4 g/L, Monopotassium phosphate: 2.2 g/L
5. Kanamycin (1000X stock) or an alternative antibiotic, depending on the resistance gene of the expression plasmid (Gold Biotechnology)
6. Isopropyl ß-D-1-thiogalactopyranoside (IPTG) (1 M Stock) (Gold Biotechnology)
7. MilliQ Ultrapure H_2O
8. HEPES (1 M pH 8.0 stock) (See Buffer Appendix)

9. KCl (Sigma-Aldrich)
10. Glycerol (Sigma-Aldrich)
11. TCEP-HCl (Sigma-Aldrich)
12. Imidazole (Sigma-Aldrich)
13. MF-Millipore Membrane Filter, 0.22 μm, 47 mm diameter, mixed cellulose esters (MCE) membrane, hydrophilic filter papers (Sigma-Aldrich) (See Buffer Appendix)
14. Lysis buffer (See Buffer Appendix)
15. MF-Millipore Membrane Filter, 0.45 μm, 47 mm diameter, mixed cellulose esters (MCE) membrane, hydrophilic filter papers (Sigma-Aldrich) (See Buffer Appendix)
16. Buffer filtration system (Nalgene™)
17. 37°C incubator
18. 500 mL baffled culture flasks
19. 2.5 L Ultra Yield® Flasks (Thomson Instrument Company; cat. no. 931136-B); essential for achieving high yields
20. Refrigerated incubator shaker
21. Photometer or spectrometer to measure optical density at 600 nm (OD_{600}). e.g., Cell Density Meter CO8000 (Biowave)
22. High-capacity, refrigerated centrifuge e.g., Beckman Avanti J-HC with JA-4.2 rotor (Beckman Coulter)

2.2 Purification of ABE and CBE enzymes

2.2.1 Cell lysis

1. MilliQ Ultrapure H_2O
2. HEPES (1 M pH 8.0 stock)
3. KCl (Sigma-Aldrich)
4. Glycerol (Sigma-Aldrich)
5. TCEP-HCl (Sigma-Aldrich)
6. Imidazole (Sigma-Aldrich)
7. cOmplete™, EDTA-free Protease Inhibitor Cocktail (Roche)
8. RQ1 DNase (Promega)
9. Lysozyme, Egg White (powder) (Sigma-Aldrich)
10. TritonX-100 (Sigma-Aldrich)
11. Lysis buffer (See Buffer Appendix)
12. MF-Millipore Membrane Filter, 0.22 μm, 47 mm diameter, mixed cellulose esters (MCE) membrane, hydrophilic filter papers (Sigma-Aldrich) (See Buffer Appendix)
13. Buffer filtration system (Nalgene™)

14. Ultrasonic processor e.g., Q700 Sonicator®
15. Oak Ridge High-speed 30 mL centrifuge tubes (Nalgene™)
16. High-speed, refrigerated centrifuge e.g., Beckman Avanti J-E with fixed angle JA-20 (Beckman Coulter)
17. Ultracentrifuge (optional) (Beckman Coulter)
    a. Beckman Type 45 Ti Rotor Assembly, Fixed-Angle Rotor, Titanium
    b. Beckman UltraCentrifuge (e.g., Optimal LE-80K)

2.2.2 Column chromatography using FPLC system

1. HEPES (1 M pH 8.0 stock)
2. KCl (Sigma-Aldrich)
3. NaOH (0.5 M Stock)
4. Glycerol (Sigma-Aldrich)
5. TCEP-HCl (Sigma-Aldrich)
6. Imidazole (Sigma-Aldrich)
7. $MgCl_2$ (Sigma-Aldrich)
8. LiCl (Sigma-Aldrich)
9. Dithiothreitol (DTT) (Gold Biotechnology)
10. Tris-HCl (1 M pH 8.0 stock)
11. MilliQ Ultrapure H_2O
12. MF-Millipore Membrane Filter, 0.22 μm, 47 mm diameter, mixed cellulose esters (MCE) membrane, hydrophilic filter papers (Sigma-Aldrich) (See Buffer Appendix)
13. Protein Ladder e.g., Unstained Protein Standard, Broad Range (New England Biolabs)
14. Low-binding RNAse-DNAse free 1.7 mL Eppendorf tubes
15. 10X Tris Glycine Running Buffer (See Buffer Appendix)
16. 4X SDS-Page Loading Dye (See Buffer Appendix)
17. SDS-Page Coomassie Blue Stain (See Buffer Appendix)
18. Peristaltic pump
19. FPLC system with the ability to detect two wavelengths (280 and 260 nm) simultaneously and equipped with a fraction collector, e.g., ÄKTA pure™ 25 M (Cytiva)
20. HisTrap™ HP, 5 mL column (Cytiva)
21. Superloop™, 50 mL, M6 fitting (Cytiva)
22. 5 mL PEEK loop (Cytiva)
23. HiPrep™ 26/10 Desalting column (Cytiva)
24. HiTrap™ Heparin HP, 5 mL (Cytiva)
25. HiLoad® 16/600 Superdex® 200 prep grade (Cytiva)

26. Mini-PROTEAN Tetra Cell (Bio-Rad)
27. PowerPac™ Basic Power Supply (Bio-Rad)
28. 10 % Gels for protein SDS-PAGE either made in-house following this protocol (Protocols, 2015), or purchased commercially e.g., 4–20 % Mini-PROTEAN® TGX™ Precast Protein Gels, 15-well, 15 μL (Bio-Rad)
29. Refrigerated tabletop centrifuge e.g., Eppendorf centrifuge 5920 R
30. Amicon® Ultra 15 mL Centrifugal Filters 100 kDa MWCO (Millipore)
31. Spectrophotometer e.g., NanoDrop One (Thermo Fisher Scientific)

2.3 In vitro transcription (IVT) of sgRNA

2.3.1 In vitro transcription

1. Dry block incubator or thermocycler
2. Forward and Reverse DNA Template for guide RNA (100 μM stock) (IDT)
3. Diethyl pyrocarbonate (DEPC)-treated water (See Buffer Appendix), or nuclease-free water
4. 5X Transcription Buffer (See Buffer Appendix)
5. NTP solutions (ATP, CTP, GTP, UTP) (100 mM preferred) (Sigma-Aldrich)
   a. If making in-house
      i. ATP disodium salt powder
      ii. CTP disodium salt powder
      iii. GTP sodium salt hydrate powder
      iv. UTP sodium salt powder
      v. 5 N NaOH
      vi. 1 N NaOH
      vii. Potassium pH strips (between pH 6-9)
         1. DTT (1 M stock) (Gold Biotechnology)
         2. Triton X-100 (10 % stock) (Sigma-Aldrich)
         3. Thermostable inorganic pyrophosphatase (TIPP) (New England Biolabs)
         4. RNase inhibitor e.g. RNAse Out (Thermo Fisher Scientific)
         5. T7 polymerase (1 mg/mL) either made in-house using this protocol (Rio, 2013) or purchased commercially
         6. RNase-free RQ1 DNase (Promega)

2.3.2 sgRNA purification by preparative polyacrylamide gel electrophoresis

1. 70 % Ethanol
2. RNase-surface decontaminant (e.g., RNase™ Away)
3. Siliconizing reagent Sigmacote® (Sigma-Aldrich)

4. Glass Plates 35 cm × 23 cm for RNA purification, 14 cm × 20 cm for analytical gel (Ellard Instrumentation)
5. 35 cm and 14 cm aluminum diffuser plates (Ellard Instrumentation)
6. Vertical Slab Gel Electrophoresis Units V3523 for RNA purification, V1420 for analytical gel (Ellard Instrumentation)
7. PowerPac HV High-Voltage Power Supply (Bio-Rad)
8. Binder Clips
9. Thick, 3 mm comb with 1 or 2 square wells (Ellard Instrumentation)
10. Thin, 0.8-1.0 mm comb with 20 square wells (Ellard Instrumentation)
11. Thick, 3 mm Spacers (Ellard Instrumentation)
12. Thin, 0.8-1.0 mm spacers (Ellard Instrumentation)
13. Saran Wrap
14. Ammonium persulfate (APS) (40 % stock) (Sigma-Aldrich)
15. N,N,N′,N′-Tetramethylethylenediamine (TEMED) (Sigma-Aldrich)
16. 5X TBE buffer (See Buffer Appendix)
17. 30 mL syringe with hypodermic needle
18. 19:1 40 % Acrylamide Solution (Sigma-Aldrich)
19. Urea (Sigma-Aldrich)
20. 1 % (w/v) Bromophenol Blue (Sigma-Aldrich)
21. 1 % (w/v) Xylene Cyanol (Sigma-Aldrich)
22. 2X RNA Loading Dye (See Buffer Appendix)
23. 2X RNA Loading Buffer (See Buffer Appendix)
24. Sterile Medical Disposable Scalpels
25. Thin Layer Chromatography (TLC) plate
26. Handheld UV Lamp (254 nm)

2.3.3 Elution of sgRNA

1. TEN Buffer (See Buffer Appendix)
2. Spatula
3. Rocker or nutator
4. 1.2 μM Supor membrane filters low DNA and RNA binding (e.g., Cameo Nylon Syringe Filters, 1.2 μm, 30 mm) (Sigma-Aldrich)
5. Low-binding RNase-, DNase-free 2.0 mL Eppendorf tubes
6. 3 mL syringe
7. Amicon® Ultra 500 μL Centrifugal Filters 3 kDa MWCO (Millipore)
8. Ice-cold 100 % RNase-, DNase-free Ethanol
9. Glycoblue (Thermo Fisher Scientific)
10. NaOAc pH 5.2 (3 M stock) (See Buffer Appendix)
11. Refrigerated tabletop microcentrifuge (Eppendorf 5425 R)

12. Cold 70 % Ethanol
13. Dry block/heat block for 1.7/2.0 mL Eppendorf tubes
14. SYBR™ Green II RNA Gel Stain (Invitrogen)
15. Gel imaging system e.g. GelDoc (Bio-Rad)

2.4 In vitro DNA deamination assay

2.4.1 Folding of sgRNA

1. Purified sgRNA (from step 2.3.3)
2. Dilution Buffer (5X Stock) (See Buffer Appendix)
3. MF-Millipore Membrane Filter, 0.22 μm, 47 mm diameter, mixed cellulose esters (MCE) membrane, hydrophilic filter papers (Sigma-Aldrich)
4. Dry block incubator or thermocycler
5. 50 mM $MgCl_2$ stock (filtered with 0.22 μm)

2.4.2 RNP assembly

1. RNP Assembly Buffer (5X Stock) (See Buffer Appendix)
2. Folded sgRNA (from step 2.4.1)
3. Purified DNA base editor (from step 2.2)

2.4.3 Formation of 5′-fluorescently labeled double-stranded DNA substrate

1. 60 nucleotides long, fluorescently-labeled (5′-IRDye® 800) single-stranded DNA (ssDNA) containing either a single A or C (depending on which BE will be tested) (IDT). Alternatively, the DNA can be labeled with other fluorophores (e.g. FAM) or radiolabeled with ^{32}P isotope.
2. 60 nucleotides long, unlabeled ssDNA complementary to both the fluorescently labeled DNA and the spacer region of the sgRNA (IDT)
3. Dry block incubator or thermocycler

2.4.4 In vitro DNA deamination assay

1. Deamination Buffer (5X Stock) (See Buffer Appendix)
2. Reconstituted ABE or CBE RNP (from step 2.4.2)
3. Double-stranded, fluorescently-labeled DNA (from step 2.4.3)
4. Dry block incubator or thermocycler
5. Monarch PCR & DNA Purification Kit (New England Biolabs)
6. 100 % RNase-, DNase- free Ethanol
7. Tabletop microcentrifuge (Eppendorf 5425 R)
8. EndoV Mix when testing ABEs (See Buffer Appendix)
9. USER Mix when testing CBEs (See Buffer Appendix)
10. Super Buffer (See Buffer Appendix)

11. 7 M urea 15 % 29:1 acrylamide gels, either made in-house or commercial 15 % TBE-Urea Gel, 12 well (Bio-Rad)
12. Mini-PROTEAN Tetra cell Apparatus (Bio-Rad)
13. Power Supply (PowerPac if using commercial urea gel, High Voltage if using in-house made gel) (Bio-Rad)
14. 2X RNA Loading Dye (See Buffer Appendix)
15. 30 mL syringe with hypodermic needle
16. Amersham Typhoon Imager or another gel imaging system capable of detecting fluorescence from the labeled DNA

3. Methods

Here we provide a comprehensive guide for the expression and purification of TadA-based DNA base editors (ABEs and CBEs), alongside the assembly of functional RNPs. To validate the activity of the prepared RNPs, we introduce a straightforward assay leveraging commercially available polyacrylamide gels to quantify DNA deamination by reconstituted RNPs. While this protocol does not cover the design of functional sgRNAs or the selection of the appropriate Cas9-deaminase variant for specific genomic targets, these aspects are thoroughly addressed in (Huang et al., 2021).

3.1 Transformation of ABE or CBE plasmid (20 h)

1. Clone the chosen base editor into a bacterial expression vector with an antibiotic selection marker, or alternatively, acquire it from Addgene. Here, we use ABE8e (Richter et al., 2020) or CBE-T1.52 (Lam et al., 2023), both cloned into a pBR322-derived plasmid containing a kanamycin resistance gene. Both base editors feature a 6xHis tag (necessary for affinity purification) and a bipartite nuclear localization signal (BPNLS, necessary for genomic DNA editing in mammalian cells) at their N-termini followed by a single deaminase domain (TadA8e or T_{AD}C-1.52) and *Streptococcus pyogenes* nCas9. CBE-T1.52 also includes two uracil glycosylase inhibitor (UGI) domains, essential for efficient C-to-U editing in mammalian cells, and an additional BPNLS domain at its C-terminus.
2. Transform BE expression plasmid into Rosetta 2(DE3) pLysS competent cells using either your custom protocol or the manufacturer's protocol, if the competent cells were purchased. Using pLysS cells is crucial to prevent basal expression of T7 RNA polymerase and improve the yield of recombinant base editors. Additionally, Rosetta 2 cells

should be used to enhance the expression of constructs that are not codon-optimized for bacterial expression and contain codons rarely used by *E. coli.*

3.2 Expression of recombinant DNA base editors (48 h)

3.2.1 Day in advance (5 h)

1. Prepare 500 mL of LB medium in a 1 L glass bottle and autoclave. Once sterilized, allow the medium to cool to at least 37°C before using it for overnight starter culture.
2. Prepare 6 L of TB medium in 8 Ultra High Yield flasks (750 mL/flask) and autoclave. Allow the media to cool to at least 37°C before using it for expression. Using Ultra High Yield flasks improves the yield of recombinant DNA base editors.

3.2.2 Day 1: Preparation of overnight culture (12–18 h)

1. Prepare two 500 mL autoclaved flasks, each containing 100 mL of autoclaved LB for starter cultures. Add the appropriate antibiotic to each flask at the final required concentration.
2. Pick a couple of colonies from the transformation plate and inoculate each flask. Incubate for 12–18 h at 37°C with shaking at 200–250 rpm. Selecting multiple colonies reduces the risk of choosing a wrong clone. It's important to note that cultures incubated for more than 24 h may enter the stationary phase and should not be used for large-scale recombinant protein expression.

3.2.3 Day 2: Large scale expression (5–6 h)

1. Inoculate 6 L of autoclaved TB media (distributed across 8 flasks) containing the appropriate antibiotic with overnight culture. Typically, the volume of overnight culture should be 1–10 % of the total expression volume. We use 20 mL of overnight culture for every 750 mL of TB medium.
   I. Tip: Pre-warming the TB media to 37°C before inoculation decreases the time needed to reach the optimal cell density necessary for induction.
2. Incubate the cultures at 37°C with shaking at 180 rpm until an OD_{600} of ~1.2 is reached.
3. Reduce temperature to 16°C.
4. After cooling the cultures to 16°C (which can be expedited by placing the flasks on ice for 10–20 min), induce the expression of DNA base editors by adding IPTG to a final concentration of 0.5 mM.

5. Incubate the cultures at 16°C with shaking at 180 rpm for **19–20 h** after induction. We recommend not exceeding 20 h, as this may reduce base editor yields. We have assessed the yield and activity of recombinant ABE8e when harvested after 18, 20 and 24 h of expression. We observe the highest protein yield when harvesting at 20 h, with activity remaining consistent across all harvest time points (Fig. 3).

3.2.4 Day 3: Harvesting cells (3 h)

1. Pellet the cells by centrifugation at 4550 x g for 30 min at 4°C.
2. Resuspend the cell pellets in lysis buffer. To maintain cell integrity during this step, it's important that the lysis buffer does not contain any detergent or lysozyme. Typically, we use 35–40 mL of lysis buffer per pellet derived from 750 mL of culture. *At this stage, there is an option to halt the process and transfer the resuspended pellets to 50 mL tubes for freezing and storage at −80°C for later purification.*

3.3 Purification of ABE and CBE enzymes (14.5 h)

3.3.1 Day 4: Cell lysis (2.5 h for thawed cells)

If pellets were stored at −80°C, thaw them. To speed up thawing, the Falcon 50 mL tubes containing the cell pellets can be placed in room temperature water and incubated until most of the ice has melted.

1. Combine the thawed cells in a 600 mL beaker on ice (typically resulting in ~300 mL of suspension).

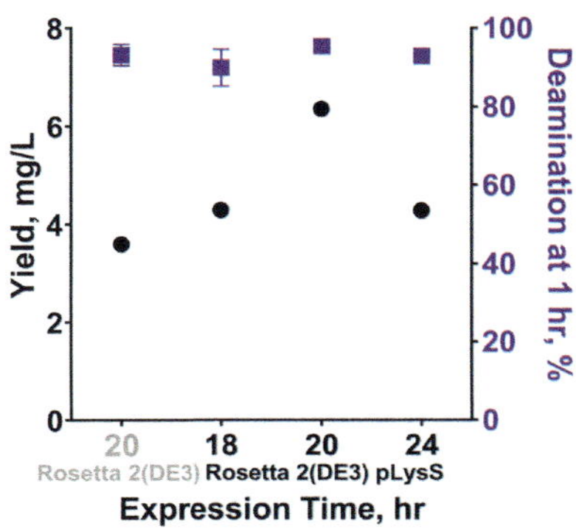

Fig. 3 Determining expression duration to maximize ABE8e yield. ABE8e was expressed at various durations (18–24 h) in Rosetta 2(DE3) (gray x-axis) and Rosetta 2(DE3) pLysS (black x-axis) *E. coli* cells. The yield (black y-axis) was the highest in cells with pLysS and when expressed for 20 h. The percentage of deamination at one hour (blue y-axis) remained constant across all expression times.

2. Next, add 100 mL of lysis buffer (bringing the final suspension volume to 400 mL), supplemented with dissolved 3 cOmplete EDTA-free protease inhibitor tablets, 400 mg of lysozyme (to achieve a final concentration of 1 mg/mL), 200 μL of RNase-free DNase1, and 2 mL of TritonX-100 (resulting in a final concentration of 0.5 %). Mix the components thoroughly, using a serological pipette or spatula, taking care to avoid foam formation. Incubate the mixture on ice for 15 min.
3. Sonicate the cell suspension on ice, being careful not to create foam, until the suspension turns fluid and thin. The lysate should have a darker green/brownish color and not be stringy. The sonication process typically takes 25–30 min. We choose sonication because it also fragments bacterial DNA, allowing DNase1 to degrade it more efficiently, which enhances the separation of base editors from nonspecific nucleic acids during purification. It is essential to prevent the lysate from heating up, as this could denature the protein or reduce its activity.
4. Transfer the lysate to Oak Ridge tubes (30 mL/tube). Ensure tubes are balanced within ±0.01 g using lysis buffer.
5. Pellet cell debris by centrifugation at 43,500 x g or higher for 45 min at 4°C.
6. Decant and save the supernatant. If the supernatant is opaque, perform an additional centrifugation at 210,000 x g for 40 min at 4°C to further clarify it. Failure to do so may clog the affinity chromatography column during the next purification step.

3.3.2 Day 4: Affinity chromatography (1.5 h)

Affinity chromatography can be performed using either Ni-NTA beads via gravity flow or a pre-packed HisTrap™ HP, 5 mL column with an FPLC system. We recommend using the latter due to its faster processing. From our experience, once the cells are lysed, it's important to proceed swiftly through the initial chromatography steps to maximize the yield of active base editors. Ensure all subsequent steps are carried out at 4°C to maintain protein stability and prevent degradation.

1. Pre-equilibrate the HisTrap™ HP, 5 mL column with 5 column volumes (CVs) of lysis buffer using a peristaltic pump or an FPLC system at 4°C.
2. Load the supernatant from step 6 onto the pre-equilibrated column using a peristaltic pump or an FPLC system.
3. Once the sample is loaded, pass 5 CVs of lysis buffer through the column. Next, perform alternating washes of 5 CVs each, switching

between lysis buffer and high salt buffer a couple of times. This step is essential for removing some of the contaminating proteins and nucleic acids. If using an FPLC system for this step, several peaks at 280 nm and 260 nm wavelengths will be observed during the change in conductivity, indicating the elution of nonspecific proteins and nucleic acids.

4. Pass 5 CV of lysis buffer through the column using FPLC system until the absorbance signal at 260 nm and 280 nm stabilizes near zero.
5. Protein elution:
   a. For ABE: use a second FPLC pump to adjust the elution buffer concentration to 10 % and pass 5 CVs through the column. Increase the elution buffer concentration to 15 % for another 5 CVs and 20 % for 5 more CVs. Finally, set up a gradient of elution buffer ranging from 20 % to 100 % over 4 min with a flow rate of 3 mL/min. Collect 20 μL samples from each fraction corresponding to the peaks observed at 280 nm and 260 nm wavelengths, and analyze them by SDS-PAGE using 10 % gel or commercial gradient gel (Fig. 4A).
   b. For TadA-based CBE: set up a gradient of elution buffer ranging from 0 % to 100 % over 4 min with a flow rate of 3 mL/min. Collect 20 μL samples from across the peaks observed at 280 nm and 260 nm wavelengths and analyze them by SDS-PAGE using 10 % gel or commercial gradient gel (Fig. 5A).
6. Pool the fractions containing the base editor.

3.3.3 Day 4: Buffer exchange (1.5 h)

In this step, the elution buffer is exchanged for a low-salt (IEXA) buffer, which is crucial for the subsequent purification step involving affinity chromatography to separate non-specifically interacting nucleic acids from the base editor.

1. Equilibrate the desalting column (we recommend using the HiPrep™ 26/10 desalting column) with 5 CVs of low salt IEXA buffer using an FPLC system. By the end of the equilibration process, the absorbance at 260 nm and 280 nm, as well as the conductivity signals, should be stable.
2. Load the protein sample onto the column using a superloop, injecting 15 mL of sample at a time if you are using the HiPrep™ 26/10 desalting column. As the protein elutes, it will be visible as a distinct peak in the absorbance at both 280 nm and 260 nm wavelengths on the chromatogram. With a sample volume of 15 mL or less, the protein peak should be well-separated from the high-salt buffer, which elutes later and is

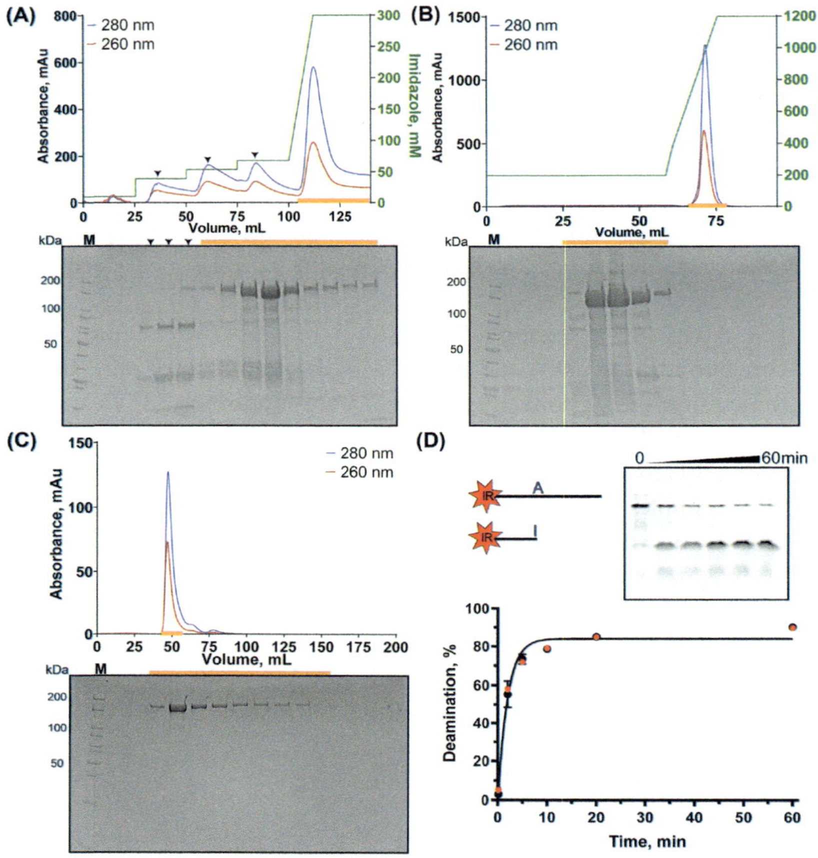

Fig. 4 Purification and single-turnover *in vitro* kinetics of ABE8e. (A) HisTrap affinity chromatography profile of ABE8e protein (top). 4–15 % gradient SDS-PAGE analysis of ABE8e purified via HisTrap affinity chromatography (bottom). The orange bar corresponds to collected fractions. ABE8e is detected in the elution fractions, as shown by the SDS-PAGE analysis, where it appears at the expected molecular weight (~182 kDa). (B) Heparin affinity chromatography profile of ABE8e protein (top). 4–15 % gradient SDS-PAGE analysis of ABE8e purified via Heparin affinity chromatography (bottom). (C) Size exclusion chromatography profile of ABE8e protein (top). 4–15 % gradient SDS-PAGE analysis of ABE8e purified via size exclusion chromatography (bottom). (D) Gel representing single-turnover deamination kinetics of ABE8e measured using fluorescently-labeled dsDNA containing a single adenine (top). The fraction of deaminated dsDNA is plotted as a function of time and fitted to a single-phase exponential equation (bottom). The extracted apparent deamination rate is $0.51 \pm 0.05\ min^{-1}$. Data are represented as the mean ± SD from three independent experiments. The red points correspond to one replicate run on a commercial gel (Section 3.5.5), whereas the black points correspond to samples run on an in-house denaturing gel (Section 3.5.4).

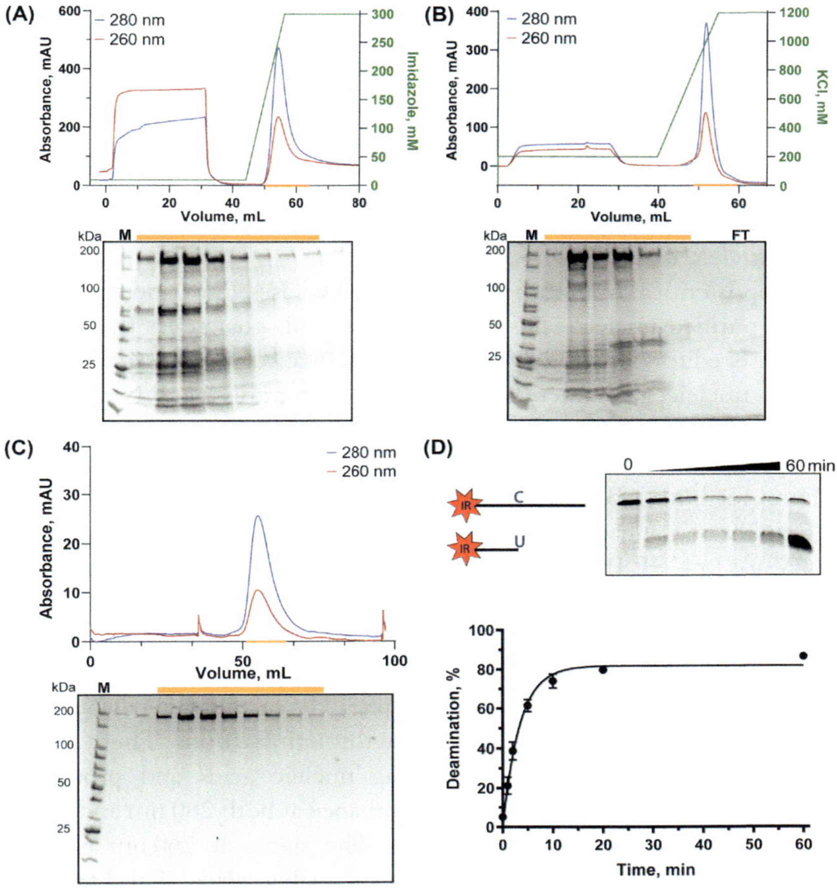

Fig. 5 Purification and single-turnover *in vitro* activity of CBET-1.52. (A) HisTrap affinity chromatography profile of CBET-1.52 protein (top). 4–15 % gradient SDS-PAGE analysis of CBET-1.52 purified via HisTrap affinity chromatography (bottom). The orange bar corresponds to collected fractions. CBET-1.52 is detected in the elution fractions, as shown by the SDS-PAGE analysis, where it appears at the expected molecular weight (~205 kDa). (B) Heparin affinity chromatography profile of CBET-1.52 protein (top). 4–15 % gradient SDS-PAGE analysis of CBET-1.52 purified via Heparin affinity chromatography (bottom). (C) Size exclusion chromatography profile of CBET-1.52 protein (top). 4–15 % gradient SDS-PAGE analysis of CBET-1.52 purified via size exclusion chromatography (bottom). (D) Gel representing single-turnover deamination kinetics of CBE-T1.52 measured using fluorescently-labeled dsDNA containing a single cytosine (top). The fraction of deaminated dsDNA is plotted as a function of time and fitted to a single-phase exponential equation (bottom). The extracted apparent deamination rate is $0.29 \pm 0.02\,\text{min}^{-1}$. Data are represented as the mean ± SD from three independent experiments.

indicated by an increase in the conductivity signal. Collect protein fractions based on the absorbance peak at 280 nm, starting slightly before the peak and continuing until the absorbance returns to baseline, ensuring you avoid collecting the high-salt buffer.

3. Repeat step 14 until the entire sample volume has been desalted.

3.3.4 Day 4: Affinity chromatography to remove contaminating nucleic acids (1 h)

This step is critical to ensure the purity of the base editor and to eliminate any co-purified nucleic acids from bacteria. This guarantees that the purified base editors are highly active. Additionally, the presence of nonspecific nucleic acids increases the risk of co-purifying RNases, which could interfere with the assembly of active and intact RNPs. For this step, we typically use affinity chromatography and HiTrap™ Heparin HP column for both ABEs and TadA-based CBEs.

1. Equilibrate the heparin column (HiTrap™ Heparin HP column) with 5 CVs of low salt IEXA buffer using an FPLC system. By the end of the equilibration process, the absorbance at 260 nm and 280 nm, as well as the conductivity signals, should be stable.
2. Load the sample, obtained by pooling the fractions collected from the desalting column (step 3.3.3), onto HiTrap™ Heparin HP column using the superloop.
3. Immediately begin collecting the flowthrough fractions. These fractions will typically contain contaminating nucleic acids and proteins, as indicated by distinct peaks in the absorbance at both 280 nm and 260 nm wavelengths on the chromatogram. The signal at 260 nm is usually higher, reflecting the predominance of nucleic acids. The base editor will bind to the heparin column in the presence of the low-salt IEXA buffer, allowing it to be separated from these contaminants.
4. To elute the protein, set up a gradient of high-salt elution buffer (IEXB) ranging from 0 % to 100 % over 6 min with a flow rate of 3 mL/min. Collect 20 μL samples from each fraction corresponding to the peaks in the absorbance at both 280 nm and 260 nm wavelengths on the chromatogram. Analyze these samples using SDS-PAGE using 10 % gel or commercial gradient gel (Figs. 4B and 5B).
5. Combine the elution fractions containing the base editor based on the absorbance peak at 280 nm. At this stage, the signal at 280 nm should be approximately twice as high as the signal at 260 nm, indicating that the protein sample is free of nucleic acids. *This is a good point to pause the purification and proceed with the remaining steps the next day.*

6. Concentrate the protein using a pre-equilibrated spin column in IEXB buffer with an appropriate molecular weight cut-off (MWCO) for the protein. For TadA-based base editors, use a 100 kDa MWCO Amicon® Ultra 15 mL filter. Centrifuge at 4000 x g for 10-minute intervals at 4–10°C until the sample volume is reduced to 5 mL or less. After each spin, mix the sample by pipetting to minimize protein aggregation. If aggregation is observed, halt the concentration process and proceed to the next purification step.
7. Transfer the sample to 2 mL Eppendorf tubes and centrifuge at 17,000 x g for 4 min at 4°C. Carefully transfer the supernatant to a new tube. This step is crucial for removing any aggregated or precipitated proteins before loading the sample onto the size exclusion chromatography (SEC) column.

3.3.5 Day 5: Size exclusion chromatography (4 h)

This step is crucial for polishing the final base editor prep by separating proteins based on their size, effectively removing contaminants and aggregates. We recommend using the HiLoad® 16/600 Superdex® 200 prep grade column, which enables the separation of molecules from 10 kDa to 600 kDa.

1. Equilibrate the size-exclusion chromatography column using an FPLC system with 2 CVs of gel filtration buffer. By the end of the equilibration process, the absorbance at 260 nm and 280 nm, as well as the conductivity signals, should be stable.
2. Inject the sample (from step 3.3.4) onto the column using a pre-equilibrated 5 mL sample loop.
3. Collect fractions from approximately 40 mL to 100 mL after injection. Base editors typically elute between 50-60 mL, corresponding to 180–210 kDa proteins based on molecular weight standards, when using a HiLoad® 16/600 Superdex® 200 prep grade column.
4. Take 20 µL samples from each fraction corresponding to the absorbance peaks at 280 nm and 260 nm wavelengths and analyze them using SDS-PAGE using 10% gel or precast gradient gel (Figs. 4C and 5C).
5. Combine fractions that contain the base editor, prioritizing those with the least amount of contaminants.

3.3.6 Day 5: Preparing sample for long-term storage (2 h)

1. Concentrate the sample to ~90 µM using a pre-equilibrated spin column in gel filtration buffer with an appropriate MWCO for the

protein. For TadA-based base editors, use a 100 kDa MWCO column. Centrifuge at 4000 x g for 10-minute intervals at 4–10°C until the sample volume is reduced to 200 μL or less.
2. Measure the base editor concentration using a UV-Vis spectrophotometer. Verify the absence of nucleic acid contamination by checking that the 260/280 ratio is approximately 0.5–0.6.
3. Make 50 μL aliquots in PCR strip tubes. Flash freeze the protein in liquid nitrogen to preserve its activity, and then store the aliquots at −80°C.

3.4 *In vitro* transcription (IVT) of sgRNA (30 h)

3.4.1 Designing sgRNA

The sgRNA can be designed by replacing the spacer region with a sequence that is complementary to the target strand of the dsDNA target of interest. For more information on sgRNA design see (Huang et al., 2021). Here, we provide guidance for designing dsDNA templates used in the *in vitro* transcription of sgRNA.

1. Design and order two DNA strands (forward and reverse) to prepare dsDNA template for *in vitro* transcription of sgRNA:
   a. The forward strand (5′–3′) should contain a T7 polymerase promoter (TAATACGACTCACTATAG) followed by the sgRNA sequence (spacer followed by the scaffold).
   b. The reverse strand (3′ to 5′) should be complementary to the forward strand and will act as the template for T7 polymerase to transcribe the sgRNA sequence.
2. Alternatively, sgRNA can be purchased and allows skipping to step 3.5.1

3.4.2 Day 1: Preparing 10 μM dsDNA template (15 min)

1. Combine the forward and reverse strands to achieve a final concentration of 10 μM each in DEPC-treated (or RNase/DNase free) H_2O.
2. Heat the mixture at 95°C for 3 min, then allow it to cool gradually at room temperature. Use immediately or store at −20°C for future use.

3.4.3 Day 1: Preparing of NTP stocks (1.5–2 h)

1. Add 500 μL of DEPC-treated (or RNase/DNase-free) H_2O to 80 mg of each NTP in a 1.7 mL Eppendorf tube.
2. Adjust the pH of each NTP solution to between 7.0 and 7.4 by adding 5 μL of 5 N NaOH, followed by ~1-5 μL of 1 N NaOH or 1–2 μL of 5 N NaOH as needed until the desired pH is reached. Monitor the pH

by spotting a few drops of the solution on pH Indicator Strips with a range of 6.5–10 to conserve the sample. Typically, ATP will require the most adjustment, while UTP will need the least.

3. Prepare a 500-fold dilution of each NTP and measure the absorbance using the appropriate wavelength for each nucleotide to determine its concentration.
   ATP λ_{max} 259 nm, ε_{max} 1.59×10^4 cm^{-1} M^{-1}
   CTP λ_{max} 271 nm, ε_{max} 0.9×10^4 cm^{-1} M^{-1}
   GTP λ_{max} 253 nm, ε_{max} 1.37×10^4 cm^{-1} M^{-1}
   UTP λ_{max} 259 nm, ε_{max} 1.59×10^4 cm^{-1} M^{-1}
4. Calculate concentrations using Beer's Law and the molar absorption coefficient for each NTP.
5. Aliquot the NTPs into 50 μL portions to avoid repeated freeze-thaw cycles, which can lead to degradation. Store the aliquots at −20°C.

3.4.4 Day 1: Setting up IVT reaction (16–18 h)

1. Make two *in vitro* transcription reactions, 100 μL each on ice in PCR tubes (see Table 1).
2. Incubate at 37°C overnight (16–18 h) in a thermocycler or dry block incubator.
3. This protocol typically yields ~50 μL of sgRNA at ~100 μM (about 5 nmol of sgRNA from 0.2 nmol of template). This quantity is generally sufficient for most biochemical applications. However, the reaction can be scaled up as needed to meet your requirements. Additionally, this protocol can be used to produce guide RNAs (pegRNAs) for prime editors with similar yields.

3.4.5 Day 2: sgRNA purification by preparative polyacrylamide gel electrophoresis (~4 h)

1. Add 5 units (5 μL) of RNase-free DNase (Promega) per 100 μL of IVT reaction directly to each reaction. Incubate at 37°C for 30 min in the thermocycler. This step ensures that the DNA template is degraded.
2. Add an equal volume (105 μL) of 2X RNA Loading Buffer directly to each reaction. *At this point, you can optionally pause the process by storing these quenched reactions at −80°C.*
3. Clean the glass plates with 70% ethanol and RNase Away solution then coat them with 200 μL of Sigmacote®. This treatment prevents the gel from sticking to the glass plates during polymerization and ensures it can be removed without tearing after electrophoresis.

Table 1 *In vitro* transcription reaction.

Components	Final concentration
DEPC-treated or nuclease-free H_2O (to 100 μL total)	
5X Transcription Buffer ATP	1X 5 mM
CTP	5 mM
GTP	5 mM
UTP	5 mM
DTT	10 mM
Hybridized DNA template	1 μM
Inorganic pyrophosphatase[a]	10 mM
RNase inhibitor[a]	0.4 U/μL
T7 polymerase[a]	100 μg/mL

[a]Mix by pipetting at this step.

4. Clean the comb (featuring one small well for the molecular weight marker and one large well for the IVT reaction) and thick spacers with 70 % ethanol and RNase Away solution. Then, coat the comb with 100 μL of Sigmacote®. This treatment ensures that the comb can be removed more easily from the gel after polymerization, minimizing the risk of tearing the wells.
5. Assemble the glass plates with side spacers and secure them with binder clips. Tape around the edges of the glass plates to ensure a tight seal (Table 2).
6. Prepare 500 mL of a solution for a 10 % (19:1) denaturing polyacrylamide gel 400 mL is needed for a big gel (35 × 25 cm) and 150 mL is needed for a small gel (14 cm × 20 cm) (see Table 3). For sgRNAs of approximately 100 nucleotides, this 10 % gel effectively separates the desired RNA length from n + 1 and n−1 products, which may result from non-templated nucleotide addition by T7 polymerase, premature termination of T7 polymerization, or incomplete synthetic template oligos. Optimization of acrylamide concentration and gel running time is required for purifying RNAs of different lengths (see Table 2).

Table 2 PAGE fragment resolution: denaturing conditions (7 M urea).

% Acrylamide	Fragment size (nt)	Bromophenol blue (nt)	Xylene cyanol (nt)
30	2–8	6	20
20	8–25	8	28
10	25–35	12	55
8	35–45	19	75
6	45–70	26	105
5	70–300	35	130
4	100–500	~50	~230

Table 3 Denaturing 7 M urea polyacrylamide components for 10 %, 12.5 %, and 15 % acrylamide gels.

Components	Final concentration		
5X TBE	0.5X		
40 % Acrylamide (19:1)[a]	10 %	12.5 %	15 % (v/v)
Urea	7 M		
MilliQ H_2O (to 500 mL)			

[a](29:1) Solution is used when running *in vitro* activity assays.

7. Filter the solution through a 0.45 μm PVDF membrane to remove small particles and dust contaminants.
8. Store the prepared gel solution at 4°C. Cold gel solution will polymerize more slowly, allowing more time for casting the gel.
9. Before casting the gel, add 500 μL 40 % APS and 300 μL TEMED to 400 mL of the polyacrylamide-urea solution. Mix thoroughly and pour the gel immediately. Insert the comb and allow the gel to polymerize for at least 1 h.
10. Prepare the denaturing gel for electrophoresis by carefully removing the comb and the tape.
11. Assemble the electrophoresis apparatus, placing an aluminum diffuser plate on the outside of the gel to act as a heat sink.
12. Fill the top and bottom buffer chambers of the electrophoresis apparatus with 0.5X TBE buffer.

13. Pre-run the gel for 30–45 min at a constant power of 40 W (30 W for smaller gels) to heat up the gel before loading samples.
14. During the pre-running of the gel, urea leaches into the buffer and gel wells. Being denser than the buffer, it forms a cushion at the bottom of the wells and impedes sample migration into the gel. To ensure optimal results, rinse the wells with 0.5X TBE buffer using a syringe fitted with a hypodermic needle to thoroughly remove the urea just before loading the gel.
15. Denature the quenched IVT reactions (from step 14) at 95°C for 5 min. This step ensures that the RNA is single-stranded and unstructured, allowing it to migrate solely based on molecular weight and charge rather than its three-dimensional shape.
16. Load the IVT reactions (approximately 400 μL total) into the wide well, and add 50 μL of 2X RNA Loading Dye to the small well to monitor sgRNA migration through the gel in real time during electrophoresis (see Table 2).
17. Run the denaturing gel at a constant power of 40 W (30 W for small gel for around 1.5 h) or until the bromophenol blue band (bottom band) has migrated 3/4 of the gel.
18. Once the run is complete, carefully remove the gel from the apparatus and separate it from the glass plates. Wrap the gel in plastic wrap, taking care to avoid any wrinkles.
19. Place the plastic-wrapped gel on a TLC plate in a dark room and shine UV light (254 nm) over it. The TLC plate will fluoresce in areas where RNA is absent. A distinct, dark band should be visible slightly above the xylene cyanol band from the marker, indicating the location of the sgRNA. Trace the band with a fine-tipped Sharpie to clearly delineate the area containing the sgRNA.
20. Excise the band through the plastic wrap using a sterile razor or scalpel. Peel off the top plastic wrap, slice the gel into small pieces, and transfer the pieces into a 15 mL Falcon tube.

3.4.6 Days 2–3: Gel extraction and precipitation of sgRNA (~21–23 h)

1. Crush the gel slices into the smallest pieces possible using a spatula that has been cleaned with 70 % ethanol and RNase Away solution.
2. Add an equal volume of TEN buffer, or just enough to achieve a slurry consistency that allows for smooth movement during rocking.
3. Elute the RNA overnight by rocking at 4°C at a high enough speed to agitate the solution.

4. Filter the sgRNA solution from the gel pieces using a pre-equilibrated low DNA/RNA binding 1.2 μm Supor membrane syringe filter and a 3 mL syringe. Transfer the thick TEN buffer-gel slurry (using a P1000 pipette with the tip cut off to facilitate pipetting of the thick slurry) into the 3 mL syringe with the attached filter. Slowly push the syringe plunger to filter the sgRNA and TEN buffer into low-binding 2 mL Eppendorf tubes. Repeat this step until you have filtered all of the slurry.
5. Ethanol precipitate the sgRNA by adding 1/10 vol 3 M NaOAc (pH 5.2), 1 μL of GlycoBlue, and 3 volumes of cold 100 % ethanol.
6. Incubate at −80°C for at least 1 h (typically overnight).
7. Centrifuge sgRNA at maximum speed (17,000 x g) for 40 min at 4°C.
8. Carefully aspirate the supernatant, taking care not to disturb the sgRNA pellet.
9. Add 300 μL of −20°C 70 % ethanol to the sgRNA pellet. Avoid disturbing the pellet and centrifuge at 17,000 x g for 10 min.
10. Carefully aspirate the supernatant, taking care not to disturb the sgRNA pellet. Leave the sample tube open on a benchtop, covered with a Kimwipe to evaporate the remaining ethanol until dry.
11. Resuspend the sgRNA pellet in 50 μL of DEPC-treated or nuclease-free water. Measure the absorbance at 260 nm using a UV spectrophotometer to determine the concentration.
12. Store sgRNA at −80°C.

3.5 *In vitro* DNA deamination assay (7-8 h)

3.5.1 *Assembly of RNP complex (1 h)*

1. Prepare the sgRNA for folding by diluting the purified sgRNA (from Step 3.4.6) or purchased sgRNA to the desired concentration using 5X Dilution Buffer to a final concentration of 1X.
2. Incubate the solution at 70°C for 5 min, then allow it to cool slowly to room temperature on the bench top.
3. Add $MgCl_2$ to a final concentration of 1 mM and mix thoroughly by pipetting.
4. Incubate at 50°C for 5 min, then allow the mixture to cool slowly to room temperature on the benchtop.
5. Add 5X RNP Assembly Buffer to the folded sgRNA sample to achieve a final 1X concentration. Then, add the purified base editor (from step 3.3.6) to the folded sgRNA at a 1:1.5 molar ratio of editor to sgRNA.
6. Incubate at room temperature for 20-30 min, then place on ice. The RNP solution can be stored at 4°C for ~5 days.

3.5.2 Single-turnover activity assay (2.5 h)

1. Form the dsDNA target by mixing together the fluorescently labeled non-target strand (NTS) with the unlabeled target strand (TS) DNA at a 1:1.2 molar ratio and incubating at 95°C for 3 min. Allow the dsDNA target to slowly cool to room temperature. It is essential to keep the mixture covered from light.
2. Mix the RNP and dsDNA target in 5X Deamination Buffer (final concentration 1X) to achieve final concentrations of 1 μM for the base editor and 10 nM for the dsDNA target.
3. Incubate the reactions at 37°C. At various time points (for TadA-based base editors, collect samples at 0, 1, 2, 5, 10, 20, and 60 min), take 20 μL aliquots and quench the reaction by adding 30 μL of heated DEPC-treated or nuclease-free water. Incubate at 95°C for 3 min, then transfer the samples to ice.
4. Purify the DNA from the quenched samples using the Monarch® PCR & DNA Cleanup Kit according to the manufacturer's instructions. Elute the DNA in 15 μL of the DNA Elution Buffer provided with the kit. *At this point, the assay can be paused, and the samples can be stored at −20°C.*

3.5.3 EndoV and USER reactions (1.5 h)

To quantify the deamination activity of ABEs, the dsDNA products (from step 3.5.2) must be processed with EndoV. In the case of CBEs, use USER enzyme.

1. Take 10 μL of sample from step 3.5.2 and add 5 μL of either EndoV or USER mix, depending on the base editor used (see Buffer Appendix).
2. Incubate the samples at 37°C for 1 hr.
3. Quench reaction by adding 15 μL of Super Buffer. *At this point, the assay can be paused, and the samples can be stored at −20°C.*
4. Next, follow either Section 3.5.4 or Section 3.5.5.

3.5.4 Denaturing polyacrylamide gel electrophoresis (1.5–2 h)

1. Prepare small or large glass plates depending on the number of samples to be processed following the steps described in 3.4.5.
2. Cast a 15 % (29:1) polyacrylamide urea gel following the steps described in 3.4.5.
3. Pre-run the gel for 30–45 min at a constant power of 40 W (30 W for smaller gels) to heat up the gel before loading samples.

4. During the pre-running of the gel, urea leaches into the buffer and gel wells. To ensure optimal results, rinse the wells with 0.5X TBE buffer using a syringe fitted with a hypodermic needle to thoroughly remove the urea just before loading the gel.
5. Denature the EndoV- or USER-treated samples from step 3.5.3 at 95°C for 1-2–min. This step ensures that the DNA is single-stranded and unstructured, allowing it to migrate solely based on molecular weight and charge rather than its three-dimensional shape.
6. Load 15 μL of the sample per well and 10 μL of the 2X RNA Loading Dye into one of the wells to monitor DNA migration through the gel in real time during electrophoresis (see Table 2).
7. Run the denaturing gel at a constant power of 40 W (30 W for small gel for around 1.5 h) or until the bromophenol blue band (bottom band) has migrated 1/3–1/2 of the gel.
8. Once the run is complete, carefully remove the gel from the apparatus, separate it from the glass plates and image on Amersham Typhoon Imager or another type of gel imaging system (Fig. 4D).

3.5.5 Denaturing polyacrylamide gel electrophoresis using commercial gel (1.5–2 h)

1. Use 1X TBE buffer as a running buffer.
2. Prepare the 15% TBE-Urea Gel (Bio-Rad) for electrophoresis following manufacturer instructions.
3. Assemble the Mini-PROTEAN Tetra cell apparatus (Bio-Rad) and carefully insert the gel into the system.
4. Pre-run gels at 300 V for 15 min.
5. Rinse the wells with 1X TBE using a syringe fitted with a hypodermic needle.
6. Denature the EndoV- or USER-treated samples from step 3.5.3 at 95°C for 1–2 min.
7. Load 10 μL of 2X RNA Loading Dye into one of the wells as a molecular weight marker, and 10 μL of each sample into the remaining wells.
8. Run the denaturing gel at a constant voltage of 300 V for 15 min. The run-time may need to be optimized due to target length to achieve proper band separation (see Table 2).
9. Once the run is complete, carefully remove the gel from the apparatus, separate it from the glass plates and image on Amersham Typhoon Imager or another type of gel imaging system (Figs. 4D and 5D).

4. Summary and conclusion

This protocol outlines the steps for the expression, purification, RNP reconstitution, and *in vitro* activity assessment of TadA-based BEs. It is valuable for both biochemical and structural characterization of BEs, as well as for preparing active BE RNPs for genome editing applications in vivo. Our lab, along with others, has conducted a range of successful studies using BE RNPs prepared following this protocol. For example, we used this protocol to prepare ABE8e and ABE7.10 RNPs for comparing their DNA deamination rates *in vitro* (Richter et al., 2020), and elucidating the molecular mechanism of ABE8e-mediated DNA adenine deamination through a combination of biochemical, biophysical, and structural approaches (Arantes et al., 2024; Lapinaite et al., 2020).

The protocol is designed to achieve high purity across a range of TadA-based BEs. It features a detailed activity assay using commercial denaturing gels, which is a convenient method using standard molecular biology equipment. This allows for rapid quality control testing of reconstituted BE RNPs before proceeding with more costly and time-consuming in vivo genome editing experiments.

While there are a couple of protocols available for the expression and purification of APOBEC-based CBEs using either human cell expression systems or *E. coli* (Jang et al., 2021; Rees et al., 2017), CBEs purified using these protocols demonstrate reduced activity in vivo genome editing applications (Jang et al., 2021; Rees et al., 2017). CBEs contain UGI domains derived from *Bacillus subtilis* bacteriophage PBS1, which directly bind to human uracil-DNA glycosylase (UNG), inhibiting the removal of uridine from genomic DNA and facilitating efficient C-to-U conversion. Given that human UNG is homologous to the *E. coli* UNG (EcoUNG) (Olsen, Aasland, Wittwer, Krokan, & Helland, 1989), it is likely that the CBEs expressed in *E. coli* co-purify with EcoUNG (Bennett, Schimerlik, & Mosbaugh, 1993), potentially preventing the interaction between UGI and human UNG during genome editing in human cells. However, the multi-step protocol outlined here ensures the production of highly pure CBEs, with no detectable EcoUNG (MW ~25.7 kDa) by SDS-PAGE analysis (Fig. 5C), which likely improves the in vivo editing efficiency of *in vitro* reconstituted CBEs.

Overall, this protocol offers a comprehensive framework for preparing high-purity and highly active BE RNPs and provides a straightforward method for testing their activity, empowering more laboratories to generate tailored BE RNPs for diverse *in vitro* and in vivo applications.

Acknowledgement

We thank Aaron Feller for his technical assistance with the purification of ABE8e.

Funding

This work was supported by the National Institutes of Health DP2GM149550 grant to A.L. The authors acknowledge support to the Gavin Herbert Eye Institute at the University of California, Irvine from an unrestricted grant from Research to Prevent Blindness and from NIH core grant P30 EY034070.

Data availability

Data and plasmids encoding ABE8e and CBET-1.52 used in this study are available from the corresponding author upon reasonable request.

5. Buffer Appendix

5.1 Purification buffers

Prepare all buffers using 0.22 μm or 0.45 μm membrane filters and buffer filtration system listed in the materials section. Allow to cool to 4°C before use.

Lysis Buffer (1.5 L).

Components	Final concentration
HEPES pH 8*	25 mM
KCl	1 M
Glycerol	10% (v/v)
TCEP	2 mM
Imidazole	10 mM
MilliQ H_2O	
Filter to or below 0.45 μm, store 4°C	

*HEPES pH 8 was added from 1 M stock adjusted with KOH or HCl to avoid the addition of ions made following https://www.aatbio.com/resources/buffer-preparations-and-recipes/hepes-buffer-ph-7-5

High Salt Wash (200 mL).

Components	Final concentration
HEPES pH 8*	25 mM
KCl	1 M
LiCl	2 M
Glycerol	10% (v/v)

TCEP	2 mM
Imidazole	10 mM
MilliQ H_2O	
Filter to or below 0.45 μm, store 4°C	

*HEPES pH 8 was added from 1 M stock adjusted with KOH or HCl to avoid the addition of ions made following https://www.aatbio.com/resources/buffer-preparations-and-recipes/hepes-buffer-ph-7-5

Elution Buffer (250 mL).

Components	Final concentration
HEPES pH 8*	25 mM
KCl	0.5 M
Glycerol	10% (v/v)
TCEP	2 mM
Imidazole	300 mM
MilliQ H_2O	
Filter to or below 0.45 μm, store 4°C	

*HEPES pH 8 was added from 1 M stock adjusted with KOH or HCl to avoid the addition of ions made following https://www.aatbio.com/resources/buffer-preparations-and-recipes/hepes-buffer-ph-7-5

IEXA (Low-Salt) Buffer (500 mL).

Components	Final concentration
HEPES pH 8*	25 mM
KCl	200 mM
Glycerol	10% (v/v)
TCEP	1 mM
MilliQ H_2O	
Filter to or below 0.45 μm, store 4°C	

*HEPES pH 8 was added from 1 M stock adjusted with KOH or HCl to avoid the addition of ions made following https://www.aatbio.com/resources/buffer-preparations-and-recipes/hepes-buffer-ph-7-5

IEXB (High-Salt) Buffer (250 mL).

Components	Final concentration
HEPES pH 8*	25 mM
KCl	1.2 M
Glycerol	10% (v/v)
TCEP	1 mM
MilliQ H_2O	
Filter to or below 0.45 μm, store 4°C	

*HEPES pH 8 was added from 1 M stock adjusted with KOH or HCl to avoid the addition of ions made following https://www.aatbio.com/resources/buffer-preparations-and-recipes/hepes-buffer-ph-7-5

Gel Filtration Buffer (1 L).

Components	Final concentration
Tris-HCl pH 8*	20 mM
KCl	200 mM
Glycerol	10% (v/v)
TCEP	1 mM
MilliQ H_2O	
Filter to or below 0.45 μm, store 4°C	

*Tris-HCl pH 8 was added from 1 M stock adjusted with KOH or HCl to avoid the addition of ions made following https://doi.org/10.1101/pdb.rec8747 *Cold Spring Harb Protoc* 2006.

10X SDS Running Buffer (2 L).

Components	Final concentration
Trizma base	250 mM
Glycine	1.92 M
SDS	1.00% (w/v)
MilliQ H_2O	
Filter to or below 0.22 μm, store 25°C	
Check pH with pH strip (should be 8.3). Do not autoclave. Dilute to 1X before use.	

4X SDS-PAGE Loading Dye (10 mL).

Components	Final concentration
Tris-HCl pH 6.8*	200 mM
SDS	8% (w/v)
Glycerol	40% (v/v)
β-mercaptoethanol	588 mM
EDTA	50 mM
Bromophenol blue	0.08% (w/v)
MilliQ H_2O	
Store −20°C	

*Tris-HCl pH 6.8 was added from 0.5 M stock adjusted with KOH or HCl to avoid the addition of ions made following https://doi.org/10.1101/pdb.rec8747 Cold Spring Harb Protoc 2006.

SDS-PAGE Coomassie Blue Gel Stain (1 L).

Components	Final concentration
Coomassie Blue	0.125% (w/v)
Methanol*	50% (v/v)
Glacial acetic acid*	10% (v/v)
MilliQ H_2O	
Store 25°C	

*Methanol and acetic acid can be diluted in water before adding. Add acetic acid last. Adapted from https://doi.org/10.1101/pdb.rec073924 Cold Spring Harb Protoc 2013

***In Vitro* Transcription of sgRNA Buffers.**
DEPC-Treated Water (1 L).

Components	Final concentration
Diethyl pyrocarbonate (DEPC)	0.1% (v/v)
MilliQ H_2O	
Mix well and set in the fume hood at 25°C for 2 h	
Autoclave at 121°C*, store 25°C	

*This step is critical for inactivation of the remaining DEPC.

5X Transcription Buffer (10 mL).

Components	Final concentration
Tris-HCl pH 8.1*	150 mM
$MgCl_2$	125 mM
Triton X−100	0.05% (v/v)
Spermidine	10 mM
DEPC-treated or nuclease-free water	
Filter to or below 0.22 μm, store −20°C	

*Tris-HCl pH 8.1 was added from 1 M stock adjusted with KOH or HCl to avoid the addition of ions made following https://doi.org/10.1101/pdb.rec8747 Cold Spring Harb Protoc 2006.

5X TBE Buffer (1 L).

Components	Final concentration
Trizma base	450 mM
Boric acid	450 mM
EDTA*	10 mM
MilliQ H_2O	
Adjust pH with HCl until 8.3. Autoclave at 121°C, store 25°C	

*EDTA was added from 0.5 M stock made following https://doi.org/10.1101/pdb.rec8030 Cold Spring Harb Protoc 2006.

TEN Buffer (50 mL).

Components	Final concentration
Tris-HCl pH 8.8*	10 mM
NaCl	300 mM
EDTA**	1 mM
DEPC-treated or nuclease-free water	
Filter to or below 0.22 μm, store −20°C	

*Tris-HCl pH 8.8 was added from 1 M stock adjusted with NaOH or HCl to avoid the addition of ions made following https://doi.org/10.1101/pdb.rec8747 Cold Spring Harb Protoc 2006. **EDTA was added from 0.5 M stock made following https://doi.org/10.1101/pdb.rec8030 Cold Spring Harb Protoc 2006.

NaOAc pH 5.2 (100 mL).

Components	**Final concentration**
Sodium acetate (NaOAc)	3 M
MilliQ H_2O	
Adjust pH with glacial acetic acid until 5.2. Filter to or below 0.2–2 μm, store 25°C	

Activity Assay Buffers.
5X Dilution Buffer (50 mL).

Components	**Final concentration**
HEPES pH 7.5*	100 mM
KCl	750 mM
Glycerol	50% (v/v)
DTT	10 mM
DEPC-treated or nuclease-free H_2O	
Filter to or below 0.22 μm, store −20°C	

*HEPES pH 7.5 was added from 1 M stock adjusted with KOH or HCl to avoid addition of ions made following https://www.aatbio.com/resources/buffer-preparations-and-recipes/hepes-buffer-ph-7-5

5X RNP Assembly Buffer (50 mL).

Components	**Final concentration**
Tris-HCl pH 7.5*	100 mM
KCl	1 M
$MgCl_2$	25 mM
Glycerol	25% (v/v)
DTT	10 mM
DEPC-treated or nuclease-free H_2O	
Filter to or below 0.22 μm, store −20°C	

*Tris-HCl pH 7.5 was added from 1 M stock adjusted with NaOH or HCl to avoid the addition of ions made following https://doi.org/10.1101/pdb.rec8747 Cold Spring Harb Protoc 2006.

5X Deamination Buffer (50 mL).

Components	Final concentration
Tris-HCl pH 7.5*	100 mM
KCl	500 mM
Glycerol	25% (v/v)
$MgSO_4$	12.5 mM
DTT	10 mM
DEPC-treated or nuclease-free H_2O	
Filter to or below 0.22 µm, store −20°C	

*Tris-HCl pH 7.5 was added from 1 M stock adjusted with NaOH or HCl to avoid the addition of ions made following https://doi.org/10.1101/pdb.rec8747 Cold Spring Harb Protoc 2006.

EndoV Mix.

Components	Final concentration
DEPC-treated or nuclease-free H_2O	50% (v/v)
NEBuffer™ 4 (NEB)	30% (v/v)
Endonuclease V*	20% (v/v)
Make Fresh	
*Mix by pipetting	

USER Mix.

Components	Final concentration
DEPC-treated or nuclease-free H_2O	50% (v/v)
rCutSmart™ Buffer (NEB)	30% (v/v)
USER® Enzyme* (NEB)	20% (v/v)
Make Fresh	
*Mix by pipetting	

2X RNA Loading Buffer (30 mL).

Components	Final concentration
Formamide	95% (v/v)
EDTA	1 mM
SDS*	0.03% (v/v)
DEPC-treated or nuclease-free H_2O	
Filter to or below 0.22 μm, store −20°C	

*SDS was added from 10% stock made following https://doi.org/10.1101/pdb.rec10480 Cold Spring Harb Protoc 2006.

2x RNA Loading Dye (30 mL).

Components	Final concentration
Formamide	95% (v/v)
EDTA	1 mM
SDS*	0.025% (v/v)
Bromophenol Blue**	0.025% (v/v)
Xylene Cyanol**	0.01% (v/v)
DEPC-treated or nuclease-free H_2O	
Filter to or below 0.22 μm, store −20°C	

*SDS was added from 10% stock made following https://doi.org/10.1101/pdb.rec10480 Cold Spring Harb Protoc 2006. **Bromophenol blue and xylene cyanol were added from 1% (w/v) stocks

Super Buffer (600 μL).

Components	Final concentration
2X RNA Loading Buffer	95.20% (v/v)
EDTA	15 mM
SDS*	0.10% (v/v)
Heparin, sodium salt**	0.02% (w/v)
Filter to or below 0.22 μm, store −20°C	

*SDS was added from 10% stock made following doi:10.1101/pdb.rec10480 Cold Spring Harb Protoc 2006. **Heparin was added from 2.5% (w/v) stock

References

Adler, B. A., Trinidad, M. I., Bellieny-Rabelo, D., Zhang, E., Karp, H. M., Skopintsev, P., ... Shi, H., ... Doudna, J. A. (2024). CasPEDIA Database: A functional classification system for class 2 CRISPR-Cas enzymes. *Nucleic Acids Research, 52*(D1), D590–D596.

Alves, C. R. R., Ha, L. L., Yaworski, R., Sutton, E. R., Lazzarotto, C. R., Christie, K. A., ... Kleinstiver, B. P. (2024). Optimization of base editors for the functional correction of SMN2 as a treatment for spinal muscular atrophy. *Nature Biomedical Engineering, 8*(2), 118–131.

Anzalone, A. V., Randolph, P. B., Davis, J. R., Sousa, A. A., Koblan, L. W., Levy, J. M., ... Liu, D. R. (2019). Search-and-replace genome editing without double-strand breaks or donor DNA. *Nature, 576*(7785), 149–157.

Arantes, P. R., Chen, X., Sinha, S., Saha, A., Patel, A. C., Sample, M., ... Palermo, G. (2024). Dimerization of the deaminase domain and locking interactions with Cas9 boost base editing efficiency in ABE8e. *Nucleic Acids Research.* https://doi.org/10.1093/nar/gkae1066.

Barrangou, R., Fremaux, C., Deveau, H., Richards, M., Boyaval, P., Moineau, S., ... Horvath, P. (2007). CRISPR provides acquired resistance against viruses in Prokaryotes. *Science.* https://doi.org/10.1126/science.1138140.

Bennett, S. E., Schimerlik, M. I., & Mosbaugh, D. W. (1993). Kinetics of the uracil-DNA glycosylase/inhibitor protein association. Ung interaction with Ugi, nucleic acids, and uracil compounds. *The Journal of Biological Chemistry, 268*(36), 26879–26885.

Chen, L., Zhu, B., Ru, G., Meng, H., Yan, Y., Hong, M., ... Li, D. (2023). Re-engineering the adenine deaminase TadA-8e for efficient and specific CRISPR-based cytosine base editing. *Nature Biotechnology, 41*(5), 663–672.

Chen, X., McAndrew, M. J., & Lapinaite, A. (2023). Unlocking the secrets of ABEs: The molecular mechanism behind their specificity. *Biochemical Society Transactions, 51*(4), 1635–1646.

Chiesa, R., Georgiadis, C., Syed, F., Zhan, H., Etuk, A., Gkazi, S. A., ... Group, Base-Edited C. A. R. T. (2023). Base-edited CAR7 T cells for relapsed T-cell acute lymphoblastic leukemia. *The New England Journal of Medicine, 389*(10), 899–910.

Cong, L., Ran, F. A., Cox, D., Lin, S., Barretto, R., Habib, N., ... Zhang, F. (2013). Multiplex genome engineering using CRISPR/Cas systems. *Science, 339*(6121), 819–823.

Doman, J. L., Raguram, A., Newby, G. A., & Liu, D. R. (2020). Evaluation and minimization of Cas9-independent off-target DNA editing by cytosine base editors. *Nature Biotechnology, 38*(5), 620–628.

Doudna, J. A., & Charpentier, E. (2014). Genome editing. The new frontier of genome engineering with CRISPR-Cas9. *Science, 346*(6213), 1258096.

Gaudelli, N. M., Komor, A. C., Rees, H. A., Packer, M. S., Badran, A. H., Bryson, D. I., & Liu, D. R. (2017). Programmable base editing of A•T to G•C in genomic DNA without DNA cleavage. *Nature, 551*(7681), 464–471.

Grünewald, J., Zhou, R., Garcia, S. P., Iyer, S., Lareau, C. A., Aryee, M. J., & Joung, J. K. (2019). Transcriptome-wide off-target RNA editing induced by CRISPR-guided DNA base editors. *Nature, 569*(7756), 433–437.

Gu, et al. (2021). Base editors: Expanding the types of DNA damage products harnessed for genome editing. *Gene and Genome Editing, 1*, 100005.

Hooper, A. J., Tang, X. L., & Burnett, J. R. (2024). VERVE-101, a CRISPR base-editing therapy designed to permanently inactivate hepatic PCSK9 and reduce LDL-cholesterol. *Expert Opinion on Investigational Drugs, 33*(8), 753–756.

Hou, Z., Zhang, Y., Propson, N. E., Howden, S. E., Chu, L.-F., Sontheimer, E. J., & Thomson, J. A. (2013). Efficient genome engineering in human pluripotent stem cells using Cas9 from Neisseria meningitidis. *Proceedings of the National Academy of Sciences, 110*(39), 15644–15649.

Huang, T. P., Newby, G. A., & Liu, D. R. (2021). Precision genome editing using cytosine and adenine base editors in mammalian cells. *Nature Protocols, 16*(2), 1089–1128.

Jang, H.-K., Jo, D. H., Lee, S.-N., Cho, C. S., Jeong, Y. K., Jung, Y., ... Bae, S. (2021). High-purity production and precise editing of DNA base editing ribonucleoproteins. *Science Advances,* 7(35), https://doi.org/10.1126/sciadv.abg2661.

Jinek, M., Chylinski, K., Fonfara, I., Hauer, M., Doudna, J. A., & Charpentier, E. (2012). A programmable dual-RNA-guided DNA endonuclease in adaptive bacterial immunity. *Science, 337*(6096), 816–821.

Jinek, M., East, A., Cheng, A., Lin, S., Ma, E., & Doudna, J. (2013). RNA-programmed genome editing in human cells. *eLife*. https://doi.org/10.7554/eLife.00471.

Kim, S., Kim, D., Cho, S. W., Kim, J., & Kim, J.-S. (2014). Highly efficient RNA-guided genome editing in human cells via delivery of purified Cas9 ribonucleoproteins. *Genome Research, 24*(6), 1012–1019.

Knipping, F., Newby, G. A., Eide, C. R., McElroy, A. N., Nielsen, S. C., Smith, K., ... Osborn, M. J. (2022). Disruption of HIV-1 co-receptors CCR5 and CXCR4 in primary human T cells and hematopoietic stem and progenitor cells using base editing. *Molecular Therapy: The Journal of the American Society of Gene Therapy, 30*(1), 130–144.

Komor, A. C., Kim, Y. B., Packer, M. S., Zuris, J. A., & Liu, D. R. (2016). Programmable editing of a target base in genomic DNA without double-stranded DNA cleavage. *Nature, 533*(7603), 420–424.

Lam, D. K., Feliciano, P. R., Arif, A., Bohnuud, T., Fernandez, T. P., Gehrke, J. M., ... Gaudelli, N. M. (2023). Improved cytosine base editors generated from TadA variants. *Nature Biotechnology, 41*(5), 686–697.

Lapinaite, A., Knott, G. J., Palumbo, C. M., Lin-Shiao, E., Richter, M. F., Zhao, K. T., ... Doudna, J. A. (2020). DNA capture by a CRISPR-Cas9-guided adenine base editor. *Science, 369*(6503), 566–571.

Makarova, K. S., Wolf, Y. I., Iranzo, J., Shmakov, S. A., Alkhnbashi, O. S., Brouns, S. J. J., ... Koonin, E. V. (2019). Evolutionary classification of CRISPR–Cas systems: A burst of class 2 and derived variants. *Nature Reviews. Microbiology, 18*(2), 67–83.

Mali, P., Yang, L., Esvelt, K. M., Aach, J., Guell, M., DiCarlo, J. E., ... Church, G. M. (2013). RNA-guided human genome engineering via Cas9. *Science, 339*(6121), 823–826.

Marraffini, L. A., & Sontheimer, E. J. (2008). CRISPR interference limits horizontal gene transfer in staphylococci by targeting DNA. *Science*. https://doi.org/10.1126/science.1165771.

Mojica, F. J. M., Díez-Villaseñor, C., García-Martínez, J., & Soria, E. (2005). Intervening sequences of regularly spaced prokaryotic repeats derive from foreign genetic elements. *Journal of Molecular Evolution, 60*(2), 174–182.

Neugebauer, M. E., Hsu, A., Arbab, M., Krasnow, N. A., McElroy, A. N., Pandey, S., ... Liu, D. R. (2023). Evolution of an adenine base editor into a small, efficient cytosine base editor with low off-target activity. *Nature Biotechnology, 41*(5), 673–685.

Nishida, K., Arazoe, T., Yachie, N., Banno, S., Kakimoto, M., Tabata, M., ... Kondo, A. (2016). Targeted nucleotide editing using hybrid prokaryotic and vertebrate adaptive immune systems. *Science, 353*(6305), https://doi.org/10.1126/science.aaf8729.

Olsen, L. C., Aasland, R., Wittwer, C. U., Krokan, H. E., & Helland, D. E. (1989). Molecular cloning of human uracil-DNA glycosylase, a highly conserved DNA repair enzyme. *The EMBO Journal, 8*(10), 3121–3125.

Protocols, C. S. H. (2015). SDS-PAGE gel. *Cold Spring Harbor Protocols*. https://cshprotocols.cshlp.org/content/2015/7/pdb.rec087908.full?sid=1ba12de3-23be-41d5-b8c0-cb384f7fa403.

Ran, F. A., Hsu, P. D., Wright, J., Agarwala, V., Scott, D. A., & Zhang, F. (2013). Genome engineering using the CRISPR-Cas9 system. *Nature Protocols, 8*(11), 2281–2308.

Rees, H. A., Komor, A. C., Yeh, W.-H., Caetano-Lopes, J., Warman, M., Edge, A. S. B., & Liu, D. R. (2017). Improving the DNA specificity and applicability of base editing through protein engineering and protein delivery. *Nature Communications, 8*, 15790.

Richter, M. F., Zhao, K. T., Eton, E., Lapinaite, A., Newby, G. A., Thuronyi, B. W., ... Liu, D. R. (2020). Phage-assisted evolution of an adenine base editor with improved Cas domain compatibility and activity. *Nature Biotechnology, 38*(7), 883–891.

Rio, D. C. (2013). Expression and purification of active recombinant T7 RNA polymerase from E. coli. *Cold Spring Harbor Protocols, 2013*(11) db.prot078527.

Song, B., Yang, S., Hwang, G.-H., Yu, J., & Bae, S. (2021). Analysis of NHEJ-based DNA repair after CRISPR-mediated DNA cleavage. *International Journal of Molecular Sciences, 22*(12), 6397.

Xue, Greene (2021). DNA repair pathway choices in CRISPR-Cas9-mediated genome editing. *Trends in Genetics: TIG, 37*(7), 639–656.

Zetsche, B., Gootenberg, J. S., Abudayyeh, O. O., Slaymaker, I. M., Makarova, K. S., Essletzbichler, P., ... Zhang, F. (2015). Cpf1 Is a single RNA-guided endonuclease of a class 2 CRISPR-cas system. *Cell, 163*(3), 759–771.

Zhang, E., Neugebauer, M. E., Krasnow, N. A., & Liu, D. R. (2024). Phage-assisted evolution of highly active cytosine base editors with enhanced selectivity and minimal sequence context preference. *Nature Communications, 15*(1), 1697.

Zuo, E., Sun, Y., Wei, W., Yuan, T., Ying, W., Sun, H., ... Yang, H. (2019). Cytosine base editor generates substantial off-target single-nucleotide variants in mouse embryos. *Science, 364*(6437), 289–292.

Zuris, J. A., Thompson, D. B., Shu, Y., Guilinger, J. P., Bessen, J. L., Hu, J. H., ... Liu, D. R. (2015). Cationic lipid-mediated delivery of proteins enables efficient protein-based genome editing in vitro and in vivo. *Nature Biotechnology, 33*(1), 73–80.

CHAPTER THIRTEEN

Bacterial directed evolution of CRISPR base editors

Reilly Q. Mach and Shannon M. Miller*
The Scripps Research Institute
*Corresponding author. e-mail address: smiller@scripps.edu

Contents

1. Introduction and background 318
   1.1 Introduction to genome editing proteins 318
   1.2 Previous engineering efforts 320
   1.3 Benefits and limitations of base editing compared to other genome editing methods 321
   1.4 Benefits and limitations of plate-based selection compared to other directed evolution methods 321
   1.5 Experimental design 322
   1.6 Reagents 322
   1.7 Equipment 325
   1.8 Software 326
   1.9 Primers 327
   1.10 Plasmids 327
2. Optimization of selection strategy 327
   2.1 Identifying compatible antibiotic resistance gene mutations for base editing 327
   2.2 Designing base editor constructs 329
   2.3 Mock selection to validate system and determine starting points 330
   2.4 Tuning stringency of the selection 331
3. Methods to generate diversity 332
   3.1 Site saturation mutagenesis 332
   3.2 Error-prone PCR 334
   3.3 DNA shuffling 336
4. Performing plate-based selection 338
   4.1 Transforming into the selection strain and performing selection 339
   4.2 Determining library size and coverage 340
5. Analyzing outcomes 341
   5.1 Genotyping of individual variants 341
   5.2 Determining success of a selection 342
   5.3 Repeating selection iteratively and sub-cloning 343
   5.4 Human cell base editing 344
   5.5 Determining the end of evolution 344
6. Conclusions 345

Methods in Enzymology, Volume 712
ISSN 0076-6879, https://doi.org/10.1016/bs.mie.2025.01.003

Appendix A. Supporting information 345
References 345

Abstract

Base editing and other precision editing agents have transformed the utility and therapeutic potential of CRISPR-based genome editing. While some native enzymes edit efficiently with their nature-derived function, many enzymes require rational engineering or directed evolution to enhance the compatibility with mammalian cell genome editing. While many methods of engineering and directed evolution exist, plate-based discrete evolution offers an ideal balance between ease of use and engineering power. Here, we describe a detailed method for the bacterial directed evolution of CRISPR base editors that compounds technical ease with flexibility of application.

1. Introduction and background

1.1 Introduction to genome editing proteins

CRISPR-Cas9 facilitates DNA targeting through base pairing of a single guide RNA (sgRNA) (Cong et al., 2013; Jinek et al., 2012; Jinek et al., 2013; Mali et al., 2013), leading to efficient reprogrammable genome editing following activity by the Cas9 nuclease or tethered DNA modifying proteins. CRISPR gene editing has demonstrated promising clinical relevance recent years (Doudna, 2020), culminating in the first CRISPR therapy clinical approval in 2023. Recent precision genome editing technologies, such as base and prime editing, catalyze the precise conversion of one sequence to another without the formation of double-stranded breaks (DSBs) (Anzalone et al., 2019; Gaudelli et al., 2017; Komor, Kim, Packer, Zuris, & Liu, 2016). Precision genome editing technologies are therefore highlighted for therapeutic potential given their avoidance of DSB formation and clean product formation, leading to increased safety, specificity, and efficiency (Anzalone, Koblan, & Liu, 2020).

Base editing enables direct base conversion within a target stretch of ssDNA exposed by Cas9 binding at which a tethered deaminase directly catalyzes transition mutations. Two major types of base editors (BEs) have been described: cytosine BEs (CBEs) that catalyze C•G to T•A transitions, and adenine BEs (ABEs) that catalyze A•T to G•C transitions (Gaudelli et al., 2017; Komor et al., 2016) (Fig. 1A). Base editing requires the target base to exist within an exposed ssDNA R-loop which, for SpCas9 and many other Cas9 orthologs, translates broadly to positions 2–8 in the protospacer, with the PAM starting at position 21.

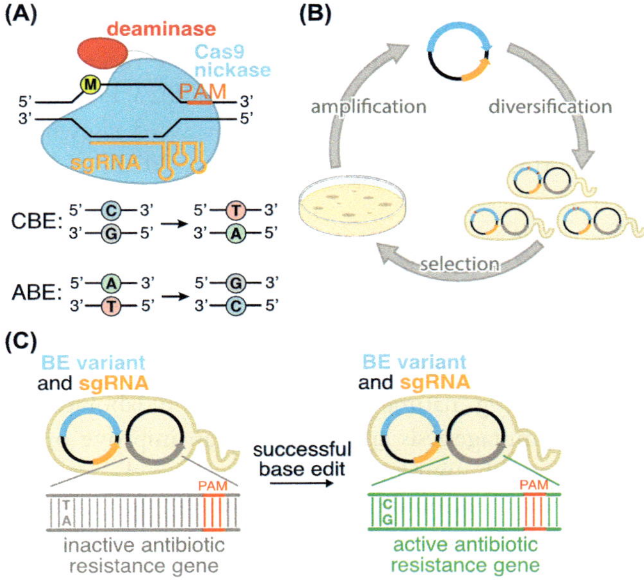

Fig. 1 Overview of plate-based selection involving a base editor. (A) Schematic of a base editor, depicted as a Cas9 nickase tethered to a deaminase enzyme. Cas9 binds to an sgRNA-designated genomic locus before revealing a target nucleotide (M=A or C) within a ssDNA R-loop. The deaminase then functions on its cognate nucleotide and the Cas9 nick on the unedited strand prompts cellular repair to incorporate the edited strand, resulting in cytosine base editing performing C•G to T•A base conversions and adenine base editing performing A•T to G•C base conversions. (B) A cycle of plate-based directed evolution, in which iterative rounds of diversification, amplification, and selection enable isolation of active BE variants. Diversification involves mutagenesis of the BE expression plasmid which is transformed into host cells containing selection plasmids, selection takes place through BE-mediated antibiotic resistance turn-on for survival on a selection plate, and amplification via clonal growth on selection media and plasmid isolation. (C) Schematic depicting BE-mediated antibiotic resistance turn-on, where a BE library member acts on a selection plasmid featuring an inactive antibiotic resistance gene. Upon introduction and induction of the BE expression plasmid, bacterial cells harboring active variants will perform targeted base edits leading to antibiotic resistance marker expression and cell survival.

Because single-nucleotide polymorphisms (SNPs) comprise the largest class of known human disease variations, these technologies have a great therapeutic potential (Nelson et al., 2004). Base editing technologies have been used in both in vivo and *ex vivo* animal models to treat various genetic diseases such as Hutchinson-Gilford progeria syndrome (Koblan et al., 2021), sickle cell disease (Newby et al., 2021), spinal muscular atrophy (Arbab et al., 2023), hypercholesterolemia (Musunuru et al., 2021), and others. Base editing therapeutics have also more recently progressed to clinical trials, with four ongoing clinical trials as of 2023. Details of these trials and further applications of base editing are reviewed elsewhere (Porto & Komor, 2023; Porto, Komor, Slaymaker, & Yeo, 2020).

1.2 Previous engineering efforts

Natural enzymes often need to be tailored using protein engineering to take advantage of their full functionality as they are repurposed as tools towards synthetic applications. Directed evolution can fine-tune these proteins by mimicking natural evolution in the laboratory, in which iterative cycles of mutagenesis and selection can improve or create novel biological function with unprecedented ease and speed. Directed evolution has been used to engineer binding proteins, nucleic acids, genetic circuits, biochemical pathways, and whole genomes (Wang et al., 2021).

Many groups have demonstrated the power of directed evolution for reprogramming Cas9 activity. To date, CRISPR proteins have been evolved to change PAM compatibility (Hu et al., 2018; Huang et al., 2023; Miller, Wang, et al., 2020; Schmidheini et al., 2024), improve human cell efficiency (Ruta et al., 2024), alter sgRNA binding (Bush et al., 2023), and increase on-target specificity (Lee et al., 2018; Spencer & Zhang, 2017). In addition, effector proteins for precision gene editing are also often evolved to improve base editing efficiency (Gaudelli et al., 2020; Neugebauer et al., 2022; Richter et al., 2020; Thuronyi et al., 2019; Zhang, Neugebauer, Krasnow, & Liu, 2024), alter the deamination substrate (Gaudelli et al., 2017; Lam et al., 2023) or deamination result (Chen et al., 2024), or improve prime editing reverse transcription (Doman et al., 2023; Weber et al., 2024). Though rational engineering is an equally explored area (Kleinstiver, Prew, Tsai, Nguyen, et al., 2015; Kleinstiver, Prew, Tsai, Topkar, et al., 2015; Nishimasu et al., 2018; Rees, Wilson, Doman, & Liu, 2019; Walton, Christie, Whittaker, & Kleinstiver, 2020; Zhou et al., 2019; Zuo et al., 2020), directed evolution frequently creates variants that outperform their rationally designed counterparts, likely due to the ability to identify residues that are not positioned near any rational interface and/or that contribute to improved biomolecule properties like solubility (Miller, Wang, & Liu, 2020). Together, these engineering efforts greatly expand the applications for base editing, tailoring editing properties depending on the specific target context.

1.3 Benefits and limitations of base editing compared to other genome editing methods

Precision genome editing technologies like base editing have an advantage over homology-directed repair (HDR) due to the avoidance of DSB formation. BEs also facilitate editing in both dividing and non-dividing cells since they are not reliant on cell-cycle dependent repair pathways, unlike HDR (Chen & Liu, 2023). HDR is also typically less efficient than base editing (O'Brien, Wilson, Burgio, & Bauer, 2019; Paquet et al., 2016), favors the formation of insertions and deletions (indels) over the desired HDR product (Inui et al., 2014), and can induce cytotoxic and oncogenic cell damage (Haapaniemi, Botla, Persson, Schmierer, & Taipale, 2018). Precision genome editors benefit from easier delivery compared to HDR, as they can be delivered entirely as RNA without the requirement of a DNA donor. Base editing can also offer advantages over prime editing if a desired edit is a transition mutation that falls in a targetable editing window, such as generally higher editing efficiencies and fewer indels (Anzalone et al., 2019). The base editor complex is also smaller than a prime editor, making delivery easier.

Base editing is limited to the presence of a suitable PAM to expose an editing window which features the target nucleotide. Engineering efforts often aim to enhance base editing activity to overcome these limitations, but a broadly active deaminase means that the likelihood of bystander editing and Cas-independent off-targets increases (Rees & Liu, 2018). Bystander editing—editing occurring on non-target cytosines or adenines within the editing window—can cause detrimental sequence mutagenesis near the target site. Cas-independent DNA and RNA off-target deamination are known to occur especially for increased-activity deaminases (Doman, Raguram, Newby, & Liu, 2020; Gaudelli et al., 2020; Grünewald et al., 2019; Yu et al., 2020; Zuo et al., 2019). Though some effort has been made towards engineering context-specific BEs (Gehrke et al., 2018; Liu et al., 2020; Zhang et al., 2024) and reduced DNA/RNA off-targets (Doman et al., 2020; Grünewald et al., 2019; Kim et al., 2017; Thuronyi et al., 2019; Yu et al., 2020; Zhou et al., 2019), prime editing is preferred when high product purity is essential (Chen & Liu, 2023).

1.4 Benefits and limitations of plate-based selection compared to other directed evolution methods

Plate-based selection is a form of non-continuous (discrete) evolution (Fig. 1B). Here we describe a robust selection for BE-mediated antibiotic resistance turn-on (Gaudelli et al., 2017) to select for variants with high desired activity. This selection can be adapted to evolve either the deaminase or Cas enzyme, creating tailored BEs for a desired application. In this selection, a mutagenized BE targets a selection plasmid with a disabled antibiotic resistance cassette. Bacterial cells harboring active variants will enable BE activity, leading to antibiotic resistance marker expression and cell survival after plating on media containing selection antibiotic (Fig. 1C). One major benefit of this selection strategy is that it allows exogenous supply of molecules like synthetic sgRNA and small molecules, which would not be possible with continuous methods of directed evolution.

In contrast to discrete evolution, continuous directed evolution methods allow fast evolution times in bacteria, yeast, and mammalian cells in which mutagenesis and selection happen simultaneously without researcher intervention (Badran & Liu, 2015; Berman et al., 2018; Esvelt, Carlson, & Liu, 2011; Miller et al., 2020; Ravikumar, Arzumanyan, Obadi, Javanpour, & Liu, 2018). While these systems offer fast evolution times with minimal researcher intervention, they require specialized instrumentation, complex genetic circuits, are restricted to fast-acting proteins, and feature size limitations on the evolving protein of interest. Plate-based selections are, in contrast, much more flexible towards the type of activity evolved and are more user-friendly to allow use by researchers new to directed evolution without specialized instrumentation or skills.

1.5 Experimental design

The overall flow of the procedure is shown in Fig. 2.

1.6 Reagents

General reagents.

- MilliQ filtered water
- Ethanol

Molecular cloning and library preparation

- RepliQa HiFi ToughMix (2X) (QuantaBio, cat. no. 95200-500)
- QIAquick Gel Extraction Kit (Qiagen, cat. no. 28704)
- QIAEX II Gel Extraction Kit (Qiagen, cat. no. 20021)
- QIAprep Spin Miniprep Kit (Qiagen, cat. no. 27106)
- Plasmid Plus Midi Kit (Qiagen, cat. no. 12945)
- TempliPhi Amplification Kit (Cytiva, cat. no. 25-6400-10)
- GeneMorph II Random Mutagenesis Kit (Agilent, cat. no. 200550)
- Agarose (Gold Biotechnology, cat. no. A-201-500)

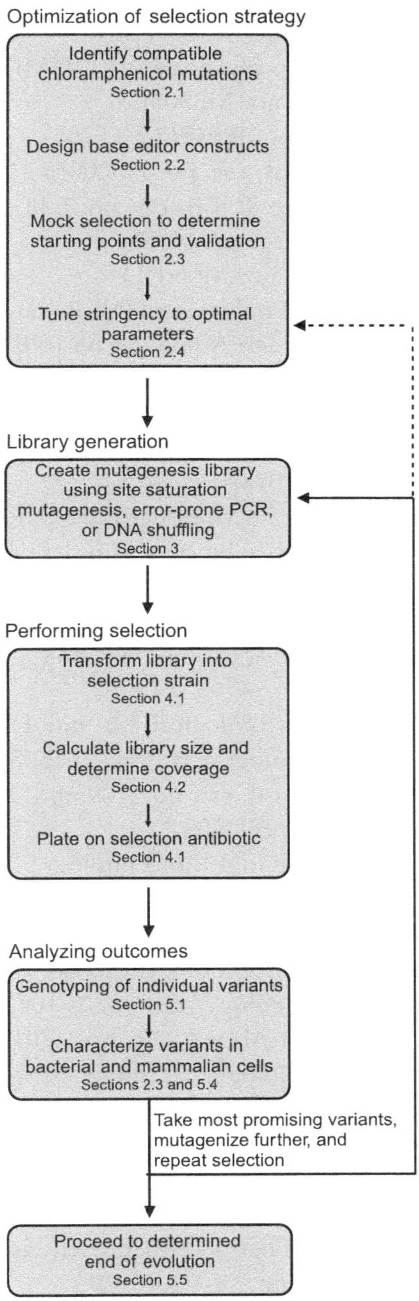

Fig. 2 Workflow schematic for base editor plate-based selections. Directed evolution is performed proceeding through the following steps: optimization of selection strategy, library generation, performing the selection, and analyzing outcomes. At any step, the selection may be iterated by proceeding to the indicated selection optimization or library generation step.

- DMSO (Corning, cat. no. 25-950-CQC)
- TAE Buffer, 10X (Corning, cat. no. 46-010-CM)
- USER enzyme (NEB, cat. no. M5505L)
- Dpn1 enzyme (NEB, cat. no. R0176L)
- 1 kb Plus DNA ladder (NEB, cat. no. N3200L)
- Gibson Assembly master mix (NEB, cat. no. E2611L)
- KLD enzyme mix (NEB, cat. no. M0554S)
- T5 exonuclease (NEB, cat. no. M0663L)
- Phusion high-fidelity DNA polymerase (NEB, cat. no. M0530L)
- Q5U hot start high-fidelity DNA polymerase (NEB, cat. no. M0515S)
- Taq DNA ligase (NEB, cat. no. M0208L)
- ETSSB (NEB, cat. no. M2401S)
- Deoxynucleotide (dNTP) solution mix (NEB, cat. no. N0447L)
- NAD+ (NEB, cat. no. B9007S)
- Taq DNA polymerase with ThermoPol Buffer (NEB, cat. no. M0267S)
- Magnesium sulfate (MgSO4) solution (NEB, cat. no. B1003S)
- Deoxynucleotide (dNTP) solution set (NEB, cat. no. N0446S)
- dUTP solution (NEB, cat. no. N0459S)
- Vent DNA polymerase (NEB, cat. no. M0254S)

Bacterial cell culture.

- 2x YT broth (United States Biological, cat. no. T9200)
- LB broth (United States Biological, cat. no. L1505)
- Agar (United States Biological, cat. no. A0930)
- SOC media (NEB, cat. no. B9020S)
- Carbenicillin or ampicillin (Gold Biotechnology, cat. no. C-103)
- Kanamycin (Gold Biotechnology, cat. no. K-120-50)
- Spectinomycin (Gold Biotechnology, cat. no. S-140-50)
- Chloramphenicol (Gold Biotechnology, cat. no. C-105)
- Tetracycline (Gold Biotechnology, cat. no. T-101)
- Anhydrotetracycline (Sigma-Aldrich, cat. no. 27919)
- L-arabinose (Gold Biotechnology, cat. no. A-300)
- Bleach (Essendant, cat. no. KIKBLEACH6)

Mammalian cell culture.

- HEK293T cell line (ATCC, cat. no. CRL-3216)
- DMEM medium (Corning, cat. no. 10-013-CV)
- Fetal bovine serum (FBS) (Thermo Fisher Scientific, cat. no. 16000044)
- DPBS (1X) (Corning, cat. no. 21-030-CVR)
- TrypLE Express Enzyme (1X), phenol red (Thermo Fisher Scientific, cat. no. 12605036)

- Trypan blue stain (0.4% (wt/vol)) (Thermo Fisher Scientific, cat. no. T10282)
- Proteinase K (NEB, cat. no. P8107S)
- SF Cell Line 4D-Nucleofector X Kit S (Lonza, cat no. V4XC-2032) or Amaxa SF Cell Line 96-well Nucleofector Kit (Lonza, cat. no. V4SC-2960)

Next-generation sequencing analysis.

- RepliQa HiFi ToughMix (2X) (QuantaBio, cat. no. 95200-500)
- QIAquick Gel Extraction Kit (Qiagen, cat. no. 28704)
- Qubit double-stranded DNA High-Sensitivity Assay Kit (Thermo Fisher Scientific, cat. no. Q32854)
- KAPA Library Quantification Kit (Roche, cat. no. 07960140001)
- MiSeq Reagent Kit v2 (300 cycles) (Illumina, cat. no. MS-103-1002—Micro Kit, ~4 million reads)

1.7 Equipment

- Filtered sterile pipette tips (e.g., Rainin, VWR)
- Serological pipettes, assorted (Corning)
- Standard microcentrifuge tubes, 1.5 ml (VWR, cat. no. 10025-724)
- Standard PCR strip tubes, 8 tubes/strip, 0.2 ml (VWR, cat. no. 20170-004)
- Conical centrifuge tubes, 15 ml (VWR, cat. no. 21008-216)
- Conical centrifuge tubes, 50 ml (VWR, cat. no. 89039-656)
- Vacuum filter/storage system, 0.22-um pore (Corning, cat. no. 431097)
- 96-well PCR plate (VWR, cat. no. 89049-178)
- PCR plate aluminum foil seals (VWR, cat. no. 60941-126)
- LightCycler 480 Multiwell Plate 96, white (Roche, cat. no. 04729692001)
- LightCycler 480 Sealing Foil (Roche, cat. no. 04729757001)
- Inoculating loop (Corning, cat. no. OS10B-02)
- L-shape cell spreader (VWR, cat. no. 76207-748)
- Disposable pipetting reservoir, 50 ml (VWR, cat. no. 89094-676)
- Petri dish, non-tissue culture-treated, 94/16 mm, standard with vents (Greiner Bio-One, cat. no. 633181)
- Petri dish, square with grid, 100mmx100mm (Fisher Scientific, cat. no. FB0875711A)
- Square large non-treated sterile culture dishes, 245 × 245 mm (Corning, cat. no. 431301)
- Polypropylene round bottom culture tube, 14 ml (Corning, cat. no. 352059)
- 96-well deep well culture plate (VWR, cat. no. 76397-576)
- Porous films for biological cultures (VWR, cat. no. 60941-086)

- Disposable plastic spectrophotometer cuvettes (VWR, cat. no. 97000-586)
- MicroPulser electroporation cuvettes, 0.1 cm or 0.2 cm gap (Bio-Rad, cat. no. 1652083—0.1 cm gap or cat. no. 1652082—0.2 cm gap)
- 96-well tissue culture-treated microplate with lid (VWR, cat. no. 10062-900)
- Cell culture flask with vented cap, 75 cm^2 (VWR, cat. no. 10062-860)
- Countess cell counting chamber slides (Thermo Fisher Scientific, cat. no. C10314)
- Gel casting system (Bio-Rad, cat. no. 1704412—caster; and Bio-Rad, cat. no. 1704416—gel tray)
- Gel electrophoresis system (Bio-Rad, cat. no. 1704401)
- Power supply for gel electrophoresis (Bio-Rad, cat. no. 1645070)
- PCR thermocycler (Bio-Rad, cat. no. 1851197)
- LightCycler 480 Instrument II (Roche, cat. no. 05015278001 or comparable system)
- Benchtop microcentrifuge (Eppendorf, cat. no. 5405000441)
- Tabletop centrifuge with rotor fitting 50-ml conical tubes (Eppendorf, cat. no. 022623508 or comparable system)
- High-speed floor centrifuge (Beckman Coulter Avanti J-E or comparable system)
- BioMate 160 UV-Vis Spectrophotometer (Thermo Fisher Scientific, cat. no. 840-301000 or comparable system)
- MicroPulser Electroporator (Bio-Rad, cat. no. 1652100)
- Qubit 4 fluorometer (Thermo Fisher Scientific, cat. no. Q33238)
- Countess II cell counter (Thermo Fisher Scientific or comparable system)
- Lonza 4D Nucleofector system with X Unit (Lonza, cat. no. AAF-1002X and AAF-1002B)
- 37 C, humidity- and CO_2-regulated incubator (Thermo Fisher Scientific, cat. no. 51030284 or comparable system)
- Benchtop vortexer (Fisher Scientific, cat. no. 02-215-414 or comparable system)
- Blue-light transilluminator for gel cutting (VWR, cat. no. 76151-834 or comparable system)
- Gel imaging system (Bio-Rad ChemiDoc or comparable system)
- MiSeq instrument (Illumina, cat. no. SY-410-1003)

1.8 Software

- CRISPResso2 (https://github.com/pinellolab/CRISPResso2)
- Docker (https://www.docker.com/products/docker-desktop/)

1.9 Primers

Primers used in this study can be found in Supplementary Tables 1–2.

1.10 Plasmids

Selection plasmid and BE expression plasmids (ABE and CBE) have been deposited to Addgene under this reference. Mammalian expression plasmids for ABE and CBE can be found on Addgene.

- pSM165 (selection plasmid, deposited under this reference)
- pRM015 (ABE expression plasmid, deposited under this reference)
- pRM016 (CBE expression plasmid, deposited under this reference)
- pCMV-ABE8e (Addgene plasmid #138489)
- pCMV-BE4max (Addgene plasmid #112093)

2. Optimization of selection strategy

2.1 Identifying compatible antibiotic resistance gene mutations for base editing

In the plate-based selection described here, a BE acts on a disabled antibiotic resistance gene to enable cell survival. The first step in designing a selection strategy is identifying compatible mutations on a given resistance gene. Editing sites must feature an accessible PAM which places a target A or C within the editing window of the protospacer. Sequences for previously validated sites for SpCas9 ABE or CBE activity on various antibiotic resistance genes are shown in Table 1 (Gaudelli et al., 2017). Using this system, BE activity is linked to cell survival on chloramphenicol through correction of the antibiotic mutation. Beyond the mutations listed in Table 1, which have been previously validated, many sites within resistance genes can be mutated to a stop codon, rendering the resistance gene inactive. In this case, only when the BE is active will the resistance gene be active, leading to cell survival. In order to create these stop codon mutants, any tryptophan (TGG), arginine (CGA), or glutamine (CAG or CAA) residue can be mutated to a stop codon (TGA, TAA, or TAG) for correction with ABE. This strategy is depicted schematically in Fig. 3. While this strategy is only compatible with ABE, a CBE selection scheme could be developed in which the introduction of a stop codon into a toxic gene, like *ccdB* or *SacB* (Kleinstiver, Prew, Tsai, Topkar, et al., 2015; Miller, Wang, et al., 2020), will render the toxic gene inactive and facilitate bacterial survival. In this case, CBE could be used to convert glutamine

Table 1 Antibiotic selection plasmids compatible with base editing. Table listing validated antibiotic resistance genes compatible with either ABE or CBE, suggested working concentrations of each antibiotic, the respective target sequence (with target nucleotide highlighted in red), inactivating mutation, and position of the target nucleotide in the protospacer. Beyond these validated sites, users can design new selection sites compatible with plate-based BE selections following guidelines described in Sections 2.1 and 2.2.

Resistance gene	Base editor	Target sequence	Inactivating mutation	Position of target in protospacer
CmR	CBE	TACGGCGTCGTGCACCTGGA	H193R	9
CmR	ABE	TACGGCGTAGTGCACCTGGA	H193Y	9
KanR	ABE	ATCTTATTCGATCATGCGAA	Q4*	6
KanR	ABE	GCTTAGGTGGAGCGCCTATT	W15*	5
KanR	ABE	TTCATTAACTGTGGCCGGCT	D208N	7
SpectR	ABE	CAATGATGACTTCTACAGCG	T89I	6

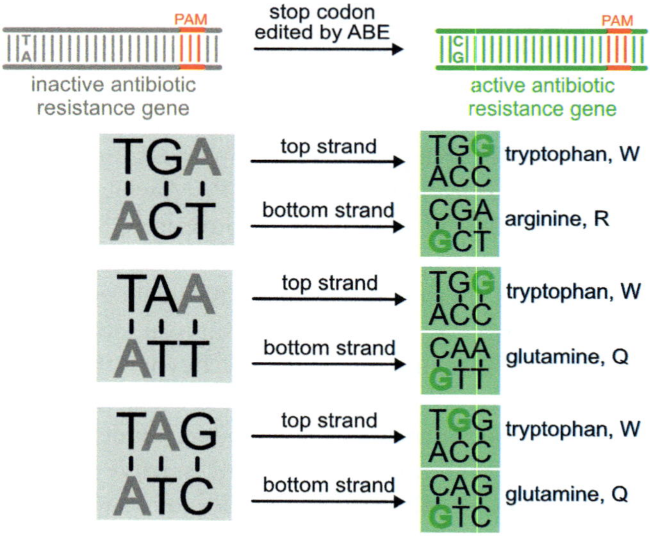

Fig. 3 ABE targeting of stop codons for antibiotic resistance gene turn-on. Sites within an antibiotic resistance gene can be manipulated by the user to a stop codon, rendering the gene inactive. Any tryptophan (TGG), arginine (CGA), or glutamine (CAG or CAA) residue can be mutated to a stop codon (TGA, TAA, or TAG) for correction with ABE. In this case, only when ABE is active will the resistance gene be expressed, linking ABE activity to cell survival on the selection antibiotic.

(CAG or CAA), arginine (CGA), or tryptophan (TGG) into stop codons to render the toxic gene inactive.

Inactivating mutations should be cloned into the selection plasmid using a preferred cloning method (we suggest Gibson assembly or USER assembly) to test the dynamic range of the selection. This can be done by plating serial dilutions on plates containing maintenance antibiotic. Ensure that both the original selection plasmid sequence and the mutated sequence are plated to draw comparisons. Stringency of the selection can be increased by including multiple inactivating sites to be corrected, which is discussed further below.

2.2 Designing base editor constructs

Candidate sgRNAs should be designed to exactly match the identified target sequence 5'-to-3', comprising the 20 base pairs 5' of the NGG PAM for SpCas9. This might change depending on the CRISPR protein chosen (Huang, Newby, & Liu, 2021). These sgRNAs can then be prepared as DNA plasmids using a one-step cloning strategy. If only switching out the target protospacer from a BE expression plasmid (deposited on Addgene), we recommend a one-step cloning method like the kinase-ligase-dpn1 (KLD) strategy. Depending on the user's evolution needs, the deaminase and/or Cas protein can be swapped to generate tailor-made novel base editing constructs using any method of cloning, including Gibson and USER cloning strategies according to previously described methods (Gibson et al., 2009; Huang et al., 2021; Thomas, Maynard, & Gill, 2015).

2.2.1 Switching out sgRNA cassette with KLD cloning

This strategy should be used when introducing a new protospacer within the BE expression plasmid, for which an existing cassette expressing the SpCas9 sgRNA scaffold is available.

1. Generate fragments via PCR from a BE expression plasmid (deposited on Addgene)
   a. Use primers RM-01 and RM-02
   b. Amplify with RepliQa (or similar) using manufacturer's protocol
2. Confirm desired band size via 1 % agarose gel electrophoresis (9.3 kb for SpCas9 ABE 7.10, 10.1 kb for SpCas9 CBE)
3. Purify the PCR reaction via column purification according to manufacturer's instructions, elute in a minimal (~10 μl) volume of water
4. Set up the KLD reaction according to manufacturer's protocols
   a. 5 μl KLD buffer, 2x

   b. 1 µl KLD enzyme mix, 10x
   c. 4 µl purified PCR template from the previous step
5. Incubate the KLD reaction at room temperature for 5 min, place on ice for immediate transformation
6. Transform the assembly into 10 µl chemicompetent *E. coli* Mach1 cells (or similar)
   a. Incubate the cell-DNA mixture on ice for 10 min
   b. Heat shock at 42 °C for 30 s
   c. Place back on ice for 1 min
7. Add 100 µl SOC medium and plate directly on a plate containing LB-agar and 50 µg/ml carbenicillin
8. Incubate overnight at 37 °C
9. Pick single colonies and prepare either colony PCR reaction (primers RM-15 and RM-16) or miniprep
10. Submit colony reactions for sequencing (whole plasmid recommended) to confirm insertion of desired protospacer sequence
11. Transfer sequence-verified colonies to a 125 ml Erlenmeyer flask with 50 ml 2xYT media and 50 µg/ml carbenicillin
12. Incubate at 37 °C, 220 rpm overnight
13. Isolate plasmid DNA using midiprep kit
   a. Wash > 3x with PE buffer
   b. Spin-dry for 3 min
   c. Elute in 100 µl prewarmed water

2.3 Mock selection to validate system and determine starting points

The identified system must then be tested and validated via mock selection in which wild-type BE and sgRNA are subjected to selection protocol and survival is measured via serial dilution. With any new selection strategy, this validation is crucial to ensure that the selection strategy works well and displays appropriate selection stringency and background for a directed evolution campaign. Survival is quantified using a serial dilution plate, which allows calculation of surviving colony forming units (CFU) on the selection antibiotic. Background survival should be determined with a negative control, like a no sgRNA control construct. Example data demonstrating the survival of SpCas9-ABE(7.10) with a sgRNA targeting a CmR H193Y mutation is shown in Fig. 4. To perform a mock selection to validate the system, follow Section 4.1 omitting step 17.

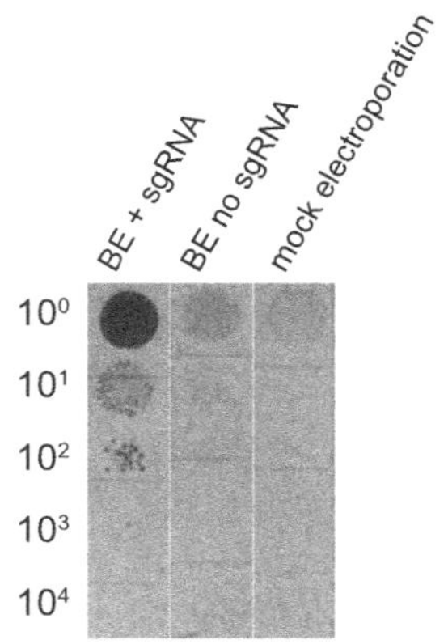

Fig. 4 Mock selection results to validate a selection. Mock selection results demonstrating survival of SpCas9-ABE(7.10) after editing using a sgRNA targeting CmR(H193Y). Protocol followed is described in Section 4.1. Controls include ABE + no sgRNA and mock electroporation. Survival of both controls is 0 colony forming units (CFU), indicating undetectable background survival. Survival of the experimental condition (BE + sgRNA) is 5 x 10^4 CFU (one colony in the 4th dilution) according to the equation found in Section 4.2. The difference between survival of the experimental condition compared to negative controls indicates a good dynamic range with appropriate selection stringency. Stringency of the selection can be further manipulated if desired, discussed in Section 2.4.

2.4 Tuning stringency of the selection

The stringency and background of the selection can be modulated depending on the goals of the evolution and the results of the mock selection. Stringency represents how difficult it is to pull out an active variant from a selection; a higher stringency represents a more rigorous selection, and a lower stringency represents a more lenient selection. The suitable selection stringency varies based on individual selections but will exist somewhere between selection background and efficiency of a positive control (i.e. SpCas9 BE on a validated guide). We suggest a starting point in which survival is ~10-fold detectable above background; lower survival might enrich non-functional constructs and higher may enrich high-performing and low-performing variants equally. It is also common in directed evolution campaigns to alter selection stringency throughout evolution; lower stringency is often required at the beginning of a campaign and then increased as more active variants are isolated.

There are several ways to tune stringency of a selection. These include selection antibiotic concentration, dosing of the sgRNA, promoters, ribosome-binding sites (RBS), induction time, and incubation (outgrowth) time. Increasing the concentration of the selection antibiotic increases stringency, as more selection plasmid copies must be edited to express sufficient enzyme to nullify the antibiotics. Increasing concentration of the sgRNA and/or the BE reduces stringency, as more active BE ribonucleoprotein formed per cell increases the likelihood that editing will occur. sgRNA and BE expression can be modulated by increasing promoter strength (Davis, Rubin, & Sauer, 2011), RBS (Ringquist et al., 1992), and/or plasmid copy number (Friehs, 2004). More information on various promoters, RBS, and copy numbers can be found online at parts.iGem.org. Concentrations of the selection antibiotic, sgRNA, and BE should be optimized in initial experiments to tune stringency to the desired point. Starting point recommendations for selection antibiotic working concentrations are listed in Table 1. Besides design, timing of the protocol can alter stringency. Induction time (duration of BE expression) and incubation time (outgrowth of the system) both have an effect on the stringency of the system. Increasing these time points will lead to lower stringency as the BE has more time to make the desired edit. Ultimately, optimal parameters that define the relationship between target activity and BE activity can help achieve desired outcomes for the system.

3. Methods to generate diversity

Diversity generation is a key component in any directed evolution campaign. To generate these mutant libraries, any appropriate cloning strategy can be used if it is compatible with electroporation. USER assembly (Cavaleiro, Kim, Seppälä, Nielsen, & Nørholm, 2015), for example, is not compatible with this method due to instability of the unligated sticky ends during electroporation. However, Gibson assembly (Gibson et al., 2009; Thomas et al., 2015), in-fusion cloning (Berrow, Alderton, & Owens, 2009), and Golden Gate cloning (Engler, Kandzia, & Marillonnet, 2008) are all good options with high speed and flexibility, allowing the introduction of many inserts. Here, we detail Gibson cloning optimization. We recommend using the Gibson Assembly master mix for these reactions, but a homemade mix can be used as a cheaper alternative.

3.1 Site saturation mutagenesis

Site saturation mutagenesis holds great potential as a tool in directed evolution, as it can introduce targeted sequence variations at virtually any region (Acevedo-Rocha, Reetz, & Nov, 2015). When generating site saturation libraries, it is crucial to limit the formation of premature stop codons. NNK degenerate codons (where N = A/C/G/T, K = G/T) are used to reduce the frequency of prematurely truncated variants. To insert these site saturation libraries into a desired position, primers encoding NNK nucleotides are used.

3.1.1 Primer design

NNK primers for Gibson assembly feature one primer encoding NNK nucleotides at the desired site and at least 17nt of annealing to the template with a T_m ~60 °C. For the other primer, follow standard primer design strategies (greater than 17nt, GC 40–60 %, T_m ~60 °C). The Gibson junction should be designed such that a 20–25 bp overlap will occur between the two amplified fragments. This design should be followed for all desired NNK sites in a library. Design strategies for these primers are depicted in Fig. 5A.

3.1.2 Site saturation mutagenesis library generation

1. Amplify fragments via 100 μl PCR reactions using primers designed in Step 3.1.1
   a. Larger or smaller reaction volumes can be used depending on efficiency of amplification or desired library size

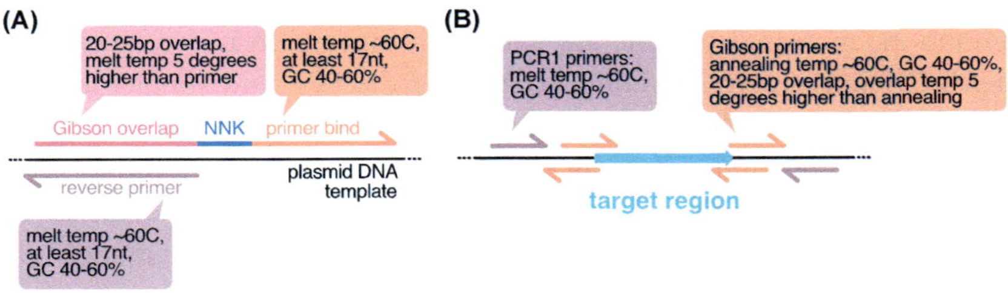

Fig. 5 Primer design parameters for mutagenesis strategies. (A) Gibson primer design strategy for generating NNK libraries, where target residue(s) are manipulated to an NNK codon within the primer. This primer features annealing portions upstream and downstream of the target residue and a Gibson overlap junction with the reverse primer to be designed within the indicated parameters. (B) Primer design strategies for error-prone PCR and DNA shuffling, where the first primer set (for PCR1) amplifies the target region of mutagenesis or uracil-incorporation PCR. After proceeding through the indicated steps, Gibson primers amplify mutagenized fragments and unmutagenized backbone for Gibson assembly.

2. Dpn1 digest each fragment at 37 °C for at least one hour to remove plasmid template
   a. Add 1 μl of Dpn1 per 50 μl PCR reaction directly to an unpurified completed PCR reaction
3. Cleanup each fragment via column purification
4. Quantify DNA concentration (i.e. nanodrop or Qubit)
5. Assemble 200 μl Gibson reactions according to manufacturer's protocol
   a. If using more than 1 fragment, ensure a 1:1 molar ratio of each fragment
   b. Total DNA should not exceed 2pmol for 2-3 fragments or 4pmol for 4 + fragments
   c. Up to 100 μl of purified fragments can be added to a 200 μl reaction
6. Incubate at 50 °C for 10 h, then cool to 4 °C until retrieved
7. Purify assembled DNA via column purification
   a. Wash > 3x with buffer PE
   b. Spin dry for > 5 min
   c. Elute DNA in 20 μl of prewarmed nuclease-free water
8. Calculate concentration of DNA (i.e. nanodrop or Qubit)
9. Proceed to Section 4.1

3.2 Error-prone PCR

Random mutagenesis of a desired gene can be achieved through error-prone PCR (epPCR). epPCR differs from standard PCR conditions through engineered polymerases, increased Taq DNA polymerase concentration, increased polymerase extension time, increased concentration of $MgCl_2$ ions, increased concentration of dNTPs, and supplementation with $MnCl_2$ ions (McCullum, Williams, Zhang, & Chaput, 2010). As a result, mutations can be introduced to any piece of the gene during PCR. We detail epPCR using the GeneMorph II Random Mutagenesis Kit, however epPCR can be performed using Taq DNA polymerase with reaction alterations (McCullum et al., 2010).

3.2.1 *Primer design*

epPCR requires a PCR1 and PCR2 reaction with different primers for each reaction. PCR1 will use a mutagenesis-prone polymerase to mutagenize the region of interest. PCR1 primers should span the region to be mutated and can follow standard primer design strategies (greater than 17nt, GC 40–60 %, T_m ~60 °C). PCR2 will amplify fragments (epPCR fragment and backbone fragment) for Gibson assembly in two separate reactions with different primer pairs. The Gibson junctions should be designed such that a 20–25 bp overlap will occur between the two amplified fragments. Design strategies for these primers are depicted in Fig. 5B and are expanded upon in Section 3.2.2.

3.2.2 Error-prone PCR library generation

1. Design primers for epPCR according to Section 3.2.1
   a. Primers RM-03 and RM-04 can be used for randomization of *E. coli* TadA, while primers RM-05 and RM-06 can be used for randomization of SpCas9
2. Set up 50 μl error-prone PCR reaction using GeneMorph II Random Mutagenesis Kit according to manufacturer's instructions
   a. Mutagenesis frequency can be lowered by lowering the number of cycles and by increasing amount of input DNA, or vice versa if higher mutation rates are desired
3. Run on 1 % agarose gel to confirm desired band size (764 bp for E. coli TadA, 4193 bp for SpCas9)
4. Dpn1 digest epPCR product at 37 °C for at least one hour
   a. Add 1 μl of Dpn1 per 50 μl epPCR reaction directly to an unpurified completed epPCR reaction
5. Purify epPCR reaction using column
6. Set up PCR reactions (200 μl per fragment) to amplify the error-prone fragment and backbone fragment for Gibson assembly
   a. Add 1 μl epPCR purified fragment (concentration 40-60 ng/μl, from Step 5) as template for epPCR amplification reaction
      i. Primers RM-07 and RM-08 can be used for amplification of *E. coli* TadA, while primers RM-09 and RM-10 can be used for amplification of SpCas9
   b. Add 10-20 ng plasmid DNA as template for backbone amplification reaction
      i. Primers RM-11 and RM-12 can be used for amplification of *E. coli* TadA backbone, while primers RM-13 and RM-14 can be used for amplification of SpCas9 backbone
   c. Run these PCR reactions under desired cycling conditions
      i. Limit to 20-25 cycles to reduce PCR bias of randomized library
7. Run on 1 % agarose gel to confirm desired band sizes
8. Add 1 μl of Dpn1 per 50 μl PCR reaction directly to an unpurified completed PCR reaction
9. Purify reactions using column

10. Quantify DNA concentration (i.e. nanodrop or Qubit)
11. Assemble 200 μl Gibson reactions according to manufacturer's protocol
    a. If using more than 1 fragment, ensure a 1:1 molar ratio of each fragment
    b. Total DNA should not exceed 2pmol for 2–3 fragments or 4pmol for 4 + fragments
    c. Up to 100 μl of purified fragments can be added to a 200 μl reaction
12. Incubate at 50 °C for 10 h, then cool to 4 °C until retrieved
13. Purify Gibson reaction using column
    a. Wash > 3x with buffer PE
    b. Spin dry for > 5 min
    c. Elute DNA in 20 μl of prewarmed nuclease-free water
14. Quantify DNA concentration (i.e. nanodrop or Qubit)
15. Proceed to Section 4.1

3.3 DNA shuffling

DNA shuffling is another recombination-based mutagenesis method where DNA sequences are randomly fragmented and then reassembled into full-length, chimeric sequences (Coco et al., 2001; Stemmer, 1994). DNA shuffling can be performed on a library of related genes to enable family shuffling diversification (Kikuchi, Ohnishi, & Harayama, 2000), used as a method to cross multiple variants isolated during an evolution campaign, or used to remove non-essential "hitchhiker" mutations via random reversion by back-crossing with the wild-type sequence.

3.3.1 DNA shuffling fragment amplification protocol

1. Design primers according to the same guidelines as Section 3.2.1 (Fig. 5B)
2. Prepare DNA templates
    a. For family shuffling: pool an equimolar amount of each isolated homolog
    b. For post-evolution shuffling: purify selected library via midiprep kit
    c. For reversion: mix purified library (as in 2b) or isolated clone with wild-type plasmid in a 10:1 ratio. Ratio can be increased/decreased if more/less reversion is desired. Using this method on a purified library will enable simultaneous shuffling and reversion.
1. Perform uracil-incorporation PCR
    a. Add together the following ingredients
        - 20 μl 5x Q5U buffer
        - 2 μl Q5U DNA polymerase

- 4 μl each 10 mM dATP, dCTP, dGTP (400 μM final)
- 1.2 μl 10 mM dUTP (120 μM final)
- 2.8 μl 10 mM dTTP (280 μM final)
- 1 μl each 100uM primer (1 μM final)
- 40–60 ng template DNA
- Water to 100 μl

b. Amplify according to manufacturer's instructions

c. Reaction volumes can be increased depending on amplification efficiency and desired library sizes

3. Run on 1 % agarose gel to confirm desired band sizes

a. Gel extraction may be required if off-target amplification occurs

4. Dpn1 digest PCR product at 37 °C for at least one hour

a. Add 1 μl of Dpn1 per 50 μl PCR reaction directly to an unpurified completed PCR reaction

5. Purify the fragment via column

a. Elute in 40 μl water

6. Digest the purified product with 2 U of USER enzyme per 40 μl in 1x CutSmart Buffer at 37 °C

a. Monitor digestion by 2 % agarose gel electrophoresis until starting material is no longer observed; digestion products may appear on the gel as a smear

b. More digestion will enable higher resolution shuffling but will decrease the efficiency of assembly

c. Digestion of larger products may require overnight incubation

d. Additional USER enzyme may be supplemented every 3–4 h to improve digestion

7. Purify the digested material with QiaexII gel extraction kit according to manufacturer's protocol

a. Follow section "Protocol: Desalting and Concentrating DNA Solutions" from manufacturer's protocol

b. Elute in 50 μl pre-warmed water

3.3.2 DNA shuffling reassembly protocol

1. Prepare 50 μl DNA shuffling reaction

a. Add together the following ingredients

- 25 μl eluted digested DNA fragments (from step 3.3.1)
- 2 μl Vent Polymerase
- 4 μl 10 mM dNTP solution mix (800 μM final)
- 1 μl Taq DNA polymerase

- 5 μl 10x ThermoPol Buffer supplemented
- 0.25 μl 100 mM $MgSO_4$ (0.5 mM final)
- Water to 50 μl

2. Run the reassembly reaction
   a. 94 °C for 3 min; [30 s at 92 °C; 60 s at (30 °C +1 °C per cycle); (60 s + 4 s per cycle) at 72 °C] x 40;
      10 min at 72 °C
   b. It is unlikely that a clear band will be visible via electrophoresis
3. Amplify the reassembled product by PCR using RepliQa according to manufacturer's instructions
   a. Combine 15 μl unpurified assembly (from the previous step) in a 200 μl reaction
   b. Primers RM-07 and RM-08 can be used for amplification of *E. coli* TadA, while primers RM-09 and RM-10 can be used for amplification of SpCas9
   c. Long assembly products may require increased extension times. We recommend 20 s per kb using RepliQa.
4. Amplify backbone by PCR using RepliQa according to manufacturer's instructions
   a. Combine 10–20 ng template plasmid in a 200 μl reaction
   b. Primers RM-11 and RM-12 can be used for amplification of *E. coli* TadA backbone, while primers RM-13 and RM-14 can be used for amplification of SpCas9 backbone
5. Run on 1 % agarose gel to confirm desired band sizes
6. Dpn1 digest PCR product at 37 °C for at least one hour
   c. Add 1 μl of Dpn1 per 50 μl PCR reaction directly to an unpurified completed PCR reaction
   d. Gel extraction may be required if off-target amplification occurs
7. Assemble via the Gibson cloning method according to manufacturer's instructions
8. Cleanup Gibson assembly via column purification
9. Proceed to Section 4.1

4. Performing plate-based selection

Assembled mutant libraries will then be taken through a plate-based selection, where the variants which are performing the desired activity will be pulled out from the library. Prior to selection, libraries must be transformed into the selection strain (2060 + selection plasmid) via electroporation. It is best practice to prep fresh electrocompetent cells on the day of the electroporation. Large libraries may necessitate more electroporations per selection; the following protocol may be scaled as such.

4.1 Transforming into the selection strain and performing selection

1. Inoculate 5 ml 2xYT supplemented with appropriate maintenance antibiotics with the selection strain from glycerol stock or fresh plasmid transformation and grow overnight at 37 °C, 220 rpm
   a. Saturated cultures can be stored at 4 °C for 2-3 days
2. Back dilute saturated culture into a total volume of 10 ml per electroporation of culture with appropriate maintenance antibiotic
   a. Ensure that the OD is less than 0.15 after back dilution, roughly a 1:50 dilution
3. Grow the cells to mid-log phase (OD 0.45-0.55)
4. Pellet cells by centrifugation for 10 min at 8000 x g, 4 °C
   a. Discard media
5. Pre-warm SOC recovery media to 37 °C
   a. Prepare 500 µl per electroporation
6. Prepare electrocompetent bacterial cells
   a. Resuspend pellet in 1 vol ice-cold 10 % glycerol (v/v) in dH_2O
   b. Pipette to resuspend, ensuring no remaining cell clumps
   c. Centrifuge for 10 min, 8000 x g, 4 °C
   d. Discard media, careful not to disturb pellet
   e. Repeat steps 6a-d 2x
   f. Resuspend pellet in 75 µl per electroporation ice-cold 10 % glycerol (v/v) in dH_2O
   g. Transfer to individual Eppendorf tubes per electroporation
7. Add library DNA
   a. Greater than 500 ng purified DNA per electroporation
   b. Volume of DNA must not exceed 10 % total electroporation volume (8 µl)
   c. Mix gently by pipetting
8. Transfer to pre-chilled electroporation cuvette(s) with a 1 mm gap
   a. Avoid bubbles when pipetting into cuvette
9. Electroporate at 1.8 kV
10. Immediately following electroporation, add 500 µl of pre-warmed SOC recovery media to cuvette and transfer to an Eppendorf tube

11. Recover for 1 hr at 37 °C, 220 rpm
12. After 1 hr recovery, take 50 µl of the culture and plate serial dilutions on maintenance antibiotics to determine library size and coverage (discussed further in Section 4.2)
    a. To characterize pre-selection clones for successful mutagenesis, proceed to Section 5.1
13. Transfer the remainder of the culture to a pre-warmed outgrowth culture: 10 ml 2xYT media with appropriate maintenance antibiotics and inducer (do not include selection antibiotic)
    a. Inducer: L-arabinose at 100 mM or anhydrotetracycline at 5 µg/ml, depending on inducible promoter
14. Place in shaking incubator (37 °C, 220 rpm) overnight (16-20 hr)
15. After outgrowth, spin down 10 ml culture for 10 min at 8000 x g, 4 °C
16. Resuspend pellet in 500 µl water, reserve 50 µl for serial dilutions
17. Plate the remainder of the 500 µl mixture on a 245 x 245mm LB agar plate supplemented with appropriate antibiotics (including selection antibiotic)
    a. Spread mixture across plate with sterile beads, ensuring culture is spread evenly across the plate
18. Perform serial dilutions (10x dilution of culture at each step) and plate on LB agar plate containing appropriate antibiotics
    a. Pipette 50 µl of resuspended culture to one well of a PCR strip tube
    b. Pipette 45 µl of dH_2O to the other 7 wells in the strip
    c. Take 5 µl from one well to the next, creating a 10-fold dilution to each tube
    d. Be sure to mix very thoroughly by pipetting between dilutions

4.2 Determining library size and coverage

After mutant libraries have been assembled, it is important to determine library size and coverage. This is to ensure that the pre-selection variant library is representative of as many variants as possible before selecting on said library. Theoretical library size can be calculated based on how many variants could result from a library. For an NNK library, one NNK library represents 32 possible codons. Theoretical library sizes for multi-site NNK libraries can thus be calculated by multiplying 32^n, where n = # of NNK sites. As a result, increasing the number of NNK sites increases the theoretical library size exponentially. Within the directed evolution community, 5-fold coverage of the theoretical library is optimal because it gives 99 % probability of having any particular member in the library present (Bosley & Ostermeier, 2005).

Observed library size can be calculated using serial dilution plates of transformation culture. When plating transformants, 50 μl of culture should be taken to perform serial dilutions, where each step is a 1:10 dilution of the previous. Plating these dilutions then allows an accurate picture of colony forming units, or library size. Library size can be calculated as follows:

$$\#colonies \times dilution\ factor \times \frac{volume\ of\ culture\ taken\ from}{volume\ of\ culture\ plated}$$

Successful mutagenesis of a library can be confirmed by sequencing pre-selection clones. Sequencing protocols for this purpose are discussed in Section 5.1.

5. Analyzing outcomes

Following selection, surviving clones must be screened to determine genotypes of resulting variants. Converging mutations are often indicative of a successful selection. The most promising variants will be identified and sub-cloned for further characterization to confirm their fitness advantage in the context of the selection. The selection can be repeated iteratively until a desired level is achieved. At this point, final variants must be thoroughly characterized in a variety of contexts, discussed further below.

5.1 Genotyping of individual variants

After performing selection, individual variants are isolated as surviving colonies picked from post-selection plates. Variants are genotyped through gene isolation and amplification by PCR, rolling circle amplification (RCA), or plasmid DNA isolation and sequencing. Whole plasmid sequencing following plasmid DNA isolation enables a more expansive view of these individual variants. At least 8 colonies per directed evolution round should be analyzed to ensure sampling of the top genotypes. In all instances, cultures of individually isolated clones should be saved for sub-cloning and evaluation post-sequencing. Promising genotypes that are the most represented within the population and/or contain mutations converged between variants should be further characterized through bacterial (Section 4.1) or mammalian (Section 5.4) assays.

5.1.1 PCR amplification protocol

1. Assemble colony PCR reaction
   a. Assemble 25 µl RepliQa PCR reactions per clone
   b. Touch colonies with the tip of a p10 pipette tip and deposit the tip in the PCR reaction before pipetting to mix
   c. Primers RM-15 and RM-16 can be used to amplify the entire BE cassette
2. Run PCR according to manufacturer's instructions
3. Confirm correct band via 1 % agarose gel electrophoresis
   a. Using primers RM-15 and RM-16, Apobec-SpCas9 CBE is 5.7 kb and TadA(7.10)-SpCas9 ABE is 4.9 kb
4. Submit to sequencing. Many commercial Sanger and whole-plasmid sequencing sources accept unpurified PCR reactions, but purification may be required

5.1.2 RCA protocol

1. Assemble RCA reactions using the TempliPhi Amplification Kit
2. Touch colonies with the tip of a p10 pipette tip and deposit the tip into 4 µl DNA amplification kit sample buffer, pipetting well to mix
3. Denature at 95 °C for 3 min, then place on ice
4. Add 5 µl of DNA amplification kit reaction buffer and 0.2 µl of DNA amplification kit enzyme mix
   a. Make a master mix for the number of reactions
   b. Add 5.2 µl of the master mix to each sample
5. Incubate at 37 °C for at least 4 h
6. Denature at 65 °C for 10 min
7. Add 90 µl of water and mix well
8. Send reactions for sequencing

5.2 Determining success of a selection

Evaluating the success of a directed evolution campaign is critical, as false positives can occur post-selection. Thorough characterization is essential to confirm selection outcomes. Early indicators, such as survival on the selection plate, can provide insights. Lack of survival often signals excessive stringency in the selection protocol, which can be adjusted to improve outcomes (Section 2.4). High dynamic range—low background survival with high target survival—may suggest a successful selection, but further characterization is necessary to confirm this. Genotyping individual variants (Section 5.1) provides the most reliable assessment of selection success.

Converging mutations across multiple clones strongly suggest that these variants are linked to the desired activity. Converging variants should be sub-cloned and tested experimentally in both bacterial and mammalian systems (Sections 5.3 and 5.4) to validate improved performance over the wild-type protein. If such variants show enhanced activity, the round can be considered successful, though further rounds of evolution may still be needed. Absence of converging mutations does not automatically imply failure. If no convergence is observed, the library can be subjected to additional selection rounds, particularly under high stringency, to enrich for beneficial variants. If convergence remains absent after multiple rounds, reconsidering the selection parameters or target may be necessary.

5.3 Repeating selection iteratively and sub-cloning

Many successful directed evolution campaigns require multiple rounds of diversification and selection to generate variants of desired efficiency. The most highly active variant(s) after each round of selection should be further mutagenized and selected upon by repeating Sections 3 and 4. If selection stringency (Section 2.4) is not adequately high, sub-cloning and selection can be performed without mutagenesis to further enrich active variants within the same pool. Sub-cloned variants can be used to evaluate survival efficiency in a mock selection (Section 2.3), serving as a proxy for bacterial base editing activity which is thought to mirror editing activity in human cells.

5.3.1 Sub-cloning

1. Set up PCR reactions (200 μl per fragment if library, 50 μl per fragment if not) to amplify the two fragments to reassemble the BE expression plasmid
   a. Add 1 μl purified library to be subcloned (concentration 40-60 ng/μl) as template for both reactions
   b. Reaction 1 primers: RM-15/RM-16
   c. Reaction 2 primers: RM-17/RM-18
      i. These primer sets will subclone most BE variants
2. Run these PCR reactions under standard cycling conditions
3. Run on 1 % agarose gel to confirm desired band sizes
4. Dpn1 digest PCR products at 37 °C for at least one hour
   a. Add 1 μl of Dpn1 per 50 μl PCR reaction directly to an unpurified completed PCR reaction
5. Purify reactions using column

6. Quantify DNA concentration (i.e. nanodrop or Qubit)
7. Assemble 200 μl Gibson reactions (if library) or 20 μl Gibson reactions (if not) according to manufacturer's protocol
   a. Ensure a 1:1 molar ratio of each fragment
   b. Total DNA should not exceed 2 pmol (for 200 μl reaction) or 0.2 pmol (for 20 μl reaction)
   c. Up to 100 μl of purified fragments can be added to a 200 μl reaction, up to 10 μl of purified fragments can be added to a 20 μl reaction
8. Incubate at 50 °C for 10 h (if library), 15 min (if not) then cool to 4 °C until retrieved
9. Purify Gibson reaction using column
   a. Wash > 3x with buffer PE
   b. Spin dry for > 5 min
   c. Elute DNA in 20 μl of prewarmed nuclease-free water
10. Quantify DNA concentration (i.e. nanodrop or Qubit)
11. Proceed to Section 4.1 if re-enrichment of a library is desired
12. Proceed to Section 2.3 to evaluate clonal bacterial efficiency

5.4 Human cell base editing

Activity of evolved BE variants should then be characterized by human cell plasmid transfection followed by high-throughput sequencing (HTS). A thorough protocol to choose genomic sites and perform BE experiments in human cells has already been published (Huang et al., 2021). We refer readers to this reference for more information on human cell BE experiments. A collection of validated genomic loci and protospacers (Gaudelli et al., 2017; Thuronyi et al., 2019) compatible with SpCas9-mediated BE are listed in Supplementary Table 3 and respective HTS primers are listed in Supplementary Table 2.

5.5 Determining the end of evolution

Ultimately, the goal of a directed evolution campaign is to evolve to levels of activity similar to that of SpCas9 ABEs or CBEs within the desired context (normally, human cell base editing), but desired activities may differ based on intended application. Iterative cycles of mutagenesis and selection should be performed until this level is achieved. Alongside desired activity levels, sufficient experimental evidence should confirm that the evolved protein shows no significant increase in off-target editing, toxicity, or immunogenicity (Huang et al., 2021; Tsai et al., 2015; Tsai et al., 2017). Once these benchmarks have been reached, the evolution campaign can be considered complete.

6. Conclusions

Here we describe a protocol for the plate-based directed evolution of base editor enzymes. Importantly, this method proceeds using minimal specialized equipment or knowledge and can be tailored for the evolution of any CRISPR effector protein or tethered catalytic domains. In this way, a researcher can optimize existing base editor combinations or develop novel methods of precision genome editing. This method is compatible with targeted mutagenesis, random mutagenesis, or family DNA shuffling, allowing for flexible DNA inputs into the selection. Furthermore, the plate-based nature of the selection enables the addition of non-biological compounds, including small molecules and chemically modified biomolecules, which could be used to develop novel inhibitors, activators, or synthetic Cas9 or sgRNA compounds. In sum, this selection represents a simple and flexible pathway for any researcher to begin directed evolution of precision genome editing compounds.

Appendix A. Supporting information

Supplementary data associated with this article can be found in the online version at https://doi.org/10.1016/bs.mie.2025.01.003.

References

Acevedo-Rocha, C. G., Reetz, M. T., & Nov, Y. (2015). Economical analysis of saturation mutagenesis experiments. *Scientific Reports, 5*(1), 10654. https://doi.org/10.1038/srep10654.

Anzalone, A. V., Koblan, L. W., & Liu, D. R. (2020). Genome editing with CRISPR-Cas nucleases, base editors, transposases and prime editors. *Nature Biotechnology, 38*(7), 824–844. https://doi.org/10.1038/s41587-020-0561-9.

Anzalone, A. V., Randolph, P. B., Davis, J. R., Sousa, A. A., Koblan, L. W., Levy, J. M., ... Liu, D. R. (2019). Search-and-replace genome editing without double-strand breaks or donor DNA. *Nature, 576*(7785), 149–157. https://doi.org/10.1038/s41586-019-1711-4.

Arbab, M., Matuszek, Z., Kray, K. M., Du, A., Newby, G. A., Blatnik, A. J., ... Liu, D. R. (2023). Base editing rescue of spinal muscular atrophy in cells and in mice. *Science (New York, N. Y.), 380*(6642), eadg6518. https://doi.org/10.1126/science.adg6518.

Badran, A. H., & Liu, D. R. (2015). In vivo continuous directed evolution. *Current Opinion in Chemical Biology, 24*, 1–10. https://doi.org/10.1016/j.cbpa.2014.09.040.

Berman, C. M., Papa, L. J., 3rd, Hendel, S. J., Moore, C. L., Suen, P. H., Weickhardt, A. F., ... Shoulders, M. D. (2018). An adaptable platform for directed evolution in human cells. *Journal of the American Chemical Society, 140*(51), 18093–18103. https://doi.org/10.1021/jacs.8b10937.

Berrow, N. S., Alderton, D., & Owens, R. J. (2009). The precise engineering of expression vectors using high-throughput in-fusion™ PCR cloning. In S. A. Doyle (Ed.). *High throughput protein expression and purification: Methods and protocols* (pp. 75–90) Humana Press. https://doi.org/10.1007/978-1-59745-196-3_5.

Bosley, A. D., & Ostermeier, M. (2005). Mathematical expressions useful in the construction, description and evaluation of protein libraries. *Biomolecular Engineering, 22*(1), 57–61. https://doi.org/10.1016/j.bioeng.2004.11.002.

Bush, K., Corsi, G. I., Yan, A. C., Haynes, K., Layzer, J. M., Zhou, J. H., ... Sullenger, B. A. (2023). Utilizing directed evolution to interrogate and optimize CRISPR/Cas guide RNA scaffolds. *Cell Chemical Biology, 30*(8), 879–892.e875. https://doi.org/10.1016/j.chembiol.2023.06.007.

Cavaleiro, A. M., Kim, S. H., Seppälä, S., Nielsen, M. T., & Nørholm, M. H. H. (2015). Accurate DNA assembly and genome engineering with optimized uracil excision cloning. *ACS Synthetic Biology, 4*(9), 1042–1046. https://doi.org/10.1021/acssynbio.5b00113.

Chen, L., Hong, M., Luan, C., Gao, H., Ru, G., Guo, X., ... Li, D. (2024). Adenine transversion editors enable precise, efficient A•T-to-C•G base editing in mammalian cells and embryos. *Nature Biotechnology, 42*(4), 638–650. https://doi.org/10.1038/s41587-023-01821-9.

Chen, P. J., & Liu, D. R. (2023). Prime editing for precise and highly versatile genome manipulation. *Nature Reviews. Genetics, 24*(3), 161–177. https://doi.org/10.1038/s41576-022-00541-1.

Coco, W. M., Levinson, W. E., Crist, M. J., Hektor, H. J., Darzins, A., Pienkos, P. T., ... Monticello, D. J. (2001). DNA shuffling method for generating highly recombined genes and evolved enzymes. *Nature Biotechnology, 19*(4), 354–359. https://doi.org/10.1038/86744.

Cong, L., Ran, F. A., Cox, D., Lin, S., Barretto, R., Habib, N., ... Zhang, F. (2013). Multiplex genome engineering using CRISPR/Cas systems. *Science (New York, N. Y.), 339*(6121), 819–823. https://doi.org/10.1126/science.1231143.

Davis, J. H., Rubin, A. J., & Sauer, R. T. (2011). Design, construction and characterization of a set of insulated bacterial promoters. *Nucleic Acids Research, 39*(3), 1131–1141. https://doi.org/10.1093/nar/gkq810.

Doman, J. L., Pandey, S., Neugebauer, M. E., An, M., Davis, J. R., Randolph, P. B., ... Liu, D. R. (2023). Phage-assisted evolution and protein engineering yield compact, efficient prime editors. *Cell, 186*(18), 3983–4002.e3926. https://doi.org/10.1016/j.cell.2023.07.039.

Doman, J. L., Raguram, A., Newby, G. A., & Liu, D. R. (2020). Evaluation and minimization of Cas9-independent off-target DNA editing by cytosine base editors. *Nature Biotechnology, 38*(5), 620–628. https://doi.org/10.1038/s41587-020-0414-6.

Doudna, J. A. (2020). The promise and challenge of therapeutic genome editing. *Nature, 578*(7794), 229–236. https://doi.org/10.1038/s41586-020-1978-5.

Engler, C., Kandzia, R., & Marillonnet, S. (2008). A one pot, one step, precision cloning method with high throughput capability. *PLoS One, 3*(11), e3647. https://doi.org/10.1371/journal.pone.0003647.

Esvelt, K. M., Carlson, J. C., & Liu, D. R. (2011). A system for the continuous directed evolution of biomolecules. *Nature, 472*(7344), 499–503. https://doi.org/10.1038/nature09929.

Friehs, K. (2004). Plasmid copy number and plasmid stability. In T. Scheper (Ed.). *New trends and developments in biochemical engineering* (pp. 47–82) Springer Berlin Heidelberg. https://doi.org/10.1007/b12440.

Gaudelli, N. M., Komor, A. C., Rees, H. A., Packer, M. S., Badran, A. H., Bryson, D. I., & Liu, D. R. (2017). Programmable base editing of A*T to G*C in genomic DNA without DNA cleavage. *Nature, 551*(7681), 464–471. https://doi.org/10.1038/nature24644.

Gaudelli, N. M., Lam, D. K., Rees, H. A., Solá-Esteves, N. M., Barrera, L. A., Born, D. A., ... Ciaramella, G. (2020). Directed evolution of adenine base editors with increased activity and therapeutic application. *Nature Biotechnology, 38*(7), 892–900. https://doi.org/10.1038/s41587-020-0491-6.

Gehrke, J. M., Cervantes, O., Clement, M. K., Wu, Y., Zeng, J., Bauer, D. E., ... Joung, J. K. (2018). An APOBEC3A-Cas9 base editor with minimized bystander and off-target activities. *Nature Biotechnology, 36*(10), 977–982. https://doi.org/10.1038/nbt.4199.

Gibson, D. G., Young, L., Chuang, R.-Y., Venter, J. C., Hutchison, C. A., & Smith, H. O. (2009). Enzymatic assembly of DNA molecules up to several hundred kilobases. *Nature Methods, 6*(5), 343–345. https://doi.org/10.1038/nmeth.1318.

Grünewald, J., Zhou, R., Garcia, S. P., Iyer, S., Lareau, C. A., Aryee, M. J., & Joung, J. K. (2019). Transcriptome-wide off-target RNA editing induced by CRISPR-guided DNA base editors. *Nature, 569*(7756), 433–437. https://doi.org/10.1038/s41586-019-1161-z.

Haapaniemi, E., Botla, S., Persson, J., Schmierer, B., & Taipale, J. (2018). CRISPR-Cas9 genome editing induces a p53-mediated DNA damage response. *Nature Medicine, 24*(7), 927–930. https://doi.org/10.1038/s41591-018-0049-z.

Hu, J. H., Miller, S. M., Geurts, M. H., Tang, W., Chen, L., Sun, N., ... Liu, D. R. (2018). Evolved Cas9 variants with broad PAM compatibility and high DNA specificity. *Nature, 556*(7699), 57–63. https://doi.org/10.1038/nature26155.

Huang, T. P., Heins, Z. J., Miller, S. M., Wong, B. G., Balivada, P. A., Wang, T., ... Liu, D. R. (2023). High-throughput continuous evolution of compact Cas9 variants targeting single-nucleotide-pyrimidine PAMs. *Nature Biotechnology, 41*(1), 96–107. https://doi.org/10.1038/s41587-022-01410-2.

Huang, T. P., Newby, G. A., & Liu, D. R. (2021). Precision genome editing using cytosine and adenine base editors in mammalian cells. *Nature Protocols, 16*(2), 1089–1128. https://doi.org/10.1038/s41596-020-00450-9.

Inui, M., Miyado, M., Igarashi, M., Tamano, M., Kubo, A., Yamashita, S., ... Takada, S. (2014). Rapid generation of mouse models with defined point mutations by the CRISPR/Cas9 system. *Scientific Reports, 4*(1), 5396. https://doi.org/10.1038/srep05396.

Jinek, M., Chylinski, K., Fonfara, I., Hauer, M., Doudna, J. A., & Charpentier, E. (2012). A programmable dual-RNA-guided DNA endonuclease in adaptive bacterial immunity. *Science (New York, N. Y.), 337*(6096), 816–821. https://doi.org/10.1126/science.1225829.

Jinek, M., East, A., Cheng, A., Lin, S., Ma, E., & Doudna, J. (2013). RNA-programmed genome editing in human cells. *Elife, 2*, e00471. https://doi.org/10.7554/eLife.00471.

Kikuchi, M., Ohnishi, K., & Harayama, S. (2000). An effective family shuffling method using single-stranded DNA. *Gene, 243*(1), 133–137. https://doi.org/10.1016/S0378-1119(99)00547-8.

Kim, Y. B., Komor, A. C., Levy, J. M., Packer, M. S., Zhao, K. T., & Liu, D. R. (2017). Increasing the genome-targeting scope and precision of base editing with engineered Cas9-cytidine deaminase fusions. *Nature Biotechnology, 35*(4), 371–376. https://doi.org/10.1038/nbt.3803.

Kleinstiver, B. P., Prew, M. S., Tsai, S. Q., Nguyen, N. T., Topkar, V. V., Zheng, Z., & Joung, J. K. (2015). Broadening the targeting range of Staphylococcus aureus CRISPR-Cas9 by modifying PAM recognition. *Nature Biotechnology, 33*(12), 1293–1298. https://doi.org/10.1038/nbt.3404.

Kleinstiver, B. P., Prew, M. S., Tsai, S. Q., Topkar, V. V., Nguyen, N. T., Zheng, Z., ... Joung, J. K. (2015). Engineered CRISPR-Cas9 nucleases with altered PAM specificities. *Nature, 523*(7561), 481–485. https://doi.org/10.1038/nature14592.

Koblan, L. W., Erdos, M. R., Wilson, C., Cabral, W. A., Levy, J. M., Xiong, Z.-M., ... Liu, D. R. (2021). In vivo base editing rescues Hutchinson–Gilford progeria syndrome in mice. *Nature, 589*(7843), 608–614. https://doi.org/10.1038/s41586-020-03086-7.

Komor, A. C., Kim, Y. B., Packer, M. S., Zuris, J. A., & Liu, D. R. (2016). Programmable editing of a target base in genomic DNA without double-stranded DNA cleavage. *Nature, 533*(7603), 420–424. https://doi.org/10.1038/nature17946.

Lam, D. K., Feliciano, P. R., Arif, A., Bohnuud, T., Fernandez, T. P., Gehrke, J. M., ... Gaudelli, N. M. (2023). Improved cytosine base editors generated from TadA variants. *Nature Biotechnology, 41*(5), 686–697. https://doi.org/10.1038/s41587-022-01611-9.

Lee, J. K., Jeong, E., Lee, J., Jung, M., Shin, E., Kim, Y.-h, ... Kim, J.-S. (2018). Directed evolution of CRISPR-Cas9 to increase its specificity. *Nature Communications, 9*(1), 3048. https://doi.org/10.1038/s41467-018-05477-x.

Liu, Z., Chen, S., Shan, H., Jia, Y., Chen, M., Song, Y., ... Li, Z. (2020). Precise base editing with CC context-specificity using engineered human APOBEC3G-nCas9 fusions. *BMC Biology, 18*(1), 111. https://doi.org/10.1186/s12915-020-00849-6.

Mali, P., Yang, L., Esvelt, K. M., Aach, J., Guell, M., DiCarlo, J. E., ... Church, G. M. (2013). RNA-guided human genome engineering via Cas9. *Science (New York, N. Y.), 339*(6121), 823–826. https://doi.org/10.1126/science.1232033.

McCullum, E. O., Williams, B. A. R., Zhang, J., & Chaput, J. C. (2010). Random mutagenesis by error-prone PCR. In J. Braman (Ed.). *In vitro mutagenesis protocols* (pp. 103–109)(3rd ed.). Humana Press. https://doi.org/10.1007/978-1-60761-652-8_7.

Miller, S. M., Wang, T., & Liu, D. R. (2020). Phage-assisted continuous and non-continuous evolution. *Nature Protocols, 15*(12), 4101–4127. https://doi.org/10.1038/s41596-020-00410-3.

Miller, S. M., Wang, T., Randolph, P. B., Arbab, M., Shen, M. W., Huang, T. P., ... Liu, D. R. (2020). Continuous evolution of SpCas9 variants compatible with non-G PAMs. *Nature Biotechnology*. https://doi.org/10.1038/s41587-020-0412-8.

Musunuru, K., Chadwick, A. C., Mizoguchi, T., Garcia, S. P., DeNizio, J. E., Reiss, C. W., ... Kathiresan, S. (2021). In vivo CRISPR base editing of PCSK9 durably lowers cholesterol in primates. *Nature, 593*(7859), 429–434. https://doi.org/10.1038/s41586-021-03534-y.

Nelson, M. R., Marnellos, G., Kammerer, S., Hoyal, C. R., Shi, M. M., Cantor, C. R., & Braun, A. (2004). Large-scale validation of single nucleotide polymorphisms in gene regions. *Genome Research, 14*(8), 1664–1668. https://doi.org/10.1101/gr.2421604.

Neugebauer, M. E., Hsu, A., Arbab, M., Krasnow, N. A., McElroy, A. N., Pandey, S., ... Liu, D. R. (2022). Evolution of an adenine base editor into a small, efficient cytosine base editor with low off-target activity. *Nature Biotechnology*. https://doi.org/10.1038/s41587-022-01533-6.

Newby, G. A., Yen, J. S., Woodard, K. J., Mayuranathan, T., Lazzarotto, C. R., Li, Y., ... Mayberry, K. (2021). Base editing of haematopoietic stem cells rescues sickle cell disease in mice. *Nature, 595*(7866), 295–302.

Nishimasu, H., Shi, X., Ishiguro, S., Gao, L., Hirano, S., Okazaki, S., ... Nureki, O. (2018). Engineered CRISPR-Cas9 nuclease with expanded targeting space. *Science (New York, N. Y.), 361*(6408), 1259–1262. https://doi.org/10.1126/science.aas9129.

O'Brien, A. R., Wilson, L. O. W., Burgio, G., & Bauer, D. C. (2019). Unlocking HDR-mediated nucleotide editing by identifying high-efficiency target sites using machine learning. *Scientific Reports, 9*(1), 2788. https://doi.org/10.1038/s41598-019-39142-0.

Paquet, D., Kwart, D., Chen, A., Sproul, A., Jacob, S., Teo, S., ... Tessier-Lavigne, M. (2016). Efficient introduction of specific homozygous and heterozygous mutations using CRISPR/Cas9. *Nature, 533*(7601), 125–129. https://doi.org/10.1038/nature17664.

Porto, E. M., & Komor, A. C. (2023). In the business of base editors: Evolution from bench to bedside. *PLoS Biology, 21*(4), e3002071. https://doi.org/10.1371/journal.pbio.3002071.

Porto, E. M., Komor, A. C., Slaymaker, I. M., & Yeo, G. W. (2020). Base editing: Advances and therapeutic opportunities. *Nature Reviews. Drug Discovery, 19*(12), 839–859. https://doi.org/10.1038/s41573-020-0084-6.

Ravikumar, A., Arzumanyan, G. A., Obadi, M. K. A., Javanpour, A. A., & Liu, C. C. (2018). Scalable, continuous evolution of genes at mutation rates above genomic error thresholds. *Cell, 175*(7), 1946–1957.e1913. https://doi.org/10.1016/j.cell.2018.10.021.

Rees, H. A., & Liu, D. R. (2018). Base editing: Precision chemistry on the genome and transcriptome of living cells. *Nature Reviews. Genetics, 19*(12), 770–788. https://doi.org/10.1038/s41576-018-0059-1.

Rees, H. A., Wilson, C., Doman, J. L., & Liu, D. R. (2019). Analysis and minimization of cellular RNA editing by DNA adenine base editors. *Science Advances, 5*(5), eaax5717. https://doi.org/10.1126/sciadv.aax5717.

Richter, M. F., Zhao, K. T., Eton, E., Lapinaite, A., Newby, G. A., Thuronyi, B. W., ... Liu, D. R. (2020). Phage-assisted evolution of an adenine base editor with improved Cas domain compatibility and activity. *Nature Biotechnology, 38*(7), 883–891. https://doi.org/10.1038/s41587-020-0453-z.

Ringquist, S., Shinedling, S., Barrick, D., Green, L., Binkley, J., Stormo, G. D., & Gold, L. (1992). Translation initiation in Escherichia coli: Sequences within the ribosome-binding site. *Molecular Microbiology, 6*(9), 1219–1229. https://doi.org/10.1111/j.1365-2958.1992.tb01561.x.

Ruta, G. V., Ciciani, M., Kheir, E., Gentile, M. D., Amistadi, S., Casini, A., & Cereseto, A. (2024). Eukaryotic-driven directed evolution of Cas9 nucleases. *Genome Biology, 25*(1), 79. https://doi.org/10.1186/s13059-024-03215-9.

Schmidheini, L., Mathis, N., Marquart, K. F., Rothgangl, T., Kissling, L., Böck, D., ... Schwank, G. (2024). Continuous directed evolution of a compact CjCas9 variant with broad PAM compatibility. *Nature Chemical Biology, 20*(3), 333–343. https://doi.org/10.1038/s41589-023-01427-x.

Spencer, J. M., & Zhang, X. (2017). Deep mutational scanning of S. pyogenes Cas9 reveals important functional domains. *Scientific Reports,* 7(1), 16836. https://doi.org/10.1038/s41598-017-17081-y.

Stemmer, W. P. (1994). DNA shuffling by random fragmentation and reassembly: in vitro recombination for molecular evolution. *Proceedings of the National Academy of Sciences, 91*(22), 10747–10751.

Thomas, S., Maynard, N. D., & Gill, J. (2015). DNA library construction using Gibson Assembly®. *Nature Methods, 12*(11), i–ii. https://doi.org/10.1038/nmeth.f.384.

Thuronyi, B. W., Koblan, L. W., Levy, J. M., Yeh, W. H., Zheng, C., Newby, G. A., ... Liu, D. R. (2019). Continuous evolution of base editors with expanded target compatibility and improved activity. *Nature Biotechnology, 37*(9), 1070–1079. https://doi.org/10.1038/s41587-019-0193-0.

Tsai, S. Q., Nguyen, N. T., Malagon-Lopez, J., Topkar, V. V., Aryee, M. J., & Joung, J. K. (2017). CIRCLE-seq: A highly sensitive in vitro screen for genome-wide CRISPR-Cas9 nuclease off-targets. *Nature Methods, 14*(6), 607–614. https://doi.org/10.1038/nmeth.4278.

Tsai, S. Q., Zheng, Z., Nguyen, N. T., Liebers, M., Topkar, V. V., Thapar, V., ... Joung, J. K. (2015). GUIDE-seq enables genome-wide profiling of off-target cleavage by CRISPR-Cas nucleases. *Nature Biotechnology, 33*(2), 187–197. https://doi.org/10.1038/nbt.3117.

Walton, R. T., Christie, K. A., Whittaker, M. N., & Kleinstiver, B. P. (2020). Unconstrained genome targeting with near-PAMless engineered CRISPR-Cas9 variants. *Science (New York, N. Y.), 368*(6488), 290–296. https://doi.org/10.1126/science.aba8853.

Wang, Y., Xue, P., Cao, M., Yu, T., Lane, S. T., & Zhao, H. (2021). Directed evolution: Methodologies and applications. *Chemical Reviews, 121*(20), 12384–12444. https://doi.org/10.1021/acs.chemrev.1c00260.

Weber, Y., Böck, D., Ivaşcu, A., Mathis, N., Rothgangl, T., Ioannidi, E. I., ... Schwank, G. (2024). Enhancing prime editor activity by directed protein evolution in yeast. *Nature Communications, 15*(1), 2092. https://doi.org/10.1038/s41467-024-46107-z.

Yu, Y., Leete, T. C., Born, D. A., Young, L., Barrera, L. A., Lee, S.-J., ... Gaudelli, N. M. (2020). Cytosine base editors with minimized unguided DNA and RNA off-target events and high on-target activity. *Nature Communications, 11*(1), 2052. https://doi.org/10.1038/s41467-020-15887-5.

Zhang, E., Neugebauer, M. E., Krasnow, N. A., & Liu, D. R. (2024). Phage-assisted evolution of highly active cytosine base editors with enhanced selectivity and minimal sequence context preference. *Nature Communications, 15*(1), 1697. https://doi.org/10.1038/s41467-024-45969-7.

Zhou, C., Sun, Y., Yan, R., Liu, Y., Zuo, E., Gu, C., ... Yang, H. (2019). Off-target RNA mutation induced by DNA base editing and its elimination by mutagenesis. *Nature, 571*(7764), 275–278. https://doi.org/10.1038/s41586-019-1314-0.

Zuo, E., Sun, Y., Wei, W., Yuan, T., Ying, W., Sun, H., ... Yang, H. (2019). Cytosine base editor generates substantial off-target single-nucleotide variants in mouse embryos. *Science (New York, N. Y.), 364*(6437), 289–292. https://doi.org/10.1126/science.aav9973.

Zuo, E., Sun, Y., Yuan, T., He, B., Zhou, C., Ying, W., ... Yang, H. (2020). A rationally engineered cytosine base editor retains high on-target activity while reducing both DNA and RNA off-target effects. *Nature Methods, 17*(6), 600–604. https://doi.org/10.1038/s41592-020-0832-x.

CHAPTER FOURTEEN

Genome editing with programmable base editors in human cells

Nicola R.B. Osgood[a,1], Natalie M. Zawalick[a,1], Courtney B. Sawyer[b], Quinn T. Cowan[a], Sifeng Gu[a], S.J. Mawson[a], Brodie L. Ranzau[a], Lehan Li[a], Melissa Gymrek[b,c], Alon Goren[b], and Alexis C. Komor[a,d,e,*]

[a]Department of Chemistry and Biochemistry, University of California San Diego, La Jolla, CA, United States
[b]Department of Medicine, University of California San Diego, La Jolla, CA, United States
[c]Department of Computer Science and Engineering, University of California San Diego, La Jolla, CA, United States
[d]Moores UCSD Cancer Center, University of California San Diego, La Jolla, CA, United States
[e]Sanford Stem Cell Institute, University of California San Diego, La Jolla, CA, United States
*Corresponding author. e-mail address: akomor@ucsd.edu

Contents

1. Key resources table 352
2. Introduction 355
3. Selecting a base editor 357
   3.1 Method 1 – selecting a base editor 361
4. Designing guide RNAs (gRNAs) 362
   4.1 Method 2a – designing a gRNA: PAM-first approach 363
   4.2 Examples (Fig. 3) 364
   4.3 Method 2b – designing a gRNA: edit-first approach 365
   4.4 Examples (Fig. 4) 366
   4.5 Method 3 – anneal and phosphorylate oligonucleotides 369
   4.6 Method 4 - golden gate cloning (Fig. 5B) 370
   4.7 Method 5 – transformation, inoculation, and sequencing 371
5. Transfecting mammalian cells and harvesting DNA 373
   5.1 Method 6a – forward transfection of adherent mammalian cells in 48-well plate 375
   5.2 Method 6b – reverse transfection of adherent mammalian cells in 48-well plate 377
   5.3 Method 6c – electroporation of suspension mammalian cells in a 12-well plate 379
   5.4 Method 6d – piggyBac™ BE transposition and gRNA transduction of human iPSCs in a 6-well plate 380
   5.5 Method 7 – lysing adherent mammalian cells following transfection in a 48-well plate 384

[1] These authors contributed equally to this work.

Methods in Enzymology, Volume 712
ISSN 0076-6879, https://doi.org/10.1016/bs.mie.2025.01.001

6. Next-generation sequencing to quantify base editing efficiency 386
6.1 Method 8 – next generation sequencing sample preparation 388
7. Enriching for cells with high BE expression or activity 397
7.1 Method 9a – FACS of adherent mammalian cells for base editing enrichment 399
7.2 Method 9b - Namocell® sorting for cell line generation with adherent mammalian cells using base editing 401
Acknowledgments 403
Competing interests 403
Appendix A. Supporting information 403
References 403

Abstract

Genome editing has garnered significant attention over the last decade, resulting in a massive expansion of the genome engineering toolbox. Base editors encompass a class of tools that enable installing single-nucleotide changes in genomic DNA without the use of double-strand breaks. With the ever-increasing development of new and/or improved base editor systems, it is easy to be overwhelmed by the abundance of options. Here, we provide clear guidance to facilitate the selection of a base editor and to design guide RNAs (gRNAs) to suit various needs. Additionally, we describe in detail how to generate gRNA plasmids, transfect various mammalian cell types, and evaluate editing efficiencies. Finally, we give alternative methods and troubleshooting tips for some common pitfalls encountered during base editing.

1. Key resources table

Resources will vary by protocol. This table provides a summary of key resources needed to conduct every protocol included within this document.

Reagent or Resource	Source	Identifier
Bacterial and Virus Strains		
Mix&Go *E. Coli* DH5α	Zymo	T3009
Biological Samples		
KOLF 2.1J iPSC cell line (Pantazis et al., 2022)	Pantazis et al., 2022	N/A
HEK293T cell line	ATCC	CRL–3216
K562-s cell line	ATCC	CRL–3343

Chemicals, Peptides, and Recombinant Proteins		
2xYT bacterial medium and agar plates		
Tryptone	Fisher Chemical	BP1421–500
Yeast extract	VWR	97063–370
Sodium chloride	Macron Fine Chemicals	7532–06
Carbenicillin	GoldBio	C–103–100
Agar	VWR	J637–1KG
Polybrene	MilliporeSigma™	TR-1003G
Sodium Dodecyl Sulfate (SDS), 10%	VWR	0227–500G
Agarose gel		
Agarose	VWR	97062–250
TAE Buffer		
Tris	Thermo Scientific™	17926
Acetic acid	Fisher Chemical	A38SI–212
EDTA	Thermo Scientific™	17892
Gibco™ Penicillin-Streptomycin (10,000 U/mL)	Life Technologies	15140122
Critical Commercial Assays/Reagents		
T4 ligase buffer	New England Biolabs	B0202S
T4 PNK	New England Biolabs	M0201S or M0201L
T4 ligase	New England Biolabs	M0202S or M0202L
BsmBI-v2 Restriction Enzyme	New England Biolabs	R0739S or R0739L
ZymoPURE™ Plasmid Miniprep	Zymo	11–550B
Lipofectamine™ 2000	Invitrogen™	11668019
Gibco™ Opti-MEM™	Thermo Scientific™	31985062
Gibco™ DMEM	Thermo Scientific™	10566024
Gibco™ Fetal Bovine Serum (FBS)	Thermo Scientific™	A5256701
Gibco™ Phosphate Buffered Saline (PBS)	Thermo Scientific™	10010049

Neon™ Transfection System, 10 μL Kit	Thermo Scientific™	MPK1096
Accumax	Innovative Cell Technologies	AM105
BioCoat® Matrigel® 6-well Plate	Corning	354671
ACCUTASE™	STEMCELL™ Technologies	07920
mTeSR™ Plus	STEMCELL™ Technologies	100–0276
Lipofectamine™ Stem	Invitrogen™	STEM00003
Proteinase K	New England Biolabs	P8107S
Phusion® High-Fidelity DNA Polymerase (Kit)	New England Biolabs	M0530S or M0530L
Phusion® High-Fidelity DNA Polymerase		
DMSO		
Phusion® HF Buffer Pack		
Phusion® GC Buffer Pack		
Deoxynucleotide (dNTP) Solution Mix	New England Biolabs	N0447S or N0447L
Monarch® Spin DNA Gel Extraction Kit	New England Biolabs	T1120S or T1120L
Oligonucleotides		
Forward and reverse oligonucleotides with corresponding spacer sequence (see Table 2)	IDT	https://www.idtdna.com/pages/products/custom-dna-rna/dna-oligos/custom-dna-oligos
Primers with NGS adapters (see Table 4)	IDT	
Illumina-specific barcode primers (see Supplementary Table 2)	IDT	
Plasmids		
See Supplementary Table 1 for BE and gRNA plasmids	Addgene	Multiple – see Supplementary Table 1
Super piggyBac™	System Biosciences	PB210PA–1
Lentivirus packaging		
pMD	Addgene	12259
psPAX2	Addgene	12260

Other		
HyPure™ Cell Culture Grade Water	Cytiva	SH30529.02
Neon™ Electroporation System	Thermo Scientific™	MPK5000
NanoDrop™ 2000c UV-Vis Spectrophotometer	Thermo Scientific™	E112352
S3e Cell Sorter	BioRad	145–1006
Namocell® Pala™ Cell Sorter and Single Cell Dispenser	Bio-Techne	NI006
Falcon™ Round-Bottom Polystyrene Test Tubes with 35 μm Cell Strainer Snap Cap	Corning	352235

2. Introduction

Since the mechanistic characterization of CRISPR-Cas9 systems as programmable nucleases in 2012, (Jinek et al., 2012) the field of genome editing has quickly expanded, with an abundance of new tools developed over the past decade. One such new class of tools is base editors (BEs), which take advantage of the programmability of CRISPR-Cas systems, and use deaminase enzymes to install single nucleotide changes in genomic DNA. Applications of BEs include, but are not limited to, disease modeling, evaluation of the impact of genetic variants, and the correction of disease-causing mutations.

Programmable DNA binding by Cas9 is facilitated by a short piece of RNA called a single guide RNA (sgRNA) or, more simply, guide RNA (gRNA). In the natural bacterial or archaeal CRISPR systems, the gRNA consists of two separate RNA molecules, the CRISPR RNA (crRNA) and the trans-activating CRISPR RNA (tracrRNA, pronounced "tracer" RNA). To facilitate a more streamlined use of the system, the 3′ end of the crRNA was joined to the 5′ end of the tracrRNA by a four-nucleotide linker sequence to produce the sgRNA construct that is used today (Jinek et al., 2012). The Cas9 protein first searches the genome for protospacer adjacent motifs (PAMs), which are short sequence motifs specific to the Cas9 homolog being used (Sternberg, Redding, Jinek, Greene, & Doudna, 2014). The most commonly used Cas9 homolog is the *Streptococcus pyogenes* Cas9 (Sp-Cas9), which natively recognizes an NGG (N = A, C, G, or T) PAM. Once the PAM sequence is located, the Cas9 protein unwinds the

DNA directly 5′ to the PAM (called the "protospacer", Fig. 1) and checks for sequence complementarity with the 20-nucleotide sequence at the 5′ end of the gRNA (called the "spacer", Fig. 1). While wild-type Cas9 is an endonuclease that cleaves the DNA backbone on both strands, BE systems are fusions between nucleic acid editing enzymes and a partially catalytically inactivated Cas9 (called nickase Cas9, or nCas9), which cleaves only the DNA strand that is base-paired with the gRNA (Fig. 1), which prompts the cell to replace this strand and use the opposite as a template.

There are two main classes of BEs: cytosine base editors (CBEs) and adenine base editors (ABEs) (Gaudelli et al., 2017; Komor, Kim, Packer, Zuris, & Liu, 2016). CBEs facilitate the conversion of C·G base pairs to T·A

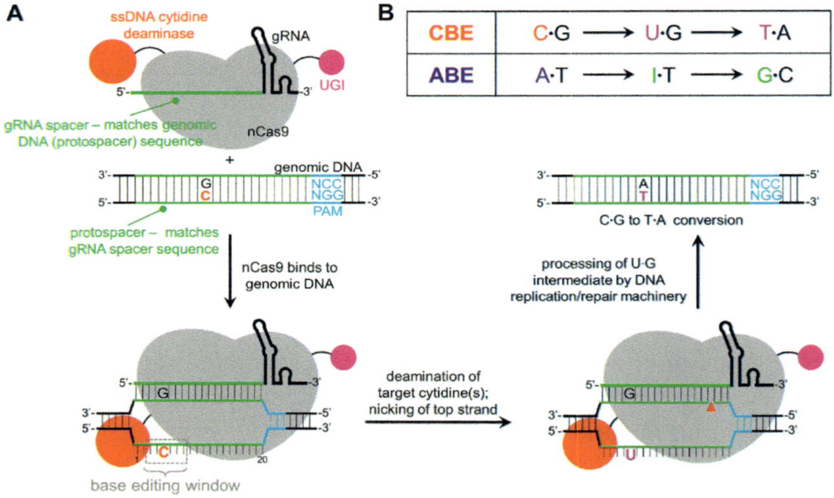

Fig. 1 Overview of base editing. (A) Binding of the BE to genomic DNA is facilitated by sequence complementarity between the gRNA spacer and the genomic DNA (the protospacer). A protospacer adjacent motif (PAM) must also be present next to the protospacer (in this case, the PAM is NGG). Once the BE is bound to the DNA, a stretch of nucleotides on the PAM-distal end of the protospacer (the "base editing window", shown in gray) is exposed to the ssDNA deaminase enzyme (a cytidine deaminase for CBEs, an adenosine deaminase for ABEs). The base editing window depends on the nCas9 homolog and the deaminase, see Fig. 2B for editing windows of specific BEs. The bases within the protospacer are numbered 1 (PAM-distal end) to 20 (PAM-proximal end) according to the illustration. A U·G intermediate (in the case of ABEs this is an I·T intermediate) is introduced into the genomic DNA following deamination of target nucleotides within the base editing window and nCas9-induced nicking of the unedited strand. This intermediate is processed by the cell to result in an overall C·G to T·A conversion (in the case of ABEs this is an A·T to G·C conversion). A uracil glycosylase inhibitor (UGI) component is also included with CBEs to increase editing efficiencies. (B) Overview of mutations facilitated by ABEs and CBEs, with intermediates shown.

(Fig. 1) and generally consist of nCas9, a cytidine deaminase enzyme, and a uracil DNA glycosylase inhibitor (UGI, Fig. 1). After nCas9 binds to the genomic DNA, the cytidine deaminase enzyme converts target cytidines to uracils (U), which are read as thymine by DNA replication and repair machinery (Fig. 1). The cytidine deaminase enzyme can only deaminate cytidines within accessible single-stranded DNA, thus the "base editing window" is located on the PAM-distal end of the DNA strand that lacks a base-pairing partner (Fig. 1A). It is important to note that the base editing window is generally a stretch of four to eight nucleotides (See Fig. 1A, bottom left panel). However, the enzyme may edit outside of this window in low frequencies. If multiple Cs are present within this window (the target C as well as neighboring "bystander" Cs), all may become edited concurrently in a process called "bystander editing". This will be important later on for judicious gRNA design. The UGI component of CBEs increases the editing efficiency by preventing the endogenous DNA repair protein uracil N-glycosylase (UNG) from excising the uracil intermediate. ABEs work analogously to CBEs but use an adenosine deaminase enzyme, and therefore convert A·T base pairs to G·C through an inosine (I) intermediate (Fig. 1). Furthermore, the use of ABEs does not require an equivalent UGI component due to inefficient inosine excision by the cell's DNA repair machinery. The use of nCas9 to nick the gRNA-paired, unmodified DNA strand promotes the cell to preferentially replace this strand and, in the process, use the U or I as a template, increasing overall base editing yields.

Whether using base editing to model a genetic disease, to study the consequences of genetic variants, or to develop a therapeutic, it is valuable to navigate through the suite of tools available and best practices when employing them. In this protocol, we discuss the process for selecting the appropriate BE and designing the most suitable gRNAs for a given target of interest. Additionally, we describe approaches for reliable generation and delivery of gRNAs (together with BEs) into various cell types. Finally, we provide guidelines for increasing base editing efficiencies and analyzing base editing outcomes.

3. Selecting a base editor

Selecting the appropriate BE for a given target is the first step in designing a base editing experiment. However, the choice may seem overwhelming given the repertoire of BEs; there are currently over 700 BE constructs on Addgene for use in mammalian cells. Here we will simplify the process of selecting a BE to enable even non-specialists to design their own experiments (Method 1). Construct maps of the basic architectures of most ABEs and CBEs are shown in Fig. 2A, and consist of two main components – the Cas protein and the deaminase. Many variations exist for each of these two components, resulting in a large number of combinations. We will discuss each component separately and distill all the apparent options down to 8 recommended BEs (4 CBEs and 4 ABEs). Note the two basic architectures that we recommend (BE4 for CBEs and ABE8 for ABEs) are comprised of an N-terminal deaminase fused to the Cas enzyme, with both N- and C-terminal nuclear localization signals (NLS). These basic architectures are among the most well-characterized, have optimal nuclear import, and have well-defined editing windows, described below. The BE4 architecture also employs two copies of UGI, which helps to increase editing efficiencies and minimize C·G to non-T·A editing outcomes when using CBEs, which can occur if intracellular UNG enzymes excise the U intermediate efficiently enough (Koblan et al., 2018). We also include a "P2A-GFP" motif in our constructs, in which GFP is

A

B

Sequence Preferences

Deaminase	Favors	Disfavors	Editing Window	
evoAPOBEC	B**C**	A**C**	3-8	CBEs
ancAPOBEC	T**C**	G**C**	3-9	
TadA8e	T**A**, C**A**	-	2-11	ABEs
TadA8.20	T**A**	-	2-9	

C
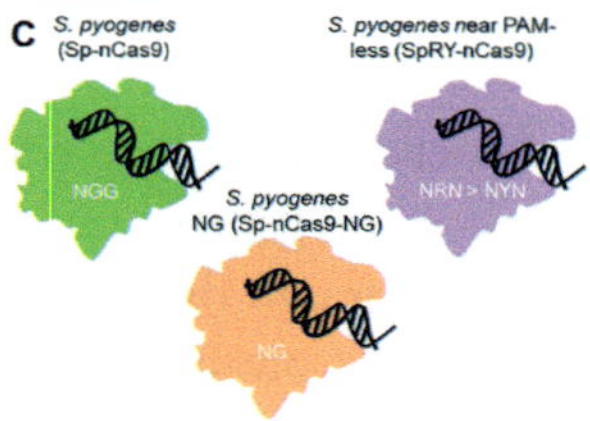

Fig. 2 Recommended Base Editors. (A) General base editor construct maps. "BE4" refers to the most optimized CBE architecture, while "ABE8" refers to the most efficient ABE constructs. The Cas component can be replaced with any nCas9 or dCas12 variant, and the deaminase can be replaced with a variety of deaminases. The P2A-GFP enables visualization of BE expression. (B) Recommended deaminase options with sequence motif preferences and editing windows when used in conjunction with a Sp-Cas9 variant (Arbab et al., 2020; Cowan et al., 2024; Gaudelli et al., 2020; Koblan et al., 2018; Richter et al., 2020; Thuronyi et al., 2019; Xiao, Wu, & Tang, 2024). The target nucleotide is bolded and underlined. B = C, T, or G. Highest editing occurs in the middle of the editing window. Position 1 is furthest from the PAM and position 20 is closest to the PAM (illustrated in Fig. 1). (C) Recommended nCas9 variants and their protospacer adjacent motifs (PAMs) (Walton, Christie, Whittaker, & Kleinstiver, 2020). PAMs are in white.

co-expressed with the BE via the self-cleaving P2A sequence. This allows for evaluation of transfection efficiencies and selection of transfected cells via cell sorting methods.

The choice of Cas protein for a BE will have a major impact on the size of the base editing window and the number of potential protospacers that can be designed (via the presence of appropriate PAMs). Cas proteins evolved from bacteria and archaea as natural defense systems, and a multitude of Cas9 and Cas12 homologs have been identified and characterized for genome editing applications. Cas12 enzymes function similarly to Cas9 enzymes. While certain Cas12 enzymes are slightly smaller than some Cas9 enzymes, they cannot be mutated to cleave the appropriate DNA strand in the context of base editing. All Cas12-derived BEs must utilize completely catalytically dead Cas12 (dCas12) which does not nick the DNA strand that is base-paired with the gRNA. We thus do not recommend the use of Cas12-derived BEs, as their efficiencies are much lower than nCas9-derived systems. Of the many nCas9 enzymes, the Sp-nCas9 homolog is the most commonly used nCas9 protein for base editing due to its PAM flexibility, high efficiency in mammalian cells, and precise base editing window. Furthermore, researchers have engineered the "PAM-relaxed" Sp-nCas9 variants Sp-nCas9-NG (recognizes NG PAMs) and SpRY-nCas9 (recognizes NR [R = G or A], and to a lesser extent, NY [Y = C or T] PAMs), (Nishimasu et al., 2018; Walton et al., 2020) which offer the greatest flexibility for gRNA design (Fig. 2C).

For the purposes of this protocol piece, we will exclusively recommend BEs derived from Sp-nCas9-NG and SpRY-nCas9, with the choice between the two depending on the protospacer sequence (detailed in the next section). The use of BEs derived from wild-type Sp-nCas9 is also recommended, but identifying an appropriate BE gRNA that is compatible with an NGG PAM is less likely, and thus we have not included plasmid constructs for these BEs with this work. Sp-nCas9-derived BEs will generally have a base editing window of positions 3–9 within the protospacer, in which the PAM-distal end is 1 and the base directly next to the PAM is 20 (Fig. 1), with slight variations depending on the choice of deaminase. This editing window offers, in our opinion, an optimal balance between the likelihood of an optimally-positioned PAM and minimizing potential bystander editing. However, nCas9 homologs from other species have been repurposed as BEs and do offer certain benefits. For example, BEs derived from the KKH variant of *Staphylococcus aureus* (Sa, NNNRRT PAM) nCas9 homolog (Kleinstiver et al., 2015) have a smaller size overall (which provides cloning and delivery benefits), but have a more restrictive PAM and much wider editing windows (typically, positions 3 to 14), increasing the likelihood of bystander editing. While this may be a drawback for certain applications (for instance, in cases where editing of only a single A or C is desired), in some scenarios a wider editing window is either necessary or preferable.

In addition to the continued discovery of various Cas9 enzymes, there has been extensive work to discover, engineer, or evolve deaminases for use in base editing. The choice of deaminase for a BE will have a major impact on editing efficiency for a given target sequence (as most deaminase enzymes have certain motif preferences and aversions) and the size of the base editing window. The adenosine deaminases used in ABEs were evolved from an *E. coli* tRNA modifying enzyme, as no naturally occurring ssDNA adenosine deaminase enzymes exist. As such, the deaminase options for ABEs are limited and we recommend either the TadA8e deaminase (which will have an editing window of positions 2 through 11 when used with Sp-nCas9) or the TadA8.20 deaminase (which will have a narrower editing window of positions 2 through 9 when used with Sp-nCas9) (Cowan et al., 2024; Gaudelli et al., 2020; Richter et al., 2020). While neither of these two evolved TadA enzymes has a particular sequence motif aversions, both prefer As that are present within TA sequence motifs. Further, TadA8e is slightly more efficient at editing As within CA motifs when compared to TadA8.20 (Fig. 2B) (Xiao et al., 2024). Cytidine deaminase enzymes are a naturally occurring family of proteins, and as such, there are many more potential options for deaminases in CBEs than ABEs. We have found the most ubiquitous cytidine deaminases for use in base editing to be the "ancestral APOBEC" variant (ancAPOBEC, which will have an editing window of positions 3 through 9 when used with Sp-nCas9) and the "evolved APOBEC" (evoAPOBEC), which will have an editing window of positions 3 through 8 when used with Sp-nCas9 (Koblan et al., 2018; Thuronyi et al., 2019). evoAPOBEC has no dominant motif preference, but has a slight aversion to Cs within AC motifs (Arbab et al., 2020), while ancAPOBEC prefers Cs within TC motifs and has an aversion to editing Cs within GC contexts (Thuronyi et al., 2019). A brief summary of editing windows and sequence preferences is shown in Fig. 2B, and the eight BEs that we recommend are summarized in Table 1.

As mentioned previously, when designing a base editing experiment, potential bystanders must always be identified and considered. Bystanders are A or C bases next to (generally, within 5 base pairs of) the target A or C

Table 1 Plasmid information for recommended BEs. All plasmids have been deposited on Addgene.

Type of BE	Construct name	Deaminase	nCas9
CBE	evoBE4-NG	evoAPOBEC	Sp-nCas9-NG
	evoBE4-SpRY	evoAPPOBEC	SpRY-nCas9
	ancBE4-NG	ancAPOBEC	Sp-nCas9-NG
	ancBE4-SpRY	ancAPOBEC	SpRY-nCas9
ABE	ABE8e-NG	TadA8e	Sp-nCas9-NG
	ABE8e-SpRY	TadA8e	SpRY-nCas9
	ABE8.20-NG	TadA8.20	Sp-nCas9-NG
	ABE8.20-SpRY	TadA8.20	SpRY-nCas9

base, which may get edited concurrently (or instead of) the desired base. This is because certain deaminases (such as TadA8e) are particularly processive and will deaminate multiple bases at a time. It is important to identify all potential bystander bases near the target editing site, recognize if such bystander editing is problematic or not (e.g., certain silent mutations may be acceptable), and note their sequence contexts as compared to the intended target base.

3.1 Method 1 – selecting a base editor

1. Identify which nucleotide conversion is desired: C•G to T•A or A•T to G•C
   a. C•G to T•A: use a CBE

      Note: If the mutation is listed as a G to A mutation on the coding strand, this is equivalent to a C to T mutation on the template strand, and a CBE can be used.

   b. A•T to G•C: use an ABE

      Note: If the mutation is listed as a T to C mutation on the coding strand, this is equivalent to a A to G mutation on the template strand, and an ABE can be used.
2. Note the sequence motif of the nucleotide to be edited. If this motif is one of the "disliked" motifs of a given deaminase, eliminate that deaminase as an option.
3. Identify potential bystander edits, their sequence motifs, and whether or not it is acceptable if they are edited.

4. If possible, identify a deaminase that prefers the motif of the target C or A and has an aversion to the motif of potential bystanders. If this is not possible, simply choose a deaminase that prefers or does not have an aversion to the motif of the target C or A.
5. Move to Method 2 (designing gRNAs) with both BEs (NG and SpRY) derived from the deaminase(s) of choice. The selection between NG and SpRY will depend on the protospacer(s).

Fig. 2 covers a minimal number of BEs, but includes those that are the most universal and thus will simplify the process of designing an initial base editing experiment, while maximizing the chances of successful genome editing. These eight BEs have been deposited to Addgene (Table 1, **Supplementary Table 1**). For a more comprehensive list of BEs, we recommend the 2020 review by Anzalone, Koblan, and Liu (Anzalone, Koblan, & Liu, 2020). New BEs are regularly in development and additional literature may be beneficial to determine whether another editor has been developed for specific needs not addressed by the more commonly used BEs.

4. Designing guide RNAs (gRNAs)

Designing a suitable gRNA for a given target is the next step. The gRNA consists of a constant 80-nt 3′ end, with a variable 20-nt 5′ end termed the spacer. The spacer must be designed based on the desired edit of interest, and is identical to the protospacer sequence within the DNA. While some tools exist to aid in spacer design, most were not developed specifically for BE gRNA design and thus we describe here processes for identifying and designing BE gRNA spacers manually, using either a "PAM-first" approach (Method 2a, Fig. 3) or an "edit-first" approach (Method 2b, Fig. 4). In the "PAM-first" approach, the identification of efficient PAMs (such as NG and NA) is prioritized, while in the "edit-first" approach, optimal positioning of the target C or A within the protospacer (positions 5, 6, and 7) is prioritized. These two methods may therefore identify slightly different gRNAs (as demonstrated below). While manually designing gRNAs for a manageable number of targets is practical, if a large library of gRNAs is desired, we recommend the "Base editor design tool" code written by Mudra Hegde and Ruth Hanna (https://github.com/mhegde/base-editor-design-tool) (Hanna et al., 2021).

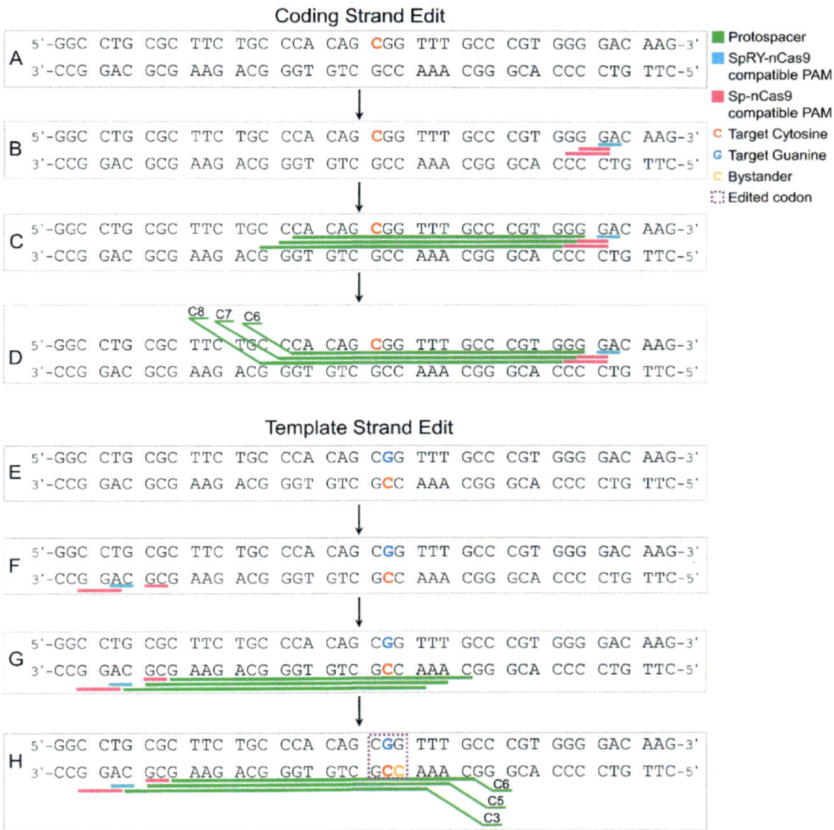

Fig. 3 PAM-first approach for gRNA design (Method 2a).

4.1 Method 2a – designing a gRNA: PAM-first approach

1. Identify the strand on which the target C or A occurs (Fig. 3A and E).
2. Using a molecular biology visualization tool (e.g., Benchling), annotate the target C or A. Be sure to annotate the appropriate DNA strand (Fig. 3A and E).
3. Scan the bases that are 9 to 18 bases away from the target C or A on the 3′ side and locate and annotate all PAMs (NGG, NG, or NA motifs are preferred, but NC and NT can be used with SpRY-BEs, Fig. 3B and F).
4. For each identified PAM, annotate the 20 bases on the 5′ side of the PAM as a potential protospacer (Fig. 3C and G).

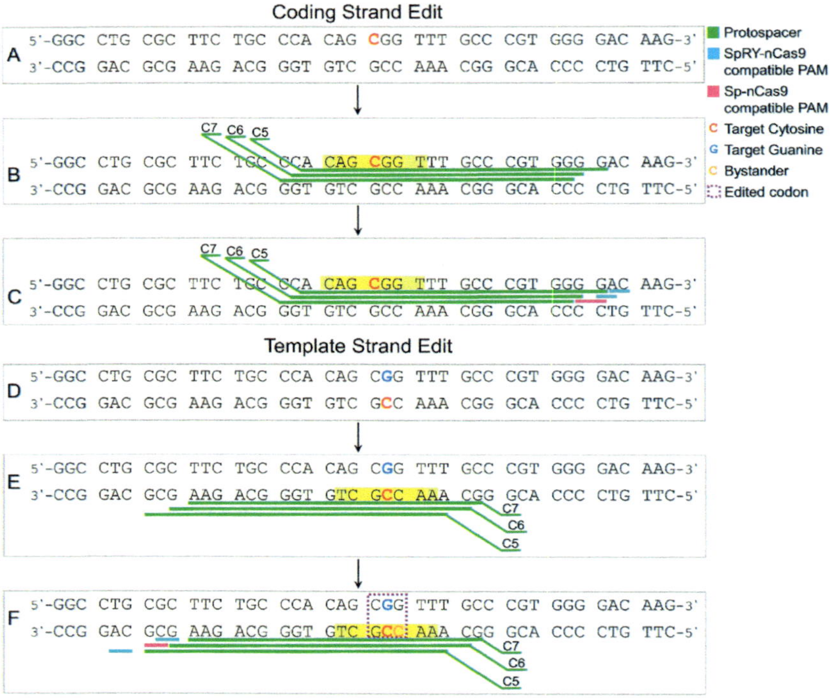

Fig. 4 Edit-first approach for gRNA design (Method 2b).

5. For each potential protospacer, determine at which position the target C or A is, and whether this is within the base editing window for the BE(s) that was chosen. Discard the protospacers in which the target C or A is outside of the editing window (Fig. 3D and H).
6. Pair each protospacer with the appropriate BE based on its PAM sequence (i.e., use either an NG- or SpRY-BE for any with NG PAMs, and use a SpRY-BE for any with NA PAMs) and the deaminase(s) that was identified previously (Fig. 3D and H).

4.2 Examples (Fig. 3)

See Fig. 3 to follow along with these two examples. In Fig. 3A, the target nucleotide for a C to T edit on the coding strand is shown in red. Note that the motif of the target C (G<u>C</u>) is a disliked motif of the ancAPOBEC deaminase. Thus, only evoAPOBEC-derived CBEs will be considered for this target. In Step 3 (Fig. 3B), PAMs are identified on the appropriate strand, 3′ of the target nucleotide. Preferred PAMs are underlined here (NGG, NG, and NA), with PAMs corresponding to Sp-nCas9 (NGG) and Sp-nCas9-NG (NG) underlined pink, and PAMs corresponding to SpRY-nCas9 (NA) underlined blue. Next, per Step 4, 20-nt protospacers for each PAM are shown in green (Fig. 3C). Finally, the position of the target nucleotide within each protospacer is identified (Step 5). As can be seen in Fig. 3D, the target nucleotide lies in position 8, 7, or 6 of the identified protospacers, which are all compatible with evoAPOBEC-derived CBEs (Fig. 2B). The C8 protospacer would be paired with evoBE4 (which uses wild-type nCas9) or evoBE4-NG, the C7 protospacer would be paired with evoBE4-NG, and the C6 protospacer would be paired with evoBE4-SpRY.

In Fig. 3E, the target nucleotide for a G to A edit on the coding strand is shown in blue. The C on the template strand (shown in red) that is base paired with the G is the actual target of the CBE that will yield the desired G•C to A•T conversion. Note that the motif of the target C (C**C**) is NOT a disliked motif of either CBE deaminases. Thus, both ancAPOBEC and evoAPOBEC-derived CBEs will be considered for this target. In Step 3 (Fig. 3F), PAMs are identified on the appropriate strand, 3′ of the target nucleotide. Preferred PAMs are underlined here (NGG, NG, and NA), with PAMs corresponding to Sp-nCas9 (NGG) and Sp-nCas9-NG (NG) underlined pink, and PAMs corresponding to SpRY-nCas9 (NA) underlined blue. Next, per Step 4, 20-nt protospacers for each PAM are shown in green (Fig. 3G). Finally, the position of the target nucleotide within each protospacer is identified (Step 5). As can be seen in Fig. 3H, the target nucleotide is in position 3, 5, or 6 of the identified protospacers, which are all compatible with both ancAPOBEC and evoAPOBEC-derived CBEs (Fig. 2B). In this example, there is a likely bystander edit, shown in orange, immediately 5′ of the target cytosine. However, this will result in a silent mutation, as both CAA and CAG code for Gln, which is the desired amino acid change in this example (the codon is indicated by a purple dotted outline). Furthermore, the motif of this bystander C (A**C**) is a disliked motif of the evoAPOBEC deaminase, and it is outside of the editing window in the C3 protospacer. It is therefore recommended to use evoAPOBEC-NG with the C6 protospacer, evoAPOBEC-SpRY with the C5 protospacer, and either evoAPOBEC-NG or ancAPOBEC-NG with the C3 protospacer.

4.3 Method 2b – designing a gRNA: edit-first approach

1. Identify the strand on which the target C or A occurs (Fig. 4A and D).
2. Using a molecular biology visualization tool (e.g., Benchling), annotate the target C or A. Be sure to annotate the appropriate DNA strand (Fig. 4A and D).

3. Design an initial protospacer in which the target C or A is at position 6: annotate the five bases 5′ of the target C or A, the target C or A, and the 14 bases 3′ of the target C or A (Fig. 4B and E).
4. Identify the PAM for this protospacer. This will determine which BE must be used. Be aware of less favorable PAMs, such as NC, which may work with SpRY-nCas9 but are not optimal (Fig. 4C and F).
5. Repeat steps 3 and 4, annotating additional protospacer sequences that place the target C or A within the editing window of potential BEs by shifting the initial protospacer one nucleotide to the 5′ or 3′ side at a time.
6. Pair each protospacer with the appropriate BE based on its PAM sequence (i.e. use either an NG- or SpRY-BE for any with NG PAMs, and use a SpRY-BE for any with NA PAMs) and the deaminase(s) that was identified previously. If no NG or NA PAMs exist that place the edit in a favorable position, SpRY BE may still be successful.

4.4 Examples (Fig. 4)

See Fig. 4 to follow along with these two examples. In Fig. 4A, the target nucleotide for a C to T edit on the coding strand is shown in red. Note that the motif of the target C (G<u>C</u>) is a disliked motif of the ancAPOBEC deaminase. Thus, only evoAPOBEC-derived CBEs will be considered for this target. In Step 3 (Fig. 4B), protospacers that place the target nucleotide in positions 5, 6, or 7 of the editing window are identified. These are underlined in green. This step could be repeated to find protospacers with the target nucleotide in additional positions within the editing window, determined by the deaminase being used (Fig. 2B), if needed. Next, per Step 4, PAMs for the spacers are identified (Fig. 4C). PAMs compatible with SpRY-nCas9 are underlined in blue, while those compatible with Sp-nCas9-NG are in pink. The PAM for the C5 protospacer is NC, which is a less favorable option as the SpRY-nCas9 variant is less efficient at binding to these targets. The C7 protospacer would be paired with evoBE4-NG, and the C6 and C5 protospacers would be paired with evoBE4-SpRY.

In Fig. 4D, the target nucleotide for a G to A edit on the coding strand is shown in blue. The C on the template strand (shown in red) that is base paired with the G is the actual target of the CBE that will yield the desired G•C to A•T conversion. Note that the motif of the target C (C<u>C</u>) is NOT a disliked motif of either CBE deaminases. Thus, both ancAPOBEC and evoAPOBEC-derived CBEs will be considered for this target. In Step 3 (Fig. 4B), protospacers that place the target nucleotide in positions 5, 6, or 7 of the editing window are identified. These are underlined in green. This step could be repeated to find protospacers with the target nucleotide in additional positions within the editing window, determined by the deaminase being used (Fig. 2E), if needed. Next, per Step 4, PAMs for the spacers are identified (Fig. 4F). PAMs compatible with SpRY-nCas9 are underlined in blue, while those compatible with S-npCas9-NG are in pink. The PAM for the C7 protospacer is NC, which is a less favorable option as the SpRY-nCas9 variant is less efficient at binding to these targets. As can be seen in Fig. 4F, there is a likely bystander edit, shown in orange, that will occur to the cytosine immediately 5′ of the target cytosine. However, this will result in a silent mutation, as both CAA and CAG code for Gln, which is the desired amino acid change in this example (this codon is indicated by a purple dotted outline). Furthermore, the motif of this bystander C (A**C**) is a disliked motif of the evoAPOBEC deaminase. It is therefore recommended to use evoAPOBEC-NG with the C6 protospacer, and evoAPOBEC-SpRY with the C5 and C7 protospacers.

After a potential protospacer is selected, it is important to make several assessments. First, confirm that the target nucleotide is on the appropriate DNA strand (the target C or A should be found within the protospacer sequence, and not its reverse complement), and is within the editing window of the BE that it is being paired with (Fig. 2B). Next, identify if there are nearby bystanders within the editing window that, if edited, would be unacceptable (e.g., non-silent mutations, splice site mutations). Lastly, if there are unacceptable bystanders, consider using an alternative protospacer that positions the bystander bases in a less accessible position. For each target base, there may be multiple BE:gRNA options. We recommend trying all options. If there are none or only a single potential protospacer, repeat the process and look for NC (preferred) or NT PAMs, to use with a SpRY-BE variant.

Once the protospacers are selected, the corresponding spacer sequence should be cloned into a plasmid backbone with the rest of the gRNA backbone. A variety of gRNA-expressing plasmids are available from Addgene, including pU6-GG-gRNA (detailed in Fig. 5A), which we recommend. pU6-GG-gRNA contains a dual Type IIS restriction enzyme (BsmBI) site between the U6 promoter and the gRNA backbone, also called a Golden Gate (GG) site. This enables custom spacer sequences to be easily ligated at this site (Fig. 5B). Type IIS restriction enzymes cleave outside of their recognition sequence, allowing for scarless restriction enzyme cloning known as Golden Gate cloning (Engler, Kandzia, & Marillonnet, 2008). To facilitate assembly into the digested backbone,

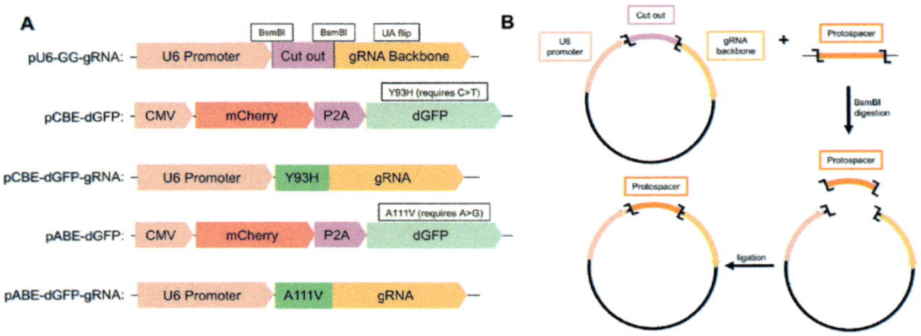

Fig. 5 Construct maps of plasmids recommended for gRNA cloning (top) and for enriching for cells with high BE activity (bottom four) are shown in (A). pU6-GG-gRNA is recommended for gRNA cloning. It contains a dual BsmBI site for restriction enzyme digestion and ligation of custom spacer sequences, as shown in (B). It also contains a "UA flip" in the gRNA backbone which reduces polymerase stalling during transcription. pCBE-dGFP and pABE-dGFP are plasmids with non-functional GFP (dGFP) which are converted to functional GFP when a critical mutation (Y93H or A11V) is successfully base edited back to wild type. Matching gRNA expressing plasmids that target the appropriate BE to the deactivating mutation (pCBE-dGFP-gRNA and pABE-dGFP-gRNA, respectively) are also shown.

custom overhangs must be appended on either side of the spacer sequence. Note that the U6 promoter requires a 5′G to initiate transcription. If your spacer sequence naturally starts with a G, this is sufficient for intracellular gRNA transcription. If your spacer does not, an extra 5′G must be added. For pU6-GG-gRNA, the forward and reverse oligonucleotides in Table 2 must be ordered for each custom spacer.

The U6 promoter utilizes RNA polymerase III (RNAPIII) for expression of high levels of the gRNA. RNAPIII recognizes poly-T stretches of 4–6 thymines as termination signals (Artsimovitch & Belogurov, 2015), which may cause issues with certain spacers. If one or more of the designed spacer sequences contains a poly-T stretch, it may cause issues with expression in cells. To overcome this, we recommend the use of synthetic gRNAs, which can be purchased commercially (e.g., from IDT, Synthego, or Thermo Fisher Scientific). These can be transfected into cells alongside BE-encoding mRNA (optimally) or a BE-encoding plasmid. Notably, the natural Sp-gRNA backbone sequence has a poly-T stretch immediately 3′ of the spacer sequence. Therefore, to improve intracellular gRNA transcription, the gRNA backbone sequence in pU6-GG-gRNA incorporates two single-nucleotide changes known as the "UA flip" (Chen et al., 2016).

Table 2 pU6-GG-gRNA oligonucleotide sequences. The adapters for the overhangs of the Golden Gate cloning are given and must be included when ordering the oligonucleotides for the spacer insert.

	Oligonucleotide	Sequence
If spacer sequence starts with G	Forward	CACC[spacer sequence]
	Reverse	AAAC[reverse complement of spacer sequence]
If spacer sequence starts with A, C, or T	Forward	CACCG[spacer sequence]
	Reverse	AAAC[reverse complement of spacer sequence]C

4.5 Method 3 – anneal and phosphorylate oligonucleotides

4.5.1 Materials and equipment

- Thermocycler
- Micropipettes
- Ice bucket
- 0.2 mL PCR tubes
- Forward and reverse oligonucleotides (25 nmol scale, standard desalting purification specifications from IDT are recommended)
- Sterile water
- T4 ligase buffer
- T4 PNK

4.5.2 Protocol

1. Resuspend any new, lyophilized oligonucleotides to a working concentration of 100 μM in sterile water.
2. On ice, mix the following, per reaction (i.e., per protospacer):

Component/Reagent	Volume (μL)
Sterile water	16.5
10X T4 ligase buffer*	2
T4 PNK	0.5
Forward oligo (100 μM)	0.5
Reverse oligo (100 μM)	0.5
Total Volume	**20**

**T4 ligase buffer is used instead of T4 PNK buffer as it includes ATP, which is necessary for the reaction to proceed.*

3. Using a thermocycler, incubate the reaction for 30 min at 37 °C, followed by 95 °C for 1 min. Ramp the temperature down to either 12 °C or 4 °C by 0.1 °C per second.
4. This product can be stored at 4 °C for up to two weeks or at –20 °C for up to a year.

4.6 Method 4 - golden gate cloning (Fig. 5B)

4.6.1 Materials and equipment

- Thermocycler or heat blocks set for 42 °C and 60 °C
- Micropipettes
- Ice bucket
- 0.2 mL PCR tubes or 1.5 mL microcentrifuge tubes
- Annealed and phosphorylated oligonucleotides from Method 3
- Sterile water
- T4 ligase buffer
- T4 ligase
- Backbone plasmid with Golden Gate site (pU6-GG-gRNA recommended)
- Compatible Type IIS restriction enzyme (BsmBI if using pU6-GG-gRNA)

4.6.2 Protocol

1. On ice, mix the following, per reaction (i.e., per protospacer):

Component/Reagent	Volume (μL)
Sterile water	to a total volume of 10
10X T4 ligase buffer	1
T4 ligase	0.5
BsmBI-v2 (or other Type IIS restriction enzyme)	0.5
Backbone	0.01 pmol
Insert (from Method 3)	0.05 pmol
Total Volume	**10**

2. Incubate the reaction for 30 min at 42 °C.

   Note: If there are issues with efficiency, the reaction can be cycled on a thermocycler as follows: 42 °C for 5 min, 16 °C for 5 min, 4–6 cycles.

3. Transfer the reaction to 60 °C for 10 min.
4. This product can be stored between 4 °C and −20 °C overnight, depending on the restriction enzyme used. For best results proceed to the transformation step immediately.

4.7 Method 5 – transformation, inoculation, and sequencing

4.7.1 Materials and equipment

- Micropipettes
- Ice bucket
- 0.2 mL PCR tubes or 1.5 mL microcentrifuge tubes
- Heat block
- Bacterial loops or spreaders
- Bacterial incubator set to 37 °C
- Bacterial incubator/shaker set to 37 °C
- Competent bacterial cells (E. Coli DH5α are recommended for most plasmids)
- Bacterial medium (LB or 2XYT) with appropriate antibiotics (ampicillin or carbenicillin for pU6-GG-gRNA)
- Agar plate(s) with appropriate growth medium (LB or 2XYT) and antibiotics (ampicillin or carbenicillin for pU6-GG-gRNA)
- Miniprep kit (e.g., ZymoPURE™ Plasmid Miniprep)

4.7.2 Protocol

This protocol uses Zymo's Mix&Go! competent *E. Coli* DH5α cells. A heat shock step might be necessary if other types of competent cells are used.

1. Thaw cells on ice.
2. Add 5 μL of Method 4 product to 50 μL cells and gently flick the tube to mix.
   Notes:
   i. *If the transformation efficiency is too high, less Golden Gate product can be added. However, it is not recommended to add less than 1 μL of product.*
   ii. *Instead of reducing the volume added, the product can be diluted prior to transformation.*
3. Incubate the cells on ice for 5 min if using Zymo Mix&Go! cells.
   Note: If using competent cells other than Zymo Mix&Go!, increase incubation time to 10–20 min.
4. If necessary, heat shock cells for 75 s at 55 °C, and incubate on ice for another 2 min after.

5. Spread the cells on a pre-warmed bacterial plate with 100 μg/mL carbenicillin or ampicillin.

   Notes:

   i. *The type of antibiotic used may vary depending on the backbone. pU6-GG-gRNA contains ampicillin resistance.*
   ii. *Recovery may be necessary if using an antibiotic other than carbenicillin or ampicillin.*
   iii. *If your transformation efficiency is too high, dilute the transformed bacteria prior to plating or plate a smaller volume.*
6. Place in 37 °C incubator for 12–20 h.
7. Upon successful cloning and transformation, select two or more colonies and inoculate each in 5 mL media with appropriate antibiotics.
8. Place cultures in a shaker at 37 °C for 12–16 h.
9. Isolate the plasmids from the cells using a miniprep kit or equivalent.
10. Send an aliquot of the isolated plasmid for sequencing to ensure the cloning was successful.

    Notes:

    i. *If using a backbone for the first time or that has been in storage for a while, whole plasmid sequencing is recommended to ensure there are no unexpected mutations.*
    ii. *If using a backbone that has been recently used and sequence-verified, only the protospacer region needs to be sequence-verified. Design a primer that is more than 100 bp away from the beginning of the protospacer to use for Sanger sequencing. The suggested Sanger sequencing primer for pU6-GG-gRNA is 5′-TAC GTG ACG TAG AAA GTA AT-3′.*
    iii. *If needing to sequence more than 700 bp* via *Sanger sequencing, generate tiled primers or a reverse primer to ensure robust sequence data of the area that you desire.*

It is important to ensure that any plasmid that will be used with a mammalian cell line is endotoxin-free. Most plasmid isolation kits come with some method of removing endotoxins from the isolation. It is imperative that this step is performed. If endotoxins are present, transfection efficiencies will be severely reduced and decreased cell viability is highly likely. Additionally, any results from assays performed with plasmids containing endotoxins are unreliable.

While this method is quite efficient for generating gRNA plasmids, there are some common pitfalls to be aware of. First, if the backbone:insert ratio is not optimal, both the transformation efficiency and likelihood of obtaining colonies with the desired plasmid sequence will suffer. The insert should be more plentiful than the backbone. In this case, the insert is the annealed and phosphorylated oligonucleotides from Method 3, and a backbone:insert ratio of 1:5 (molar ratio) has been reliably successful. Additional decreases in transformation efficiency may be seen if the competent bacterial cells are mishandled. Thaw competent cells on ice and do not introduce bubbles or vortex the cells. Another variable that may impact transformation efficiency is the amount of DNA introduced to the competent cells. Sometimes, adding a smaller amount of ligation product may yield greater transformation efficiencies. If efficiency is poor, vary the amount of ligation mixture that is transformed.

5. Transfecting mammalian cells and harvesting DNA

To effectively perform genome editing, the gRNA and BE selected and generated in the previous sections must be delivered into the cells of interest. The most common method for this process is using a cationic lipid-based reagent, such as Lipofectamine™, to perform a transfection. These cationic lipids can be used to perform both forward and reverse transfections (described below), and can effectively deliver plasmid DNA as well as other nucleic acids such as siRNA, mRNA, synthetic gRNA, and even purified Cas9 protein if it is first complexed with a gRNA.

We describe in detail an optimal method for the transfection of plasmids encoding BE and gRNA into HEK293T cells in a 48-well plate format using Lipofectamine™ 2000 (L2000). However, if working with other cell types, both the amount of DNA used and the ratio of L2000 to DNA should be optimized. Additionally, even if using HEK293T cells, the amount of DNA used per well does not scale linearly based on surface area and should be optimized when scaling up or down. Finally, different cationic lipid reagents should be evaluated if working with other cell types, or if delivering nucleic acids other than plasmid DNA (such as synthetic gRNA and mRNA).

An additional consideration for transfections is cell density. Ideally, cells should be experiencing logarithmic growth when transfected. This typically is when cells are approximately 80 % confluency. However, some cell types may be much more or much less dense than this to achieve optimal transfection efficiency. Optimization of the cell density upon transfection is highly recommended for new cell lines. When optimizing, we recommend that densities both well above and well below 80 % confluency are tested, as some cell types may appear more confluent than they truly are due to different morphologies and growth patterns.

Here, we present two methods for transfecting cells: both forward (Method 6a) and reverse (Method 6b) transfection. A forward transfection is when cells are plated first, and a reverse transfection is when the reagents are plated first and a cell suspension is added on top. Forward transfections are gentler on sensitive cell types, as the process of attaching to the plate while taking up plasmid (as in reverse transfections) can cause toxicity. This can typically be compensated for by plating a larger number of cells. Reverse transfections generally afford higher transfection efficiencies with adherent cells, as there is theoretically a greater amount of cell surface area with which the transfection reagents come into contact with. Additionally, they require one less day of laboratory work (as the cells are plated on the same day they are transfected), and often require less DNA input. As with all deviations from a standard protocol, we recommend optimization of the conditions for a reverse transfection. However, plating twice the density of cells required for an optimized forward transfection is a decent starting point for a reverse transfection.

Cells should not be transfected freshly from cryopreservation. Cells should be passaged a minimum of two times prior to transfection to ensure they have successfully recovered from cryopreservation. While cells may be transfected sooner, transfection efficiency and cell viability will decrease significantly.

Some cell types may be challenging to transfect utilizing cationic lipids. This may be due to a particularly delicate cell type, or extreme resistance to transfection via cationic lipids. In this case, alternate methods exist to deliver nucleic acid payloads into cells. Electroporation is the next most common method, which uses a transient electric pulse to briefly open the cell membrane, allowing cargo to pass through. Here, we present a method for electroporating K562 suspension cells (Method 6c). Finally, certain sensitive cell types such as stem cells exhibit extreme toxicity when exposed to plasmid DNA. Further, long-term expression of BEs can also be toxic in these cell types. We therefore offer a method for genomically encoding a small molecule-inducible BE construct into induced pluripotent stem cells (iPSCs) via piggyBac™ transposition and delivering the gRNA via lentiviral transduction, to facilitate base editing in stem cells (Method 6d).

5.1 Method 6a – forward transfection of adherent mammalian cells in 48-well plate

5.1.1 Materials and equipment

- 1.5 mL microcentrifuge tubes

- 0.2 mL PCR tubes
- Tissue culture treated plate(s)
- 15 mL conical tubes
- 50 mL conical tubes
- Micropipettes
- Centrifuge
- Tissue culture incubator set to optimal conditions for cell type
- Microscope with fluorescence imaging capabilities
- Lipofectamine™ 2000
- Opti-MEM™
- Antibiotic-free medium
- PBS
- Plasmid DNA encoding gRNA
- Plasmid DNA encoding BE

5.1.2 Protocol

5.1.2.1 Day one

1. Plate cells in antibiotic-free medium such that they will be experiencing logarithmic growth upon transfection (around 80 % confluency for many cell types).
2. Ensure wells are included for a no-color control, and color control(s). A color control should be present for each potential fluorescent marker used. For example, if only using a plasmid with GFP, include one well for a no-color control and one well for a GFP+ control. If using GFP alongside another marker such as mCherry, three control wells are necessary: one no-color, one GFP+ , and one mCherry+.

   Notes:

   i. *Some cell types and their optimal plating densities can be found in* Table 3.

   ii. *It is highly recommended to use antibiotic-free medium at this point, as antibiotics in the medium will cause excessive cell death upon transfection.*

5.1.2.2 Day two

1. Bring Opti-MEM™ or similar serum-reduced transfection medium to room temperature.
2. For each transfection condition, there will be two tubes - one will contain DNA and the other will contain the L2000.
   a. Tube 1 – combine the appropriate volume of gRNA and BE plasmids. Dilute with Opti-MEM™ (or equivalent) to 12.5 μL total volume. For transfecting HEK293T cells in a 48-well plate, we recommend using 750 ng of BE plasmid with 250 ng of gRNA plasmid.

Table 3 Recommended seeding densities for a 48-well plate. Additional optimization may be necessary if cells appear to be growing at atypical rates. Cells must be experiencing logarithmic growth when transfected. These values provide starting densities for potential further optimization by the user.

Cell type	**Seeding density (cells per well of 48-well plate)**	
	Forward transfection	**Reverse transfection**
HEK293T	50,000	100,000
HEK293A	35,000	70,000
RPE1	55,000	110,000
AC16	40,000	80,000
HeLa	30,000	60,000

b. Tube 2 – combine 1.5 μL L2000 and 11 μL Opti-MEM™ (or equivalent).

3. Mix the contents of Tube 1 and Tube 2 for a total volume of 25 μL per well. Incubate for 15 min at room temperature.

Notes:

i. *These dilutions and mixes can be done in either 0.2 mL PCR tubes or 1.5 mL microcentrifuge tubes, depending on the final volume needed.*

ii. *If transfecting multiple wells, make a master mix with extra volume to account for error.*

4. After incubation, add the DNA/L2000 mix dropwise to the well of cells.

5. Return the plate to the incubator.

5.1.2.3 Day three

1. Observe the cells using a fluorescence microscope. If the BE used has a fluorescent marker, fluorescence will be visible if the transfection was successful (Fig. 6A–C).

2. Some cationic lipid transfection reagents can be toxic to some cell types. If cells are more sensitive or delicate, medium should be changed at this time.

a. Carefully aspirate medium and gently add 150 μL PBS.

b. Carefully aspirate PBS and add 250 μL fresh, prewarmed medium. *Note: Medium can contain antibiotics at this point.*

3. Return the plate to the incubator.

5.1.2.4 Day five

1. Harvest the cells via either bulk lysis (see **Method 7**) or flow cytometry (see Enriching Cells section).

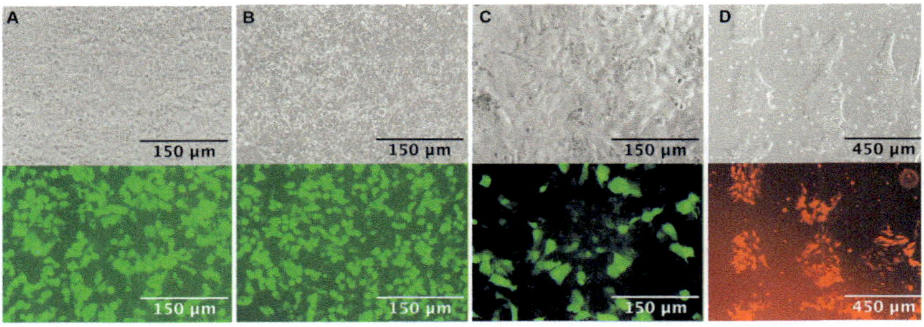

Fig. 6 Fluorescence imaging of cells after a successful transfection. (A–B) HEK293T cells were transfected using the forward (A) and reverse (B) methods with a gRNA and a base editor with a GFP marker, and images were taken after 48 h of incubation. (C) AC16 cells were transfected using the forward method with a gRNA, base editor, and dGFP turn-on plasmid. Images were taken after 48 h of incubation. (D) hiPSCs with a genomically integrated base editor (using the piggyBac transposition method, in which the BE is co-expressed with mCherry) and transduced gRNA were treated with doxycycline for 48 h to induced base editor expression.

5.2 Method 6b – reverse transfection of adherent mammalian cells in 48-well plate

5.2.1 Materials and equipment

- 1.5 mL microcentrifuge tubes
- 0.2 mL PCR tubes
- Tissue culture treated plate(s)
- 15 mL conical tubes
- 50 mL conical tubes
- Micropipettes
- Centrifuge
- Tissue culture incubator set to optimal conditions for cell type
- Microscope with fluorescence imaging capabilities
- Lipofectamine™ 2000
- Opti-MEM™
- Antibiotic-free medium
- PBS
- Plasmid DNA encoding gRNA
- Plasmid DNA encoding BE

5.2.2 Protocol

5.2.2.1 Day one

1. Bring Opti-MEM™ or similar serum-reduced transfection medium to room temperature.

2. Detach cells from the culture vessel. Spin the cells down and aspirate the medium. Resuspend in a minimal amount of prewarmed, antibiotic-free medium.

Note: For a fully confluent T25 flask, 1 mL of medium is sufficient.

3. While detaching cells, generate the transfection mixtures. Ensure wells are included for a no-color control, and color control(s). See Method 6a, Day One, Step 2 for additional details.

4. For each transfection condition, there will be two tubes – one will contain DNA and the other will contain the L2000.

a. Tube 1 – combine the appropriate volume of gRNA and BE plasmids. Dilute with Opti-MEM™ (or equivalent) to 12.5 µL total volume. For transfecting HEK293T cells in a 48-well plate, we recommend using 750 ng of BE plasmid with 250 ng of gRNA plasmid.

b. Tube 2 – combine 1.5 µL L2000 and 11 µL Opti-MEM™ (or equivalent).

5. Mix the contents of Tube 1 and Tube 2 for a total volume of 25 µL per well. Incubate for 15 min at room temperature.

Notes:

i. *These dilutions and mixes can be done in either 0.2 mL PCR tubes or 1.5 mL microcentrifuge tubes, depending on the final volume needed.*

ii. *If transfecting multiple wells, make a master mix with extra volume to account for error.*

6. While the DNA/L2000 mix incubates, dilute the cells such that the number of cells per well is double what they would be for a forward transfection. Note that 250 µL of cell mixture will be added to each well.

Note: See Table 3 *for some cell types and their optimal plating densities.*

7. After the DNA/L2000 mix has finished incubating, add it dropwise to the empty plate well.

8. Gently add 250 µL of cells to each well.

9. Return the plate to the incubator.

5.2.2.2 Day two

1. Proceed per **Method 6a**, *Day Three.*

5.2.2.3 Day four

1. Proceed per **Method 6a**, *Day Five.*

5.3 Method 6c – electroporation of suspension mammalian cells in a 12-well plate

5.3.1 Materials and equipment

- Electroporator (e.g., Neon™)
- 1.5 mL microcentrifuge tubes
- 0.2 mL PCR tubes
- Tissue culture treated plate(s)
- 15 mL conical tubes
- 50 mL conical tubes
- Micropipettes
- Centrifuge
- Tissue culture incubator set to optimal conditions for cell type
- Microscope with fluorescence imaging capabilities
- Antibiotic-free medium
- Plasmid DNA encoding gRNA
- Plasmid DNA encoding BE
- PBS
- Neon™ Transfection System 10 μL Kit

5.3.2 Protocol

This protocol uses the Neon™ Electroporation System.

5.3.2.1 Day one

1. Warm appropriate antibiotic-free medium and PBS to 37 °C.
2. Aliquot 1 mL of media into each well of a 12-well plate. Place the plate in the incubator.
3. Prepare the Neon™ tube by inserting it into the pipette station. Add 3 mL of Buffer E to the Neon™ tube.
4. Add the DNA mixture (250 ng gRNA plasmid and 750 ng BE plasmid) to a 1.5 mL centrifuge tube.
5. Count cells and transfer the desired number of cells per transfection into a 1.5 mL centrifuge tube. For K562-s cells, 200,000 cells per well for a 12-well plate format is sufficient.
6. Centrifuge the cells to generate a cell pellet. For K562-s cells, centrifuge at 200 xg for 8 min. Aspirate the media and add 1 mL of pre-warmed PBS to wash the cells.
7. Centrifuge the cells as before and aspirate the PBS.
8. Resuspend the cells in Buffer R so that there are 10,000 cells per microliter. That is, for every 200,000 cells, add 20 μL.

9. Transfer the 20 μL of cells into the previously prepared 1.5 mL centrifuge tube with the DNA mixture and gently pipette to mix.
10. Load the 10 μL Neon™ tip into the Neon™ pipette.
11. Slowly aspirate the cell/DNA mixture, being careful to avoid generating bubbles, as bubbles will prohibit adequate electroporation.
12. Submerge the pipette tip in the Buffer E within the Neon™ tube and dock the pipette. A click sound indicates appropriate docking.
13. Run the protocol on the instrument. For K562-s cells, use 1700 V for 20 ms and 1 pulse.
14. Transfer the cells to the pre-warmed plate. Repeat steps 11–13 using the same tip for the remaining cell-payload mixture.
15. Repeat for as many wells as needed, then return the plate to the incubator.

5.3.2.2 Day two

1. Observe the cells using a fluorescence microscope. If the BE used has a fluorescent marker, fluorescence will be visible if the transfection was successful.

5.3.2.3 Day four

1. Harvest the cells via either bulk lysis or flow cytometry (see Enriching Cells section).

In some cases, it may be necessary to split or passage cells after transfection. This is advised when cells are too confluent or if the incubation period will be longer than 48 h. This may also be useful if two identical populations of transfected cells are desired for different assays.

5.4 Method 6d – piggyBac™ BE transposition and gRNA transduction of human iPSCs in a 6-well plate

5.4.1 Materials and equipment

- 1.5 mL microcentrifuge tubes
- 0.2 mL PCR tubes
- Tissue culture treated plate(s), coated with Matrigel®
- 15 mL conical tubes
- 50 mL conical tubes
- Micropipettes
- Centrifuge
- Tissue culture incubator set to optimal conditions for cell type
- Human induced pluripotent stem cell (hiPSC) line (this protocol is optimized for the KOLF 2.1 J line)

- piggyBac™ cargo plasmid with BE and fluorescent marker (see Supplementary Table 1; the constructs that we provide use mCherry as a marker for BE expression)
- piggyBac™ transposase plasmid (this protocol uses Super piggyBac™: PB210PA-1, System Biosciences)
- Lentivirus containing gRNA with fluorescent marker (see Supplementary Table 1 and the protocol available at https://primer3.ut.ee/20)
- PBS
- ACCUTASE™
- mTeSR™ Plus
- ROCK inhibitor
- Lipofectamine™ Stem
- Opti-MEM™
- Polybrene

5.4.2 Protocol – transposition

5.4.2.1 Day one

1. Aspirate medium from iPSC culture. Wash once with 1 mL PBS.
2. Add 1 mL of ACCUTASE™ and incubate at 37 °C for 10–12 min, until cells appear in single or 2-cell clumps.
3. Add 1 mL medium and mix well to break up any remaining clumps.
4. Measure the cell concentration using a hemocytometer or equivalent.
5. Centrifuge cells for 5 min at 100 × g to pellet them.
6. Aspirate media and resuspend cells in mTeSR™ Plus with 5–10 μM ROCK inhibitor to a final concentration of 250,000 cells/mL.
7. Add 2 mL of the cell mixture to one well of a 6-well plate, to transfect 500,000 cells.

   Note: the number of cells transfected may be varied and can be optimized by the user as necessary.
8. In 1.5 mL microcentrifuge tubes, prepare the DNA mixture to add to the cells. Include the following conditions:
   a. Negative control: transposase only (600 ng)
   b. Experimental: cargo plasmid (1800 ng) + transposase (600 ng)

      Note: a 3:1 ratio of cargo to piggyBac™ transposase is recommended. However, some optimization may be necessary and desired.
9. Mix the DNA mixture with OptiMEM™ to total volume of 100 μL.
10. Make a Lipofectamine™ Stem (LipoStem) master mix by adding 5 μL to 95 μL OptiMEM™ for a total volume of 100 μL per sample.

11. For each sample, add 100 μL of the LipoStem mixture to the 100 μL DNA mixture and vortex. Incubate the LipoStem/DNA mixture for 10 min at room temperature.
12. Add the 200 μL of the LipoStem/DNA mixture to the cells in the 6-well plate. Swirl or tap to ensure even distribution within the medium.
13. Incubate at 37 °C overnight.

5.4.2.2 Day two

1. Replace with fresh mTeSR™ Plus medium.

5.4.2.3 Day three

1. Once cells are fully confluent, replace medium with mTeSR™ Plus medium containing 50 μg/mL hygromycin. Continue replacing the medium with fresh medium containing hygromycin every other day until all negative control cells are dead.
2. If cells are not yet fully confluent, continue replacing the medium every other day with fresh medium without antibiotics until cells are confluent. Then, switch to medium with hygromycin and proceed as described previously.

5.4.2.4 Day ten to fourteen

1. Once all negative control cells are dead, bank cells by expanding cell culture up to two 10 cm dishes and cryopreserving 20–25 vials of cells. *Note: Cell banking is required prior to doxycycline induction of the BE. After induction commences, we do not recommend ceasing induction and re-using the cells, as silencing of the BE locus often occurs.*

5.4.2.5 After cell banking

1. Perform a doxycycline titration to determine the optimal concentration of doxycycline for BE induction. We recommend testing concentrations ranging from 100 ng/mL to 2000 ng/mL. These concentrations are the final concentrations in the well, so a single well of a 6-well plate at the lowest concentration would have 200 ng doxycycline. See Fig. 6D for an example of successful BE expression induction.

5.4.3 Protocol – transduction

5.4.3.1 Prior to commencing transduction

1. Generate lentivirus containing the desired gRNA construct using preferred lentivirus generation protocol.

2. Freeze lentivirus in single-use aliquots, as freeze-thaw cycles reduce lentivirus efficacy.
3. Perform a function titration experiment to determine virus titer required for desired multiplicity of infection (MOI).

   Notes:

   i. *To do this, transduce variable amounts of virus, keeping cell count the same. This experiment will yield a range of MOIs and inform you of how much virus to add to a certain number of cells to achieve a specific MOI. These values scale linearly.*

   ii. *For example, if adding 2 μL virus to 500,000 cells results in 50 % of the cells expressing GFP, this is an MOI of 0.5. To achieve the same MOI in 2000,000 cells, add 8 μL of virus.*

   iii. *If a single integrant per cell is desired, it is recommended to strive for a MOI of 0.2–0.4.*

5.4.3.2 Day one

1. Aspirate medium from iPSC culture. Wash once with 1 mL PBS.
2. Add 1 mL of ACCUTASE™ and incubate at 37 °C for 10–12 min, until cells appear in single or 2-cell clumps.
3. Add 1 mL medium and mix well to break up any remaining clumps.
4. Measure the cell concentration using a hemocytometer or equivalent.
5. Centrifuge cells for 5 min at 100 g to pellet them.
6. Aspirate media and resuspend cells in mTeSR™ Plus with 5–10 μM ROCK inhibitor and 8-10 μg/mL polybrene to a final concentration of 250,000 cells/mL.
7. Add 2 mL of the cell mixture to one well of a 6-well plate, to transduce 500,000 cells.

   Note: the number of cells transduced may be varied and can be optimized by the user as necessary.

8. Add the amount of virus necessary to reach the intended multiplicity of infection (MOI).

   Notes:

   i. *First, measure the titer of your virus through a functional titration experiment, as described previously.*

   ii. *Every time you freeze and thaw lentivirus, it loses efficacy. It is recommended to freeze lentivirus in single-use aliquots.*

9. Incubate at 37 °C overnight.

5.4.3.3 Day two

1. Replace with fresh mTeSR™ Plus medium.

5.4.3.4 Day three

1. Induce base editor expression with doxycycline at the concentration determined by the previous doxycycline titration experiment (see Fig. 6D for an example).
2. Maintain doxycycline on the cells for at least five days.
   Note: fluorescent marker expression peaks on day six or seven.

5.4.3.5 Day four or more

1. Once cells are confluent, expand or split cells as desired.
   Note: ensure you maintain doxycycline during the expansion or splitting process if prior to day eight.
2. Bank cells for potential future use.

5.4.3.6 Day six or seven

1. Confirm successful induction and integration using a fluorescent microscope (Fig. 6) flow cytometry.
   Note: this assumes both the BE and gRNA are tagged with fluorescent proteins. It is highly recommended that fluorescent proteins are used as the selection markers. If lentiviral selection markers instead confer antibiotic resistance (e.g., *puromycin), modify protocol accordingly.*
2. Confirm successful base editing using **Methods 7–9**.

5.5 Method 7 – lysing adherent mammalian cells following transfection in a 48-well plate

5.5.1 Materials and equipment

- Thermocycler
- 1.5 mL microcentrifuge tubes
- 0.2 mL PCR tubes
- Micropipettes
- Microscope with fluorescence imaging capabilities
- PBS
- Sterile water
- 1 M Tris
- 10 % SDS
- Proteinase K

5.5.2 Protocol

1. 48–72 h post-transfection, observe the cells using a fluorescence microscope. If the BE used has a fluorescent marker, fluorescence will be visible in successful transfections. Note the confluency of each well.

2. Prepare an appropriate amount of fresh lysis buffer:
   a. Scale the amount of lysis buffer to the confluency of each well. For a 100 % confluent well, use 100 μL of lysis buffer, scale linearly down (i.e. 50 % confluent well requires 50 μL lysis buffer).

      Note: If between two percentages, round down. It is better to add less and be able to add additional later, rather than add too much lysis buffer to begin with.
   b. Prepare fresh lysis buffer the day of lysis and keep at 4 °C until use. For 200 μL of buffer, use the below numbers and scale linearly as needed.

Component/Reagent	Volume (μL)
Sterile water	196.75
1 M Tris	2
10% SDS	1
Proteinase K	0.25

   Note: it is not feasible nor recommended to pipette the volume of Proteinase K needed for a final total volume less than 200 μL.
3. Carefully aspirate growth medium and gently add 150 μL PBS.
4. Carefully aspirate PBS, ensuring all liquid has been removed from the well.
5. Add the appropriate amount of lysis buffer to each well.
6. Let cells incubate with lysis buffer for 1–5 min at room temperature, then resuspend thoroughly. Be sure to scrape the tip of the pipette across the bottom of the well to remove all cells. Lysis material should be viscous.

   Note: If lysis is not viscous, either too much lysis buffer was added, or the PBS wash was too aggressive and removed a significant portion of the cells.
7. Transfer resuspension into 0.2 mL PCR tubes.
8. Using a thermocycler, incubate the resuspension at 37 °C for 1 h and 80 °C for 30 min. Ramp the temperature down to either 12 °C or 4 °C by 0.1 °C per second.
9. Store lysis at 4 °C for up to 1 year, but ideally no more than 6 months.

6. Next-generation sequencing to quantify base editing efficiency

All genome editing experiments require a method for quantifying editing efficiency. Next Generation Sequencing (NGS), also known as High Throughput Sequencing (HTS), provides information on different sequence outcomes (alleles), along with quantitative editing efficiencies for each allele (the lower limit of detection is roughly 0.1 %). This is useful for sites with low editing percentages, protospacers with multiple editable bases (i.e. bystanders, in which some alleles may have only one base edited, while others may have more than one edited, see Fig. 7C), or for isogenic cell line characterization at sites that have multiple copies of the gene of interest. Here, we provide guidance for the preparation of samples for a 300-cycle paired-end NGS sequencing run (a 2 ×150 cycle run specifically) with amplicons of 200- to 250-bp in length (Fig. 7A). However, this protocol can be adjusted for different sequencing configurations. When quantifying editing efficiencies from "bulk samples" (i.e. those that are not clonally expanded and thus a heterogeneous mixture of genotypes), we do not recommend using Sanger sequencing as the relative peak heights of Sanger sequencing data at mixed-base sites (such as SNVs) are not necessarily a quantitative representative of the proportion of each base (the incorporation rate of ddNTP terminators is highly dependent on template concentrations and the sequence context surrounding the SNV location) (Carr et al., 2009). For isogenic cell lines (which are clonal and thus a homogeneous genotype), Sanger sequencing may be appropriate for

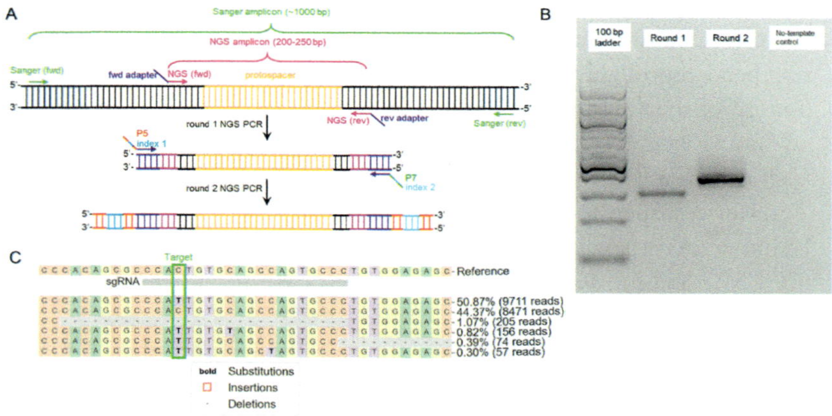

Fig. 7 Sequencing preparation and analysis examples. (A) Example schematic for designing amplicons and primers for Sanger and NGS analysis (top), and work flow of the two rounds of PCR for NGS preparation (bottom). (B) Agarose gel showing the round 1 and 2 PCR products with an increase in size between the rounds due to the Illumina adapters added in round 2 amplification. (C) CRISPResso2 analysis of allele-specific editing data. The target edited sequence is shown to have the highest prevalence, while the wildtype is the second most commonly identified allele. Other alleles, such as large deletions and bystander edits are undesired outcomes.

genotyping. However, if a particular clone is heterozygous, and the cell line is known to have an unusual or unstable karyotype, NGS is recommended for genotyping.

To quantify base editing efficiencies with NGS, the genomic locus surrounding the protospacer is amplified from the lysis obtained from **Method 7**, **Method 9a**, or **Method 9b**. First, design primers to amplify the protospacer plus additional sequence on either side to produce an amplicon of 200- to 250-bp with Illumina adapter sequences on either side (round 1 PCR, Fig. 7A and B). 200- to 250-bp provides optimal clustering, but a maximum amplicon size of 285-bp is possible. Specifically, we recommend designing the amplicon such that the forward and reverse reads have at least 10-bp of overlap between them in the middle of the amplicon, as this overlap is required for some commonly used software that merge paired-end reads. The annealing regions of these primers should be designed using a primer design software (e.g., Benchling's Primer3), (Untergasser et al., 2012) as PCR of genomic DNA lysate can be quite challenging. Round 1 NGS primer adapters can be found in Table 4. After this initial amplification, Illumina-specific barcodes (index 1 and 2 sequences) and P5/P7 sequences are added (round 2 PCR) to enable pooling of multiple different samples in the same NGS run (Fig. 7A). These allow for de-multiplexing of the samples after the NGS run, separating the different samples into separate FASTQ files. Note that if performing an NGS run that differs from a 2 × 150 cycle run, the amplicon size will need to correlate with the sequencing run. For example, if running a 2 × 75 cycle run, your amplicon should be under 135-bp (to ensure at least 10-bp of overlap between the paired-end reads). After the NGS run is completed, the data can be processed using CRISPResso2 (Clement et al., 2019) software or custom NGS analysis codes.

Table 4 Round 1 primer adapter sequences. These sequences must be added to the 5′ end of the annealing region of the primers designed for amplification. Illumina-specific round 2 primers can be found in Supplementary Table 2.

Primer	Sequence
NGS fwd primer	ACACTCTTTCCCTACACGACGCTCTTCCGATCTNNNN [fwd annealing region]
NGS rev primer	TGGAGTTCAGACGTGTGCTCTTCCGATCT [rev annealing region]

6.1 Method 8 – next generation sequencing sample preparation

6.1.1 Materials and equipment

- Thermocycler
- UV-Vis spectrophotometer (e.g., NanoDrop™)
- Micropipettes
- Ice bucket
- Razor blade
- 0.2 mL PCR tubes
- Sterile water
- HF buffer
- GC buffer
- dNTPs
- Phusion®
- DMSO
- Cell lysis
- Primers with NGS adapters
- Illumina-specific barcode primers
- 2 % agarose gel
- Gel extraction kit (e.g., Monarch® DNA Gel Extraction Kit)

6.1.2 Protocol

6.1.2.1 Round 1 preparation

1. Make a 20 μM mix of the found 1 forward and reverse NGS primers.
2. Mix per reaction in a 0.2 mL PCR tube:

Component/Reagent	Volume (μL)
Sterile water	8.17
GC green buffer	2.50
dNTPs	0.25
DMSO (3%)	0.375
Primer mix (20uM)	0.08
Phusion®	0.125
Cell lysate (DNA)	1.00
Total Volume	**12.5**

Notes:

i. *Be sure to add the Phusion® or equivalent polymerase last and keep the reaction mixture on ice after the addition of the polymerase.*

ii. *This can be scaled up to make a master mix where all components except the cell lysate are added, and the individual cell lysates are added after the master mix is distributed.*

1. Add the sample(s) to a thermocycler and cycle per the protocol below:

Cycles	Time	Temperature	Purpose
1 cycle	1 min	98 °C	initial denaturation
22–28 cycles	10 s	98 °C	denaturation
	20 s	65 °C	annealing
	10 s	72 °C	extension
1 cycle	5 min	72 °C	final extension
1 cycle	forever	4–12 °C	ramp to hold

Note: it is important to run the fewest possible number of cycles to avoid PCR bias. This may require optimization when working with a new set of fwd/rev NGS primers.

1. Run 1–3 μL of each sample on a 2 % agarose gel using a 100-bp ladder. It is recommended to run at 115–125 V for 15–25 min to resolve any potential primer dimer bands.

Note: A clear, single band that is 66 bp longer than the amplicon should be observed. This added length accounts for the NGS adapter sequences. Refer to the troubleshooting sub-section if multiple bands are observed, no bands are observed, or the primer dimer band density outweighs the on-target band.

6.1.2.2 Round 2 preparation

1. Make a 1 μM mix of the round 2 forward and reverse Illumina adapter primers. Each sample should have its own unique fwd/rev combination.

2. Mix per reaction in a 0.2 mL PCR tube:

Component/Reagent	Volume (μL)
Sterile water	6.34
HF green buffer	2.40
dNTPs	0.24
Primer mix (1uM)	2.40
Phusion®	0.12
Round 1 PCR product	0.50
Total Volume	**12**

Notes:

i. *Primers can be added individually at 1.2 µL per sample. It is recommended to pool either the forward or reverse primer (if one is held constant across all samples) in the master mix to make pipetting simpler and reduce error.*

ii. *Be sure to add the Phusion® or equivalent polymerase last and keep the reaction mixture on ice after the addition of the polymerase.*

iii. *This can be scaled up to make a master mix where all components with the exception of the Round 1 PCR product and one of the primers (if one is held constant across all samples) are added, with these components being added after the master mix is distributed.*

1. Add the sample(s) to a thermocycler and cycler per the protocol below:

Cycles	Time	Temperature	Purpose
1 cycle	1 min	98 °C	Initial denaturation
8–16 cycles	10 s	98 °C	Denaturation
	20 s	65 °C	Annealing
	10 s	72 °C	Extension
1 cycle	5 min	72 °C	Final extension
1 cycle	forever	4–12 °C	Ramp to hold

1. Run 1–3 µL of each sample on a 2 % agarose gel using a 100-bp ladder. It is recommended to run at 115–125 V for 8–20 min to resolve any potential primer dimer bands.

Note: A clear, single band that is 140 bp longer than the amplicon should be observed. This addition accounts for both the NGS adapter sequences from Round 1 and the Illumina-specific barcode additions from Round 2. Refer to the troubleshooting sub-section if multiple bands are observed, no bands are observed, or the primer dimer band density outweighs the on-target band.

6.1.2.3 Pooling and gel extraction

1. Pool all Round 2 PCR products that are a similar size together depending on the intensity of their band on the gel in step 8. Product sizes should vary no more than 50 base pairs, i.e. a 300 bp product and 350 bp product can be pooled.

Notes:

i. *For bands brighter than the average intensity, pool less (i.e. half) than the normalized amount. For bands dimmer than the average intensity, pool more (i.e. double) than the normalized amount.*

   ii. *Ensure enough of each of the Round 2 samples remain to feasibly repeat the pooling should a mistake be made past this step. If your library is small (< 10 samples) consider doubling the scale of your round 2 reaction to allow for multiple attempts with a larger pooling volume.*

2. Run the entire pooled library on a 2% agarose gel using a wide comb. Run at 100–110 V for 35–45 min for the best resolution of potential primer dimer bands.

   Note: Run as much volume as feasible in a well of the gel. Well size will vary depending on gel or gel caster.

3. Perform a gel extraction of the on-target band. Use a NanoDrop™ spectrophotometer or equivalent to measure the concentration and assess DNA quality before proceeding.

   Notes:

   i. *The minimum mass of DNA required to run during NGS is around 100 ng total DNA. The A260/A280 should be around 1.80 and A260/A230 should be around 2.20. However, if the peak is clean, minor deviations from these parameters are acceptable. If there is a shoulder on the peak, then an additional PCR cleanup of the product is recommended.*

   ii. *Expect to lose half of the total yield of DNA with each additional cleanup.*

4. Quantify the library using a Qubit dsDNA HS assay kit. This will quantify the actual amount of dsDNA in the sample, ignoring any other contaminants. This value should be used for dilution calculations when preparing to run on NGS and is considered more reliable than a measurement from a NanoDrop™ spectrophotometer or equivalent.

5. Set up the NGS according to the Illumina protocols for the instrument, or submit this library to a sequencing core. To obtain robust, reliable, and quantitative sequencing data, we recommend obtaining at least 10,000 sequencing reads per sample.

6. The instrument will demultiplex the samples according to the forward and reverse barcode sequence combinations, with all of the reads from a given sample in separate fastq files. Those who are familiar with NGS datasets and analysis may prefer to analyze their fastq files for editing efficiencies with custom code. For those not familiar with NGS datasets, we recommend using CRISPResso2, which has a simple user interface and can analyze targeted amplicon sequencing data for genome editing efficiencies. To use the web version:

   a. Under "Editing tools", select "Base editors".

b. Under "Sequencing design", select "Paired end reads" unless a single end read run was completed.
c. Under "Fastq file R1", upload your read 1 (R1) fastq file, and under "Fastq file R2", upload your read 2 (R2) fastq file.
d. Under "Amplicon", input the entire amplicon sequence (this will include the annealing regions of the forward and reverse round 1 primers, plus the intervening sequence). Note that CRISPResso2 requires 10-bp of overlap between the R1 and R2 reads for paired end reads.
e. Under "sgRNA", enter the 20-bp protospacer sequence (do not include the PAM). This will result in the program quantifying editing efficiencies within this region and providing an allele plot focused on this region as well.
f. Under "Optional parameters", select the appropriate base editor "target base" and "result base". For CBE, this will be a target base of C and a result base of T. For ABE, this will be a target base of A and a result base of G.
g. The remainder of the default parameters should be sufficient.
h. Click on "Submit!" to analyze. See Fig. 7C for an example output.
i. Alternatively, if there are many fastq files to analyze, it may be desired to download the source program from github (https://github.com/pinellolab/CRISPResso2) and run the analysis in "batch mode", which allows for analysis of multiple fastq files at once (the web version can only handle up to 100 MB of data at a time).

Editing efficiency can vary depending on multiple factors. Some factors to consider include the protospacer selected (e.g., position of the edit and of bystanders), the base editor used (e.g., editing window, processivity, sequence motifs), the delivery method for the gRNA and base editor (e.g., electroporation, cationic lipid, transduction), the cell line and cell type (e.g., some cell lines are more challenging to transfect and edit in than others), and the collection and harvesting method (e.g., bulk lysis, FACS, enriching for transfection vs. enriching for BE expression). Therefore, we have provided detailed information from the transfections covered in this work in Table 5. The necessary level of editing for an experiment to be deemed a success may differ based on desired outcomes. However, if the goal is to generate an isogenic cell line, we recommend that editing efficiency be $\geq 10\%$.

Table 5 Editing data for transfections performed in this work. Additional editing data for experiments not performed in this work, but by these authors, is also provided. The protospacer sequence and corresponding gene are given in column 1. Expected editing efficiency is determined by the allele-specific target base conversion percentage provided by NGS analysis, in technical triplicate, unless indicated (*) which is in technical duplicate.

Protospacer sequence (5′–3′) and gene target	Editor used	Delivery method	Cell type	Harvest method (bulk, FACS for BE expression, etc.)	Expected editing efficiency
GGCCCAGACTGAGCACGTGA (HEK site 3)	ancBE4 - NG	Method 6d	iPSC (KOLF 2.1J)	FACS for GFP (gRNA)	~50 %
CAAACCGCTGTGGGCAGAAG (ERCC2)	ancBE4-NG	Method 6a	HEK293T	Method 7: Bulk	~10 %
GGCTCGGTGCTCTCCTCCGT (SCN5A)	ancBE4-NG	Method 6a	HEK293T	Method 7: Bulk	~15 %
CCACTGTGCAGCCAGTGCCC (MUTYH)	ancBE4-NG	Method 6a	HEK293T	Method 7: Bulk	~15 %
CAAACCGCTGTGGGCAGAAG (ERCC2)	ancBE4-NG	Method 6a	HEK293T	Method 9a: enriched FACS	~25 %
GGCTCGGTGCTCTCCTCCGT (SCN5A)	ancBE4-NG	Method 6a	HEK293T	Method 9a: enriched FACS	~35 %*
CCACTGTGCAGCCAGTGCCC (MUTYH)	ancBE4-NG	Method 6a	HEK293T	Method 9a: enriched FACS	~35 %*

(continued)

Table 5 Editing data for transfections performed in this work. Additional editing data for experiments not performed in this work, but by these authors, is also provided. The protospacer sequence and corresponding gene are given in column 1. Expected editing efficiency is determined by the allele-specific target base conversion percentage provided by NGS analysis, in technical triplicate, unless indicated (*) which is in technical duplicate. *(cont'd)*

Protospacer sequence (5′–3′) and gene target	Editor used	Delivery method	Cell type	Harvest method (bulk, FACS for BE expression, etc.)	Expected editing efficiency
CAAACCGCTGTGGGCAGAAG (ERCC2)	ancBE4-NG	Method 6a	HEK293T	Method 9a: dGFP turn-on FACS for BE expression	~30 %
GGCTCGGTGCTCTCCTCCGT (SCN5A)	ancBE4-NG	Method 6a	HEK293T	Method 9a: dGFP turn-on FACS for BE expression	~45 %*
CCACTGTGCAGCCAGTGCCC (MUTYH)	ancBE4-NG	Method 6a	HEK293T	Method 9a: dGFP turn-on FACS for BE expression	~45 %*
CAAACCGCTGTGGGCAGAAG (ERCC2)	ancBE4-NG	Method 6b	HEK293T	Method 7: Bulk	~15 %
GGCTCGGTGCTCTCCTCCGT (SCN5A)	ancBE4-NG	Method 6b	HEK293T	Method 7: Bulk	~20 %
CCACTGTGCAGCCAGTGCCC (MUTYH)	ancBE4-NG	Method 6b	HEK293T	Method 7: Bulk	~20 %

CAAACCGCTGTGGGCAGAAG (ERCC2)	ancBE4-NG	Method 6b	HEK293T	Method 9a: enriched FACS	~25 %
GGCTCGGTGCTCTCCTCCGT (SCN5A)	ancBE4-NG	Method 6b	HEK293T	Method 9a: enriched FACS	~35 %
CCACTGTGCAGCCAGTGCCC (MUTYH)	ancBE4-NG	Method 6b	HEK293T	Method 9a: enriched FACS	~40 %
CAAACCGCTGTGGGCAGAAG (ERCC2)	ancBE4-NG	Method 6b	HEK293T	Method 9a: dGFP turn-on FACS for BE expression	~30 %
GGCTCGGTGCTCTCCTCCGT (SCN5A)	ancBE4-NG	Method 6b	HEK293T	Method 9a: dGFP turn-on FACS for BE expression	~45 %
CCACTGTGCAGCCAGTGCCC (MUTYH)	ancBE4-NG	Method 6b	HEK293T	Method 9a: dGFP turn-on FACS for BE expression	~50 %
CCACTGTGCAGCCAGTGCCC (MUTYH)	ancBE4-NG	Method 6c	K562-s	Method 7: Bulk	~10 %

6.1.3 Troubleshooting

6.1.3.1 Round 1 PCR

- No bands:
  - o Lysis may be too concentrated or too dilute. Try a gradient of lysis input from 1 μL of a 1:10 dilution of the lysis, up to 2 μL of the undiluted lysis.
    - ■ A smaller gradient of 0.5-2 μL of lysis may also be used.
  - o Primers may not be binding. If the annealing region is too long, the annealing temperature may be too high. Either redesign the primers or try a temperature gradient on the annealing step from 54 °C to 72 °C.
- *Note: Running a PCR with the annealing temperature equal to or higher than the extension temperature is not recommended as it can lead to PCR bias, but in cases of particularly tricky sites, and an inability to redesign the primers, this could help with getting any bands to move forward with in Round 2.*
  - o Primers may be forming self-dimers or heterodimers and not binding to the target. Use a primer analyzer tool such as IDT's OligoAnalyzerTM tool (www.idtdna.com/calc/oligoanalyzer) to see if the primers are forming significant primer dimers or secondary structures. Include the adapter sequences in your analysis.
- Unbalanced primer dimer and on-target bands:
  - o Ideally, the on-target band should be stronger than the primer dimer band. If not, try a temperature gradient on the annealing step from 54 °C to 72 °C. Also, try using half the amount of primer, or double the amount of lysis. Finally, try adding 3–5 % DMSO to the PCR reaction mixture.
- More than one band:
  - o A primer dimer band is normal, but multiple bands in addition to a primer dimer band could be due to the primer binding at other regions in the genome. Be sure to check if the primer (including the NGS adapters) has any homology to other regions of the genome.
  - o Adding 3–5 % DMSO can also help to reduce off-target binding.

6.1.3.2 Round 2 PCR

- Unbalanced primer dimer and on-target bands:
  - o Try increasing the number of cycles up to 10–16.
  - o Try increasing the amount of Round 1 PCR input (1–2 μL).
- Having no bands or multiple Round 2 bands is rare and likely due to contamination of reagents.

6.1.3.3 Gel extraction

- The goal is to have a clean and concentrated gel extraction product. An additional PCR cleanup on the gel extraction product may be necessary. Ensuring an A260/A280 peak around 1.80 and A260/A230 peak around 2.20 on a NanoDrop™ spectrophotometer or equivalent is advised.
- Save some of the Round 2 samples so that re-pooling is possible should the gel extraction fail.

7. Enriching for cells with high BE expression or activity

If the initial base editing efficiency, as assessed from bulk samples, is low (<10 %), editing can be improved using Fluorescence-Activated Cell Sorting (FACS) (Fig. 8). The BE plasmids that we recommend here have fluorescent markers incorporated (via the self-cleaving P2A sequence), enabling users to enrich for cells with high BE expression using FACS. To improve editing efficiency further, fluorescent turn-on plasmids such as pCBE-dGFP and pABE-dGFP (Fig. 5A) can be used. Both plasmids contain a functional mCherry gene (which acts a fluorescent marker for transfection) and a GFP gene with an inactivating mutation (which we call "dGFP"). In pCBE-dGFP, a C·G to T·A mutation is required to restore fluorescence, while in pABE-dGFP, an A·T to G·C mutation is required. pCBE-dGFP-gRNA and pABE-dGFP-gRNA are the respective gRNA-expressing plasmids to facilitate GFP turn-on of these two dGFP plasmids. When employing this strategy, the dGFP and dGFP-gRNA plasmids are added to the transfection mixture in Methods 6a, 6b, or 6c in a ratio of 1:5 and 1:10 to the endogenous gRNA plasmid, respectively. The BE performs editing at both the endogenous target and the episomal dGFP target, and GFP+ cells are sorted (Fig. 9). It is important to note that pCBE-dGFP and pABE-dGFP cannot be used with BE plasmids that contain GFP fluorescent markers. Both of these strategies can greatly improve base editing efficiencies to levels that make isogenic cell line generation much more feasible (we recommend obtaining editing efficiencies of >10 % for this) (Fig. 8). If editing efficiency for a given target remains low (<5 %) even after incorporating these strategies, it is recommended to use the same gRNA with the appropriate wild-type Cas9 variant (Cas9, Cas9-NG, or SpRY-Cas9) and quantify indel introduction efficiencies at the target site. This will provide information on how efficiently the Cas9:gRNA complex

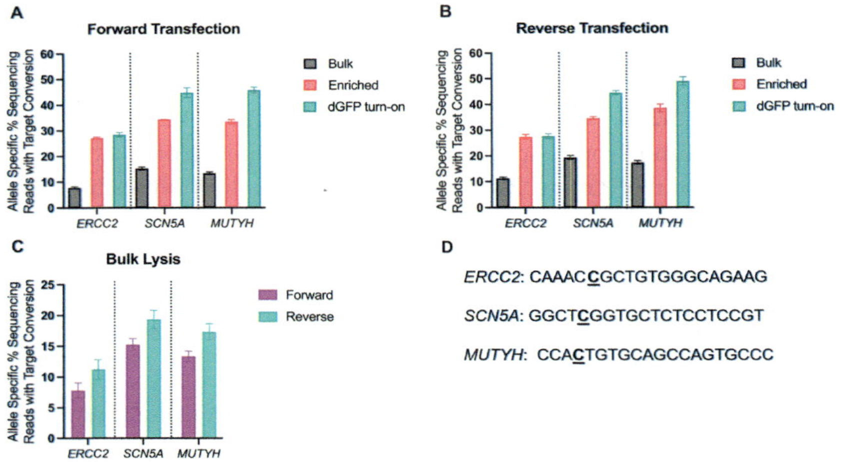

Fig. 8 Target conversion efficiencies across various conditions in HEK293T cells. (A) Under forward transfection conditions, target conversion efficiencies are shown for three genomic targets within the *ERCC2*, *SCN5A*, and *MUTYH* genes. Bulk lysis, where cells are not sorted using FACS, is shown in gray. Enriched samples, shown in pink, were transfected with a BE construct that constitutively expresses GFP. Therefore, only cells that were successfully transfected were collected, lysed, and sequenced. In teal, dGFP turn-on samples are shown. Cells were transfected with a construct containing dGFP that can be revived if base editing is active (according to the constructs shown in Fig. 5A). Thus, only cells with active base editor are collected, lysed, and sequenced. (B) Reverse transfection target conversion efficiencies are shown in the same manner as (A). Notably, some genomic loci can be more challenging to edit for a variety of reasons. (C) Cells transfected using forward and reverse transfections are compared after bulk lysis. Under bulk lysis conditions, reverse transfections yield a higher rate of target conversions at all three loci. (D) Spacer sequences for the loci. The target nucleotide is bolded and underlined. The corresponding gRNAs were combined with the ancBE4-NG editor.

can bind to the genomic locus (which is a prerequisite for base editing). If indel introduction efficiency is low, gRNA expression may be an issue, in which case we recommend trying with synthetic gRNA.

7.1 Method 9a – FACS of adherent mammalian cells for base editing enrichment

7.1.1 Materials and equipment

- Flow cytometry instrument (BioRad S3e or equivalent)
- Micropipettes
- Filtered cap FACS tubes (35 μm mesh size)
- 1.5 mL microcentrifuge tubes

- Ice
- Accumax
- PBS
- Fluorescent cells

7.1.2 Protocol

1. This protocol is to be performed after Method 6, instead of Method 7.
2. 48–72 h post-transfection, observe the cells using a fluorescence microscope to confirm that fluorescence is present.
3. Carefully aspirate growth medium and gently add 150 μL PBS.
4. Carefully aspirate PBS, ensuring all liquid has been removed from the well.
5. Add 60 μL of Accumax to each well.

   Note: Accumax is used over Trypsin detachment mediums as it is more efficient, particularly at separating clumps of mammalian cells. This is ideal for FACS applications as even small clumps of cells result in clogs on the instrument.
6. Incubate at room temperature for 3–5 min. The cells can be moved into a 37 °C incubator to shorten the incubation time to 1–2 min.
7. Resuspend each well in 140 μL of cold PBS. This decreases the probability of the cells reattaching to each other before they are run on the FACS instrument.

   Note: Keep the plate on ice for the duration of the sorting process.
8. Immediately before loading the cells onto the instrument, resuspend each well again, and push the resuspension through the filtered cap of a FACS tube.
9. Generate gates using a no-color control. Load the instrument and run the sample, accounting for cell death and singlets versus multiples.
10. Refine gates using color control(s). A single-color control is required for each color used. For example, if only using GFP, only a green color control is required. If using both mCherry and GFP, two color controls are required – one red, one green.
11. Place a 1.5 mL microcentrifuge tube containing 500 μL of cold PBS into the cell dispensing area of the FACS instrument.
12. Load the instrument with a sample well, using previously established gates from Steps 9 and 10 for either GFP+ or dual GFP+/mCherry + cells, depending on the enrichment strategy used.
    a. Note the number of cells collected. This is important for calculating the amount of lysis buffer to use, later.

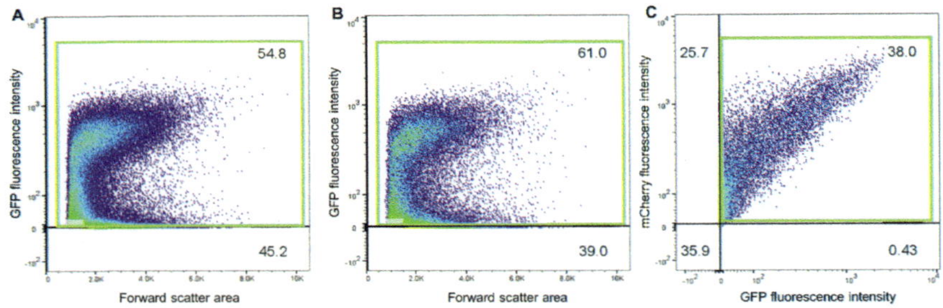

Fig. 9 Fluorescence-Activated Cell Sorting (FACS) scatter plots. (A–B) HEK293T cells were transfected with base editor and GFP as a transfection marker. Forward transfected samples (A), and reverse transfected samples (B) are shown comparing the forward scatter area (x-axis) to GFP intensity (y-axis). The forward scatter area provides information on the size of the cells. When only a single color is measured, this can be plotted against forward scatter (to assess the homogeneity of the size of the cells) or as a histogram. (C) HEK293A cells with the dGFP turn-on enrichment plasmid are shown. These cells express mCherry as a transfection control, and GFP fluorescence is restored in cells with active base editor. This graph compares mCherry (y-axis) to GFP (x-axis) fluorescence. Cells in the green boxes were gated and sorted for enrichment, then later lysed and sequenced.

b. Collect a minimum of 2000 cells.

c. Ideally, collect 10,000 or more cells.

d. See Fig. 9 for an example of gating.

13. Collect the microcentrifuge tube and invert it gently to resuspend any cells on the wall of the tube. Keep on ice until the rest of the sorting is complete.

14. Spin the microcentrifuge tubes at 300 g for 10 min, ensuring the back tab of the tube is facing the same direction on all tubes so the cell pellet location is known. Ideally, face the tab outwards.

15. Gently pipette off the PBS, being very careful not to disturb the pellet.

16. Add the lysis buffer to the tube and resuspend.

Notes:

i. *Refer to* ***Method*** *7 step 4b for lysis buffer recipe.*

ii. *The amount of lysis buffer to use is dependent on the amount of cells collected. Use 0.5 μL lysis buffer per 1000 cells collected.*

17. Add the resuspension to a 0.2 mL PCR tube.

18. Using a thermocycler, incubate the resuspension at 37 °C for 1 h and 80 °C for 30 min. Ramp the temperature down to either 12 °C or 4 °C by 0.1 °C per second.

19. Store lysis at 4 °C for up to 1 year, but ideally no more than 6 months.

20. Sequence the genomic locus of interest using Method 8.

If the goal of an experiment is to generate an isogenic cell line harboring the target mutation, individual cells must be isolated after the editing process has occurred and allowed to proliferate into a clonal population. To sort individual cells, we recommend using a FACS instrument called a Namocell®. This instrument uses single-use sterile cartridges that are loaded with cells still suspended in growth medium, and sorts cells at a far lower psi (<2 psi) compared to traditional FACS instruments (20–70 psi). Traditional FACS instruments are known to frequently contaminate cells as they pass through shared internal tubing inside the machine, and the sterile cartridges of the Namocell® eliminate this source of contamination. Furthermore, the higher psi at which cells are expelled in traditional FACS instruments is not only toxic to the cells, but also increases the possibility of multiple cells landing in a single well when sorting, which results in a non-clonal cell line comprised of a mixture of different genotypes. We have found that with the Namocell®, approximately one third of cells survive the single-cell isolation and clonal expansion process, compared to approximately one in ten with traditional FACS instruments.

7.2 Method 9b - Namocell® sorting for cell line generation with adherent mammalian cells using base editing

7.2.1 Materials and equipment

- Namocell®
- Micropipettes
- Ice bucket
- 96-well plate(s)
- Filtered cap FACS tubes (35 μm mesh size)
- Accumax
- PBS
- Cell culture media
- Optional: Penicillin Streptomyocin
- Fluorescent cells

7.2.2 Protocol

1. This protocol is to be performed after Method 6, instead of Method 7. 48–72 h post-transfection, observe the cells using a fluorescence microscope to confirm that fluorescence is present.
2. Prepare a 96-well plate with 100 μL of growth medium in each well. Put the plate(s) into the incubator to warm before sorting.

   Notes:

   i. *Add an additional 10–20 μL in edge wells to account for evaporation.*
   ii. *For HEK293T or HEK293A cells, DMEM + 25 % FBS + 1 % Penicillin Streptomycin is recommended.*
   iii. *Prepare as many plates as needed to sort. Assume a 10–40 % survival rate and be prepared to care for that many individual wells when deciding how many wells to sort into.*

3. Carefully aspirate growth medium and gently add 150 μL PBS.

4. Carefully aspirate PBS, ensuring all liquid has been removed from the well.

5. Add 60 μL of Accumax to each well of the 48-well plate.

Note: Accumax is used over Trypsin detachment mediums as it is more efficient, particularly at separating clumps of mammalian cells. This is ideal for FACS applications as even small clumps of cells result in clogs on the instrument.

6. Incubate at room temperature for 3–5 min. The cells can be moved into a 37 °C incubator to shorten the incubation time to 1–2 min.

7. Resuspend each well in 190 μL of cold growth medium. This decreases the probability of the cells reattaching to each other before they are run on the Namocell®. Either keep the resuspension in the plate, or move into a 1.5 mL microcentrifuge tube, and place on ice.

Note: Keep the plate or tube on ice for the duration of the sorting process.

8. Immediately before loading the instrument, resuspend each well again, and push the resuspension through a filtered cap of a FACS tube.

9. The Namocell® requires an incredibly dilute sample volume (~5000–20,000 cells/mL) depending on the iteration of the Namocell® in use (Hana or Pala). Dilute the sample accordingly with the appropriate volume of growth medium.

Note: A 1:10 dilution is a good place to start.

10. After starting up and washing the machine, load ~500 μL of the sample into a new sterile sorting cartridge.

11. Set gates using your sample, gating for the appropriate fluorophores for the experiment. See Method 6a Day One Step 2 for necessary controls.

12. Add the prepared and warmed plate onto the stage of the Namocell® and sort samples. Repeat as needed for additional plates.

13. Immediately place cells in the incubator after sorting. Leave them for at least 48 h before attempting to move them to a different incubator.

14. Single colonies can usually be observed after 1 week, be sure to note if multiple colonies are observed in one well.

15. After 2 weeks cells will begin to require passaging. Split all viable wells into a new "master plate" and a "sequencing plate." These plates should be identical and will serve as the numbering for each colony going forward.
16. Once wells in the "sequencing plate" are confluent, lyse them according to **Method 7**. For a 96-well plate be sure to scale the lysis buffer down to 50 μL for a fully confluent well.
17. Keep the "master plate" alive and split wells as needed until all of the "sequencing plate" wells have been sequenced.
18. Keep 3 homozygous clones, 3 heterozygous clones, and 3 wild type clones as controls. Expand the wells up to a 6-well or T25 and cryopreserve these lines.

Acknowledgments

The authors acknowledge the continual contributions towards ongoing projects cited in this chapter. We would also like to thank Lingzhi Zhang for providing expertise and helpful discussions on hiPSC culturing. This work was additionally made possible by the UC San Diego Stem Cell Program and a CIRM Major Facilities grant (FA1–00607) to the Sanford Consortium for Regenerative Medicine. This publication includes data generated at the UCSD Human Embryonic Stem Cell Core Facility, using the Namocell® Flow Cytometry Sorters. This work was supported by Cottrell Scholar Award no. 27502 (to A.C.K), the National Institute of Health (NIH) through grants no. R35GM138317 (to A.C.K.) and R56HG013535 (to A.C.K., A.G., and M.G.), and the California Institute of Regenerative Medicine through grant no. DISC0–13808 (to A.C.K., A.G., and M.G.). N.M.Z. was supported by the American Heart Association Predoctoral Fellowship no. 24PRE1201491. N.R.B.O. was supported by the Contemporary Approaches to Cancer Cell Signaling and Communication Training Grant, NIH Grant T32 CA009523–37. B.L.R. was supported by the Chemistry-Biology Interface Training Program, NIH Grant T32 GM112584. Q.T.C. was supported by the Molecular Biophysics Training grant, NIH Grant T32 GM008326.

Competing interests

A.C.K. is a member of the SAB of Pairwise Plants, is an equity holder for Pairwise Plants and Beam Therapeutics, and receives royalties from Pairwise Plants, Beam Therapeutics, and Editas Medicine via patents licensed from Harvard University. A.C.K.'s interests have been reviewed and approved by the University of California, San Diego in accordance with its conflict of interest policies. All other authors declare no competing financial interests.

Appendix A. Supporting information

Supplementary data associated with this article can be found in the online version at https://doi.org/10.1016/bs.mie.2025.01.001.

References

Anzalone, A. V., Koblan, L. W., & Liu, D. R. (2020). Genome editing with CRISPR–Cas nucleases, base editors, transposases and prime editors. *Nature Biotechnology, 38*, 824–844.

Arbab, M., et al. (2020). Determinants of base editing outcomes from target library analysis and machine learning. *Cell, 182*, 463–480.e30.

Artsimovitch, I., & Belogurov, G. A. (2015). Creative math of RNA polymerase III termination: Sense plus antisense makes more sense. *Molecular Cell, 58*, 974–976.

Carr, I. M., et al. (2009). Inferring relative proportions of DNA variants from sequencing electropherograms. *Bioinformatics (Oxford, England), 25*, 3244–3250.

Chen, B., et al. (2016). Expanding the CRISPR imaging toolset with Staphylococcus aureus Cas9 for simultaneous imaging of multiple genomic loci. *Nucleic Acids Research, 44*, e75.

Clement, K., et al. (2019). CRISPResso2 provides accurate and rapid genome editing sequence analysis. *Nature Biotechnology, 37*, 224–226.

Cowan, Q. T., et al. (2024). Development of multiplexed orthogonal base editor (MOBE) systems. *Nature Biotechnology*, 1–15. https://doi.org/10.1038/s41587-024-02240-0.

Engler, C., Kandzia, R., & Marillonnet, S. (2008). A one pot, one step, precision cloning method with high throughput capability. *PLoS One, 3*, e3647.

Gaudelli, N. M., et al. (2020). Directed evolution of adenine base editors with increased activity and therapeutic application. *Nature Biotechnology, 38*, 892–900.

Gaudelli, N. M., et al. (2017). Programmable base editing of A•T to G•C in genomic DNA without DNA cleavage. *Nature, 551*, 464–471.

Hanna, R. E., et al. (2021). Massively parallel assessment of human variants with base editor screens. *Cell, 184*, 1064–1080.e20.

Jinek, M., et al. (2012). A programmable dual RNA-guided DNA endonuclease in adaptive bacterial immunity. *Science (New York, N. Y.), 337*, 816–821.

Kleinstiver, B. P., et al. (2015). Broadening the targeting range of Staphylococcus aureus CRISPR-Cas9 by modifying PAM recognition. *Nature Biotechnology, 33*, 1293–1298.

Koblan, L. W., et al. (2018). Improving cytidine and adenine base editors by expression optimization and ancestral reconstruction. *Nature Biotechnology, 36*, 843–846.

Komor, A. C., Kim, Y. B., Packer, M. S., Zuris, J. A., & Liu, D. R. (2016). Programmable editing of a target base in genomic DNA without double-stranded DNA cleavage. *Nature, 533*, 420–424.

Nishimasu, H., et al. (2018). Engineered CRISPR-Cas9 nuclease with expanded targeting space. *Science (New York, N. Y.), 361*, 1259–1262.

Pantazis, C. B., et al. (2022). A reference human induced pluripotent stem cell line for large-scale collaborative studies. e22 *Cell Stem Cell, 29*, 1685–1702.e22.

Richter, M. F., et al. (2020). Phage-assisted evolution of an adenine base editor with improved Cas domain compatibility and activity. *Nature Biotechnology, 38*, 883–891.

Sternberg, S. H., Redding, S., Jinek, M., Greene, E. C., & Doudna, J. A. (2014). DNA interrogation by the CRISPR RNA-guided endonuclease Cas9. *Nature, 507*, 62–67.

Thuronyi, B. W., et al. (2019). Continuous evolution of base editors with expanded target compatibility and improved activity. *Nature Biotechnology, 37*, 1070–1079.

Untergasser, A., et al. (2012). Primer3—New capabilities and interfaces. *Nucleic Acids Research, 40*, e115.

Walton, R. T., Christie, K. A., Whittaker, M. N., & Kleinstiver, B. P. (2020). Unconstrained genome targeting with near-PAMless engineered CRISPR-Cas9 variants. *Science (New York, N. Y.), 368*, 290–296.

Xiao, Y.-L., Wu, Y., & Tang, W. (2024). An adenine base editor variant expands context compatibility. *Nature Biotechnology*, 1–12. https://doi.org/10.1038/s41587-023-01994-3.

CHAPTER FIFTEEN

Prime editing in bacteria with BacPE

Hongyuan Zhang[a] and Quanjiang Ji[a,b,c,*]
[a]School of Physical Science and Technology, ShanghaiTech University, Shanghai, P.R. China
[b]Gene Editing Center, School of Life Science and Technology, ShanghaiTech University, Shanghai, P.R. China
[c]Shanghai Clinical Research and Trial Center, Shanghai, P.R. China
*Corresponding author. e-mail address: quanjiangji@shanghaitech.edu.cn

Contents

1. Introduction 405
2. Before you begin: design pegRNA for genome editing in *E. coli* 406
   2.1 Materials and equipment 409
   2.2 Step-by-step method details: perform prime editing in *E. coli* 410
3. Concluding remarks 416
4. Limitations 417
5. Declaration of AI and AI-assisted technologies in the writing process 417
References 417

Abstract

Programmable genome editing technologies have revolutionized the ability of researchers to alter the genomes of microorganisms in a straightforward and efficient manner, significantly advancing the field of microbiology. To date, several CRISPR-Cas-based genome-editing systems have been developed for use in *E. coli*, including CRISPR/Cas9, base editing, and prime editing technologies. In this chapter, we describe the design and experimental application of BacPE, a variant of prime editing technology optimized for *E. coli*. BacPE facilitates the introduction of point mutations, insertions, and deletions without the need for double-strand DNA breaks. We demonstrate that BacPE is a powerful tool for genome editing in *E. coli* and highlight its potential applicability to other bacterial species.

1. Introduction

The CRISPR-Cas9 system has been widely harnessed for genome editing in diverse organisms (Anzalone, Koblan, & Liu, 2020; Cong et al., 2013; Jiang, Bikard, Cox, Zhang, & Marraffini, 2013; Jinek et al., 2012). In *E. coli*, the CRISPR-Cas9 system has been shown to facilitate accurate genome modifications, such as insertions and deletions (Jiang et al., 2013; Jiang et al., 2015;

Methods in Enzymology, Volume 712
ISSN 0076-6879, https://doi.org/10.1016/bs.mie.2025.01.026

Zhang et al., 2020). The Cas9-induced DSBs are repaired by λ-Red-mediated recombination, allowing for the incorporation of desired edits by supplying well-designed DNA donors. While the CRISPR-Cas9 system offers significant potential for various types of genetic editing, its application is frequently constrained by the requirement of donor DNA templates (Wu et al., 2020). To mitigate this limitation, base editor systems were developed, which consist of a deaminase enzyme fused to Cas9 (Banno, Nishida, Arazoe, Mitsunobu, & Kondo, 2018; Gaudelli et al., 2017; Komor, Kim, Packer, Zuris, & Liu, 2016; Rees & Liu, 2018). This system enables the precise conversion of C-to-T or A-to-G at target sites without introducing double-strand DNA breaks.

Although the base editing system enables the editing of various nucleotides, it cannot perform insertions and deletions. The recently developed prime editing system combines the advantages of both the CRISPR-Cas9 and base editing systems, allowing for all types of nucleotide edits as well as DNA insertions and deletions (Anzalone et al., 2019; Chen & Liu, 2023). Prime editors (PE) consist of a reverse transcriptase (RT) fused to Cas9 nickase and a prime editing guide RNA (pegRNA). The pegRNA includes a spacer for DNA targeting and a 3′ extension with the desired edits. The prime editing complex binds to the target site through base pairing with the pegRNA spacer and nicks the non-target strand to release a 3′ DNA end. This 3′ end then hybridizes with the pegRNA's primer binding site (PBS) to initiate the reverse transcription reaction via the engineered RT, synthesizing the desired edits from the RT template. Subsequent flap equilibration, 5′ flap cleavage, and DNA repair processes enable the integration of the edited 3′ DNA flap into a target genomic site (Chen et al., 2021).

In this chapter, we present our engineered prime editing system, BacPE, designed for rapid and versatile genome editing in *E. coli* strains (Zhang et al., 2024). We detail the protocols for the design and construction of BacPE plasmids and their applications in various types of genome editing in *E. coli*. Additionally, we outline the procedures for transformation, genome editing, and plasmid clearance. Finally, we briefly discuss the potential applications of the BacPE system in other bacterial strains.

2. Before you begin: design pegRNA for genome editing in *E. coli*

The prime editing system comprises two main components: the prime editor protein and the pegRNA. To leverage the prime editing system for genome editing in *E. coli*, we devised a three-plasmid system, referred to as BacPE (Fig. 1). The Sp_PE plasmid encodes the prime editor protein, the pBbS8c_ddCpf1_final plasmid contains a crRNA expression cassette along with ddCpf1, and the pegRNA2_tevopreQ1 plasmid features a pegRNA cloning site flanked by *Bsa*I restriction sites. To execute prime editing in *E. coli*, the pegRNA sequence was cloned into the pegRNA2_tevopreQ1 plasmid using Golden Gate assembly. Subsequently, the pegRNA2_tevopreQ1 plasmid, now containing the pegRNA sequence, was transformed into an *E. coli* strain already harboring the Sp_PE and pBbS8c_ddCpf1_final plasmids.

Next, we will demonstrate how to design a pegRNA for gene editing in *lacZ* gene. All operations are conducted in Benchling. Just like SpCas9, the PE system requires an NGG PAM sequence to target the desired location. We have marked the PAM in orange, and the nick introduced by PE is marked by a black triangle. Additionally, the prime editing range is also noted (Fig. 2A). The protospacer is marked in green, the PAM sequence is highlighted in orange, and the red box indicates the type of gene editing (Fig. 2B). In Fig. 2C, DNA sequences used for pegRNA design were highlighted with boxes. We recommend starting with a PBS and RT template length of approximately 13 nucleotides and testing different PBS and RT template lengths for optimization. The linker sequence in the pegRNA was designed using pegLIT (https://peglit.liugroup.us/). The input sequences were copied as indicated in Fig. 2C. Subsequently, all DNA sequences mentioned in Fig. 2C and the linker pattern calculated by pegLIT were entered into the "pegRNA_design.xlsx" file, which automatically generated primers for pegRNA cloning (The pegRNA design file is available online: https://github.com/CRISPRmining/Prime-editing-in-bacteria-with-BacPE.git).

The "pegRNA cloning" step is sometimes confusing, so we illustrated this step in Fig. 3.

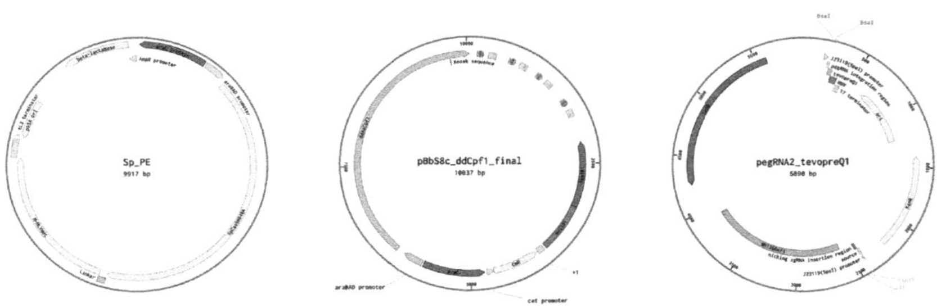

Fig. 1 BacPE plasmids for targeted genome editing in *E. coli*.

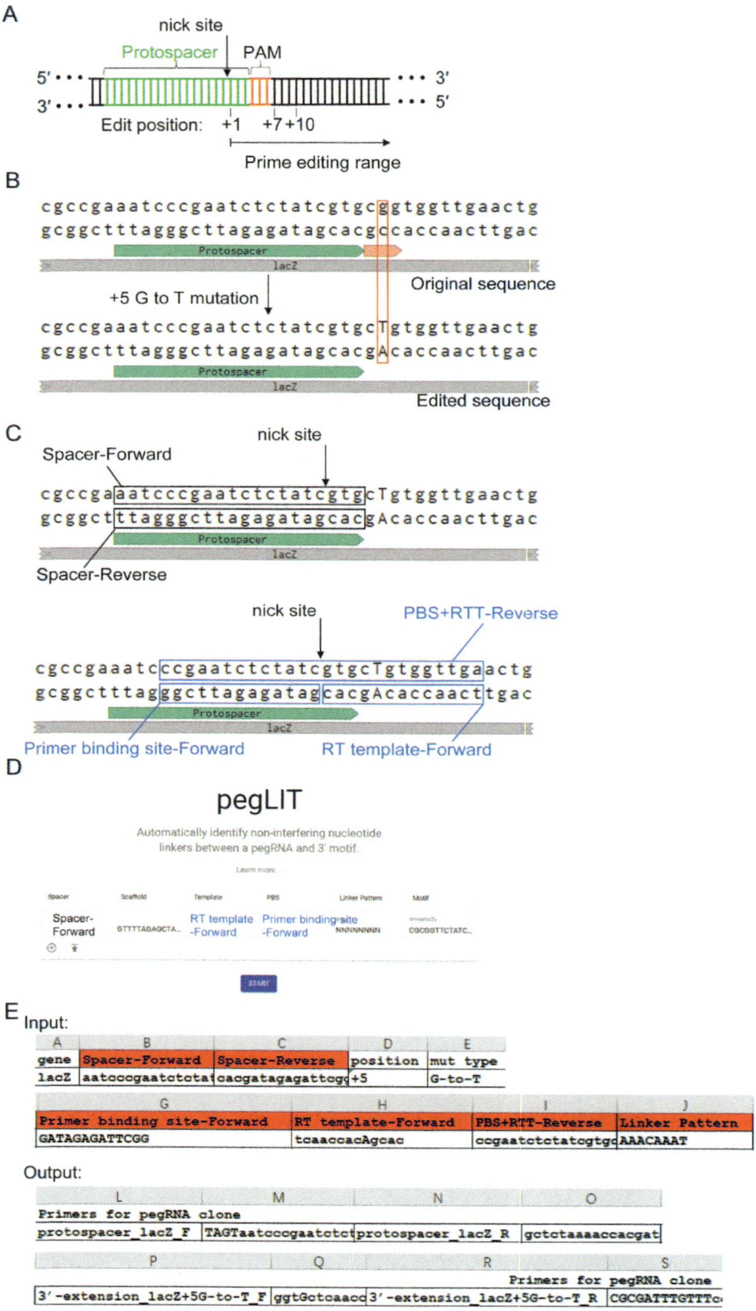

Fig. 2 Overview of pegRNA design. (A) The prime editing range of PE. (B) Introduce a mutation in *lacZ*. The red box highlights the type and position of DNA mutation.

In order to help audience to navigate through different sections, we also provided an additional workflow scheme (Fig. 4). In this figure, we outlined the technical goals for each step in brackets.

2.1 Materials and equipment

1 M Tris-Cl Buffer, pH8.5, Sangon Biotech
5 M NaCl solution, Coolaber
Nuclease-free water
1 M $MnCl_2$ solution, Coolaber
1 M $CaCl_2$ solution, Coolaber
1 M KCl solution, Coolaber
1 M $MgCl_2$ solution, Coolaber
1 PIPES solution, pH = 6.7, Coolaber
0.22 μm membrane filter, Jet Biofil
LB broth, Sangon Biotech
SOC broth, Sangon Biotech
Agar, Sangon Biotech
TB broth, MDbio
PEG3350, Sangon Biotech
DMSO, Sangon Biotech
Glycerol, Sangon Biotech
*Bsa*I, Vazyme
TIANquick midi purification kit, TIANGEN
T4 PNK, Vazyme
T4 DNA ligase, Vazyme
Kanamycin, Sangon Biotech
IPTG, Sangon Biotech
Arabinose, Sangon Biotech
Chloramphenicol, Sangon Biotech
Carbenicillin, Sangon Biotech
Sucrose, Sangon Biotech

(C) The DNA sequence file was opened in Benchling, allowing specific DNA sequences to be copied for further use. (D) The pegLIT interface for linker design. Paste the DNA sequences from Fig. 2C into the designated area and click "START." The "Linker Pattern" will be generated in approximately one minute. (E) The interface for the automated primer generation script used for pegRNA cloning. After inputting the DNA sequences, Excel automatically returns the primer sequences. These sequences can be ordered for pegRNA cloning.

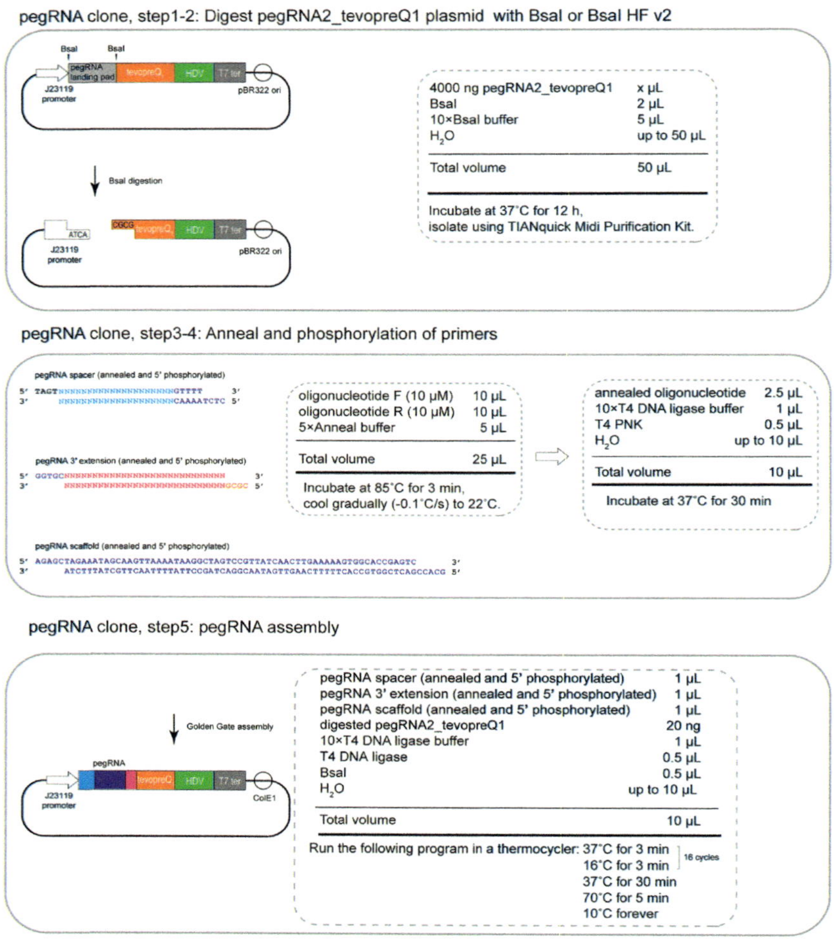

Fig. 3 Overview of pegRNA cloning step. This figure illustrated the steps related to "pegRNA cloning" section.

1 mm curvette, Bio-Rad

Gene Pulser Xcell, Bio-Rad

Refrigerated centrifuge

2.2 Step-by-step method details: perform prime editing in *E. coli*

2.2.1 Prepare materials

Annealing buffer: Nuclease-free water supplemented with 50 mM Tris-Cl at pH 8.5 and 250 mM NaCl.

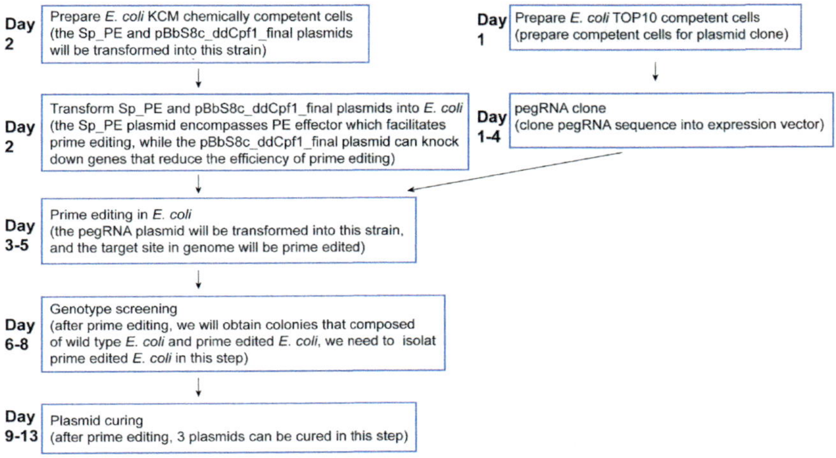

Fig. 4 Workflow scheme of BacPE system.

Inoue transformation solution: 33 mL nuclease-free water supplemented with 2.75 mL 1 M $MnCl_2$, 0.75 mL 1 M $CaCl_2$, 12.5 mL 1 M KCl and 0.5 mL 1 M PIPES (pH = 6.7). Sterilize using a 0.22-μm syringe filter and keep on ice for later use.

TSS solution: LB broth supplemented with 20 mM $MgCl_2$, 10 % wt/vol PEG3350 and 5 % vol/vol DMSO. Sterilize using a 0.22-μm syringe filter and storage at 4 °C.

5 × KCM solution: Nuclease-free water supplemented with 100 mM KCl, 30 mM $CaCl_2$ and 50 mM $MgCl_2$. Sterilize using a 0.22-μm syringe filter and storage at 4 °C.

10 % glycerol solution: Deionized water supplemented with 10 % vol/vol glycerol. Sterilize at 121 °C for 30 min and storage at 4 °C.

The pegRNA design file is available online: https://github.com/CRISPRmining/Prime-editing-in-bacteria-with-BacPE.git

All plasmid sequences are available from benchling:

Sp_PE: https://benchling.com/s/seq-tSYVS1PpE5B1lZk5sJn3?m=slm-YZDhDKBPpbIfqkrktE57

pegRNA2_tevopreQ1: https://benchling.com/s/seq-rOthM2vP1E2xZ32jf20U?m=slm-7BQKlJsvWWCEHYODYXUh

pBbS8c_ddCpf1_final: https://benchling.com/s/seq-IV0cRNa1ngRHHJC0Dlto?m=slm-I4inF2LHwLFwRXe0ZqcN

MG1655_lacZ: https://benchling.com/s/seq-Gg1sdGfHnSaRYjtpELsM?m=slm-tegaS7HzHzB4FwSCZxIc

MG1655_lacZ_G-to-T: https://benchling.com/s/seq-2AhYBCdMwsBY7qxcIC3W?m=slm-ywTrnvS7qxzN9ID4hqzJ
pegRNA2_tevopreQ1_+pegRNA for lacZ editing: https://benchling.com/s/seq-K2gIuAfSc7cyzRPPSbc1?m=slm-idWjyalYhfpNNCphYpCJ

2.2.2 Prepare E. coli TOP10 competent cells

1. Incubate several bacterial colonies in 10 mL of SOC medium at 30 °C with shaking overnight.
2. Inoculate 300 mL SOC medium with a 2 % inoculum from step1. Shake the culture at 25 °C.
3. When the OD_{600} of the culture reaches about 0.5, cool on ice for 10 min.
4. Centrifuge at 2500 g for 10 min at 4 °C. Completely discard the supernatant.
5. Gently resuspend the bacterial pellet in 80 mL transformation solution (do not use a pipette to blow, use a pipette to transfer).
6. Centrifuge at 2500 g for 10 min at 4 °C. Completely discard the supernatant.
7. Gently resuspend the bacterial pellet in approximately 10 mL transformation solution (do not use a pipette to blow, use a pipette to transfer), and measure the precise volume of the bacterial liquid. Add DMSO to achieve an accurate final concentration of 7 %, mix well.
8. Dispense in 100 μL aliquots on ice, flash freeze in liquid nitrogen, and transfer to −80 °C for long-term storage.

Note: Researchers can also buy commercial competent cells with transformation efficiency > $1 \times 10^7 \sim 1 \times 10^8$ transformants/1.

μg pUC 19 plasmid.

2.2.3 pegRNA cloning

1. Digest 2000 ng of pegRNA2_tevopreQ1 plasmid in a PCR tube. For each 20 μL reaction, add 2000 ng of pegRNA2_tevopreQ1 plasmid, 2 μL of *Bsa*I and 2 μL of 10 × Digestion Buffer II. Adjust the volume to the necessary amount with nuclease-free water. Incubate at 37 °C overnight.
2. Add 0.5 μL 5 M NaCl into the PCR tube. Purify DNA fragment with TIANquick midi purification kit.
3. Anneal oligonucleotides. For each 25 μL reaction, add 10 μL Forward oligonucleotide (10 μM), 10 μL Reverse oligonucleotide (10 μM) and 5 μL annealing buffer. In a thermocycler, heat at 85 °C for 3 min, then

cool gradually (0.1 °C/s) to 25 °C. (Note: It is necessary to anneal the spacer oligonucleotide, pegRNA scaffold oligonucleotide, and 3' extension oligonucleotide in separate reaction tubes to ensure proper annealing.)

4. Phosphorylate the annealed oligonucleotides. For each 10 μL reaction, add 1 μL 10x T4 DNA ligase buffer (Vazyme), 1 μL annealed oligonucleotides and 1 μL T4 PNK. Incubate at 37 °C for 1 h. (Note: Chemically synthesized primers often have a –OH group at the 5' end. Therefore, after phosphorylation modification, the efficiency of DNA ligation can be enhanced.)
5. Assembly pegRNA using Golden Gate assembly. For each 10 μL reaction, add 50 ng linearized pegRNA2_tevopreQ1 plasmid, 1 μL annealed and phosphorylated protospacer oligonucleotides, 1 μL annealed and phosphorylated 3'-extension oligonucleotides, 1 μL annealed and phosphorylated scaffold oligonucleotides, 0.5 μL *Bsa*I, 0.5 μL T4 DNA ligase and 1 μL 10x T4 DNA ligase buffer. Adjust the volume to the necessary amount with nuclease-free water. Cycle between 3 min at 37 °C and 3 min at 16 °C for 15 cycles, follwed by 37 °C for 30 min and 70 °C for 5 min
6. Place the TOP10 competent cells on ice and let them thaw completely for 5 min. Once fully thawed, add the products of GoldenGate Assembly to the competent cells, carefully flick the tube 15 times to mix cells and DNA. Do not vortex.
7. Place the mixture on ice for 20 min. Do not mix.
8. After heat shocking at 42 °C for 45 s, transfer the tube to ice and chill for 2 min. Following this, add 900 μL of SOC medium to the tube.
9. Recover the cells at 37 °C in a biological shaker for 1 h. After the recovery period, plate 100 μL cells onto Petri dishes containing kanamycin (50 μg/mL). Incubate the Petri dishes at 37 °C overnight.
10. Pick 2 colonies for Sanger sequencing. The constructed pegRNA expressing plasmid should be stored in a −20 °C freezer for long-term storage.

2.2.4 Prepare E. coli KCM chemically competent cells

1. Incubate several monoclonal bacterial colonies in 10 mL LB in a 37 °C biological shaker overnight.
2. Take 1 mL of the saturated cell culture and dilute it in 100 mL of LB broth. Proceed by incubating it at 37 °C on a biological shaker until it reaches an OD600 of approximately 0.4–0.6.

3. Allow these cells to cool on ice for 10 min prior to centrifuging them at 4000 g for 10 min at 4 °C.
4. Discard supernatant and resuspend the pellet by gentle stirring in 10 mL of TSS solution.
5. Divide into 100-μL portions and place in microcentrifuge tubes, then freeze in liquid nitrogen. Store at −80 °C until ready for use.

(Note: the "KCM chemically competent cells" method mentioned above can be applied to any *E. coli* strain in principle).

However, in "pegRNA clone" step, we recommend the "Prepare *E. coli* TOP10 competent cells" method, because competent cells prepared by this method (modified Inoue method) shows higher transformation efficiency.

In "Prime editing in E. coli" step, in order to get more colonies after transformation, we provided an electroporation protocol rather than KCM method or Inoue method, in general, the ranking of transformation efficiency in *E. coli* is as follows: electroporation method > Inoue method > KCM method.

Transform Sp_PE and pBbS8c_ddCpf1_final plasmids into *E. coli*.

1. Thaw a 100 μL aliquot of *E. coli* KCM chemically competent cells on ice.
2. Add 100 ng Sp_PE, 200 ng pBbS8c_ddCpf1_final plasmid and 20 μL 5 × KCM solution into KCM chemically competent cells, add nuclease-free water until the total volume reaches 200 μL. Flick the tube 15 times to mix cells and DNA.
3. Incubate on ice for 20 min
4. Heat shock at 42 °C in a water bath for 90 s
5. Transfer cells back onto ice and then immediately add 800 μL of LB media.
6. Recover the cells at 30 °C in a biological shaker for 1 h. After the recovery period, plate 100 μL cells onto Terrific Broth agar containing chloramphenicol (25 μg/mL) and carbenicillin (50 μg/mL). Incubate the Petri dishes at 30 °C overnight.

2.2.5 Prime editing in E. coli

1. For *E. coli* containing Sp_PE and pBbS8c_ddCpf1_final plasmids, five colonies were selected and grown in TB broth, supplemented with chloramphenicol (25 μg/mL) and carbenicillin (50 μg/mL). The cultures were incubated overnight at 30 °C in a biological shaker.
2. Inoculate 0.5 mL of the overnight seed culture into 100 mL of TB broth. The cultures were incubated for 2 h at 30 °C in a biological

shaker. Next, IPTG (10 mM) and arabinose (0.1 M) were added to the culture to induce the expression of effector proteins and crRNA. Chill cells on ice when OD600 of approximately 0.4. (Note: This is a key step. Sp_PE plasmid encodes PE effector protein, and pegRNA plasmid encodes pegRNA, co-transformation of these two plasmids usually lead to low transformation efficiency and low genome editing efficiency. pBbS8c_ddCpf1_final encodes CRISPRi system to knock down several genes that can reduce prime editing efficiency, however, CRISPRi system need to be induced prior prime editing. Therefore, Sp_PE and pBbS8c_ddCpf1_final plasmids need to be transformed and induced prior pegRNA plasmid transformation.)

3. Centrifuge at 4000 g for 10 min at 4 °C. Completely discard the supernatant. Resuspend cells in 40 mL of 10% glycerol solution. Repeat this step for 3 times.
4. Centrifuge at 4000 g for 10 min at 4 °C. Completely discard the supernatant. Resuspend cells in 1 mL of 10% glycerol solution.
5. Put a 1 mm curvette (Bio-Rad) on ice, then add a mix of 50 μL resuspended cells with 50 ng of pegRNA expression plasmid to the curvette.
6. Using Gene Pulser Xcell, perform electroporation with the following settings: 25 μF, 200 Ω, 1.8 kv. After electroporation, the time constant should be close to 5 ms.
7. Add 900 μL of TB broth to the cuvette, then mix by pipetting up and down twice. Transfer the cells into a 1.5 mL tube and incubate at 30 °C in a biological shaker for 1 h to recover the cells.
8. Dilute cells serially in TB broth in three 10-fold increments to yield four total samples (undiluted and 10^1-, 10^2- and 10^3-fold diluted). Transfer 10 μL of each dilution into TB agar plate containing arabinose (0.1 M), IPTG (10 mM), chloramphenicol (25 μg/mL), kanamycin (50 μg/mL) and carbenicillin (50 μg/mL).
9. Incubate at 37 °C for 1 day.

2.2.6 Genotype screening

1. Pick a single colony into 10 μL TB broth by gently touching a P10 pipette tip to the surface of the plate. Mix by pipetting up and down 10 times.
2. PCR amplify the target locus on genomic DNA by placing the tip from step 1 into a PCR reaction mixture. Perform Sanger sequencing on the PCR products. Considering that *E. coli* is a Gram-negative bacterium,

there is no need to extract the genome. The bacteria can be directly used as a PCR template for PCR amplification.

3. Meanwhile, streak the remaining bacteria from step 1 onto TB agar plates for isolation.
4. If the Sanger sequencing results from step 2 indicate that the target site has been successfully edited, you can randomly pick a certain number of colonies from the plates in step 3. Cultivate the picked colonies in TB broth, and simultaneously, take a portion of the bacteria for sequencing. (Note: For example, if Sanger sequencing indicated the prime editing efficiency is 60%, we recommend picking 3 colonies for further culturation. When 3 colonies were picked, the probability to pick a prime edited colony is: $1\text{-}(1\text{-}60\%)^3 = 95.6\%$)
5. Based on the results of Sanger sequencing, preserve the successfully gene-edited bacteria at −80 °C. When preserving bacteria, mix them with glycerol at a 4:1 ratio and flash-freeze using liquid nitrogen.

2.2.7 Plasmid curing (optimal)

1. Inoculate the genotyped bacteria into LB broth supplemented with 5% sucrose and incubate overnight at 42 °C with 250 rpm agitation.
2. A fraction of the culture can be plated on LB agar to isolate single colonies. Please note, do not add any antibiotics to the plates. Incubate overnight at 37 °C.
3. Prepare one LB agar plate without antibiotics, one LB agar plate with chloramphenicol, one LB agar plate with kanamycin, and one LB agar plate with carbenicillin. Pick a single colony into 10 μL LB broth by gently touching a P10 pipette tip to the surface of the plate. Mix by pipetting up and down 10 times. Take 2.5 microliters of bacteria and culture them separately on four different plates. Incubate overnight at 37 °C. (Note: two independent colonies from step2 should be picked for antibiotic challenging assay, this can improve the probability to get a plasmid free colony.)
4. The strains with restored antibiotic sensitivity have successfully cleared all three BacPE plasmids.

3. Concluding remarks

In previous studies, we demonstrated that the prime editing system can be used in *E. coli*, *A. baumannii*, *K. pneumoniae*, and *M. smegmatis*. We developed the BacPE system in *E. coli*, which enables prime editing in wild-type *E. coli*, by inhibiting the expression of three exonucleases through CRISPRi. However, in *A. baumannii* and *K. pneumoniae*, we do not have an appropriate CRISPRi system, making it difficult to use the prime editing system in wild-type bacteria. Interestingly, unlike in *E. coli*, *K. pneumoniae*, or *A. baumannii*, *M. smegmatis* demonstrated highly efficient prime editing without any interference with cellular genetic pathways. The absence of the key 3′→5′ ssDNA exonuclease *sbcB* may account for the high editing efficiency observed in *M. smegmatis*.

The BacPE system has been designed to enable prime editing with high efficiency in multiple *E. coli* species. To assemble the pegRNA_tevopreQ1 plasmid that target a certain gene for prime editing, the spacer, pegRNA scaffold and 3′ extension can be inserted into pegRNA_tevopreQ1 plasmid through Golden Gate assembly. Therefore, the BacPE system constitutes a significant enhancement to the toolkit available to genetic and metabolic engineers working with *E. coli*. It also offers a foundational framework for the development of prime editing systems across various bacterial species.

4. Limitations

In the pBbS8c_ddCpf1_final plasmid, four spacers are designed to target *sbcB*, *xseA*, and *exoX*, which affect prime editing efficiency. However, some *E. coli* strains may carry mutations in these genes, resulting in their inactivity. Consequently, the prime editing efficiency might be low.

5. Declaration of AI and AI-assisted technologies in the writing process

Statement: During the preparation of this work the author(s) used [GPT4/Language polishing] in order to.
[improve readability and language]. After using this tool/service, the author (s) reviewed and edited the content as needed and take(s) full responsibility for the content of the publication.

References

Anzalone, A. V., Koblan, L. W., & Liu, D. R. (2020). Genome editing with CRISPR–Cas nucleases, base editors, transposases and prime editors. *Nature Biotechnology, 38*(7), 824–844. https://doi.org/10.1038/s41587-020-0561-9.

Anzalone, A. V., Randolph, P. B., Davis, J. R., Sousa, A. A., Koblan, L. W., Levy, J. M., ... Liu, D. R. (2019). Search-and-replace genome editing without double-strand breaks or donor DNA. *Nature, 576*(7785), 149–157. https://doi.org/10.1038/s41586-019-1711-4.

Banno, S., Nishida, K., Arazoe, T., Mitsunobu, H., & Kondo, A. (2018). Deaminase-mediated multiplex genome editing in Escherichia coli. *Nature Microbiology, 3*(4), 423–429. https://doi.org/10.1038/s41564-017-0102-6.

Chen, P. J., Hussmann, J. A., Yan, J., Knipping, F., Ravisankar, P., Chen, P.-F., ... Liu, D. R. (2021). Enhanced prime editing systems by manipulating cellular determinants of editing outcomes. *Cell, 184*(22), 5635–5652.e5629. https://doi.org/10.1016/j.cell.2021.09.018.

Chen, P. J., & Liu, D. R. (2023). Prime editing for precise and highly versatile genome manipulation. *Nature Reviews. Genetics, 24*(3), 161–177. https://doi.org/10.1038/s41576-022-00541-1.

Cong, L., Ran, F. A., Cox, D., Lin, S., Barretto, R., Habib, N., ... Zhang, F. (2013). Multiplex genome engineering using CRISPR/Cas systems. *Science (New York, N. Y.), 339*(6121), 819–823. https://doi.org/10.1126/science.1231143.

Gaudelli, N. M., Komor, A. C., Rees, H. A., Packer, M. S., Badran, A. H., Bryson, D. I., & Liu, D. R. (2017). Programmable base editing of A•T to G•C in genomic DNA without DNA cleavage. *Nature, 551*(7681), 464–471. https://doi.org/10.1038/nature24644.

Jiang, W., Bikard, D., Cox, D., Zhang, F., & Marraffini, L. A. (2013). RNA-guided editing of bacterial genomes using CRISPR-Cas systems. *Nature Biotechnology, 31*(3), 233–239. https://doi.org/10.1038/nbt.2508.

Jiang, Y., Chen, B., Duan, C., Sun, B., Yang, J., & Yang, S. (2015). Multigene editing in the Escherichia coli genome via the CRISPR-Cas9 system. *Applied and Environmental Microbiology, 81*(7), 2506–2514. https://doi.org/10.1128/AEM.04023-14.

Jinek, M., Chylinski, K., Fonfara, I., Hauer, M., Doudna, J. A., & Charpentier, E. (2012). A programmable dual-RNA–guided DNA endonuclease in adaptive bacterial immunity. *Science (New York, N. Y.), 337*(6096), 816–821. https://doi.org/10.1126/science.1225829.

Komor, A. C., Kim, Y. B., Packer, M. S., Zuris, J. A., & Liu, D. R. (2016). Programmable editing of a target base in genomic DNA without double-stranded DNA cleavage. *Nature, 533*(7603), 420–424. https://doi.org/10.1038/nature17946.

Rees, H. A., & Liu, D. R. (2018). Base editing: Precision chemistry on the genome and transcriptome of living cells. *Nature Reviews. Genetics, 19*(12), 770–788. https://doi.org/10.1038/s41576-018-0059-1.

Wu, Z., Wang, Y., Zhang, Y., Chen, W., Wang, Y., & Ji, Q. (2020). Strategies for developing CRISPR-based gene editing methods in bacteria. *Small Methods, 4*(2), 1900560. https://doi.org/10.1002/smtd.201900560.

Zhang, H., Ma, J., Wu, Z., Chen, X., Qian, Y., Chen, W., ... Ji, Q. (2024). BacPE: A versatile prime-editing platform in bacteria by inhibiting DNA exonucleases. *Nature Communications, 15*(1), 825. https://doi.org/10.1038/s41467-024-45114-4.

Zhang, Y., Zhang, H., Wang, Z., Wu, Z., Wang, Y., Tang, N., ... Ji, Q. (2020). Programmable adenine deamination in bacteria using a Cas9–adenine-deaminase fusion. *Chemical Science, 11*(6), 1657–1664. https://doi.org/10.1039/C9SC03784E.

CHAPTER SIXTEEN

A quick guide to evaluating prime editing efficiency in mammalian cells

Chengfang Liu[a,1], Sifan Cheng[a,1], Junjie Zhu[a], Lina Zhou[a], and Jia Chen[a,b,*]

[a]Gene Editing Center, School of Life Science and Technology, ShanghaiTech University, Shanghai, P.R. China

[b]Shanghai Clinical Research and Trial Center, Shanghai, P.R. China

*Corresponding author. e-mail address: chenjia@shanghaitech.edu.cn

Contents

1. Introduction 420
2. Mechanism of prime editor 421
3. Construction of plasmids expressing pegRNA and nicking-sgRNA 424
   3.1 Reagents and materials 424
   3.2 Oligonucleotide sequences 424
   3.3 Protocol 425
   3.4 Notes 428
4. Transfection 429
   4.1 Methods of cell transfection 429
   4.2 Transfection based on liposomes 429
   4.3 Notes 431
5. Examination of editing efficiency 431
   5.1 Methods for assessing editing efficiency 431
   5.2 Examining of editing efficiency based on NGS 431
   5.3 Result 434
6. Conclusion 434
References 435

Abstract

According to the Clinvar database, modeling the diseases associated with pathogenic mutations requires the installation of base substitutions, small insertions or deletions. Prime editor (PE) was recently developed to precisely install any base substitutions and/or small insertions/deletions (indels) in mammalian cells and animals without requiring DSBs or donor DNA templates. PE also offers greater editing and targeting flexibility compared to other precision CRISPR editing methods because the versatile editing information is encoded in the reverse-transcription template of its prime editing guide RNA. However,

[1] These authors contributed equally to this work.

Methods in Enzymology, Volume 712
ISSN 0076-6879, https://doi.org/10.1016/bs.mie.2025.01.016

optimal PE system selection and experimental design can be complex, and there are various factors that can affect PE efficiency. This chapter serves as a rapid entry-level guideline for the application of PE, providing an experimental framework for using PE at a specific genomic locus. *RUNX1* was selected as a representative target site to illustrate the detailed methodology for constructing PE plasmids and the process of transfecting these plasmids into 293FT cells. We further examined the efficiency of PE-mediated genome editing in mammalian cells by using next-generation sequencing.

1. Introduction

Gene editing tools derived from the Clustered Regularly Interspaced Short Palindromic Repeats (CRISPR)-CRISPR associated protein (Cas) system have had a significant impact on life science research and CRISPR-based gene editing drugs (e.g., exagamglogene autotemcel for sickle cell disease and β-thalassemia) have been approved by the U.S. Food and Drug Administration (Frangoul et al., 2021). Four classes of CRISPR-derived editing tools, including CRISPR-Cas nucleases, transposases, base editors, and prime editors, are commonly used in biomedical research. However, each tool has its own limitations, including insufficient editing efficiency, limited editing capabilities, inadequate targeting scope, and poor editing specificity. Cas nucleases, such as Cas9 (Ran et al., 2013) or Cas12a (Wang et al., 2020), can bind to and cleave targeted genomic DNA in living cells. Double-strand breaks (DSBs) generated by Cas nucleases can be utilized for genome editing through homology-directed repair (HDR) to induce precise sequence replacement in the presence of an exogenous DNA template. While HDR has been successful in correcting mutations in certain cell types (Chapman, Taylor, & Boulton, 2012; Cox, Platt, & Zhang, 2015), it remains inefficient in most cell types and often leads to unwanted indels. Base editors (BEs), on the other hand, can precisely install C•G-to-T•A mutations and A•T-to-G•C mutations without requiring DSBs by fusing a Cas moiety with a nucleotide deaminase moiety (Gaudelli et al., 2017; Komor, Kim, Packer, Zuris, & Liu, 2016). However, these methods still face challenges such as bystander editing. The development of CRISPR-targeted transposases has provided inspiration for genome editing in mammalian cells but remains a challenging endeavor (Lampe et al., 2024; Tou, Orr, & Kleinstiver, 2023).

In order to address these limitations, Anzalone et al., (Anzalone et al., 2019) developed prime editing, a revolutionary genome editing technology that allows for the rewriting of genomes with almost any intended changes, including base substitutions, small indels, or combination edits, without the need for DSBs. Prime editor (PE) offers versatile and precise genome editing with minimal off-target effects (Gao et al., 2022; Yang, Yang, & Chen, 2019). Researchers have used PE to precisely edit disease-associated sites in mouse or patient-derived cell models for disease treatment and also installed disease-associated mutations to model the relevant diseases in mice (Davis et al., 2024; Ely et al., 2024; Li et al., 2023; Lin, Liu, et al., 2021; Schene et al., 2020; Sousa et al., 2024; Zhao et al., 2020). In this chapter, we provide a comprehensive guide on the application of PE in mammalian cells, including the experimental framework, plasmid construction, transfection process, and examination of editing efficiency.

2. Mechanism of prime editor

The original prime editing system consists of PE1, a Cas9 (H840A) nickase fused with the Moloney murine leukemia virus (M-MLV) reverse transcriptase (RT), and a prime editor guide RNA (pegRNA) derived from a single guide RNA (sgRNA). The PE2 utilizes an evolved RT that significantly increases editing efficiency. In the PE3 system, an additional sgRNA is introduced to guide the editor to nick the non-edited DNA strand. This additional nick biases the outcomes towards utilizing the edited strand as a replication template, leveraging the nick-directed nature of eukaryotic mismatch repair (MMR) (Komor et al., 2016), thereby enhancing the editing efficiency. PE4 and PE5 further enhance editing efficiency by incorporating a dominant-negative mismatch repair protein (MLH1dn) based on the PE2 and PE3 systems (Chen et al., 2021). Doman et al., employed protein evolution and engineering techniques to develop prime editors with reduced size and improved efficiency, known as PE6 (Doman et al., 2023). More recently, Yan et al., developed a prime editor (PE7) through the fusion to the RNA-binding N-terminal domain of La protein (Yan et al., 2024). Overall, PE offers versatile and precise genome editing with minimal off-target effects (Fig. 1). This chapter provides detailed instructions on utilizing PE in mammalian cells and outlines a specific set of processes for effective implementation.

The detailed mechanism of prime editing involves a complex series of events, influenced by the structure of the pegRNA and PE protein, as well as cellular factors. The pegRNA contains an additional 3′ extension comprising a primer-binding site (PBS), which is complementary to the target genomic site, and a reverse transcription template (RTT) that encodes the desired edit. Once

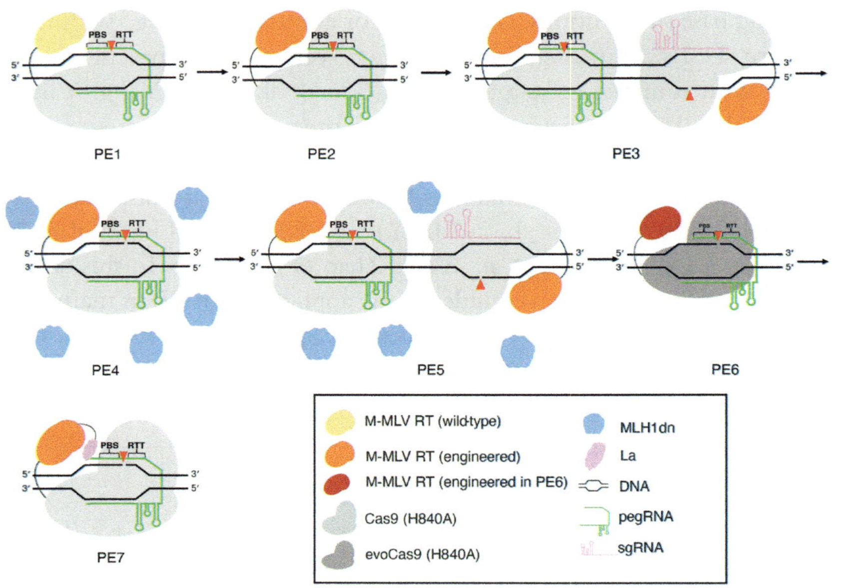

Fig. 1 Development of PE. The PE1 system comprises a Cas9 (H840A) nickase fused with the Moloney murine leukemia virus (M-MLV) reverse transcriptase (RT). It works in conjunction with a prime editing guide RNA (pegRNA) derived from a single guide RNA (sgRNA). The PE2 system incorporates an evolved RT that exhibits significantly enhanced efficiency compared to PE1. In addition, the PE3 system introduces an additional sgRNA that guides the editor to nick the non-edited DNA strand. PE4 and PE5 further enhance editing efficiency by incorporating a dominant-negative mismatch repair protein (MLH1dn), building upon the advancements of the PE2 and PE3 systems. PE6 reduces the tool size and improves efficiency through protein evolution and engineering. PE7 enhances editing efficiency by incorporating the RNA-binding N-terminal domain of the La protein and optimizing the Cas9(H840A).

the pegRNA matches the target site, the nCas9 (H840A) nicks the edited strand, exposing the 3′ end. Subsequently, the PBS of the pegRNA anneals to the 3′ end, serving as a primer, and the RT can extend the flap by synthesizing the desired sequence entirely encoded by the RTT of the pegRNA. The newly synthesized DNA 3′ end incorporates the desired edit. Through intracellular DNA repair mechanisms, both target DNA strands undergo targeted genome editing. Many enhancements have been made to the PE system to improve editing efficiency. These include engineering and evolution of the RT (PE2, PE6), incorporating additional nick sgRNA or inhibitors for the MMR pathway (PE3, PE5), engineering the pegRNA structure (Li et al., 2022; Lin, Jin, et al., 2021; Nelson et al., 2022), or stabilizing the hybridization of pegRNA and DNA strands (PE6). However, the editing efficiency of PE in cells may still be suboptimal due to various factors, including the intracellular environment and the design of PE systems (Fig. 2). Our aim is to provide clear instructions on using PE to assist researchers in applying this technology in scientific research.

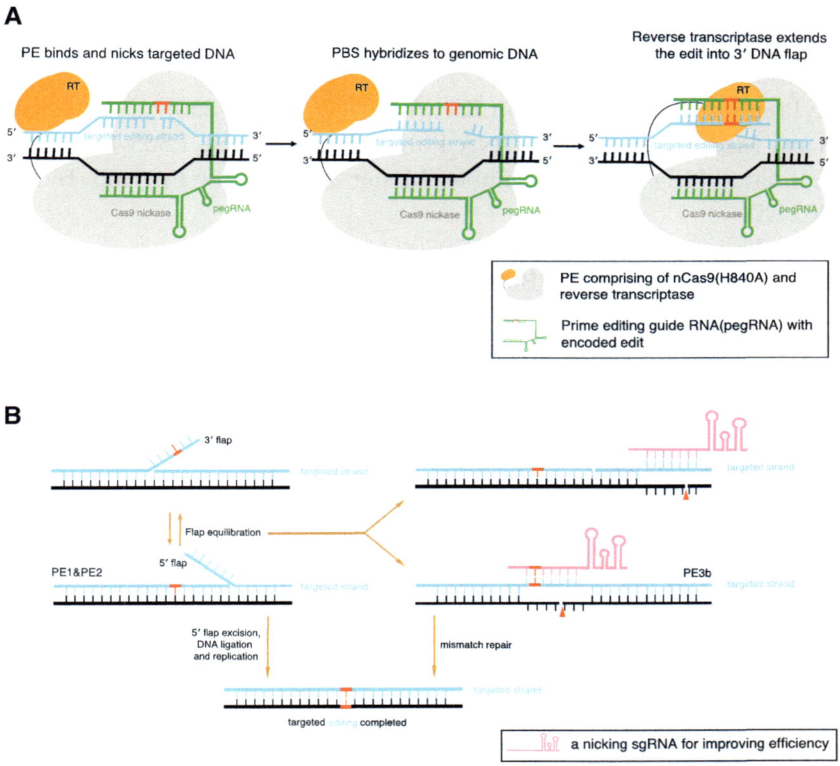

Fig. 2 Mechanism of PE. A. The putative mechanism for prime editing using various editing systems can be illustrated through the following steps. Recruitment of Cas9 nickase (gray) to the target DNA site by a prime editing guide RNA (pegRNA) (green), resulting in a nick on the targeted editing strand (blue) and generating a free 3′ end of DNA. The primer-binding site (PBS) of the pegRNA binds to the genomic DNA flap. The reverse transcriptase (RT) recognizes the hybridization between the targeted DNA and the PBS and it copies the sequence encoded by the reverse transcriptase template (RTT) of the pegRNA. B. The reverse transcription process incorporates the desired edit and extends the downstream homologous DNA. In the PE3 and PE5 systems, a nicking-sgRNA is designed to further enhance editing efficiency by manipulating cellular mismatch repair (MMR). The 5′ flap is degraded within the cell, and the edited 3′ flap is ligated into the genome, completing the editing process. This mechanism illustrates the general steps involved in prime editing using different editing systems. The single-base edit is in red.

3. Construction of plasmids expressing pegRNA and nicking-sgRNA

3.1 Reagents and materials

T4 DNA ligase (New England Biolabs, M0202L)
*Bsa*I-HFv2 (New England Biolabs, R3733L)
*Eco*RI-HF (New England Biolabs, R3101L)
Agarose (Yeasen, 10208ES60)
AxyPrep DNA Gel Extraction Kit (Axygen, AP-GX-250)
Phanta Max Super-Fidelity DNA Polymerase (Vazyme, P505-D1)
NEBuilder HiFi DNA Assembly Master Mix (New England Biolabs, E2621X)
LB medium (Sangon Biotech, A507002-0250)
LB agar medium (Sangon Biotech, A507003-0250)
Ampicillin sodium salt (ABCONE, A08483-100G)
TIANprep Midi Plasmid Kit (TIANGEN,PD106-02)
DH5α Competent Cells (Sangon Biotech, B528413-0100)

3.2 Oligonucleotide sequences

Listed below (Table 1) are the sequences for constructing the plasmids expressing pegRNA targeting the *RUNX1* site (ppegRUNX1+6G-to-C), the nicking-sgRNA (pnicking-sgRUNX1), and the sequences required for subsequent cloning validation. The edited site is indicated by the base in italics.

Table 1 Example primer sequences.

Primer name	Sequence (5′ to 3′)
pegRNA_RUNX1_FOR	CTTGTGGAAAGGACGAAACACCGGGC ATTTTCAGGAGGAAGCGAGTTTTAGA GCTAGAAA
pegRNA_RUNX1_REV	CTCGAGGTCGAGAAT TCAGGAGGAAGCGATG*C*CTTCAGACA GCATATTTGGCACCGACTCGGTGCCA CTT
RUNX1_nick_FOR	CCGGATGAAGCACTGTGGGTACGA
RUNX1_nick_REV	AAACTCGTACCCACAGTGCTTCAT
assembly-F	CGATTAGTGAACGGATCTCGACG

3.3 Protocol

3.3.1 Construction of ppegRUNX1+6G-to-C

Day 1:

1. To obtain the linearized vector, digest pGL3-U6-sgRNA-PGK-Puromycin (Addgene, #51133) with *Bsa*I and *Eco*RI enzymes. Set up a 30 μL digestion reaction (2 h at 37 °C):

Components	Volume	Amount
*Bsa*I-HFv2	1 μL	
*Eco*RI-HF	1 μL	
10 ×rCutSmart	3 μL	
pGL3-U6-psgRNA-PGK-puromycin		1 μg
ddH2O	up to 30 μL	

2. Amplify pGL3-U6-sgRNA-PGK-Puromycin using the primer set pegRNA_RUNX1_FOR/pegRNA_RUNX1_REV to obtain the scaffold fragment containing pegRUNX1+6G-to-C. Set up a 50 μL PCR reaction system:

Components	Volume	Amount
2 × Phanta Max Buffer	25 μL	
dNTP Mix (10 mM each)	1 μL	
pegRNA_RUNX1_FOR (10 μM)	2 μL	
pegRNA_RUNX1_REV (10 μM)	2 μL	
Phanta Max Super-Fidelity DNA Polymerase	1 μL	
pGL3-U6-sgRNA-PGK-Puromycin		25 ng
ddH2O	up to 50 μL	

PCR program:

Step 1	95 °C 3 min
Step 2	95 °C 15 s
	60 °C 15 s
	72 °C 20 s

	Repeat Step 2 for 29 cycles before proceeding to Step3
Step 3	72 °C 5 min
Step 4	4 °C Hold

3. Run 30 μL of the digestion product and 50 μL PCR product on 1% agarose gel and recover the digestion product using the AxyPrep DNA Gel Extraction Kit. Quantify using a spectrophotometer.
4. Clone the amplified pegRUNX1+6G-to-C scaffold fragment into *Bsa*I and *Eco*RI linearized pGL3-U6-sgRNA-PGK-Puromycin vectors using NEBuilder HiFi DNA Assembly Master Mix to generate the vector ppegRUNX1 +6G-to-C for the expression pegRUNX1+6G-to-C. Set up a 10 μL recombination reaction (30 min at 50 °C):

Components	Volume	Amount
Recommended DNA Molar Ratio		vector:insert = 1:3
NEBuilder HiFi DNA Assembly Master Mix	5 μL	
ddH2O	up to 10 μL	

5. Transform 5 μL of the recombinant product into 50 μL of DH5α competent cells by heat shock at 42 °C, recover the cells at 37 °C for 30 min, and plate them on an LA culture dish containing ampicillin. Incubate at 37 °C for 12–16 h.

Day 2:

Select two to four colonies from the culture dish and use assembly-F for Sanger sequencing to verify the correct sequence of bacterial clones.

Day 3:

1. Inoculate the verified bacterial colonies into 15 mL LB medium containing ampicillin and shake overnight at 37 °C.
2. Isolate plasmid DNA using the TIANprep Midi Plasmid Kit to obtain the expression vector ppegRUNX1+6G-to-C. Quantify using a spectrophotometer.

3.3.2 Construction of pnicking-sgRUNX1

Day 1:

1. To obtain the linearized vector, digest pGL3-U6-sgRNA-PGK-Puromycin (Addgene, #51133) with *Bsa*I enzyme. Set up a 30 μL digestion reaction (2 h at 37 °C):

Components	Volume	Amount
*Bsa*I-HFv2	1 μL	
10 ×rCutSmart	3 μL	
pGL3-U6-psgRNA-PGK-puromycin		1 μg
ddH2O	up to 30 μL	

2. Run 30 μL of the digestion product on a 1 % agarose gel and recover the digestion product using the AxyPrep DNA Gel Extraction Kit. Quantify using a spectrophotometer.
3. Anneal the oligonucleotides encoding the complementary region of the spacer. Set up a 20 μL annealing reaction:

Components	Volume
RUNX1_nick_FOR (10 μM)	10 μL
RUNX1_nick_REV (10 μM)	10 μL

PCR program:

Step 1	95 °C 3 min
Step 2	Cool to 25 °C with a rate of 0.1 °C/s
Step 3	4 °C Hold

Note: Annealed products should be prepared and used immediately.

4. Ligate the *Bsa*I-digested backbone of pGL3-U6-sgRNA-PGK-Puromycin with the annealed product using T4 DNA ligase to generate the vector pnicking-sgRUNX1 for the expression nicking-sgRUNX1. Set up a 10 μL ligation reaction (30 min at room temperature):

Components	Volume	Amount
Linearized vector Annealing products	7 μL	50 ng
T4 DNA Ligase	1 μL	
T4 DNA Ligase Buffer (10X)	1 μL	
ddH2O	up to 10 μL	

5. Transform 5 μL of the recombinant product into 50 μL of DH5α competent cells by heat shock at 42 °C, recover the cells at 37 °C for 30 min, and plate them on an LA culture dish containing ampicillin. Incubate at 37 °C for 12–16 h.
6. Refer to Section 3.3.1 for the remaining steps.

3.4 Notes

1. The efficiency of the prime editing system greatly depends on the PBS and RTT components of the pegRNA. Choosing the optimal combination of PBS and RTT for a given locus or edit may not be immediately evident. Therefore, it is crucial to optimize the lengths of PBS and RTT when utilizing the PE system. We suggest researchers start to design pegRNAs with a 13-nt PBS and an RTT longer than 10-nt (depending on the type of edits), and subsequently optimize them according to experimental results, expanding to a 9- to 15-nt PBS and a 10- to 20-nt RTT for testing. For the identification of pegRNAs that have the potential for high editing efficiency, we recommend utilizing design tools such as pegFinder (Chow, Jennifer, Shen, & Chen, 2021) (https://sidichenlab.org/pegfinder/).
2. Engineered pegRNAs, such as spegRNA, apegRNA (Li et al., 2022), and epegRNA (Nelson et al., 2022), can potentially improve the editing efficiency of pegRNAs. Different pegRNA scaffolds can be selected based on specific experimental requirements.
3. PegRNAs with additional base substitutions have the potential to induce higher expected single-base editing efficiencies (Li et al., 2022). Additional point mutations can be introduced into the RTT as needed for the particular experiment.

4. Transfection

Various transfection methods, including chemical transfection, physical transfection, and virus-mediated transfection, exhibit different mechanisms and effects. It is essential to carefully select the most appropriate transfection method based on specific experimental goals and cell types. This selection plays a critical role in enhancing gene editing efficiency and ensuring reliable results.

4.1 Methods of cell transfection

We have listed some methods for cell transfection (Table 2).

4.2 Transfection based on liposomes

Here we describe the transfection of plasmids expressing pegRUNX1+6G-to-C, nicking-sgRUNX1 and the PE3-expressing plasmid (pPE3) into 293FT cells using Lipofectamine LTX.

4.2.1 Reagents and materials

293FT (Thermo Fisher Scientific, R70007)
DMEM (Gibco/Thermo Fisher Scientific, 10566)
FBS (Gibco/Thermo Fisher Scientific, 16000044)
Lipofectamine™ LTX Reagent with PLUS™ Reagent (Thermo Fisher Scientific, 15338100)
Puromycin solution (InvivoGen, ant-pr-1)
QuickExtract™ DNA Extraction Solution (Epicentre, QE09050)

4.2.2 Protocol

Day 1:

Seed 293FT cells into a 24-well plate at a density of 0.7 × 10^5 cells per well and allow them to adhere and grow overnight.

Day 2:

Transfect each well with 100 μL of serum-free Opti-MEM containing:

Components	Volume	Amount
Lipofectamine LTX	2.6 μL	
Lipofectamine PLUS	1.3 μL	
pPE3		0.9 μg
ppegRUNX1+6G-to-C		0.3 μg
pnicking-sgRUNX1		0.1 μg

Table 2 Comparison of transfection methods.

Method	Principle	Advantages	Limitations
Lipid-Based Transfection	Uses liposomes to deliver nucleic acids via membrane fusion.	Relatively high efficiency, versatile, easily scalable.	Can cause cytotoxicity, usually results in transient expression.
Electroporation	Applies an electrical field to create pores in the cell membrane for nucleic acid entry.	High efficiency, broad applicability.	Can reduce cell viability, requires optimization.
Viral Transduction	Uses viral vectors to deliver nucleic acids into cells via infection.	Relatively high efficiency, stable integration possible.	Biosafety concerns, complex production.
Physical Methods	Direct delivery of nucleic acids via microinjection or biolistics (gene gun).	Precision, effective for various cell types.	Labor-intensive, requires specialized equipment, not suitable for high-throughput.

Note: The amount of plasmid and the ratio of transfection reagent can be adjusted to meet specific experimental needs.

Day 3:

To maximize transfection efficiency, add puromycin to the cell culture medium to a final concentration of 4 μg/mL.

Note: The final concentration of puromycin can be determined based on the growth status and density of cells.

Day 5:

Isolate genomic DNA from transfected cells using QuickExtract™ DNA Extraction Solution according to the manufacturer's instructions for subsequent sequencing analysis.

4.3 Notes

1. Experiment Optimization: When conducting similar experiments in different cell lines, it may be necessary to optimize the transfection system, including adjusting the amount of transfection reagent, medium composition, and other experimental conditions.
2. Transfection Method Selection: Depending on the specific characteristics of different cell lines, it may be necessary to choose alternative transfection methods to enhance transfection efficiency and maintain cell viability.

5. Examination of editing efficiency

Assessing the editing efficiency of target sites is vital for optimizing the experimental approach and ensuring the accuracy and reliability of the results.

5.1 Methods for assessing editing efficiency

Some methods for assessing the efficiency of genome target site editing are listed in Table 3.

5.2 Examining of editing efficiency based on NGS

Here, we describe the assessing of editing efficiency at *RUNX1*_Genomic site through NGS.

5.2.1 Reagents and materials

High fidelity DNA polymerase: Phanta Max Super Fidelity DNA Polymerase (vazyme, P505-D1)

Primer: RUNX1_G_FOR/RUNX1_G_REV

DNA template: genome from edited cells collected in Section 4.2.2

Agarose (Yeasen, 10208ES60)

Table 3 Comparison of genome editing efficiency assessing techniques.

Method	Principle	Advantages	Limitations
Sanger Sequencing and TIDE Analysis	Utilizes PCR amplification, Sanger sequencing, and chromatogram analysis	Simple and accessible, enables direct quantification	Limited sensitivity and labor-intensive
Next-Generation Sequencing (NGS)	Employs high-throughput sequencing for detailed analysis	Offers high sensitivity and precision with comprehensive analysis	Higher cost and requires sophisticated bioinformatics tools
Restriction Fragment Length Polymorphism (RFLP) Analysis	Involves restriction enzyme digestion and fragment pattern analysis	Cost-effective and allows quick assessment	Detects only specific site modifications and is less precise
Fluorescence-Based Reporter Assays	Utilizes fluorescent markers integrated into the genome	Enables real-time assessment with high throughput	Requires specific constructs and potential variability due to unrelated factors
Quantitative PCR (qPCR)	Quantifies edited versus unedited alleles through specific primer amplification	Offers high sensitivity with precise measurements	Requires careful primer design and is influenced by PCR conditions

5.2.2 Protocol

1. Primer design: Amplify the fragment containing *RUNX1* target site sequence. Always design primers that are at least 250 bps away from the 5′ and 3′ - ends of the sgRNA target site.

Primer name	Sequence (5′ to 3′)
RUNX1_G_FOR	GCTTGGGGAGTCCCAGAGGT
RUNX1_G_REV	TTCTCACGCACCGACTGAAC

Note : Ensure primers are specific to the target region and avoid regions with secondary structures or repetitive sequences.

2. PCR Reaction Setup: Establish 50 μL PCR reaction system.

Components	Volume
2 × Phanta Max Buffer	25 μL
dNTP Mix (10 mM each)	1 μL
RUNX1_G_FOR (10 μM)	2 μL
RUNX1_G_REV (10 μM)	2 μL
Phanta Max Super-Fidelity DNA Polymerase	1 μL
Template DNA	X μL
ddH2O	up to 50 μL

PCR program:

Step 1	95 °C 3 min
Step 2	95 °C 15 s
	58 °C 15 s
	72 °C 30 s
	Repeat Step 2 for 29 cycles before proceeding to Step3
Step 3	72 °C 5 min
Step 4	4 °C Hold

Note: Use high-quality and properly quantified genomic DNA to avoid inhibitors that could affect the PCR reaction.

3. Gel Electrophoresis: Run 2 μL of the PCR product on a 1.5 % agarose gel to verify the size and purity of the amplicon.

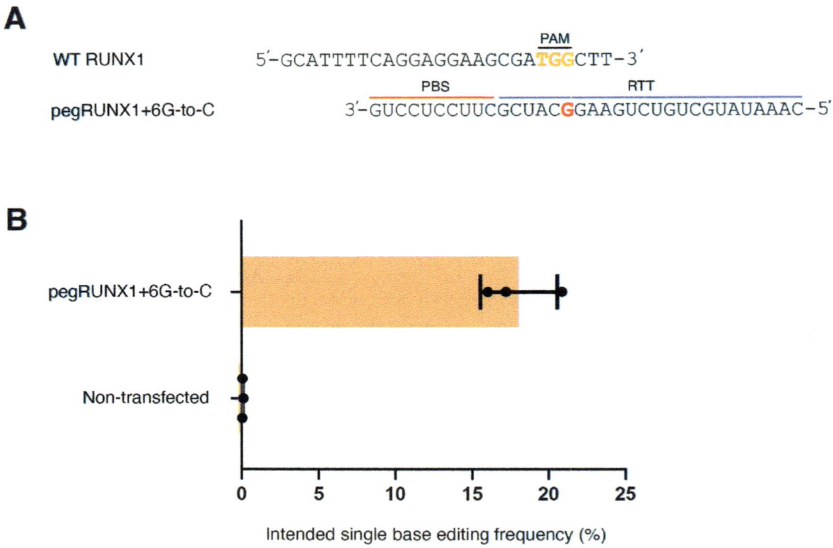

Fig. 3 Prime editing efficiency at *RUNX1* target site. A. The sequences of wild-tpye (WT) *RUNX1* target site and PBS and RTT of the pegRNA for the intended +6G-to-C editing (pegRUNX1 +6G-to-C). Intended single-base edit is in red and PAM is in orange. B. The intended single-base editing frequencies in the transfected 293FT cells. The data are presented as the mean ± s.d. from three independent experiments.

4. Data Analysis: After confirming a single sharp band on the agarose gel, proceed with PCR product gel extraction to isolate the target amplicon. Subsequently, perform NGS to examine the editing efficiency of the PE3 system at the designated target site.

5.3 Result

Here we performed NGS to determine the editing efficiency at the *RUNX1* target site. The NGS results show that the co-transfection of pPE3, ppegRUNX1+6G-to-C and pnicking-sgRUNX1 achieved average 18.06 % G-to-A editing efficiency at the *RUNX1* target site (Fig. 3).

6. Conclusion

In recent years, there have been remarkable advancements in the development of precise and versatile genome editing tools. These advancements have brought us closer to the ultimate goal of making specific sequence changes in the genomes of living cells without generating unnecessary or unintended editing products. Among these cutting-edge tools, PE stands out as a promising one as it can generate versatile editing products with high purity. We expect this protocol can provide helpful guidance for the researchers who start to use PE for scientific explorations.

References

Anzalone, A. V., Randolph, P. B., Davis, J. R., Sousa, A. A., Koblan, L. W., Levy, J. M., ... Liu, D. R. (2019). Search-and-replace genome editing without double-strand breaks or donor DNA. *Nature, 576*, 149–157.

Chapman, J. R., Taylor, M. R., & Boulton, S. J. (2012). Playing the end game: DNA double-strand break repair pathway choice. *Molecular Cell, 47*, 497–510.

Chen, P. J., Hussmann, J. A., Yan, J., Knipping, F., Ravisankar, P., Chen, P. F., ... Liu, D. R. (2021). Enhanced prime editing systems by manipulating cellular determinants of editing outcomes. *Cell, 184*, 5635–5652.e29.

Chow, R. D., Jennifer, S. C., Shen, J., & Chen, S. (2021). A web tool for the design of prime-editing guide RNAs. *Nature Biomedical Engineering, 5*, 190–194.

Cox, D. B., Platt, R. J., & Zhang, F. (2015). Therapeutic genome editing: Prospects and challenges. *Nature Medicine, 21*, 121–131.

Davis, J. R., Banskota, S., Levy, J. M., Newby, G. A., Wang, X., Anzalone, A. V., ... Liu, D. R. (2024). Efficient prime editing in mouse brain, liver and heart with dual AAVs. *Nature Biotechnology, 42*, 253–264.

Doman, J. L., Pandey, S., Neugebauer, M. E., An, M., Davis, J. R., Randolph, P. B., ... Liu, D. R. (2023). Phage-assisted evolution and protein engineering yield compact, efficient prime editors. *Cell, 186*, 3983–4002.e26.

Ely, Z. A., Mathey-Andrews, N., Naranjo, S., Gould, S. I., Mercer, K. L., Newby, G. A., ... Jacks, T. (2024). A prime editor mouse to model a broad spectrum of somatic mutations in vivo. *Nature Biotechnology, 42*, 424–436.

Frangoul, H., Altshuler, D., Cappellini, M. D., Chen, Y. S., Domm, J., Eustace, B. K., ... Corbacioglu, S. (2021). CRISPR-Cas9 gene editing for sickle cell disease and β-thalassemia. *The New England Journal of Medicine, 384*, 252–260.

Gao, R., Fu, Z. C., Li, X., Wang, Y., Wei, J., Li, G., ... Chen, J. (2022). Genomic and transcriptomic analyses of prime editing guide RNA-independent off-target effects by prime editors. *The CRISPR Journal, 5*, 276–293.

Gaudelli, N. M., Komor, A. C., Rees, H. A., Packer, M. S., Badran, A. H., Bryson, D. I., & Liu, D. R. (2017). Programmable base editing of A•T to G•C in genomic DNA without DNA cleavage. *Nature, 551*, 464–471.

Komor, A. C., Kim, Y. B., Packer, M. S., Zuris, J. A., & Liu, D. R. (2016). Programmable editing of a target base in genomic DNA without double-stranded DNA cleavage. *Nature, 533*, 420–424.

Lampe, G. D., King, R. T., Halpin-Healy, T. S., Klompe, S. E., Hogan, M. I., Vo, P. L. H., ... Sternberg, S. H. (2024). Targeted DNA integration in human cells without double-strand breaks using CRISPR-associated transposases. *Nature Biotechnology, 42*, 87–98.

Li, C., Georgakopoulou, A., Newby, G. A., Chen, P. J., Everette, K. A., Paschoudi, K., ... Lieber, A. (2023). In vivo HSC prime editing rescues sickle cell disease in a mouse model. *Blood, 141*, 2085–2099.

Li, X., Zhou, L., Gao, B.-Q., Li, G., Wang, X., Wang, Y., ... Chen, J. (2022). Highly efficient prime editing by introducing same-sense mutations in pegRNA or stabilizing its structure. *Nature Communications, 13*, 1669.

Lin, J., Liu, X., Lu, Z., Huang, S., Wu, S., Yu, W., ... Liu, Z. (2021). Modeling a cataract disorder in mice with prime editing. *Molecular Therapy Nucleic Acids, 25*, 494–501.

Lin, Q., Jin, S., Zong, Y., Yu, H., Zhu, Z., Liu, G., ... Gao, C. (2021). High-efficiency prime editing with optimized, paired pegRNAs in plants. *Nature Biotechnology, 39*, 923–927.

Nelson, J. W., Randolph, P. B., Shen, S. P., Everette, K. A., Chen, P. J., Anzalone, A. V., ... Liu, D. R. (2022). Engineered pegRNAs improve prime editing efficiency. *Nature Biotechnology, 40*, 402–410.

Ran, F. A., Hsu, P. D., Wright, J., Agarwala, V., Scott, D. A., & Zhang, F. (2013). Genome engineering using the CRISPR-Cas9 system. *Nature Protocols, 8*, 2281–2308.

Schene, I. F., Joore, I. P., Oka, R., Mokry, M., van Vugt, A. H. M., van Boxtel, R., ... Fuchs, S. A. (2020). Prime editing for functional repair in patient-derived disease models. *Nature Communications, 11*, 5352.

Sousa, A. A., Hemez, C., Lei, L., Traore, S., Kulhankova, K., Newby, G. A., ... Liu, D. R. (2024). Systematic optimization of prime editing for the efficient functional correction of CFTR F508del in human airway epithelial cells. *Nature Biomedical Engineering*.

Tou, C. J., Orr, B., & Kleinstiver, B. P. (2023). Precise cut-and-paste DNA insertion using engineered type V-K CRISPR-associated transposases. *Nature Biotechnology, 41*, 968–979.

Wang, X., Ding, C., Yu, W., Wang, Y., He, S., Yang, B., ... Chen, J. (2020). Cas12a base editors induce efficient and specific editing with low DNA damage response. *Cell Reports, 31*, 107723.

Yan, J., Oyler-Castrillo, P., Ravisankar, P., Ward, C. C., Levesque, S., Jing, Y., ... Adamson, B. (2024). Improving prime editing with an endogenous small RNA-binding protein. *Nature, 628*, 639–647.

Yang, L., Yang, B., & Chen, J. (2019). One prime for all editing. *Cell, 179*, 1448–1450.

Zhao, W., Li, J., Wang, X., Xu, W., Gao, B. Q., Xiang, J., ... Yang, B. (2023). Prime editor-mediated functional reshaping of ACE2 prevents the entry of multiple human coronaviruses, including SARS-CoV-2 variants. *MedComm, 4*(2020), e356.

CHAPTER SEVENTEEN

Using Prime Editing Guide Generator (PEGG) for high-throughput generation of prime editing sensor libraries

Samuel I. Gould[a,b] **and Francisco J. Sánchez-Rivera**[a,b,*]
[a]Department of Biology, Massachusetts Institute of Technology, Cambridge, MA, United States
[b]David H. Koch Institute for Integrative Cancer Research, Massachusetts Institute of Technology, Cambridge, MA, United States
*Corresponding author. e-mail address: fsr@mit.edu

Contents

1. Introduction 438
2. Materials 440
3. Methods 441
   3.1 Installation 441
   3.2 Mutation input formatting 441
   3.3 Reference genome loading 443
   3.4 Running PEGG to generate pegRNAs 443
   3.5 Sensor design options 445
   3.6 Exhaustive description of PEGG output 445
   3.7 Generating non- and safe-targeting pegRNAs with the library module 447
   3.8 Oligonucleotide design 448
   3.9 Visualization of pegRNAs 449
4. Notes 449
Acknowledgements 450
Competing interests 450
References 450

Abstract

Prime editing enables the generation of nearly any small genetic variant. However, the process of prime editing guide RNA (pegRNA) design is challenging and requires automated computational design tools. We developed Prime Editing Guide Generator (PEGG), a fast, flexible, and user-friendly Python package that enables the rapid generation of pegRNA and pegRNA-sensor libraries. Here, we describe the installation and use of PEGG (https://pegg.readthedocs.io) to rapidly generate custom pegRNA-sensor libraries for use in high-throughput prime editing screens.

Methods in Enzymology, Volume 712
ISSN 0076-6879, https://doi.org/10.1016/bs.mie.2025.01.006

1. Introduction

Prime editing enables the generation of nearly any small genetic variant, including all types of nucleotide substitutions, as well as insertions and deletions, across multiple organisms and mammalian cell types (Anzalone et al., 2019; Doman, Sousa, Randolph, Chen, & Liu, 2022; Nelson et al., 2022). Prime editing is a modular system wherein a prime editor protein, consisting of a Cas9 nickase fused to a reverse transcriptase, installs genomic edits based on the information encoded in a prime editing guide RNA (pegRNA). These pegRNAs include a "search" sequence, a protospacer that directs the prime editor complex to a genetic locus, and a "replace" sequence, a 3′ extension that provides a template for reverse transcription-mediated introduction of a defined genomic edit. The 3′ extension is further divided into two components: (1) a primer binding site (PBS) that binds to the exposed DNA flap introduced by the Cas9 nickase and serves as a primer for reverse transcription, and (2) a reverse transcription template (RTT) that templates the synthesis of a single strand of DNA containing the genomic edit of interest (Fig. 1A).

Compared to traditional guide RNAs (gRNAs) used for CRISPR nuclease, CRISPRi/a, or base editing, which solely consist of a variable protospacer sequence and constant scaffold region, pegRNAs are considerably more complex due to the addition of the 3′ extension that templates that DNA strand synthesized by reverse transcriptase. For a given genomic edit of interest, there often exist hundreds of possible pegRNA designs because of the vast combinatorial space that emerges from varying the choice of protospacer, as well as the sequence content and length of the PBS and RTT.

Prime editing is increasingly being used by our group and others to interrogate thousands of genetic variants in parallel in high-throughput screens (Belli, Karava, Farouni, & Platt, 2024; Cirincione et al., 2024; Erwood et al., 2022; Gould et al., 2024; Kim, Oh, Lee, & Kim, 2024; Ren et al., 2023). Due to the high variability in the editing efficiency of pegRNAs, these applications require the generation of dozens of pegRNAs for each mutation of interest. As a result, high-throughput prime editing screens necessitate designing tens or hundreds of thousands of pegRNAs. To enable the expansion of these approaches, there is a need for computational tools that are able to easily and rapidly design and rank many possible pegRNAs. To this end, we developed Prime Editing Guide Generator (PEGG), a Python package that allows users to design pegRNA libraries within user-friendly Jupyter notebooks (https://pegg.readthedocs.io).

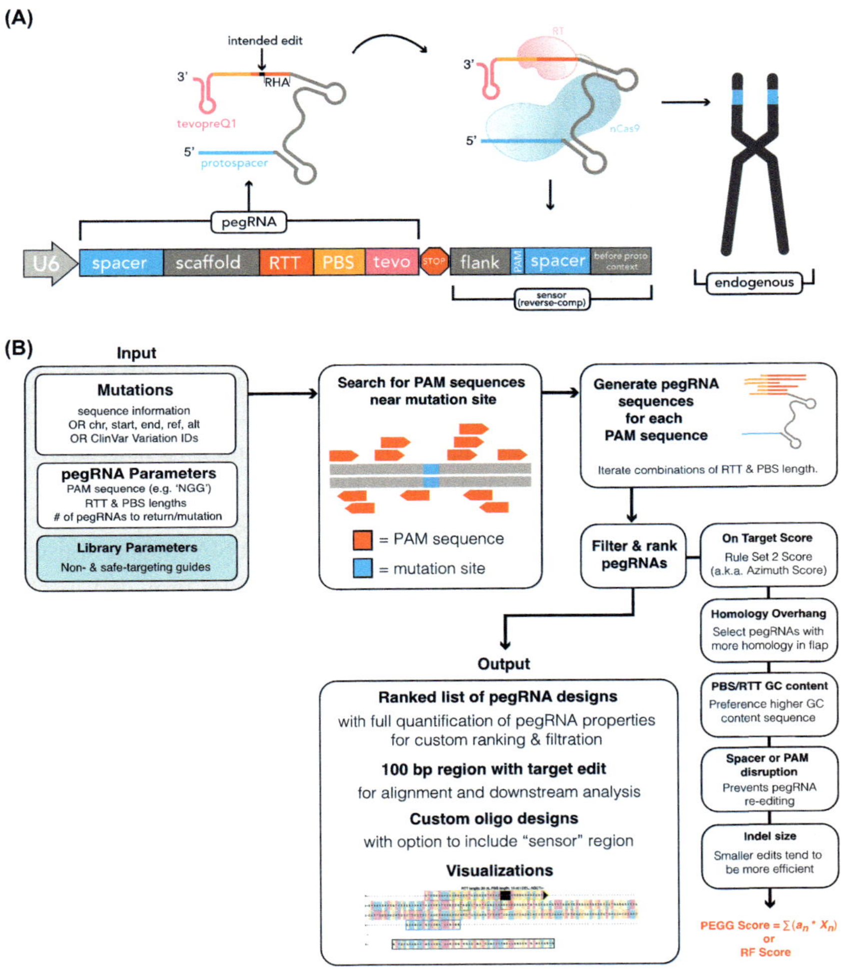

Fig. 1 Using PEGG to generate pegRNA-sensor libraries. (A) Schematic of a pegRNA-sensor construct generated by PEGG. Spacer, protospacer; RTT, reverse transcription template; PBS, primer binding site; tevo, tevopreQ1 pseudoknot motif; RHA, right homology arm; STOP, polyT sequence to stop U6 transcription; nCas9, nicking Cas9; RT, reverse transcriptase. (B) Overall schematic of the PEGG pipeline.

A number of computational tools have been created to design pegRNAs. This includes user-friendly web-based tools that facilitate the design of one or a small number of pegRNAs (Anderson, Haldrup, Thomsen, Wolff, & Mikkelsen, 2021; Chow, Chen, Shen, & Chen, 2021; Hwang et al., 2021; Li, Chen, Tsai, & Cheng, 2021; Mathis et al., 2023; Morris, Rahman, Guo, & Sanjana, 2021), as well as locally installed software solutions and command line tools that enable higher throughput design but can be difficult to use depending on the user's experience (Bhagwat et al., 2020; Hsu et al., 2021; Standage-Beier, Tekel, Brafman, & Wang, 2021). Several groups have also undertaken machine learning efforts to predict pegRNA efficiency and identify optimal pegRNAs (Kim et al., 2021; Koeppel et al., 2023; Li et al., 2021; Mathis et al., 2023; Mathis et al., 2024; Yu et al., 2023). These approaches, while important and useful, quickly become computationally intensive to run for projects that require the design of large pegRNA libraries.

What makes PEGG uniquely useful is its speed, native Python integration, flexibility of input mutation formatting, and automated sensor design (Fig. 1B). First, PEGG is fast–capable of designing tens of thousands of pegRNAs in minutes. Second, PEGG is available for installation through the Python Package Index (PyPI) and is designed to be imported and used natively within interactive Jupyter notebook environments for ease of use. Third, PEGG can take as input a range of mutation formats, including the sequence of the mutation (the standard input format for most pegRNA design software), ClinVar IDs (Landrum et al., 2018), and mutation tables provided in cBio-Portal format that provide genome coordinates for mutations of interest (Cerami et al., 2012). This flexibility allows users to generate pegRNAs for engineering mutations of interest without labor-intensive reformatting of this information, lowering the barriers to entry for pegRNA design. Fourth, PEGG automates the design of sensor sequences—synthetic copies of the endogenous target site that enable empirical evaluation of pegRNA editing efficiency and precision in a pooled format without the need to sequence each targeted endogenous locus (Gould et al., 2024; Gould, 2024) (Fig. 1A). Lastly, PEGG is the only software package that automates the process of designing pegRNA sensors, which have been recently shown by our group and others to significantly increase the signal-to-noise ratio in complex pooled prime editing screens (Cirincione et al., 2024; Gould et al., 2024).

This manuscript describes in-depth the implementation of PEGG to rapidly design pegRNA-sensor libraries for high-throughput prime editing screens.

2. Materials

1. A computer with at least 16 Gb of RAM.
2. Python version 3.9 or 3.10

3. pandas (Python package; any version)
4. *Optional*: Genome sequence of the organism that is the target of prime editing (e.g. human, mouse, etc.).

3. Methods

3.1 Installation

PEGG can be installed from the Python Package Index (PyPI). In the command line, or within a Jupyter notebook environment, use the following command:

```
> pip install pegg.
```

PEGG has been tested with python versions 3.9 and 3.10. Python versions higher than 3.10 are not compatible with the scikit-learn package version needed to compute protospacer on-target scores. To get it to install, you may need to use a conda virtual environment:

```
> conda create -n my_environment_name python= 3.9.
> conda activate my_environment_name.
> pip install pegg.
```

3.2 Mutation input formatting

PEGG accepts multiple formats for the mutations that users seek to engineer. This includes two acceptable formats that are purely sequence based (i.e. the user provides the sequences they want to prime edit), and two formats based on referencing the genome of the organism under consideration. These options are listed below:

3.2.1 WT-ALT format

In this format, the user feeds in a set of wildtype sequences, and the corresponding set of desired mutant sequences. PEGG then uses a built-in aligner to determine the mutation from these sequences, and design pegRNAs for these variants. Note: Complex INDELs will break this function because they are difficult to align. If you want to design INDELs, use one of the two formats below.

The pandas DataFrame must have at least 2 columns, with the following column names (1) 'WT' and (2) 'ALT'.

3.2.2 PrimeDesign format

In this format, users feed in a set of sequences with the desired edit in PrimeDesign format. For example, an A > G SNP would look like: "AATCG(A/G)GCTAG", a "GTT" insertion would be: "AATCG (/GTT)GCTAG", a "GTT" deletion: "AATCG(GTT/)GCTAG", and a "C > GTT" INDEL: "AATCG(C/GTT)GCTAG". It is recommended that you include at least 100 nt of context sequence on either side of the variant.

The pandas DataFrame must have at least 1 column, with the following column name: (1) 'SEQ'. This column contains the PrimeDesign formatted sequences.

3.2.3 cBioPortal format

In this format, users simply provide a set of variants with their genome coordinates and reference and alternate alleles, as well as the variant type. This format is compatible with all of the datasets in the cBioPortal (Cerami et al., 2012), which contains cancer-associated variant datasets.

If you want to build you own mutation dataset, it must have the following columns, with the associated column header names, in.csv format:

A. **Chromosome:** The chromosome where the location is located. Use integers or 'X' and 'Y'.
B. **Start_Position:** The start position of mutation in the reference genome (mutations should only be reported on the + strand)
C. **End_Position**: The end position of mutation in the reference genome
D. **Variant_Type:** The type of mutation. Options: "SNP", "ONP", "INS", "DEL", "INDEL"
E. **Reference_Allele:** The reference allele. For insertions, this should be set to "-".
F. **Tumor_Seq_Allele2:** The mutant allele (i.e. what is the mutation sequence). For deletions, this should be set to "-".

3.2.4 ClinVar IDs

PEGG also has a built-in function for translating ClinVar IDs to the "cBioPortal" input format. To do so, download a ClinVar vcf.gz file, and choose your desired Variation IDs that you wish to model. These vcf.gz files can be accessed here: https://ftp.ncbi.nlm.nih.gov/pub/clinvar/. To use this function, open a Python computing environment or Jupyter notebook, import the PEGG package, and the run the ClinVar translation function on your list of ClinVar Variation IDs:

```
> from pegg import prime.
> filepath = '.../GrCh37_clinvar_20230923.vcf.gz'.
> variation_ids = [925574, 925434, 926695, 925707,
  325626, 1191613, 308061, 361149, 1205375, 208043].
> df = prime.clinvar_VCF_translator(filepath, var-
  iation_ids).
```

The resulting pandas DataFrame can be used to generate pegRNAs.

3.3 Reference genome loading

In addition, to use the "cBioPortal" input format, you must provide a genome that PEGG can use to find the associated sequences. There is a built in genome loader function that you can use, or you can format the chromosome sequences in a dictionary, with the keys as the chromosome identifiers:

```
> from pegg import prime.
#filepath to fna.gz file containing reference
 genome (here = GrCh37).
> filepath = './GCF_000001405.25_GRCh37.p13_
  genomic.fna.gz'.
> chrom_dict, i = prime.genome_loader(filepath).
```

This function has been tested on human and mouse genomes in gzipped FastA format.

3.4 Running PEGG to generate pegRNAs

At the most basic level, users can run PEGG with the default parameters enabled to generate pegRNAs for a set of mutations in one of the aforementioned input formats. Below is an example of generating pegRNAs in PrimeDesign format, where "input" is a pandas DataFrame containing the mutations of interest. The output of this function is a pandas DataFrame containing the pegRNAs.

```
> from pegg import prime.
#options = 'cBioPortal', 'WT_ALT', 'PrimeDesign'.
> input_format = 'PrimeDesign'.
> pegRNAs = prime.run(input, input_format).
```

Alternatively, if working in "cBioPortal" format, you must also provide the reference genome that has been formatted into a dictionary by the prime.genome_loader() function (see above). Here input is a pandas DataFrame in "cBioPortal" format:

```
> from pegg import prime.
#options = 'cBioPortal', 'WT_ALT', 'PrimeDesign'.
> input_format = 'cBioPortal'.
> pegRNAs = prime.run(input, input_format, chrom_-
  dict=chrom_dict).
```

In addition, users are provided the option of altering the default parameters of the prime.run() function to change the properties of the pegRNAs that are generated, as well as the number and ranking of pegRNAs generated for each mutation. This includes choosing the PAM sequence, RTT lengths, PBS lengths, and ranking system, among other options, which are exhaustively listed below (Fig. 1A):

A. **PAM:** PAM sequence for searching. Default = "NGG". Can include any nucleic acid code (e.g. PAM = "NRCH").
B. **pegRNAs_per_mut:** How many pegRNAs to produce per mutation. Default = 'All' (all possible pegRNAs with parameters). Otherwise, choose an integer value (e.g. 5).
C. **rankby:** What pegRNA parameter to rank pegRNAs by. Options = "PEGG2_Score" (weighted linear regression of different pegRNA parameters) or "RF_Score" (random forest predictor of pegRNA efficiency).
D. **RTT_lengths:** List containing RTT lengths of designed pegRNAs. Default = [5,10,15,20,25,30].
E. **PBS_lengths:** List containing PBS lengths of designed pegRNAs. Default = [10,13,15].
F. **min_RHA_size:** Minimum size of the RHA (Right homology arm). Default = 1. pegRNAs below this value with be filtered from the output.
G. **RE_sites:** A list containing the restriction enzyme recognition sites to recognize and denote in the output (e.g. ['CGTCTC', 'GAATTC'] for *Esp*3I and *Eco*RI). Default = None (no filtration).
H. **polyT_threshold:** The length of the polyT sequence to classify as a U6 terminator and denote in the output. Default = 4.
I. **proto_size:** The length of the protospacer (excluding the appended G at the 5′ end). Default = 19 (G+19).

For example, a user that wants to generate pegRNAs with (1) an "NRCH" PAM sequence, (2) searching over RTT lengths ranging from 10 to 15 nt, (3) searching over PBS lengths equal to 12 nt, (4) with a minimum right homology arm length of 5 nt, (5) producing the top 5 ranked pegRNAs for each mutant, and (6) ranking the pegRNAs by the random forest score would execute the function as follows:

```
> from pegg import prime.
> input_format = 'cBioPortal'.
> pegRNAs =prime.run(input, input_format, chrom_-
  dict=chrom_dict, PAM="NRCH", RTT_lengths=[10–15],
  PBS_lengths=[12], min_RHA_size=5, pegRNAs_per_
  mut=5, rankby="RF_Score").
```

3.5 Sensor design options

In addition to the option to alter the parameters of the pegRNAs generated by the prime.run() function, users can also alter the design of the sensor sequence. This includes altering the length of the sensor, its orientation relative to the protospacer, as well as the length of sequence between the start of the protospacer and the end of the sensor sequence (Fig. 1A). These options are described below:

A. sensor: The options are True or False. Default = True. If True, PEGG generates sensor sequences for each pegRNA.

B. sensor_length: An integer value for the length of the sensor sequence. Default = 60.

C. sensor_orientation: The orientation of the sensor sequence relative to the protospacer. Options provided are "forward" or "reverse-complement". Default = "reverse-complement".

D. before_proto_context: Length of nucleotide context to put before the protospacer in the sensor. Default = 5.

For example that wants to generate sensors whose length is 50 nt, with 10 nt of context sequence prior to the protospacer, with the sensor in the forward orientation relative to the protospacer would use the following command:

```
> from pegg import prime.
> input_format = 'cBioPortal'.
> pegRNAs =prime.run(input, input_format, chrom_-
  dict=chrom_dict, sensor=True, sensor_length=50,
  sensor_orientation="forward", before_proto_
  context=10).
```

3.6 Exhaustive description of PEGG output

The prime.run() function produces a pandas DataFrame that contains the sequences and descriptions of the generated pegRNAs. The returned

DataFrame also automatically includes any information/columns that were present in the input DataFrame provided to the prime.run() function. The details of the output column are provided below:

A. **PAM:** Sequence of the PAM for a given pegRNA.
B. **PAM_strand:** "+ " or "-" indicating whether the PAM sequence is on the plus or minus strand of the input sequence.
C. **Protospacer_30:** The protospacer sequence with additional context to allow for on-target cutting efficiency prediction.
D. **Protospacer:** The true protospacer sequence, including a "G" appended to the 5′ end to prime U6-mediated transcription.
E. **RTT:** The sequence of the RTT in 5′ to 3′ orientation.
F. **PBS:** The sequence of the PBS in 5′ to 3′ orientation.
G. **RTT_PBS:** The sequence of the full 3′ extension (PBS and RTT) in 5′ to 3′ orientation.
H. **RTT_length:** The length of the RTT for the given pegRNA.
I. **PBS_length:** The length of the PBS for the given pegRNA.
J. **PAM_disrupted:** True/False value indicating whether the pegRNA's PAM sequence is disrupted by the introduction of the intended mutation.
K. **Proto_disrupted:** True/False value indicating whether the pegRNA's protospacer sequence is disrupted by the introduction of the intended mutation.
L. **Distance_to_nick:** The nucleotide distance between the nick site (3 nucleotides prior to the PAM sequence) and the start of the mutated sequence.
M. **RHA_size:** The size of the right homology arm (RHA). In other words, the nucleotide sequence length spanning from the end of the mutated sequence to the end of the RTT.
N. **OnTarget_Azimuth_Score:** The Rule Set 2 (Azimuth) Score for the protospacer (Doench et al., 2016).
O. **PBS_GC_content:** The GC content of the PBS sequence (as a fraction from 0 to 1).
P. **RTT_GC_content:** The GC content of the RTT sequence (as a fraction from 0 to 1).
Q. **indel_size:** The size of the indel introduced by the mutation.
R. **PEGG2_Score:** The linear PEGG2 score for the pegRNA.
S. **RF_Score**: The random forest score for the pegRNA.
T. **pegRNA_rank:** The ranking of the pegRNA relative to other pegRNAs designed to engineer the same mutation. This is based on the user's choice of the "rankby" parameter in the prime.run() function.

U. **sensor_wt:** The wildtype sensor sequence.
V. **sensor_alt:** The sensor sequence with the intended mutation.
W. **sensor_orientation:** The orientation of the sensor sequence relative to the protospacer.
X. **sensor_error:** This indicates whether there was an error in generating the sensor sequence. In certain cases where deletions are large, the sensor may be too short to accommodate the entire sequence, in which case the error is indicated here.
Y. **contains_polyT_terminator:** True/False value indicating whether the pegRNA contains a polyT terminator sequence of a length greater than or equal to that provided by the "polyT_threshold" parameter in prime.run().
Z. **contains_RE_site:** True/False value indicating whether the pegRNA contains any of the restriction enzyme sites provided by the "RE_sites" parameter in prime.run().

3.7 Generating non- and safe-targeting pegRNAs with the library module

PEGG also provides built in functions for generating non-targeting pegRNAs, whose protospacers have no match in the genome (Meier, Zhang, & Sanjana, 2017), as well as safe-targeting (Morgens et al., 2017) pegRNAs that target "safe regions" with no annotated function. These functions work for both the human and mouse genomes. Note: safe-targeting guides require human genome GRCh37 or mouse genome GRCm38.

To generate non-targeting pegRNAs, as well as safe-targeting regions that can be fed into PEGG, use the following command from PEGG's library module:

```
> from pegg import prime.
> from pegg import library.
#filepath to.gz containing reference genome (here
 = GrCh37).
> filepath
  = './GCF_000001405.25_GRCh37.p13_genomic.fna.gz'.
> chrom_dict, i = prime.genome_loader(filepath).
#generate 100 safe-targeting loci that can be fed
 into PEGG.
> num_muts = 100.
> safe = library.safe_muts(num_muts, chrom_dict,
  organism='human').
```

```
#and also generate 100 non-targeting guides for
 prime editing.
> num_guides = 100.
> nontarget  = library.nontargeting_guides(num_-
  guides, edit_type='prime').
```

3.8 Oligonucleotide design

To automatically generate oligonucleotides that contain the pegRNAs designed using PEGG, the prime.prime_oligo_generator() function provides multiple options, and produces a pegRNA oligo or epegRNA oligo (with a 3′ structural motif, tevopreQ1 or mpknot)[2] in traditional or sensor format. These oligos are designed with premade adapters for restriction cloning. These adapters, as well as the scaffold sequence, and epegRNA motif, can be customized by altering the following parameters in the prime.prime_oligo_generator() function:

A. epeg: True/False value indicating whether to include an epegRNA motif at the end of the 3′ extension.

B. epeg_motif: Which epeg motif to include at end of 3′ extension. Default = "tevopreQ1"; Other option = "mpknot". Can also input a custom motif sequence (just put in the sequence in 5′ to 3′ orientation)

C. five_prime_adapter: The sequence of the 5′ adapter to append to the 5′ end of the pegRNA-sensor oligo. The automatically provided 5′ adapter contains an *Esp*3I (BsmBI) site. The default sequence is 5′-AGCGTACACGTCTCACACC-3′ This can be swapped with a user-defined value.

D. three_prime_adapter: The sequence of the 3′ adapter to append to the 3′ end of the pegRNA-sensor oligo. The automatically provided 5′ adapter contains an *Eco*RI site. The default sequence is 5′-GAATTCTAGATCC GGTCGTCAAC-3′ This can be swapped with a user-defined value.

E. gRNA_scaff: The sequence of the gRNA scaffold. The default scaffold is 5′-GTTTTAGAGCTAGAAATAGCAAGTTAAAATAAGGCTAGTC CGTTATCAACTTGAAAAAGTGGCACCGAGTCGGTGC-3′.

An example of this function executed with the default parameters is provided below, where "pegRNA_df" is a DataFrame containing the pegRNAs generated by the prime.run() function:

```
> from pegg import prime.
> oligo_df = prime.prime_oligo_generator(pegRNA_df).
```

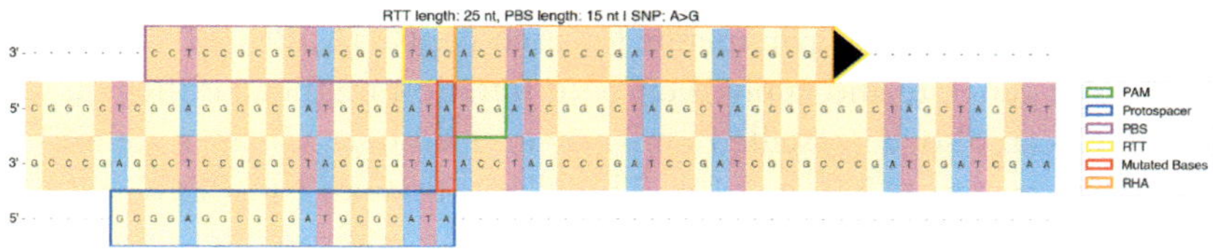

Fig. 2 | Visualization of pegRNA-sensor pair provided by built-in PEGG visualization function.

The resulting DataFrame contains the original column values with a new column, "pegRNA_oligo," that indicates the final pegRNA oligo sequence.

3.9 Visualization of pegRNAs

Finally, PEGG contains a built-in function for visualizing pegRNA-sensor pairs. Once a set of pegRNA-sensors are generated by the prime.run() function, this DataFrame can be provided to the prime.sensor_viz() function to visualize the resulting pegRNA and sensor. The user simply provides the row number of the of the pegRNA-sensor pair that they wish to visualize, and the resulting plot is returned (Fig. 2):

```
> from pegg import prime.
#visualizing pegRNA-sensor in row #4 from the
 pegRNAs produced by prime.run().
> prime.sensor_viz(pegRNAs, 4).
```

4. Notes

1. Exhaustive documentation for the possible parameters of each function included in PEGG is provided at the documentation website: https://pegg.readthedocs.io
2. PEGG also includes a "base" module for designing base editing guide RNAs, which functions similarly to the "prime" module. Further information about this can also be found at the documentation website.
3. Users can choose to filter pegRNAs based on their own parameters of interest. For example, instead of choosing the top ranked pegRNAs according the scoring metrics provided by PEGG, users could filter pegRNAs based on their disruption of the protospacer or PAM sequence, the RHA length, or the on-target score of the protospacer. By

providing all of these parameters for each pegRNA, PEGG allows users to easily generate and apply their own filtration methods if they choose.

4. PEGG can also be installed and run in a Google Colab notebook (if Python version = 3.9 or 3.10). This allows users to run PEGG without setting up their own local Python computing environment, but has not been tested exhaustively.

Acknowledgements

S.I.G. was supported by T32GM136540 from the National Institutes of Health (NIH)/ National Institute of General Medical Sciences and the Massachusetts Institute of Technology (MIT) School of Science Fellowship in Cancer Research. Work in the Sánchez-Rivera laboratory is supported by the Howard Hughes Medical Institute (Hanna Gray Fellowship, GT15656), the V Foundation for Cancer Research (V2022-028), NCI Cancer Center Support Grant P30-CA014051, the Virginia and D.K. Ludwig Fund for Cancer Research, Koch Institute Frontier Research Program, the Casey and Family Foundation Cancer Research Fund, the Michael (1957) and Inara Erdei Fund, the MIT Research Support Committee, the Upstage Lung Cancer Foundation, and a Traditional Project Award from the Bridge Project, a partnership between the Koch Institute for Integrative Cancer Research at MIT and the Dana-Farber/Harvard Cancer Center. We thank the Koch Institute Swanson Biotechnology Center Barbara K. Ostrom (1978) Bioinformatics Facility and the Genomics Facility for support.

Competing interests

F.J.S.R. has consulted for Repare Therapeutics and Ono Pharma. S.I.G. declares no competing interests.

References

Anderson, M. V., Haldrup, J., Thomsen, E. A., Wolff, J. H., & Mikkelsen, J. G. (2021). pegIT – a web-based design tool for prime editing. *Nucleic Acids Research, 49*, W505–W509.

Anzalone, A. V., et al. (2019). Search-and-replace genome editing without double-strand breaks or donor DNA. *Nature, 576*, 149–157.

Belli, O., Karava, K., Farouni, R., & Platt, R. J. (2024). Multimodal scanning of genetic variants with base and prime editing. *Nature Biotechnology*, 1–13.

Bhagwat, A. M., et al. (2020). multicrispr: gRNA design for prime editing and parallel targeting of thousands of targets. *Life Science Alliance, 3*.

Cerami, E., et al. (2012). The cBio cancer genomics portal: An open platform for exploring multidimensional cancer genomics data. *Cancer Discovery, 2*, 401–404.

Chow, R. D., Chen, J. S., Shen, J., & Chen, S. (2021). A web tool for the design of prime-editing guide RNAs. *Nature Biomedical Engineering, 5*, 190–194.

Cirincione, A., et al. (2024). A benchmarked, high-efficiency prime editing platform for multiplexed dropout screening. *Nature Methods*, 1–10.

Doench, J. G., et al. (2016). Optimized sgRNA design to maximize activity and minimize off-target effects of CRISPR-Cas9. *Nature Biotechnology, 34*, 184–191.

Doman, J. L., Sousa, A. A., Randolph, P. B., Chen, P. J., & Liu, D. R. (2022). Designing and executing prime editing experiments in mammalian cells. *Nature Protocols*. https://doi.org/10.1038/s41596-022-00724-4.

Erwood, S., et al. (2022). Saturation variant interpretation using CRISPR prime editing. *Nature Biotechnology, 40*, 885–895.
Gould, S. I., et al. (2024). High-throughput evaluation of genetic variants with prime editing sensor libraries. *Nature Biotechnology,* 1–15.
Gould, S. I. (2024). Prime editing sensors enable multiplexed genome editing. *Nature Reviews. Genetics.* https://doi.org/10.1038/s41576-024-00737-7.
Hsu, J. Y., et al. (2021). PrimeDesign software for rapid and simplified design of prime editing guide RNAs. *Nature Communications, 12*, 1034.
Hwang, G.-H., et al. (2021). PE-designer and PE-analyzer: Web-based design and analysis tools for CRISPR prime editing. *Nucleic Acids Research, 49*, W499–W504.
Kim, H. K., et al. (2021). Predicting the efficiency of prime editing guide RNAs in human cells. *Nature Biotechnology, 39*, 198–206.
Kim, Y., Oh, H.-C., Lee, S., & Kim, H. H. (2024). Saturation profiling of drug-resistant genetic variants using prime editing. *Nature Biotechnology,* 1–14.
Koeppel, J., et al. (2023). Prediction of prime editing insertion efficiencies using sequence features and DNA repair determinants. *Nature Biotechnology.* https://doi.org/10.1038/s41587-023-01678-y.
Landrum, M. J., et al. (2018). ClinVar: Improving access to variant interpretations and supporting evidence. *Nucleic Acids Research, 46*, D1062–D1067.
Li, Y., Chen, J., Tsai, S. Q., & Cheng, Y. (2021). Easy-Prime: A machine learning-based prime editor design tool. *Genome Biology, 22*, 235.
Mathis, N., et al. (2024). Machine learning prediction of prime editing efficiency across diverse chromatin contexts. *Nature Biotechnology,* 1–8.
Mathis, N., et al. (2023). Predicting prime editing efficiency and product purity by deep learning. *Nature Biotechnology.* https://doi.org/10.1038/s41587-022-01613-7.
Meier, J. A., Zhang, F., & Sanjana, N. E. (2017). GUIDES: sgRNA design for loss-of-function screens. *Nature Methods, 14*, 831–832.
Morgens, D. W., et al. (2017). Genome-scale measurement of off-target activity using Cas9 toxicity in high-throughput screens. *Nature Communications, 8*, 15178.
Morris, J. A., Rahman, J. A., Guo, X., & Sanjana, N. E. (2021). Automated design of CRISPR prime editors for 56,000 human pathogenic variants. *iScience, 24*, 103380.
Nelson, J. W., et al. (2022). Engineered pegRNAs improve prime editing efficiency. *Nature Biotechnology, 40*, 402–410.
Ren, X., et al. (2023). High-throughput PRIME-editing screens identify functional DNA variants in the human genome. *Molecular Cell, 83*, 4633–4645.e9.
Standage-Beier, K., Tekel, S. J., Brafman, D. A., & Wang, X. (2021). Prime editing guide RNA design automation using pine-cone. *ACS Synthetic Biology, 10*, 422–427.
Yu, G., et al. (2023). Prediction of efficiencies for diverse prime editing systems in multiple cell types. *Cell, 186*, 2256–2272.e23.

CHAPTER EIGHTEEN

Cloning and validating systems for high throughput molecular recording

Anqi Zhao[a,b] and Michelle M. Chan[a,b,*]
[a]Lewis-Sigler Institute for Integrative Genomics, Princeton University, Princeton, NJ, United States
[b]Department of Molecular Biology, Princeton University, Princeton, NJ, United States
*Corresponding author. e-mail address: mmchan@princeton.edu

Contents

1. Introduction 454
2. Before you begin 455
   2.1 Target site design considerations 455
   2.2 Lineage tracing vector design 457
3. Materials and equipment 457
4. Time course experiment to assess the cut site performance 459
   4.1 Plasmid construction 460
   4.2 Cell line engineering 463
   4.3 Time course experiment 464
5. Target site - triple guide plasmid generation 466
   5.1 Step 1: Triple guide cassette assembly 466
   5.2 Step 2: Target site assembly 468
   5.3 Step 3: lineage tracing vector plasmid library construction 468
6. Lineage tracing cell line construction and validation 469
7. An example of conducting in vitro lineage tracing in gastruloids 470
8. Limitations 471
References 472

Abstract

Molecular recording technologies record and store information about cellular history. Lineage tracing is one form of molecular recording and produces information describing cellular trajectories during mammalian development, differentiation and maintenance of adult stem cell niches, and tumor evolution. Our molecular recorder technology utilizes CRISPR-Cas9 barcode editing to generate mutations in genomically integrated, engineered DNA cassettes, which are read out by single-cell RNA sequencing and used to produce high-resolution lineage trees. Here, we describe optimized cloning and validation procedures to construct the molecular recorder lineage tracing system. We include information on considerations of technology design, cloning procedures, the generation of lineage tracing cell lines, and time course experiments to assess their performance.

Methods in Enzymology, Volume 712
ISSN 0076-6879, https://doi.org/10.1016/bs.mie.2025.01.015

Abbreviations

scRNA-seq single cell RNA sequencing.
gRNA guide RNA.
mESC mouse embryonic stem cells.
intBC integration barcode.
UMI unique molecular identifier.
PAM protospacer adjacent motif.
PB PiggyBac.

1. Introduction

Lineage tracing is a fundamental tool to understand how cells proliferate and dynamically undergo cell state changes. Charting the ancestral relationship between cells may reveal the processes governing development from single cell zygotes into multicellular organisms, the coordination within cellular populations to achieve homeostasis in adult stem cell niches, and the transformations normal cells undergo to become tumorigenic. Prospective lineage tracing technologies, such as the injection of dyes, multicolor Cre-reporters, and using viral infection to introduce genetic markers, have made significant advances in understanding the developmental programs of the lung, the gut, T cells, as well as many other systems but are restricted to labeling cellular clones, limiting the resolution of lineage information (Kretzschmar & Watt, 2012). Direct observation by live imaging of cell lineages is the ideal method to observe comprehensive records of fate decisions but remains challenging to study in many systems including mammalian embryogenesis due to the scale of information (McDole et al., 2018). Conversely, single cell RNA-sequencing (scRNA-seq) experiments enable high-resolution charting of cell identities for many cellular events, but the characterization of differentiation pathways requires inference without direct measurement (Mittnenzweig et al., 2021; Pijuan-Sala et al., 2019).

Several groups have constructed continuously evolving lineage tracing technologies to address the challenge of collecting high resolution lineage information at scale and have applied it to study the development in multiple model organisms and cancer models (Alemany, Florescu, Baron, Peterson-Maduro, & Van Oudenaarden, 2018; Chan et al., 2019; Choi et al., 2022; Kalhor et al., 2018; McKenna et al., 2016; Quinn et al., 2021; Raj et al., 2018; Yang et al., 2022). In this report, we will focus on the molecular recorder developed by Chan, Smith et al., which is based on CRISPR-Cas9 barcode editing and scRNA-seq. The technology utilizes genome-integrated target sites each of them containing three cut sites and a unique integration barcode. The Cas9 protein in complex with a guide RNA (gRNA) targets the target site, generates double-strand breaks, and in turn, activates the DNA repair mechanism resulting in heritable insertions or deletions (indels). The array of indel mutations accumulated over time can be used to reconstruct the lineage tree. Here, we outline the design and cloning procedure of the target site plasmids. In addition, we provide protocols for validating target site behavior and integrating the target sites into mouse embryonic stem cells (mESCs) for downstream applications.

2. Before you begin

2.1 Target site design considerations

The target site cassette comprises a 14 bp unique integration barcode (intBC) and three individual cut sites that are separated by linker sequences and flanked by PCR amplification adaptors (Fig. 1A). The intBC is used to distinguish multiple integrations of the target site cassette in the genome. This target site is specifically designed to be compatible with 10x single-cell RNA-seq and Illumina sequencing with the sequencing of 10x cell barcodes and Unique Molecular Identifiers (UMIs) on read1 alongside the target site sequencing on read2. Consequently, to accommodate the target site sequencing on the Illumina Novaseq 300 cycle platform, the maximum length of the target site cannot exceed 300 minus the combined length of the 10x cell barcode and UMI. For implementation with other sequencing technologies, the target site length should be adjusted accordingly. To save costs, we recommend that 10x single cell RNA-seq libraries are sequenced separately on 100 cycle kits from target site libraries that are sequenced on 300 cycle kits.

The selection of cut sites is crucial for the design of this technology. Three primary criteria are considered:

a. The cut sites should exhibit minimal homology to the genome of the species that will be used in experiments.

b. The indels generated by Cas9 cleavage and DNA repair should be diverse to minimize events that result in the same indel. In the ideal scenario, the diversity of indels generated from a single cut site would be as large as possible and distributed uniformly. While experimental profiling of the indel distribution for a potential cut site is recommended, in silico prediction of the indel distribution can be used when experiments are not feasible (Shen et al., 2018).

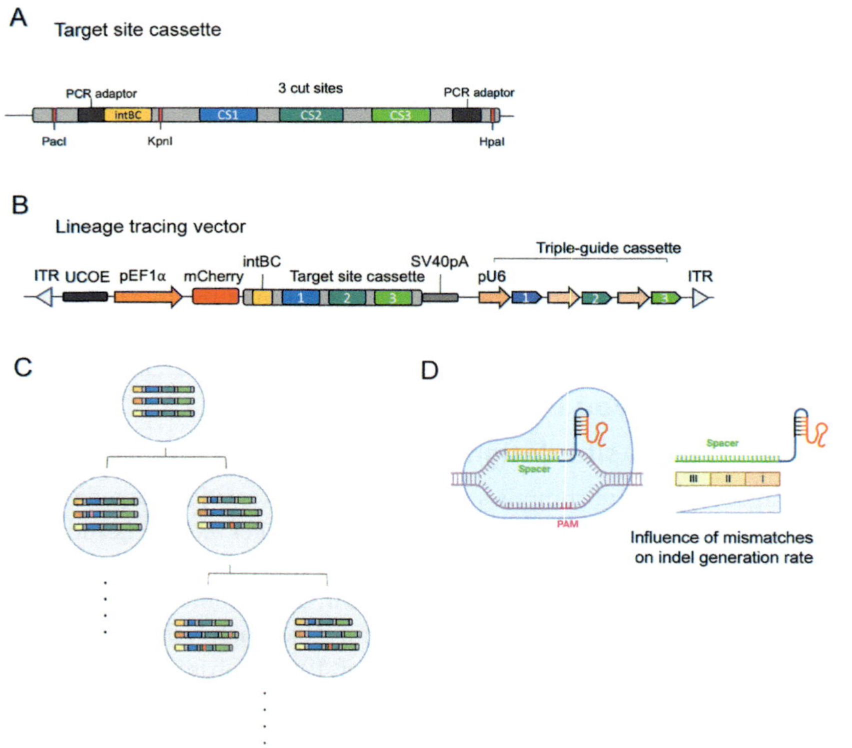

Fig. 1 (A) Target site cassette design. (B) The lineage tracing vector includes both the target site and corresponding gRNAs on a contiguous cassette. (C) An example showing the accumulation of indels. Each cell contains multiple target sites, as the cell proliferates, various heritable indels are generated. (D) Mismatches in the gRNA sequence can affect the indel generation rate. (Created with Biorender.com) In the figure, the spacer sequence is divided into three regions based on their proximity to PAM sequence. Generally, a mismatch in region I has the greatest impact on indel generation rate, followed by region II, while region III has the least influence.

c. The rate of indel generation should correspond to the time scale of the cellular process of interest. Cas9 may induce indels more rapidly than desired. To decrease the rate of indel generation, mismatches can be introduced into the gRNA sequence. Generally, mismatches in the region closer to the Protospacer Adjacent Motif (PAM) sequence will likely have a greater influence on the indel generation rate (Chan et al., 2019; Gilbert et al., 2014) (Fig. 1D). However, the precise degree of influence is largely sequence-dependent, and can only be determined through time course experiments, as demonstrated below.

2.2 Lineage tracing vector design

There are three essential components in this lineage tracing toolbox: the Cas9 protein, the target site cassette (which consists of three cut sites and an intBC), and the corresponding three gRNAs. While the Cas9 protein is often introduced separately, we combine the target site cassette with the gRNAs as a contiguous lineage tracing cassette (Fig. 1B). PiggyBac (PB) transposition is selected to integrate multiple copies of the lineage tracing cassette into the genome. The number of copies of the cassette in the genome determines the information capacity of the technology (Fig. 1C). The plasmid carrying the lineage tracing cassette (referred to as the lineage tracing vector) has inverted terminal repeats (ITR) sequences on both ends of the cassette, which are recognized by PB transposase. When the vector is co-transfected with the PB transposase-expressing vector into the cells, the PB transposase relocates the contents between the two ITR sequences into the cell genome through a "cut and paste" mechanism (Ding et al., 2005).

To aid cell line construction and ensure continuous expression of the recorder, the target site cassette is positioned within the 3′UTR of a mCherry fluorescent reporter. The universal chromatin opening element (UCOE) is inserted upstream of the EF1-α promoter to facilitate stable mCherry-target site expression during cell state transitions and differentiation events. The three gRNAs corresponding to the three cut sites are assembled into a triple-guide cassette adapted from the three-guide Perturb-seq vector (Adamson et al., 2016). The detailed plasmid construction procedure is discussed below.

3. Materials and equipment

Materials.

1. Nuclease-Free Water (not DEPC-Treated) (Invitrogen, AM9937)
2. Duplex Buffer (IDT, 11-05-01-12)
3. IDTE (1xTE solution) (IDT, 11-05-01-13)
4. Restriction enzymes from NEB: *Bst*XI (R0113), *Blp*I (R0585), *Xba*I (R0145), *Not*I-HF (R3189), *Pac*I (R0547), *Hpa*I (R0105), *Kpn*I-HF (R3142)
5. T4 DNA Ligase and T4 DNA Ligase Reaction Buffer (NEB, M0202)
6. NEB® Stable Competent *E. coli* (High efficiency) (NEB, C3040)
7. NEB® 5-alpha Competent *E. coli* (High Efficiency) (NEB, C2987)
8. LB Agar plates with Carbenicillin (Teknova, L7010)

9. SYBR™ Safe DNA Gel Stain (Invitrogen, S33102)
10. Quick-Load Purple 1 kb Plus DNA Ladder (NEB, B7025)
11. Ultra-Pure Agarose (Invitrogen, 16500-100)
12. Nucleospin Gel and PCR Clean-Up Kit (Takara, 740609)
13. NEBuilder HiFi DNA Assembly Master Mix (NEB, E2621L)
14. Endura Electrocompetent Cells (LGC, 60242-1)
15. QIAprep Spin Miniprep Kit (QIAGEN, 27106)
16. ZymoPure plasmid prep kit (Size various according to the culture input)
17. NEBNext Ultra II Q5 Master Mix (NEB, M0544)
18. AMPure XP beads (Backman, A63881)
19. RNeasy Plus Mini Kit (Qiagen, 74134)
20. High-Capacity RNA-tocDNA Kit (Applied Biosystems, 4387406)
21. Xfect mESC Transfection Reagent (Takara, 631320)
22. DPBS, w/o $MgCl_2/CaCl_2$ (Gibco, 14190144)
23. DPBS, calcium, magnesium (Gibco, 14041082)
24. TrypLE (Gibco, 12604013)
25. Gelatin 2 % solution (Sigma, G1393)
26. Super PiggyBac Transposase Expression Vector (SBI, PB210PA-1)
27. Lonza P3 Primary Cell 4D-Nucleofector® X Kit L (Lonza, V4XP-3024)
28. CF1 Mouse Embryonic Fibroblasts, irradiated (IRR) (Thermo Scientific, A34180)
29. KnockOut DMEM (Gibco, 10829-018)
30. Penicillin Streptomycin (10,000 U/ml) (Thermo Scientific, 15140122)
31. Glutamax™ Supplement (Thermo Scientific, 10439024)
32. MEM Non-Essential Amino Acids Solution (100X) (Thermo Scientific, 11140050)
33. Gibco 2-Mercaptoethanol, 55 mM solution in DPBS (Gibco, catalog number: 21985023)
34. Fetal Bovine Serum, embryonic stem cell-qualified (Thermo Fisher, 10439024)
35. Gibco Fetal Bovine Serum, Qualified (Thermo Fisher, 10437028)
36. TrypLE (Gibco, 12604013)
37. DNeasy Blood & Tissue Kit (Qiagen, 69504)
38. Murine Leukemia Inhibitory Factor (LIF) ESGRO™ (10^7U/ml) (Millipore, ESG1107)
39. Ultra-low attachment 96-well plates (Corning, Costar, CLS7007)
40. NDiff 227 medium (Takara, catalog number: Y40002)
41. CHIR99021 (Merck Millipore, Cat# SML1046)

Equipment.

1. Lonza 4D-Nucleofector Core Unit (Lonza, AAF-1003B) and X Unit (Lonza, AAF-1003X)
2. Sony MA900 Cell Sorter
3. Biorad MicroPulser Electroporator
4. Countess 3 Automated Cell Counter

Recipe.

1. Mouse embryonic fibroblast (MEF) medium
   500 ml KnockOut DMEM 55 ml **regular** FBS
   5.5 ml 100 × GlutaMAX 2.75 ml 200 × Penicillin/Streptomycin
   Sterile filter
   Aliquot in ~45 ml portions and freeze at −20 °C After thaw, store at 4 °C
2. Mouse embryonic stem cell (mESC) medium
   500 ml KnockOut DMEM
   90 ml **mESC qualified** FBS
   6 ml 100x GlutaMAX
   3 ml 200x Penicillin/Streptomycin
   6 ml NE AA
   1.21 ml bate-ME
   Sterile filter Aliquot mESC medium without LIF in 40-ml portions and freeze at −20 °C Thaw before use, then store at 4 °C Add 1:10,000 LIF immediately before use (4 μL into 40 ml medium) Store at 4 °C.

4. Time course experiment to assess the cut site performance

We quantitatively assess the performance of the cut sites, including indel diversity and generation rates with different gRNA mismatches, through a time course experiment. The performance of cut sites may depend on the specific cell line used; therefore, selecting an appropriate cell line based on downstream applications is required. Moreover, the duration of the time course experiment should align with the timescale of the targeted recording event. The time course experiment begins with generating cell lines that harbor two important components: an inducible Cas9-GFP system and the cut site – single guide cassette (Fig. 2B). Following the induction of Cas9-GFP expression, the GFP-positive population is collected every 24 h for downstream analysis. In this protocol, as the system is

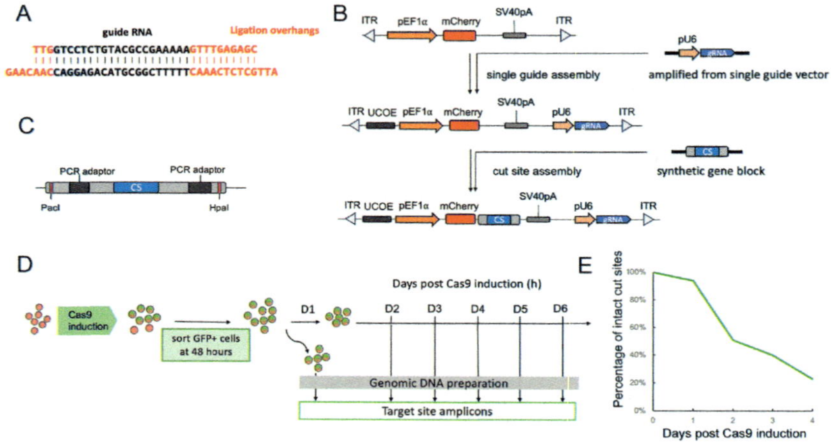

Fig. 2 (A) An example of the gRNA insert formed by annealing two ssDNA oligos with complementary protospacer sequence and ligation overhangs. (B) A schematic of the cut site-single guide vector cloning strategy. (C) The design of the cut site synthetic gene block. (D) Experimental diagram of a time course experiment. (E) An example of the cut site saturation curve.

designed to record lineage trajectories in early mouse embryogenesis, mouse embryonic stem cells are selected, and the sample collection takes place over 6–8 days.

4.1 Plasmid construction

Timing: 2–3 weeks

4.1.1 Step 1: Constructing single guide expression cassette

In this step, two single-strand DNA (ssDNA) oligonucleotides containing the complementary protospacer sequence and the ligation overhangs are annealed together and ligated onto the single guide vector backbone (Fig. 2A). These single vector backbones each contain specific assembly adaptors flanking the single guide expression cassettes with different U6 promoters to minimize the intramolecular recombination at repetitive sequences during the triple guide assembly reaction described below (Adamson et al., 2016). For cut sites that will be combined into one target site cassette, different single guide vector backbones should be selected. The step-by step cloning procedure is described below:

1. In a PCR tube, set up a 50 μL reaction containing 1 μL of each 100 μM oligonucleotides in IDT Duplex Buffer.

2. Incubate the reaction at 95 °C for 5 min in the thermocycler. Then take out the tube and place it on the bench to let it cool down to room temperature (RT), which allows the annealing of complementary oligos.
3. Ligation reaction: mix 2 μL 1:20 diluted annealed oligos with 100 ng *Bst*XI and *Blp*I digested single guide vector backbone in a 20 μL T4 ligation reaction. It is recommended to include a negative control ligation reaction.
4. Incubate the ligation reaction at RT for 1-4 h or at 16 °C overnight.
5. Transform 1 μL each ligation product into 20 μL NEB 5-alpha Competent *E. coli*:
   a. Pre-warm the plate(s) at 37 °C.
   b. Combine 20 μL cells (thaw on ice) with 1 μL ligation product. Stir gently with the pipette tip to mix. Incubate the mixture for 30 min on ice.
   c. Heat shock the tubes at 42 °C in the water bath for exactly 45 s.
   d. Place the tube immediately back on ice for ~1 min.
   e. add 100 μL pre-warmed SOC media to the transformation mixture and recover at 37 °C with shaking for 1 h. Plate 20 μL to 50 μL of the recovery culture.
   f. Incubate at 37 °C overnight.
6. Pick 2–4 colonies from the plate and sequence the plasmids with the primer oMJ0041 or other preferred sequencing primers.

NOTE: Aliquot the 10x T4 ligation buffer to avoid repeated freeze and thaw cycles.

oMJ0041: ACGGCGACTACTGCACTTAT.

4.1.2 Step 2: Single guide expression cassette assembly

1. Amplify the single guide expression cassettes from single guide vectors constructed in step 1 using the primers indicated in Table 1. Confirm the correct amplification by gel electrophoresis. Purify the amplified

Table 1 PCR forward and reverse primers for single guide expression cassette assembly.

Vector	PCR fw primer	PCR rv primer
pMJ114	DY007	AZ055
pMJ179	AZ056	AZ057
pMJ117	AZ058	DY008

fragments using PCR or gel purification with Nucleospin gel and PCR Clean-Up Kit.

2. Assemble the purified PCR fragment into *Xba*I/*Not*I digested pDY001 (Yang et al., 2022) using the NEBuilder HiFi DNA Assembly Master Mix following the manufacturer's protocol.
3. Transform the assembled product into NEB STABLE competent cells following the manufacturer's protocol. Sequence the colonies to confirm the correct assembly.
   DY007: GACTGGATTCCTTTTTTAGGGCCCATTGGTctagaCG TGACCGAGCTTGTC
   DY008: CGGGGAAAAAGCCATGTTTAAACGcggccgcctaatggatcct agtactcgaG
   AZ055: CGGGGAAAAAGCCATGTTTAAACGcggccgcctaatggatccta gCATCGTGATAGTCAGACTGG
   AZ056: GACTGGATTCCTTTTTTAGGGCCCATTGGTctagaCC AGTCTGACTATCACGATG
   AZ057: CGGGGAAAAAGCCATGTTTAAACGcggccgcctaatggatccta gGCTGAGTGTAGATTCGAGC
   AZ058: GACTGGATTCCTTTTTTAGGGCCCATTGGTctagaGC TCGAATCTACACTCAGC

4.1.3 Step 3: Cut site cassette assembly

1. The cut site cassette is ordered as a synthetic gene block from Twist Bioscience (Fig. 2C). Note that the sequence adjacent to the cut site might also influence the cut site behavior, so it is recommended to adjust the linker sequence flanking the cut site when designing the single cut site cassette.
2. Use restriction enzymes *Pac*I and *Hpa*I to digest the plasmid assembled in step 2.
3. Add the assembly arm overlapping with both ends of the gene block by PCR amplification using primers AZ019 and AZ020. Purify the amplified fragments using PCR or gel purification with Nucleospin gel and PCR Clean-Up Kit.
4. Use the standard NEBuilder HiFi DNA Assembly protocol to assemble the synthetic target site cassette onto the *Pac*I/*Hpa*I digested vector backbone with a 2:1 insert-to-vector molar ratio.
5. Transform the assembled product into NEB STABLE competent cells following the manufacturer's protocol. Sequence the colonies to confirm the correct assemble.

AZ019: gctgtacaagtaaTGAATTAATtaaGTCACGAATCCAGCTAGC
AZ020: CCGCTAATAGGTGAGCAGTTAACACCTGCAGGAGC GATGG

4.2 Cell line engineering

The inducible Cas9-GFP system was introduced into mouse embryonic stem cells as a Cre-inducible Cas9-EGFP cassette (CAG-driven Lox-STOP-Lox-Cas9-EGFP) integrated into the Rosa26 safe harbor site (Platt et al., 2014), referred to as LSL-Cas9 mESCs. (LSL-Cas9 mESCs can be derived from mouse strain #024857 from the Jackson Laboratory, or requested from our lab.) It has been validated that this cell line exhibits minimum leakage and high Cas9-EGFP expression after electroporating Cre-expressing plasmid (Bolondi et al., 2024). The procedures for integrating the cut site-single guide cassette into the LSL-Cas9 mESCs are described below:

Timing: 4–5 weeks

1. Thaw and culture the cells for at least two passages before starting the transfection. (detailed cell culture protocol could refer to Bolondi et al., 2021)
2. At the second split, transfect the cells using Xfect mESC Transfection Reagent (Bolondi et al., 2024).
   a. Wash the mESCs with DPBS (minus Calcium and Magnesium) and treat with TrypLE for 5-7 min.
   b. After neutralizing the TrypLE with culture media, further dissociate the cells into a single-cell suspension by pipetting 15 times.
   c. The cells are then pelleted at 200 rcf for 5 min, resuspended in culture media.
   d. To perform MEF depletion, sequentially plate the cells on gelatin-coated dishes for 25 min, 20 min, and 15 min. During each transfer, carefully collect the supernatant to avoid disturbing the MEF cells adhered to the bottom. Transfer it into a new gelatin coated dish and pipette it 10 times to maintain a single-cell suspension (Bolondi et al., 2021).
   e. After the final transfer, collect the supernatant that contains mostly mESCs and count.
   f. Next, seed 2×10^5 MEF-depleted mESCs in a 0.2% gelatin-coated well of a 12-well plate with culture media containing 2 x LIF. Place the plate in the incubator and allow the cells to settle undisturbed for 5 h.
   g. After 5 h incubation, add the transfection mixture containing 2.5 μg cut site-single guide vector as transposon/0.5 μg Super PiggyBac Transposase Expression Vector into the cells.

h. 3 h after adding the transfection mixture, replace it with fresh media containing feeder cells.

Expected Outcomes: The percentage of mCherry$^+$ cells 2 days post transfection is usually around 40 %. At this stage, most of the mCherry signal is from the ectopic expression. The actual percentage of cells with cassettes integrated could be quantified 6 days post transfection, which is typically lower than 5 %. Fluoresence-activated cell sorting (FACS) can help enrich these mCherry$^+$ cells.

Optimization And Trouble Shooting: If the percentage of mCherry + cells is low, increasing the amount of transposon plasmid or adjusting the molar ratio of transposon to transposase might be helpful.

4.3 Time course experiment

Timing: 3–4 weeks

Fig. 2D provides an overview for time course experiment.

1. Thaw and culture the cells for at least two passages before starting the time course experiment.

2. The time course experiment begins with activating Cas9-EGFP expression in the cell lines constructed in 4.1 through pCAG-Cre nucleofection. The nucleofection procedure is described below:

- **a.** Wash the mESCs with DPBS (minus Calcium and Magnesium) and treat with TrypLE for 5–7 min.
- **b.** After neutralizing the TrypLE with culture media, further dissociate the cells into a single-cell suspension by pipetting 15 times.
- **c.** Pellet the cells at 200 rcf for 5 min. Resuspend the pellet in media.
- **d.** Perform MEF depletion by plating the cells on gelatin-coated dishes for 25 min, 20 min, and 15 min, as described in 4.2.
- **e.** During this time, prepare the nucleofection solution:
  For a 100 μl Nucleocuvette Vessel:
  Nucleofection Solution: 82 μL
  Supplement: 18 μL
  Plasmid DNA: 10 μg (volume should not exceed 10 μL)
- **f.** After MEF depletion, count the cell number. Transfer 1 M MEF-depleted mESCs into a new tube containing 10 ml DPBS (with Calcium and Magnesium), then spin down the cells.
- **g.** Aspirate DPBS and resuspend the cell pellet in prepared nucleofection solution.

   h. Pipette the cell-nucleofection solution mixture into the cuvette. Tap to ensure no bubble in the cuvette.
   i. Immediately proceed with Nucleofection in the Lonza X-unit using program CG-104.
   j. Add 500 μL of pre-warmed medium to the cuvette and use special pipette (provided by nucleofection kit) to recover all the cells from the bottom of the cuvette.
   k. Seed nucleofected cells onto a well of 6-well plate with feeder cells.

3. 24–48 h post nucleofection, perform FACS on mCherry- and GFP-double positive cells. Half of the sorted population is frozen for genomic DNA extraction and the rest is re-plated on gelatin-coated wells.

4. After sorting, with 24-hour intervals, collect half of the cells and freeze them as pellets while re-plate the rest to allow the further accumulation of indels.

5. Extract gDNA from the frozen time course samples using the DNeasy Blood & Tissue Kit.

6. Prepare the target site amplicons as described below (protocol modified from Chan et al., 2019):
   a. Amplify the target site from genomic DNA (PCR1): Make 2 × 25 μL PCR reaction containing 200 ng gDNA, 0.5 μM forward and reverse primers (AZ021 and AZ022) and using NEBNext Ultra II Q5 Master Mix. The PCR program is: (1) 98 °C, 1 min; (2) 98 °C, 10 s; (3) 65 °C, 75 s; (repeat step 2-3 for 12 cycles) (4) 65 °C, 5 min.
   b. Purify the pooled PCR product through 0.5x-0.8x double-sided AMPure XP beads cleanup.
   c. To add the sequencing adapter, the purified PCR1 product with 0.5 μM P5 and P7 are amplified using the sample protocol in step 1 (PCR2). (PCR cycle number might vary)
   d. Purify the PCR2 product with twice 0.8x beads cleanup.

7. Sequence the target site amplicons using paired-end sequencing (300 cycles kit, 150/150) on a Miseq platform at a minimum of 1 M reads per sample.

8. Amplicon reads are aligned to the reference sequence to generate the indel tables and the percent of intact cut sites is calculated accordingly. Various software packages are available to conduct alignment including EMBOSS (Rice, n.d.) and CRISPResso (Clement et al., 2019). Fig. 2E shows an example of the cut site saturation curve.

   AZ021: TCGTCGGCAGCGTCAGATGTGTATAAGAGACAGAA TCCAGCTAGCTGTGCAGC

AZ022: GTCTCGTGGGCTCGGAGATGTGTATAAGAGACAGG CGATGGACGATTGCGGAAGACAG

Expected Outcomes: The expected percentage of GFP+ cells 48 h post nucleofection is usually higher than 80 %. The sorted GFP+ cells are usually capable to maintain Cas9-EGFP expression over time with minimum silencing effect observed. If significant silencing issues occur (for example, GFP+ cells percentage drops under 90 %), one could consider sorting on GFP+ cells every time before sample collection.

Optimization and trouble shooting

a. If cell viability is low after MEF depletion, one could consider using column-based methods, for example, using feeder depletion beads, to remove the feeder cells.

b. If Cas9 induction efficiency is low, try to operate more quickly during the nucleofection procedure and reducing the time the cells spent in nucleofection solution.

5. Target site - triple guide plasmid generation

Timing: around 4 weeks

To build a new lineage tracing vector based on pDYT001 (Yang et al., 2022), a three-step cloning procedure is required (Fig. 3A).

5.1 Step 1: Triple guide cassette assembly

1. Construct the single guide vectors as demonstrated in 4.1.1 step 1.
2. Amplify the single guide expression cassettes from the single guide vectors using the primers in Table 2. The primers contain the assembly arms for the following step.
3. Confirm the size of the amplified fragments by gel electrophoresis, and purify the fragments using PCR or gel purification with the Nucleospin Kit.
4. Assemble the three amplified single guide expression cassettes with 100 ng *Xba*I/*Not*I digested pDY001 in equal molar ratios using NEBuilder HiFi DNA Assembly Master Mix. To increase assembly efficiency, incubate the four-fragment assembly reaction at 50 °C for one hour.
5. Transform the assembly product into NEB Stable Competent Cells, following the high-efficiency transformation protocol provided by the manufacturer. Plate all the transformed cells to maximize recovery, as

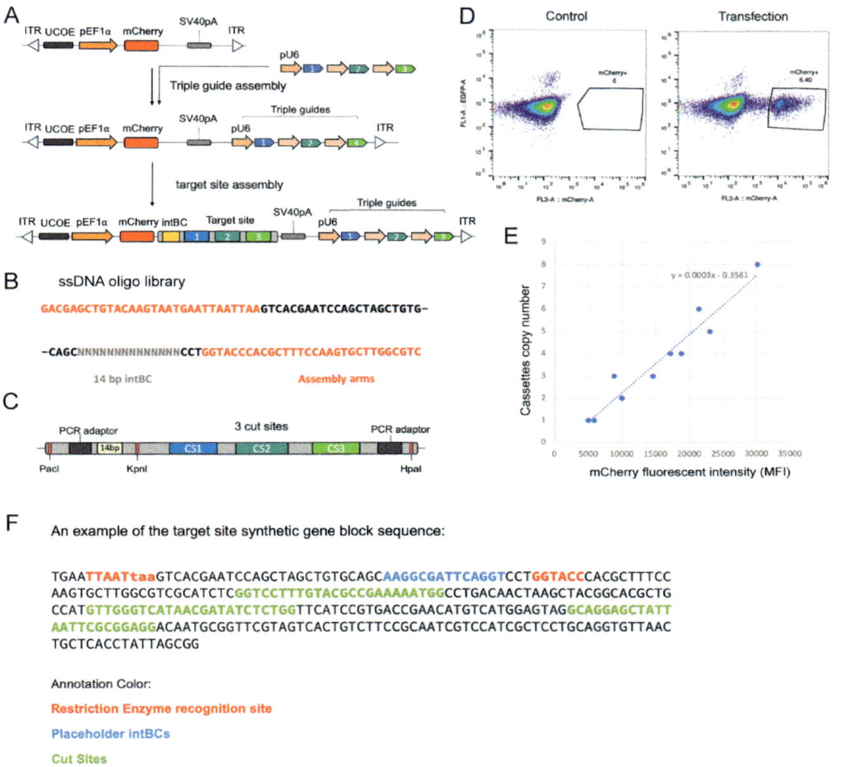

Fig. 3 (A) A schematic overview of the lineage tracing vector cloning strategy. (B) An example of the ssDNA oligo library. (C) Target site cassette design. (D) Flow cytometry analysis 6 days post-transfection. The control group was used to decide the mCherry + gate. (E) The relationship between the copy number of cassettes and median fluorescence intensity (MFI).

Table 2 PCR forward and reverse primers for triple guide assembly.

Vector	PCR fw primer	PCR rv primer
pMJ114	DY007	BA278
pMJ179	BA279	BA280
pMJ117	BA281	DY008

four-fragment assembly reactions typically have low efficiency, yielding only a few colonies from a full transformation.

6. Pick the colonies and verify the correct assembly by whole plasmid sequencing.
BA278: CATCGTGATAGTCAGACTGG
BA279: CCAGTCTGACTATCACGATG
BA280: GCTGAGTGTAGATTCGAGC
BA281: GCTCGAATCTACACTCAGC

5.2 Step 2: Target site assembly

1. Order the target site cassette with a placeholder intBC as a synthetic gene block from Twist Bioscience (Fig. 3C).

2. Follow 4.1.3 step 3 to place the target site into the *Pac*I/ *Hpa*I digested triple guide vector from step 1. Now we have the target site- triple guide vector.

5.3 Step 3: lineage tracing vector plasmid library construction

1. Remove the placeholder intBC through *Pac*I and *Kpn*I restriction enzyme digestion.

2. The random intBCs library is ordered as a single-stranded DNA (ssDNA) oligo library with random 14 bp Ns and overlapping arms from IDT (Fig. 3B).

3. Utilize the ssDNA oligo bridging method provided by NEBuilder HiFi DNA assembly master mix to place them into the placeholder removed vector. The 20 μL assembly reaction contains 5 μL 0.2 μM ssDNA oligo library (serial diluted in TE buffer with 50 mM NaCl) and 0.005 pmol digested vector. Incubate the reaction at 50 °C for one hour. It is critical to include a vector-only control reaction to access the assembly background.

4. To ensure an efficient assembly with low background, perform a test transformation by transforming 2 μL of the reaction into 50 μL NEB Stable competent cells, following the high-efficiency transformation protocol. The next day, count the number of colonies in the assembly and control plates and calculate the background by dividing the number of colonies on the assembly plate by the control plate. We consider the background acceptable when it is lower than 5 %.

5. Pick 10-20 colonies for Sanger sequencing to validate the intBCs are inserted into the correct position.

6. Once efficient and low-background assembly reaction is confirmed, purify the assembly product via the PCR purification protocol provided by the Nucelospin kit.

7. Perform large-scale transformation by electroporating 5 μL or less than 100 ng of the purified assembly product into 50 μL Endura Electrocompetent Cells on Biorad MicroPulser Electroporator with the following settings: 1 mm cuvette size, 10 μF capacitance, 600 Ohm resistance, and 1800 V voltage. It is critical to start with a relatively large amount of the purified assembly product to ensure the complexity of barcodes in the plasmid library.
8. After electroporation, incubate the cells in recovery media at 37 °C for one hour. Afterward, take 5 μL from the recovered culture and sequentially dilute to plate on LB for calculating the total number of recovered colonies. Inoculate the rest of the recovered culture overnight for plasmid preparation.
9. Perform plasmid preparation using ZymoPure plasmid prep kit.

Optimization and trouble shooting

a. Triple guide assembly is quite inefficient. If no colonies are observed, transforming the assembly reaction into more competent cells might help.
b. For electroporation, usually around 10 M cells recovered is a safe number to proceed.
c. Since the electroporation efficiency might vary, it is recommended to monitor the OD_{600} starting from 10 h post-inoculation and proceed to plasmid preparation when the OD_{600} is around 0.2-0.35 for a ten-fold diluted culture. Over-growing the culture could lead to significant decrease in plasmid yield.
d. The complexity of the library could be assessed via high-throughput sequencing following the target site amplicon preparation protocol above.

6. Lineage tracing cell line construction and validation

1. To construct the lineage tracing cell line, integrate the lineage tracing vector with a library of random intBCs into the LSL-Cas9 mESCs as described in 4.2 using piggyBac transposition.

   Note: Fig. 3D illustrates the expected mCherry fluorescent intensity distribution 6 days post-transfection. The relationship of the copy number of cassettes and mCherry median fluorescence intensity (MFI) could be described by a linear function: n intBCs = a*mCherryMFI + b, in which a and b are cell line- and FACS setting- depedent coefficients

(Fig. 3E). Therefore, the broad range of mCherry fluorescence intensity in Fig. 3D indicates the successful integration of cassettes with varying copy numbers.

2. Sort mCherry+ cells 6 days after transfection. To grow monoclonal lineage tracing cell lines, one could either sort single cells into wells of 96-well plate (with feeder cells) or seed approximately 6000 cells on a 6 cm dish with feeder cells and pick single colonies under a florencent microscope.
3. Check the number of target sites carried by each cell line through amplicon sequencing as described in 4.3. The number of unique integration barcodes detected in each amplicon library represents the number of integrated cassettes in each clone.
4. Some integrated cassettes might exhibit weak expression, making them difficult to detect via scRNA-seq. To confirm the number of actively expressing cassettes, target site amplicon sequencing from cDNA can be performed. The procedure is described below:
   a. Extract RNA from the cell pellet using RNeasy Plus Mini Kit.
   b. Perform cDNA synthesis using 1 ug RNA by the High-Capacity RNA-to-cDNA Kit.
   c. Prepare the target site amplicon library using a two-step PCR amplification as described in 4.3 with the following modifications: 100 ng of cDNA as input and 9 PCR cycles for step both 1 and 2.
5. To validate the target site performance, a similar time course experiment could be performed as described above, which includes activating Cas9-EGFP expression in selected colonies, sorting and sampling mCherry- and GFP-double positive cells over a series of time points, and preparing target site amplicon libraries from gDNA or cDNA for sequencing.
6. These lineage tracing cell lines are now capable of performing lineage tracing in various conditions.

7. An example of conducting in vitro lineage tracing in gastruloids

Timing: less than 2 weeks

Day 0:

1. Activate Cas9-EGFP expression in the lineage tracing mESCs through nucleofection as described in 4.3. In the recovering step, add 500 μL of NDiff instead of culture media into the nucleofection cuvette. Then

transfer all the cells into a 15 ml tube containing 5 ml NDiff. Spin down to obtain the cell pellet.

2. Resuspend the cell pellet in 1 ml NDiff. Count the cells using the Countess cell counter.
3. Calculate and dilute the cell suspension as needed. For example, to seed x cells per well in 40 μL, dilute the cell suspension to 25x cells/ml.
4. Use multi-channel pipet to transfer 40 μL to each well of an ultra-low attachment 96-well plate.
5. Allow the aggregation to happen undisturbed for 48 h.

Day 2:

6. At 48 h post aggregation, Use a multichannel pipet to add 150 μL NDiff supplemented with 3 μM CHIR99021 to each well of the plate containing the aggregates.

Day 3 and Day 4:

7. At 72 and 96 h post aggregation, perform media change. Carefully remove 150 μL from each well using multichannel pipette without disturbing the structure. Then replace it with 150 μL fresh NDiff.

Day 5:

8. At 120 h post aggregation, examine the structures for expected morphology, such as a clear elongation tip. Further validations including morphology measurements, immunofluorescent staining, bulk sequencing analysis could be performed according to application (Bolondi et al., 2024). High-quality lineage tracing gastruloids can be picked for scRNA-seq.

Note.

1. Cells post nucleofection are fragile, handle with care.
2. Pre-equilibrate NDiff in a 10 cm dish in the incubator for at least 20 min before use.

Optimization and troubleshooting

Optimal seeding number is cell line dependent. It is recommended to test various seeding number to determine the range that yields expected morphology.

8. Limitations

We cannot rule out the potential cellular toxicity introduced by Cas9 editing. Due to the stochastic nature of indel generation, the incomplete capture and recovery during scRNA-seq, and the potential out-growth competition of Cas9-negative cells after induction, the resulting lineage tree may not fully reconstruct the complete lineage history. Despite these limitations, this lineage tracing system is still valuable for unbiasedly uncovering differentiation trajectories during cellular events.

References

Adamson, B., Norman, T. M., Jost, M., Cho, M. Y., Nuñez, J. K., Chen, Y., ... Weissman, J. S. (2016). A multiplexed single-cell CRISPR screening platform enables systematic dissection of the unfolded protein response. *Cell, 167*(7), 1867–1882.e21. https://doi.org/10.1016/j.cell.2016.11.048.

Alemany, A., Florescu, M., Baron, C. S., Peterson-Maduro, J., & Van Oudenaarden, A. (2018). Whole-organism clone tracing using single-cell sequencing. *Nature, 556*(7699), 108–112. https://doi.org/10.1038/nature25969.

Bolondi, A., Haut, L., Gassaloglu, S., Burton, P., Kretzmer, H., Buschow, R., ... Veenvliet, J. (2021). Generation of mouse pluripotent stem cell-derived trunk-like structures: An in vitro model of post-implantation embryogenesis. *Bio-protocol, 11*(11), https://doi.org/10.21769/BioProtoc.4042.

Bolondi, A., Law, B. K., Kretzmer, H., Gassaloglu, S. I., Buschow, R., Riemenschneider, C., ... Chan, M. M. (2024). Reconstructing axial progenitor field dynamics in mouse stem cell-derived embryoids. *Developmental Cell, 59*(12), 1489–1505.e14. https://doi.org/10.1016/j.devcel.2024.03.024.

Chan, M. M., Smith, Z. D., Grosswendt, S., Kretzmer, H., Norman, T. M., Adamson, B., ... Weissman, J. S. (2019). Molecular recording of mammalian embryogenesis. *Nature, 570*(7759), 77–82. https://doi.org/10.1038/s41586-019-1184-5.

Choi, J., Chen, W., Minkina, A., Chardon, F. M., Suiter, C. C., Regalado, S. G., ... Shendure, J. (2022). A time-resolved, multi-symbol molecular recorder via sequential genome editing. *Nature, 608*(7921), 98–107. https://doi.org/10.1038/s41586-022-04922-8.

Clement, K., Rees, H., Canver, M. C., Gehrke, J. M., Farouni, R., Hsu, J. Y., ... Pinello, L. (2019). CRISPResso2 provides accurate and rapid genome editing sequence analysis. *Nature Biotechnology, 37*(3), 224–226. https://doi.org/10.1038/s41587-019-0032-3.

Ding, S., Wu, X., Li, G., Han, M., Zhuang, Y., & Xu, T. (2005). Efficient transposition of the piggyBac (PB) transposon in mammalian cells and mice. *Cell, 122*(3), 473–483. https://doi.org/10.1016/j.cell.2005.07.013.

Gilbert, L. A., Horlbeck, M. A., Adamson, B., Villalta, J. E., Chen, Y., Whitehead, E. H., ... Weissman, J. S. (2014). Genome-scale CRISPR-mediated control of gene repression and activation. *Cell, 159*(3), 647–661. https://doi.org/10.1016/j.cell.2014.09.029.

Kalhor, R., Kalhor, K., Mejia, L., Leeper, K., Graveline, A., Mali, P., & Church, G. M. (2018). Developmental barcoding of whole mouse via homing CRISPR. *Science (New York, N. Y.), 361*(6405), eaat9804. https://doi.org/10.1126/science.aat9804.

Kretzschmar, K., & Watt, F. M. (2012). Lineage tracing. *Cell, 148*(1–2), 33–45. https://doi.org/10.1016/j.cell.2012.01.002.

McDole, K., Guignard, L., Amat, F., Berger, A., Malandain, G., Royer, L. A., ... Keller, P. J. (2018). In toto imaging and reconstruction of post-implantation mouse development at the single-cell level. *Cell, 175*(3), 859–876.e33. https://doi.org/10.1016/j.cell.2018.09.031.

McKenna, A., Findlay, G. M., Gagnon, J. A., Horwitz, M. S., Schier, A. F., & Shendure, J. (2016). Whole-organism lineage tracing by combinatorial and cumulative genome editing. *Science (New York, N. Y.), 353*(6298), aaf7907. https://doi.org/10.1126/science.aaf7907.

Mittnenzweig, M., Mayshar, Y., Cheng, S., Ben-Yair, R., Hadas, R., Rais, Y., ... Stelzer, Y. (2021). A single-embryo, single-cell time-resolved model for mouse gastrulation. *Cell, 184*(11), 2825–2842.e22. https://doi.org/10.1016/j.cell.2021.04.004.

Pijuan-Sala, B., Griffiths, J. A., Guibentif, C., Hiscock, T. W., Jawaid, W., Calero-Nieto, F. J., ... Göttgens, B. (2019). A single-cell molecular map of mouse gastrulation and early organogenesis. *Nature, 566*(7745), 490–495. https://doi.org/10.1038/s41586-019-0933-9.

Platt, R. J., Chen, S., Zhou, Y., Yim, M. J., Swiech, L., Kempton, H. R., ... Zhang, F. (2014). CRISPR-Cas9 knockin mice for genome editing and cancer modeling. *Cell, 159*(2), 440–455. https://doi.org/10.1016/j.cell.2014.09.014.

Quinn, J. J., Jones, M. G., Okimoto, R. A., Nanjo, S., Chan, M. M., Yosef, N., ... Weissman, J. S. (2021). Single-cell lineages reveal the rates, routes, and drivers of metastasis in cancer xenografts. *Science (New York, N. Y.), 371*(6532), eabc1944. https://doi.org/10.1126/science.abc1944.

Raj, B., Wagner, D. E., McKenna, A., Pandey, S., Klein, A. M., Shendure, J., ... Schier, A. F. (2018). Simultaneous single-cell profiling of lineages and cell types in the vertebrate brain. *Nature Biotechnology, 36*(5), 442–450. https://doi.org/10.1038/nbt.4103.

Rice, P. (n.d.). EMBOSS: The European Molecular Biology Open Software Suite.

Shen, M. W., Arbab, M., Hsu, J. Y., Worstell, D., Culbertson, S. J., Krabbe, O., ... Sherwood, R. I. (2018). Predictable and precise template-free CRISPR editing of pathogenic variants. *Nature, 563*(7733), 646–651. https://doi.org/10.1038/s41586-018-0686-x.

Yang, D., Jones, M. G., Naranjo, S., Rideout, W. M., Min, K. H. J., Ho, R., ... Weissman, J. S. (2022). Lineage tracing reveals the phylodynamics, plasticity, and paths of tumor evolution. *Cell, 185*(11), 1905–1923.e25. https://doi.org/10.1016/j.cell.2022.04.015.

CHAPTER NINETEEN

Delivery of genome editors with engineered virus-like particles

Christopher Lu[a,b], Yuanhang Li[b,c], Jacob Ryan Cummings[a,b], and Samagya Banskota[a,b,*]

[a]Department of Biomedical Engineering, Boston University, Boston, MA, United States
[b]Biological Design Center, Boston University, Boston, MA, United States
[c]Department of Mechanical Engineering, Boston University, Boston, MA, United States
*Corresponding author. e-mail address: samagya@bu.edu

Contents

1. Introduction 476
 1.1 Therapeutic gene editing technologies 476
 1.2 Delivery of genome editors and comparisons of various delivery methods 480
 1.3 VLPs as a delivery platform 481
2. Materials and methods 490
 2.1 Key resources – materials and equipment 490
 2.2 eVLP production and purification 492
 2.3 Cellular assays for eVLP validation 499
 2.4 eVLP characterization 502
3. Optimization and troubleshooting 507
 3.1 Transfection 507
 3.2 Characterization assays 508
4. Summary 509
Acknowledgements 510
Declaration of interests 510
References 510

Abstract

Genome editing technologies have revolutionized biomedical sciences and biotechnology. However, their delivery *in vivo* remains one of the major obstacles for clinical translation. Here, we introduce various emerging genome editing systems and review different delivery systems have been developed to realize the promise of *in vivo* gene editing therapies. In particular, we focus on virus-like particles (VLPs), an emerging delivery platform and provide in depth analysis on recent advancements to improve VLPs delivery potential and highlight opportunities for future improvements. To this end, we also provide detail workflows for engineered VLP (eVLP) selection, production, and purification, along with methods for characterization and validation.

Methods in Enzymology, Volume 712
ISSN 0076-6879, https://doi.org/10.1016/bs.mie.2025.01.007

1. Introduction

The ability to precisely edit the mammalian genome *in vivo* represents a new era of precision medicine. Here, we provide a brief overview of CRISPR-based genome editing technologies and various *in vivo* delivery methods, with a focus on engineered virus-like particles (eVLPs), which is an emerging class of delivery modality. We present detailed methods for using eVLPs to deliver genome editors and discuss their capabilities and limitations.

1.1 Therapeutic gene editing technologies

1.1.1 Nucleases

Zinc-finger nucleases (ZFNs) and transcription activator-like effector nucleases (TALENs) are regarded as early-generation technologies for programmable gene editing (Bibikova et al., 2002; Cermak et al., 2011; Joung & Sander, 2013). However, the extensive protein engineering required to generate these technologies with high sequence specificity for each new genomic target makes the design process time-consuming and labor-intensive. Thus, the discovery and subsequent development of Clustered Regularly Interspaced Short Palindromic Repeats (CRISPR) and CRISPR-associated (Cas) proteins revolutionized the field of gene editing as the CRISPR/Cas nucleases achieve accurate genomic target recognition through simple RNA:DNA base-pairing interactions between the target DNA and CRISPR RNA (crRNA) to introduce double-strand breaks (DSBs) with high precision. The canonical CRISPR/Cas9 system which adapts the Cas9 protein and a single guide RNA (sgRNA), a fusion of trans-activating RNA (tracrRNA) and crRNA, simplifies the system while maintaining high programmability. By just changing the sgRNA sequence, the researchers can direct Cas9 nuclease to target and cleave any genomic site adjacent to a specific short sequence called the protospacer adjacent motif (PAM) as shown in Fig. 1. This simplified approach, which does not require any protein engineering to target and cleave a specific DNA sequence, has significantly advanced the development of CRISPR technologies.

Currently, numerous clinical trials are underway using the CRISPR/Cas system for a wide range of diseases including cancer, hematological disorders and respiratory conditions (Zhang et al., 2023). Recently, CASGEVY, the first CRISPR/Cas9-based gene editing therapy for sickle cell disease and β-Thalassemia was approved by the FDA, marking a new era of gene editing therapeutics (Philippidis, 2024; U.S. Food & Drug Administration, 2023). However, while nuclease-based gene editing holds great therapeutic potential

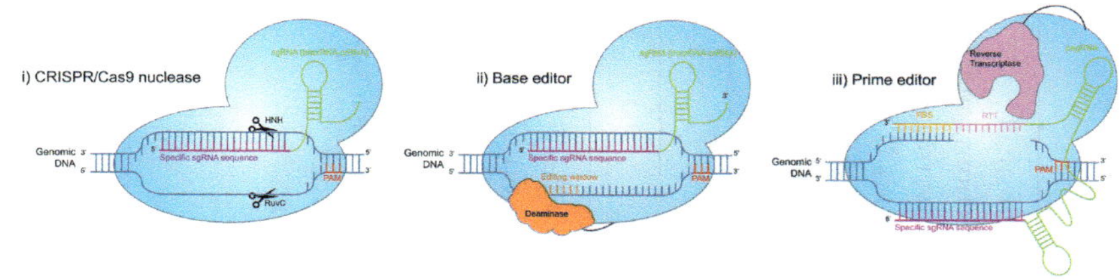

Fig. 1 Overview of genome editing platforms. The diagram highlights three distinct genome-editing methods with schematic depictions of CRISPR/Cas9 nuclease, base editor, and prime editor. (i) CRISPR/Cas9 nuclease targeted by the sgRNA sequence induces DSBs facilitated by the RuvC and HNH domains of Cas9 nuclease. (ii) Base editors targeted by the sgRNA sequence introduce point mutations using either a cytidine deaminase (C to T) or adenosine deaminase (A to G) without causing DSBs. (iii) Prime editors targeted by the sgRNA sequence of the pegRNA alter specific sequences with reverse transcriptase facilitated by the primer-binding site (PBS) and reverse transcriptase template (RTT) without causing DSBs or requiring donor DNA templates.

for gene disruption, the introduction of undesired insertions or deletions at the DSB site may not be ideal for precision gene editing outcomes as detailed in Table 1. Moreover, DSBs generated by nucleases can trigger large-scale deletions, chromosomal translocations, or other abnormalities, presenting significant safety challenges for further clinical applications (Song et al., 2020; Kosicki et al., 2018; Stadtmauer et al., 2020; Turchiano et al., 2021; Leibowitz et al., 2021; Alanis-Lobato et al., 2021).

1.1.2 Base editing

Base editors (BEs) consist of DNA-binding proteins fused with nucleotide-modifying enzymes. Specifically, the most efficient version of BE incorporates Cas9 nickase (nCas9) with a D10A mutation, fused with cytidine deaminase or adenosine deaminase to achieve cytidine base editing (CBE) or adenine base editing (ABE), respectively. To perform base editing, programmable nCas9, under the guidance of sgRNA, binds to a target gene locus and nicks the non-edited strand without generating DSBs as shown in Fig. 1. R-loop formation by nCas9-sgRNA complex exposes unwound single-stranded DNA (ssDNA), allowing the fused deaminase domain to perform deamination (Komor et al., 2016; Nishida et al., 2016; Gaudelli et al., 2017). Within this editing window, the deamination process is constrained to only several nucleotides on the unwound ssDNA and the size of the editing window is dependent on different combinations of Cas proteins and deaminase variants (Komor et al., 2016; Richter et al., 2020).

Table 1 Comparisons of CRISPR/Cas9-based gene editing technologies.

Strategies	Nucleases	Base editing	Prime editing
Mechanism	Introduction of double-strand breaks (DSBs)	Conversion of single nucleotides without DSBs using nCas9 fused with DNA deaminases	Introduction of small insertions, deletions, and nucleotide conversions using nCas9 fused with RT and pegRNA; adding a secondary ngRNA biases desired editing outcomes
DSB Generation	Yes	No	No
Specificity	Moderate	High	Very High
Editing Outcomes	Insertions, deletions, and large-scale genomic alterations	Single-nucleotide substitutions (C-G to T-A, C-G to G-C, and A-T to G-C)	All possible single-nucleotide substitutions, small DNA insertions and deletions
Limitations	• DNA off-target affecting DSBs can lead to large deletions and chromosomal translocations • Limited to precise gene disruption	• Limited in possible single base conversions • Bystander editing • DNA and RNA off-target editing	• Relatively low efficiency • Requires further development

Notably, the invention of BE technology addresses some of the drawbacks of nuclease-based gene editing by enabling more accurate conversions of single base pairs without generating DSBs as detailed in Table 1.

In CBE, the deamination of C•G base pairs creates U•G base pair mismatch, which is protected by the expression of uracil glycosylase inhibitor (UGI) fused to the C terminus of nCas9. The U•G base pair ultimately is read as a T•A base pair following biased endogenous DNA repair mechanisms (Komor et al., 2016). In absence of UGI, the cytosine deamination process can result in C•G to G•C base pair conversion, enabling the further development of C•G to G•C BE (CGBE). By fusing uracil DNA N-glycosylase (UNG) or other DNA repair proteins, CGBEs with higher editing purity have been identified (Koblan et al., 2021). ABEs, on the other hand, were developed by fusing a lab-evolved deoxyadenosine deaminase enzyme derived from an *E. coli*-originated tRNA adenosine deaminase (TadA) to nuclease impaired nCas9 (Gaudelli et al., 2017). Since the first report of base editing, more than 100 BE variants have been developed with different properties in terms of sequence contexts, specificity, editing window or activity. BEs have been applied extensively both *ex vivo* and *in vivo* for therapeutic genome editing to rescue disease phenotypes. In fact, base editing-based therapies have already been used in humans for various clinical trials (Horie & Ono, 2024). While BEs are attractive for correcting single point mutations or installing single-nucleotide variants, current generations of BEs cannot perform all possible single-nucleotide conversions or achieve targeted small insertions or deletions. BEs can also lead to undesired bystander edits if the same nucleotide as the target base is present within the optimal editing window, limiting their therapeutic potential (Huang et al., 2021).

1.1.3 Prime editing

Prime editing overcomes most of the limitations of nuclease and base editing technologies by enabling all possible single base conversions, small insertions, small deletions, and combinations thereof, without generating DSBs as detailed in Table 1. The prime editor (PE) system consists of a reverse transcriptase (RT) enzyme fused to nCas9 domain with an H840A mutation along with a prime editing guide RNA (pegRNA) and a nicking sgRNA (ngRNA). During prime editing, nCas9 domain nicks the non-targeted strand (NTS) while the fused RT initiates reverse transcription on the NTS primed by the 3' extension region of the pegRNA to install the edit of interest that is encoded on the pegRNA as shown in Fig. 1. To bias the DNA repair to the desired edited outcome, a secondary nick on the unedited strand is introduced by the ngRNA (Anzalone et al., 2019).

Recent efforts focused on engineering both PE enzymes as well as pegRNA have resulted in different versions of PE with improved editing efficiencies. For example, a pseudoknot-structured RNA motif was added on the 3' end of pegRNA to prevent degradation by exonucleases, resulting in engineered pegRNA (epegRNA) that substantially improved editing in multiple cell types as well as *in vivo* (Nelson et al., 2022; Davis et al., 2024). In addition, PE protein architecture improvements and manipulation of endogenous host factors including DNA mismatch repair pathway and small RNA-binding proteins further improved prime editing outcomes (Chen et al., 2021; Yan et al., 2024). These improvements have led to several examples of *in vivo* prime editing to correct disease-causing mutations or to install variants that prevent or rescue disease phenotype (Davis et al., 2024; Liu et al., 2021; Böck et al., 2022; Zhi et al., 2022). Recently, analogous editing methods to prime editing have also been developed wherein instead of RT, a DNA-dependent DNA polymerase was used for targeted genome editing with higher editing specificity and fidelity (Liu et al., 2024).

1.2 Delivery of genome editors and comparisons of various delivery methods

Current state-of-the-art gene editing technologies employ either non-viral systems, such as lipid nanoparticles (LNPs), or viral vectors, such as lentiviruses, adenoviruses, and adeno-associated viruses (AAVs) for *in vivo* delivery as shown in Fig. 2. LNPs are the most advanced non-viral delivery

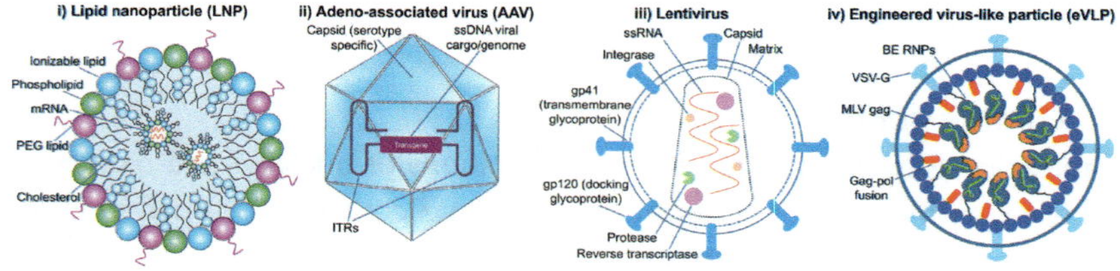

Fig. 2 Overview of various delivery systems for genome editors. The diagram highlights distinct delivery systems. (i) LNPs have variable expression levels depending on cargo type with high packaging capacity for mRNA and RNP delivery limited by high liver delivery levels. (ii) AAVs have long-term transgene expression levels but a strict 4.7 kb packaging capacity for ssDNA with potential for oncogenic integration and pre-existing immunity. (iii) Lentivirus have stable expression levels with a 10 kb optimal packaging limit for ssRNA with potential for genomic integration and off-target editing. (iv) eVLPs have high, transient expression levels capable of mRNA and RNP delivery with less immunogenicity than common viral vectors; eVLPs show promise for *in vivo* delivery applications due to their versatility, targeting capability, rapid clearance, and minimal integration risks.

system in terms of clinical use. But they also face significant challenges due to rapid blood clearance, toxic liver accumulation, and poor endosomal escape of cargo, limiting their broader application as detailed in Table 2 (Zu & Gao, 2021; Sioson et al., 2021). Conversely, viral vectors like lentiviruses, adenoviruses, and adeno-associated viruses offer high delivery efficiency and can target tissues beyond the liver. However, they present challenges including safety concerns, potential genomic integration, and limited cargo capacity as detailed in Table 2 (Zu & Gao, 2021; Sioson et al., 2021; He et al., 2022; Lukashev & Zamyatnin, 2016). Altogether, these platforms restrict efficient and broader *in vivo* delivery of genome editors to target cells (He et al., 2022). Recently, great progress has been made in developing virus-like particles (VLPs) that combine the advantages of viral and non-viral delivery as a promising alternative for *in vivo* delivery of genome editors as detailed in Table 2 (He et al., 2022; Mangeot et al., 2019; Gee et al., 2020; Lyu et al., 2019; Banskota et al., 2022). In section 1.3, we discuss VLP technology, their capabilities and limitations.

1.3 VLPs as a delivery platform

1.3.1 Introduction of VLPs designed from viruses (mRNA packaging and protein/ribonucleoprotein packaging)

VLPs are non-infections macromolecular complexes composed of viral proteins that resemble the natural structure of viruses, however, without the viral genome. Due to the lack of viral pathogenicity, these complexes are instead engineered to incorporate diverse biomolecular cargoes including mRNAs, proteins or even ribonucleoproteins (RNPs), thus making VLPs a novel and versatile delivery platform. VLPs adapt some of the natural properties of viruses such as endosomal escape and specific cellular specificity, leading to efficient and transient delivery of gene editing components. Besides, by transiently delivering cargoes like proteins, RNPs and mRNAs, VLPs also eradicate the risk of viral genome integration to the host and dramatically limit off-target effects in both genomic and transcriptomic editing (Banskota et al., 2022).

Currently, the most efficient VLPs for mRNA and protein delivery of genome editors are derived from retroviruses (RVs) due to their unique natural features in terms of their size and structural geometry (Raguram et al., 2022). First, RV particles are spherical and commonly lack rigid structural symmetry, allowing more flexibility for cargo packaging and particle engineering (Zhang et al., 2015). Second, RV particles have large diameters (100–200 nm), providing more physical space for loading large

Table 2 Consideration metrics of different delivery systems for genome editors.

Delivery system	Expression levels	Packaging capacity, cargo delivery, and immunogenicity	Potential disadvantages	Citations
LNP	Variable expression levels based on cargo type and genomic locus without potential for genomic integration	• High packaging capacity capable of mRNA delivery • Delivery to dividing and non-dividing cells • Low immunogenicity	• Limited delivery to the liver • Short circulation time • Hepatotoxicity • Limited examples for RNP delivery	Kedmi et al. (2010), Kanasty et al. (2013), Li et al. (2019), Liu et al. (2019), Di et al. (2022), Godbout and Tremblay (2022), Semple et al. (2022), Hajiaghapour Asr et al. (2023), Herrera-Barrera et al. (2023)
Lentivirus	Stable expression with potential for genomic integration	• <10 kb optimal packaging of single-stranded RNA (ssRNA) • Delivery to dividing and non-dividing cells • High potential for immunogenicity	• Potential for insertional mutagenesis • Prolonged expression leads to off-target editing, oncogenicity, and toxicity	Montini et al. (2006), Cieślar-Pobuda et al. (2017), Bushman (2020), Dong and Kantor (2021), Kalidasan et al. (2021), Arjomandnejad et al. (2023)

AAV	Long-term transgene expression with low potential for genomic integration	• <4.7 kb optimal packaging of ssDNA • Delivery to dividing and non-dividing cells • Less immunogenic than lentiviruses	• Limited packaging capacity • Potential for oncogenic integration and insertional mutagenesis • Pre-existing immunity restricts therapeutic efficacy	Banskota et al. (2022), Srivastava et al. (1983), Alexander et al. (1994), Hirata et al. (2002), Grieger and Samulski (2005), Herzog (2019), Kenjo et al. (2021)
Adenovirus	High transgene expression levels but shorter-term expression with few indications of genomic integration	• Up to 36 kb packaging of dsDNA • Delivery to dividing and non-dividing cells • High potential for immunogenicity	• Widespread pre-existing immunity leads to high immunogenicity	Advani et al. (2003); Bollard et al. (2010); Bulcha et al. (2021); Crystal (2014); Fausther-Bovendo and Kobinger (2014); Herzog (2019); Lee et al. (2019)
VLP	High expression levels for transient expression with few indications of genomic integration	• Large cargo capacity for mRNA and RNPs • Delivery to dividing and non-dividing cells • Potential for immunogenicity but less than viruses containing genetic material	• Potential for nonspecific co-packaging of cellular components during production • Requires optimization for large-scale production • Potential for immunogenicity from viral components	An et al. (2024); Banskota et al. (2022); Lyu and Lu (2022); Nooraei et al. (2021)

cargoes such as Cas9-based RNPs (Zhang et al., 2015). Third, RVs exhibit high modularity due to the distinct domains responsible for cellular targeting and cargo packaging, wherein cellular targeting specificity is achieved through diverse envelope glycoproteins, while the packaging process is mediated by the gag-pro-pol structural polyprotein complex. This modularity allows for both robust packaging of different cargoes by fusion to the gag protein and highly programmable targeting to various cell types; targeting can be achieved by using diverse naturally occurring viral glycoproteins or combining various targeting moieties on the surface (Michels et al., 2022; Zhang et al., 2015). By combining the effective targeting capabilities of viral delivery methods with the transient expression of non-viral methods, RV-based VLPs simultaneously increase potency and decrease unwanted off-target activity (Zhang et al., 2015).

1.3.2 mRNA delivery

VLPs can be engineered to deliver mRNA into target cells. RVs naturally package RNA using a packaging signal (Ψ) that is recognized by the nucleocapsid domain of retroviral gag protein (Jouvenet et al., 2011). While the Ψ signal can be used to package mRNA into VLPs derived from RVs, it enables packaging of only two copies of mRNA per particle (Galla et al., 2004). Therefore, alternative methods using the interaction between the bacteriophage MS2 coat protein (MS2cp) and the bacteriophage MS2 aptamer (MS2apt) have been developed for improved mRNA packaging into VLPs (Prel et al., 2015).

Using this method, several research groups have utilized VLPs to deliver Cas9 mRNA for gene editing applications, demonstrating the versatility and potential of this approach. Galla and colleagues delivered SpCas9 mRNA and sgRNA into human embryonic kidney 293T (HEK293T) and Jurkat cells by fusing two copies of MS2cp to the C-terminus of the murine leukemia virus (MLV) gag protein and adding two copies of MS2apt to the 3′ UTR of both SpCas9 mRNA and sgRNA (Knopp et al., 2018). While promising, their method showed limited gene editing efficiencies due to insufficient delivery of the sgRNA. Atala and colleagues addressed this limitation by packing SaCas9 mRNA with MS2ap using MS2cp fused to lentiviral nucleocapsid (Lu et al., 2019). However, the sgRNA was encoded by a separate integrase-deficient lentiviral vector (IDLV) genome. More recently, Cai and colleagues extended the lentiviral mRNA with IDLV sgRNA expression approach to *in vivo* applications (Ling et al., 2021). Their study not only demonstrated efficient packaging and co-delivery of SpCas9 mRNA and sgRNA within the IDLV genome in various cell types, including HEK293T, NIH3T3, K562, and Jurkat cells, but also, in a mouse model of age-related macular degeneration induced by VEGF-A, resulting in a 44% knockout of the *VEGFA* gene, highlighting its potential for therapeutic applications, as shown in Fig. 3 (Ling et al., 2021). These studies collectively demonstrate the potential of VLP-mediated delivery of mRNA for both *in vitro* and *in vivo* gene editing applications, paving the way for further advancements in this field.

While VLPs offer several advantages for mRNA delivery, including transient expression of the delivered mRNA which minimizes the duration of gene-editing activity and reduces potential off-target effects, there are limitations. Co-packaging of sgRNA and Cas9 mRNA within the same VLP remains

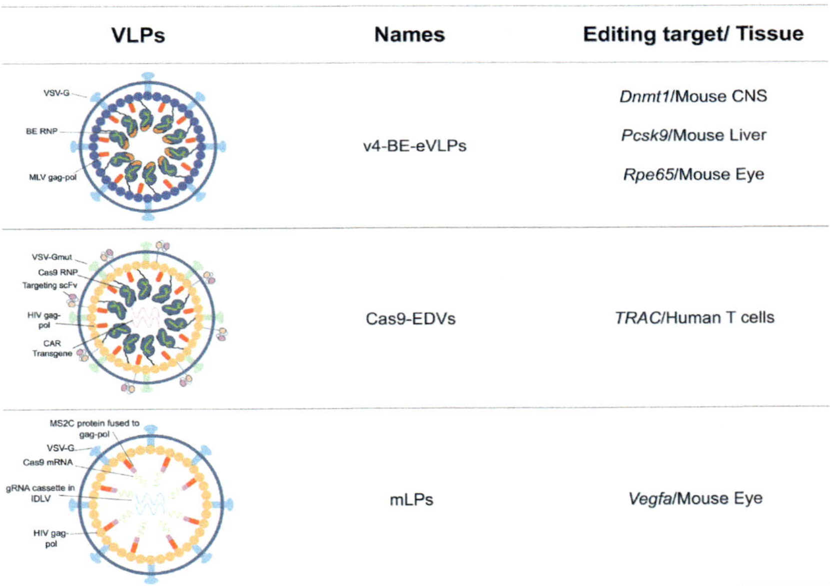

Fig. 3 Advancements in VLP-based protein or mRNA delivery for genome editors. (i) v4 BE-eVLPs enable efficient delivery of base editors and guide RNA *in vivo*, resulting in *Dnmt1* editing in mouse CNS, *Pcsk9* editing in mouse liver and *Rpe65* editing in mouse retinal pigment epithelium (RPE) tissue (Banskota et al., 2022). (ii) Cas9-packaging enveloped delivery vehicles (Cas9-EDVs) result in disruption of *TRAC* gene of T cells in humanized mice (Hamilton et al., 2021). (iii) a single subretinal injection of mRNA-carrying lentiviral particles (mLPs) can mediate efficient knockout of *Vegfa* in mouse RPE tissues of a mouse model of wet age-related macular degeneration (Ling et al., 2021).

challenging due to packaging stoichiometry and the rapid degradation of unmodified sgRNA before it can complex with the Cas9 proteins in transduced cells. The use of IDLVs to express sgRNA can mitigate this limitation, but IDLV cassettes remain as episomal DNA in transduced cells and have a low yet detectable frequency of genomic integration, which increases the associated risks (Kymäläinen et al., 2014; Wang et al., 2010).

1.3.3 Protein/RNP delivery

In addition to mRNA cargo, VLPs can also package and deliver proteins or RNPs. Protein cargo is incorporated into VLPs by fusing the cargo to the gag structural proteins. This fusion ensures the assembly and packaging of the protein cargo within the VLPs during production. A protease-cleavable linker sequence, recognized by the co-encapsulated viral protease, is introduced between the cargo and gag fusion. This allows the viral protease to release the packaged cargo during particle maturation, enabling the cargo to freely access its target upon delivery into the transduced cells (Cai et al., 2014; Kaczmarczyk et al., 2011; Voelkel et al., 2010).

This approach has been employed to package and deliver various protein cargoes within VLPs, including genome editors. Ricci and colleagues adapted the MLV VLP system for delivery of Cas9 RNP. They fused Cas9 protein to the C-terminus of gag using MLV protease-cleavable linker and expressed sgRNAs from a plasmid during VLP production. Co-expression of Cas9 and sgRNA in the producer cell led to direct packaging of Cas9 RNPs during VLP assembly (Mangeot et al., 2019). Similarly, Doudna and colleagues demonstrated that SpCas9 RNP can be delivered using HIV VLPs. In this system, the Cas9 protein is fused to HIV gag with a HIV protease-cleaved peptide linker while sgRNA expression was done either using a co-packaged IDLV genome or co-expressed using a non-lentiviral plasmid during VLP production in the producer cells (Hamilton et al., 2021). This design resulted in up to 90 % editing efficiency in Jurkat cells and 70 % in primary human T cells (Hamilton et al., 2021).

However, while most of these studies showed efficient editing *in vitro,* they showed limited or no editing *in vivo*. Consequently, recent efforts have focused on enhancing their delivery potency for *in vivo* applications. eVLPs based on Moloney murine leukemia virus (MMLV) have shown therapeutic levels of editing across various organs in mice (Banskota et al., 2022). Liu and colleagues identified several bottlenecks limiting the performance of VLPs for *in vivo* delivery and engineered solutions to each of these bottlenecks as detailed in Table 3. First, the cleavage of protease-

Table 3 Selection of BE-eVLPs from various versions.

BE-VLP system	Components	Uses
v1 BE-VLP	Friend murine leukemia virus (FMLV) gag-ABE8e fusion construct, WT FMLV gag-pro-pol polyprotein, VSV-G envelope glycoprotein, and sgRNA	Not recommended; initial BE-VLP construct designed from the FMLV-based VLPs that packaged and delivered Cas9 RNPs (Mangeot et al., 2019; Banskota et al., 2022)
v2.4 BE-eVLP	MMLV gag-ABE8e fusion construct, MMLV gag-pro-pol polyprotein, VSV-G envelope glycoprotein, and sgRNA	Not recommended; v2.4 yields lower editing than v3.4 and v4 BE-eVLP but optimized linker-cleavage efficiency dictated by protease-substrate specificity by switching retroviral scaffold from FMLV to MMLV (Banskota et al., 2022)
v3.4 BE-eVLP	MMLV gag-3xNES-ABE8e fusion construct, MMLV gag-pro-pol polyprotein, VSV-G envelope glycoprotein, and sgRNA	v3.4 contains nuclear export signals (NESs) that localize cargo to the cytoplasm of producer cells for improved packaging. NESs are cleaved following VLP maturation liberating NLS-flanked ABE8e to improve import of eVLP into the nuclei of transduced cells (Banskota et al., 2022)
v4 BE-eVLP	MMLV gag-3xNES-ABE8e fusion construct, MMLV gag-pro-pol polyprotein, VSV-G envelope glycoprotein, and sgRNA	v4 contains optimized gag-cargo:gag-pro-pol stoichiometry by balancing plasmid transfection ratio (25 % gag-cargo); this decreased BE protein molecules/particles but maintained the same amount of sgRNA molecules as v3.4 indicating more RNPs/particles. v4 showed higher levels of editing at common loci (Banskota et al., 2022)
v5 BE-eVLP	MMLV gag-3xNES-ABE8e fusion construct with mutations in gag domain, MMLV gag-pro-pol polyprotein with mutations in gag domain, VSV-G envelope glycoprotein, and sgRNA	v5 contains capsid mutations (Gag^{Q226P}-pro-pol, Gag^{C507V}-cargo) that improved RNP cargo packaging and release along with subsequent changes in capsid structure of eVLPs. v5 showed 2–4-fold higher levels of editing compared to v4 *in vitro* (An et al., 2024)

linkers between the MMLV gag and the genome editor cargo was optimized. This optimization aimed to minimize pre-mature cleavage before eVLP assembly while maximizing cleavage and cargo release from gag after particle maturation and delivery in the target cells (Banskota et al., 2022). Second, a nuclear export sequence (NES) was introduced into gag-Cas9 fused protein sequences to increase the cytosolic localization of the cargo and thus improving packaging efficiency (Banskota et al., 2022). Finally, the ratio of MMLV structural protein to cargo was optimized to enhance both packaging and editing outcomes (Banskota et al., 2022). As a result, eVLPs successfully delivered base editing RNP complexes *in vivo*, enabling 5 % bulk editing and 60 % editing of cells enriched for VLP transduction in the brain, 20–30 % editing in mouse retinal pigment epithelium cells for ocular function, and 63 % editing of *Pcsk9* gene in mouse liver, as shown in Fig. 3 (Banskota et al., 2022). More recently, eVLPs have also achieved transient delivery of RNP machinery for prime editing *in vitro* and *in vivo*, leading to therapeutic levels of editing and rescue of disease phenotype in mouse models of genetic blindness (An et al., 2024).

1.3.4 Recent advances in VLP engineering

Recent advancements in VLP engineering have focused on expanding cell and tissue targeting while reducing viral components to minimize immunogenicity. The canonical VLP design utilizes the broadly tropic vesicular stomatitis virus glycoprotein (VSV-G), which employs the low-density lipoprotein receptor (LDL-R) for cellular entry. Ricci and colleagues demonstrated that incorporating baboon retroviral envelope alongside VSV-G in Cas9 VLPs enhanced targeting of hematopoietic stem cells (HSCs), as quiescent HSCs do not express LDL-R (Mangeot et al., 2019). Similarly, VLPs have been pseudotyped with envelope proteins from HIV to preferentially target genome editing to $CD4^+$ T cells (Hamilton et al., 2021).

More recently, targeting domains such as antibodies and antibody fragments (scFvs) that bind to cell type-specific surface ligands have been employed to program particle targeting. Doudna and colleagues showed that scFvs can be displayed on VLP surfaces by overexpressing them with a transmembrane domain tag in the cell line used for VLP production. Combining scFvs with a viral fusogen, a viral glycoprotein with an inactivated targeting domain but retained membrane fusion ability, enables selective delivery of gene editing machinery to specific cell types. This envelope design successfully delivered Cas9 nuclease to primary T cells *ex vivo* (Hamilton et al., 2024), albeit with low activity *in vivo*, as shown in Fig. 3. A limitation of this system is the requirement for engineering scFvs for each individual receptor of interest.

Building on this platform, Zhang and colleagues developed a cell type-specific delivery platform named DIRECTED, which combines viral fusogen with protein AG (Strebinger et al., 2023). Protein AG binds to the Fc domain of antibodies, creating a universal platform that does not require optimization and engineering for each targeting domain. Future work on re-engineering VLP targeting should focus on optimizing and improving this system for *in vivo* gene editing.

The sequencing of eukaryotic genomes has revealed that a majority of DNA comprises retroelements, including retroviruses, that were integrated into the genome throughout evolution. These endogenous retroviral sequences encode core structural genes (gag) as well as envelope (env) genes. Zhang and colleagues utilized genome sequencing data to identify several gag homologs from endogenous retroviruses capable of assembling into VLPs (Segel et al., 2021). One LTR retrotransposon homolog, PEG10, was shown to assemble into a VLP-like structure while preferentially binding and encapsulating mRNA sequences with a specific packaging signal (Segel et al., 2021). This system, called SEND, successfully delivered Cas9 mRNA to various cell lines *in vitro*, establishing the use of endogenous elements to create VLP particles for delivery with potentially lower immunogenicity (Segel et al., 2021). However, these systems have yet to be tested *in vivo*, and their low immunogenicity remains to be validated. The scalability of both the conventional and next generation VLPs for clinical translation remain a significant challenge. However, recent advancements in using directed evolution of the gag structural protein of VLPs to enhance cargo packaging and production offer a promising solution (Raguram et al., 2024). These challenges represent potential areas for future exploration and improvement in VLP engineering for gene editing delivery applications.

1.3.5 Advantages and disadvantages of VLP-based delivery platforms

The VLP delivery system offers many advantages over traditional plasmid- and virus-based delivery methods for gene editing applications. Notably, VLP-based delivery has demonstrated significantly reduced off-target editing both *in vitro* and *in vivo*, addressing a critical concern for gene editing technologies. Furthermore, the primary cargo for VLP delivery—mRNA and proteins—eliminates the risk of viral genomic integration and insertional mutagenesis, enhancing the safety profile of this delivery system (Mangeot et al., 2019; Banskota et al., 2022).

Despite these promising attributes, several challenges remain in the development and translation of VLP-based delivery systems into the clinic. Current VLP production, primarily done using transient transfection of HEK293T cells, requires significant optimization for large-scale manufacturing (Fuenmayor et al., 2017). While the modularity of VLPs enables targeting of various cell types, achieving specificity for a single cell type *in vivo* remains challenging, necessitating further exploration of different binding moieties and their combinations (Raguram et al., 2022). Additionally, the cargo packaging process of VLPs inadvertently includes non-cargo host proteins from producer cells, warranting thorough investigation before potential clinical applications (Mangeot et al., 2019). Finally, the use of viral structural proteins in VLPs raises concerns about potential immunogenicity *in vivo*. Therefore, future research should focus on scaling up production and improving yield of VLPs and minimizing immune responses by exploring alternatives to viral proteins (Hamilton et al., 2021; Raguram et al., 2022). Addressing these challenges will be crucial in realizing the full potential of VLP-based delivery for gene editing and other therapeutic applications.

2. Materials and methods

2.1 Key resources – materials and equipment

2.1.1 Key resources notes

1. The development of the eVLP system for efficient genome editor delivery involved multiple rounds of iterative engineering, with key design improvements summarized in Table 3. For Cas9 and base editor delivery, we recommend using either the v4 BE-eVLP or v5 BE-eVLP design. For prime editors, the v3 PE3-eVLP or v3b PE3-eVLP designs are recommended. For users interested in developing new constructs, we recommend considering both mechanistic and stoichiometric factors, focusing on cargo packaging efficiency (NESs/NLSs), stoichiometric optimization of gag-pro-pol and gag-cargo constructs, protease specificity, sgRNA loading, and payload stability.
2. Plasmids for eVLP production are available on Addgene and we have listed them below. While the following protocol is optimized for v4 BE-eVLPs and includes the transfection outline for v3 PE-eVLP and v3b PE-eVLPs, the same methodology can be applied to other genome editors.

3. Our method uses Takara's Gesicle Producer 293T Cell Line. This cell line is a derivative of HEK293T cell line engineered to efficiently produce cell-derived nanovesicles. After thawing, we recommended passaging the cells at least twice prior to eVLP production.
4. All cell culture experiments, plasmid transfection, and eVLP purification steps (other than centrifugation) should take place inside a sterile tissue culture hood.
5. Common lab equipment that is needed to carry out the eVLP production, purification, and characterizations steps as shown in Fig. 4 that are not highlighted in the table include:
   a) single-channel pipettes
   b) multi-channel pipettes
   c) serological pipette controller
   d) Centrifuge/Microcentrifuge
   e) PCR machine/qPCR machine
   f) Gel electrophoresis system
   g) Milli-Q® system
   h) Thermal mixer
   i) Hotplate stirrer
   j) Orbital shaker
   k) microbiological shaking incubator and incubator
   l) pH meter
   m) Dry bead/Hot water bath set to 37 °C
   n) Cell culture incubator set to 37 °C
   o) Inverted microscope/Fluorescent microscope
   p) 4 °C fridge/−20 °C freezer/−86 °C freezer

2.1.2 Key resources tables

Resource or reagent	Source	Identifier
Transfection Plasmids		
pCMV-MMLVgag-3xNES-Cas9	Addgene	181752
pCMV-VSV-G	Addgene	8454
pBS-CMV-gag pol	Addgene	35614
pCMV-MMLVgag-3xNES-ABE8e	Addgene	181751
gag-MCP-pol	Addgene	211370
gag-PE	Addgene	211371

gag-COM-pol	Addgene	211373
gag-P3-pol	Addgene	211374
P4-PE	Addgene	211375
Bacterial and Virus Strains		
NEB® 5-alpha Competent *E. coli*	New England Biolabs	C2987H
Biological Samples (Cell Lines)		
Gesicle Producer 293T Cell Line	Takara	632617
Instruments (equivalents can be used)		
Invitrogen™ Countess™ 3 Automated Cell Counter	Thermo Fisher	AMQAX2000
Optima™ XPN-90 – IVD	Beckman Coulter	A99833
Type 45 Ti Fixed-Angle Titanium Rotor	Beckman Coulter	339160
Tube Topper Kit, 60 Hz	Beckman Coulter	358312
SW 32 Ti Swinging-Bucket Rotor	Beckman Coulter	369650
MiSeq™ System	Illumina	SY−410 −1003
BioTek Synergy H1 Multimode Reader	Agilent	
DynaPro™ Dynamic Light Scattering (DLS) Plate Reader Instrument	DynaPro	
NanoSight Pro	Malvern Panalytical	
Applied Biosystems® StepOnePlus System	Thermo Fisher	

2.2 eVLP production and purification

2.2.1 Materials and equipment

Cell culture and eVLP production components (media, transfection, *etc.*)		
Phosphate-buffered saline (DPBS, 1X), Dulbecco's formula, without calcium, without magnesium	Thermo Fisher	J67802. K2
Corning® 100 mL Trypsin EDTA 1X	Corning	25-052-CI
DMEM, high glucose, GlutaMAX™ Supplement, pyruvate	Thermo Fisher	10569069
Gibco™ Fetal Bovine Serum, certified, heat inactivated, United States	Thermo Fisher	10082147

Opti-MEM™ I Reduced Serum Medium	Thermo Fisher	31985070
Trypan Blue Solution, 0.4%	Thermo Fisher	15250061
jetPRIME®, DNA and siRNA transfection reagent	VWR	89129-920
PEG-it™ Virus Precipitation Solution (5 ×)	System Biosciences	LV825A-1
Chemicals		
Polyethylene Glycol 8000 (PEG), Fisher BioReagents™	Fisher Scientific	BP233-100
Sodium chloride, extra pure, Thermo Scientific Chemicals	Thermo Fisher	194090010
Sodium hypochlorite, 10–15% active chlorine, Thermo Scientific Chemical	Fisher Scientific	AC219255000
Sucrose, 99%	Thermo Fisher	A 15583.0 C
Thermo Scientific™ Glycerol	Thermo Fisher	T17904K2
UltraPure™ DNase/RNase-Free Distilled Water	Thermo Fisher	10977023
Consumables		
Corning® 500 mL Bottle Top Vacuum Filter, 0.22 μm Pore PES Membrane, Sterile	Corning	431097
Corning® 250 mL Bottle Top Vacuum Filter, 0.22 μm Pore PES Membrane, Sterile	Corning	431096
Corning™ U-Shaped Cell Culture Flasks	Fisher Scientific	07-202-000
Countess™ Cell Counting Chamber Slides	Thermo Fisher	C10228
Fisherbrand™ Sterile Syringes for Single Use, 10 mL	Fisher Scientific	14-955-459
Basix™ Syringe Filters, PES, Sterile	Fisher Scientific	13-100-107
Kimberly-Clark Professional™ Kimtech Science™ Kimwipes™ Delicate Task Wipers, 1-Ply	Fisher Scientific	06-666
Nunc™ Serological Pipettes	Thermo Fisher	
xTIP4 LTS Compatible Pipette Tips	Biotix	

Nunc™ 15 mL and 50 mL Conical Sterile Polypropylene Centrifuge Tubes	Thermo Fisher	339651
Snap Cap Low Retention Microcentrifuge Tubes, 1.5 mL graduated, sterile (polypropylene)	Thermo Fisher	3451
31.5 mL, Open-Top Thinwall Polypropylene Konical Tube, 25 x 89 mm	Beckman Coulter	358126
Instruments		
Invitrogen™ Countess™ 3 Automated Cell Counter	Thermo Fisher	AMQAX2000
Optima™ XPN–90 – IVD	Beckman Coulter	A99833
Type 45 Ti Fixed-Angle Titanium Rotor	Beckman Coulter	339160
Tube Topper Kit, 60 Hz	Beckman Coulter	358312
SW 32 Ti Swinging-Bucket Rotor	Beckman Coulter	369650

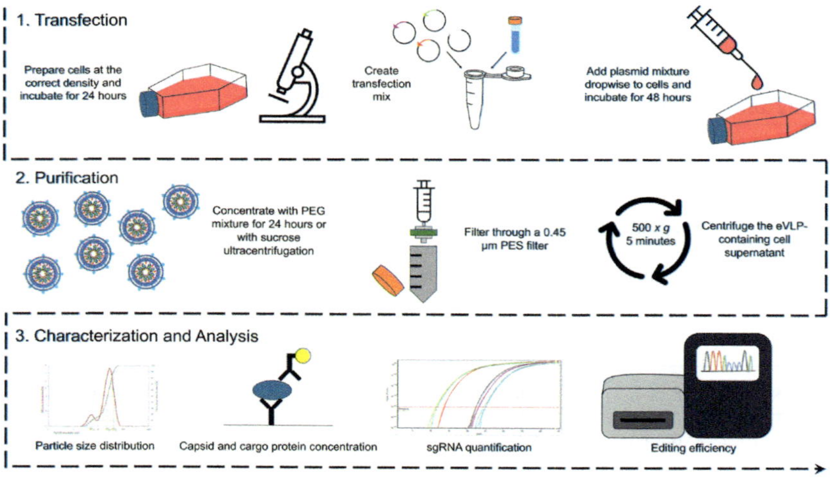

Fig. 4 Workflow of eVLP production, purification, and characterization. Depiction of the essential steps for the (1) production *via* transfection of Gesicle Producer 293T cells in a T75 flask format, (2) purification *via* PEG concentration or sucrose ultracentrifugation, (3) characterization and analysis of produced eVLPs for validation of particle size, capsid and cargo protein concentration, sgRNA levels, and genome editing efficiency.

2.2.2 Preparation of cell culture media

1. Remove 50 mL of media from a 500 mL stock bottle of DMEM using a sterile serological pipette and replace with 50 mL of FBS.
2. Secure cap and gently rotate the cell culture media bottle to mix components to avoid introducing foam and filter the media through a Corning® 500 mL Bottle Top Vacuum Filter, 0.22 μm Pore PES Membrane.
3. Store the cell culture media at 4 °C for up to four weeks or use immediately for cell culturing and passaging.

2.2.3 Cell seeding at optimal density for transfection

Timing: Day 0

1. Prior to passaging cells, warm the culture media, DPBS, and Trypsin-EDTA to 37 °C using a bead bath or water bath.
2. Passage Gesicle Producer 293T cells in a 75 cm^2 cell culture flask to 70–80 % confluency, confirming the cell coverage using a microscope.
3. Aspirate the cell culture media, being careful not to disturb the adherent cell layer on the bottom surface of the flask.
4. Gently add 6 mL of DPBS to the flask and rock side-to-side to wash off residual spent cell culture media, aspirating the DPBS afterwards. Do not pipette directly onto the cell layer to prevent any detachment of the adherent cells.
5. Add 3 mL of Trypsin-EDTA to the cell culture flask, ensuring adequate coverage of the entire bottom surface of the flask, and return the flask to the cell culture incubator at 37 °C, 5 % CO_2 for 3–5 min.
6. Deactivate the Trypsin-EDTA by adding 9 mL of cell culture media to the flask. Break up any cell clumps by repetitively pipetting the cell media over the side of the flask which had the cells adhered, and, if necessary, pipette flush against the back wall of the cell culture flask to ensure no cell clumps remain.
7. Transfer the cell culture media containing the resuspended cells to a clean 15 mL conical tube.
8. Count the number of viable cells suspended in media using a Invitrogen™ Countess™ 3 Automated Cell Counter or other appropriate equipment. If using Invitrogen™ Countess™ 3 Automated Cell Counter, briefly, mix 10 μL of resuspended cell-containing media with 10 μL of Trypan Blue 0.4 %, pipette mix the solution to confirm even dispersion of cells throughout the mixture.

9. Add 10 μL of the solution to a Countess™ Cell Counting Chamber Slide and insert the slide into the machine. Follow the user manual to record the concentration of viable cells per mL and then dilute the resuspended cells to 5×10^5 cells per mL with fresh cell culture media.
10. Add 10 mL of diluted cells (total of 5 million cells) to a clean 75 cm^2 cell culture flask and return to the cell culture incubator at 37 °C, 5 % CO_2 for 24 h.

2.2.4 Transfect cells with plasmids

Timing: Day 1 (18–24 h after seeding).

1. In a sterile 1.5 mL polypropylene microcentrifuge tube, combine 3.3375 μg of pBS-CMV-gagpol, 1.125 μg of pCMV-MMLVgag-3xNES-ABE8e, 0.4 μg of pCMV-VSV-G, and 4.4 μg of sgRNA plasmid with 500 μL of jetPRIME®, DNA and siRNA transfection reagent buffer.
   a. If making PE-eVLPs, combine the following plasmids with 500 μL of jetPRIME®, DNA and siRNA transfection reagent buffer, which has been previously detailed by Liu and colleagues (An et al., 2024):
      v3 PE-eVLPs
      i. 0.4 μg pCMV-VSV-G
      ii. 2.813 μg pBS-CMV-gagpol
      iii. 1.125 μg gag-MCP-Pol
      iv. 0.563 μg gag-PE
      v. 4.4 μg MS2-epegRNA, or 3.52 μg MS2-epegRNA and 0.88 μg MS2-ngRNA
      v3b PE-eVLPs
      vi. 0.4 μg pCMV-VSV-G
      vii. 2.813 μg pBS-CMV-gagpol
      viii. 2.0 μg gag-COM-Pol
      ix. 0.422 μg gag-P3-Pol
      x. 0.422 μg P4-PE
      xi. 4.4 μg COM-epegRNA, or 3.52 μg COM-epegRNA and 0.88 μg COM-ngRNA
   b. For very concentrated stock solutions of plasmids, it may be beneficial to first dilute the plasmids in sterile UltraPure™ water before making the appropriate mix of plasmids to avoid pipetting very small volumes.
2. Vortex the plasmid mix well to ensure homogenous distribution of DNA in solution.
3. Add 20 μL of jetPRIME®, DNA and siRNA transfection reagent to the plasmid mixture and quickly pipette-mix the components together.

4. Allow the mixture to incubate at room temperature for 10-15 min, then slowly add the entire volume dropwise to the cell culture flask prepared the day prior.
5. Return the cell flask to the cell culture incubator at 37 °C, 5 % CO_2 and wait 48 h to harvest eVLPs.

2.2.5 Purify virus-like particles in cell supernatant

Timing: Day 3 (48 h after transfection) – Day 4.

1. Use a serological pipette to transfer the entire supernatant from the cell culture flask into a sterile 15 mL polypropylene conical tube. When pipetting, draw the liquid from the bottom corner of the cell culture flask to avoid dislodging the adhered cells.
2. Centrifuge the cell supernatant at 500 *x g* for five minutes to separate eVLPs from cellular debris.
3. Filter the clarified supernatant through a 0.45 μm pore size Basix™ PES Syringe Filter attached to a 10 mL syringe. Briefly, draw the clarified supernatant into a sterile syringe, attach the syringe filter, then, at a steady rate, dispense the supernatant through the syringe filter into a new sterile 15 mL conical tube.

   Note: The following two methods can be used to concentrate virus-like particles in the cell supernatant. Concentrating with PEG is a non-specific method that separates matrix components based on their solubility. The 100-fold PEG concentration method is suitable for *in vitro* work and does not require ultracentrifugation, although it may be less effective in purifying eVLPs from cell supernatant compared to a sucrose density gradient. For *in vivo* applications of eVLPs, method B is recommended for a higher-purity product. Additionally, using a sucrose gradient for purification allows for a higher titer of virus-like products in comparison to a PEG concentration method, with an achievable concentration factor of 1000–5000-fold depending on the downstream use.

Method A: Concentrate with PEG mixture

4. First, create 4x PEG concentration solution by dissolving 80 g of Polyethylene Glycol 8000 (PEG) and 14 g of sodium chloride into a mixture of 80 mL of Milli-Q® water and 20 mL of 10x PBS, RNase free.
5. Adjust the pH to between 7.0-7.2 and bring the solution to a final volume of 200 mL, then filter the solution through a Corning® 250 mL Bottle Top Vacuum Filter, 0.22 μm Pore PES Membrane and store at 4 °C or use immediately.

6. Dilute the 4x PEG solution to 1x in the filtered supernatant and store at 4 °C for 24 h.
    a. **Alternative:** use the 5x PEG-iT™ solution. Dilute the 5x PEG-iT™ solution to 1x in the filtered supernatant and store at 4 °C for 24 h.
7. Centrifuge the supernatant and PEG mixture at 1500 *x g* for 30 min at 4 °C. A small beige pellet should be visible at the bottom of the conical tube
8. Slowly pour the media into 10–15 % sodium hypochlorite and use a clean Kimwipe® to wipe the inner walls of the conical tube to remove any traces of media and place the conical tube inverted on clean Kimwipes®. This helps ensure that any leftover media is removed from the tube while minimizing disturbance or resuspension of any of the pelleted eVLPs.
9. Resuspend the eVLP pellet in 1/100th of its original volume using 4 °C Opti-MEM™ buffer. For example, if there were 10 mL of cell supernatant and PEG solution initially, resuspend the pellet in 100 μL of Opti-MEM.
10. Aliquot the eVLPs and store at −80 °C, or directly transduce cells and/or run analytical tests if prepared.

Method B: Concentrate with sucrose ultracentrifugation.

1. Prepare 20 % w/v sucrose solution by dissolving 20 g of sucrose in 100 mL of 1x PBS solution.
2. Add 25–27 mL of filtered supernatant to the polypropylene Konical tubes designed for ultracentrifugation.
3. Add 2–3 mL of 20 % sucrose to the bottom of the polypropylene Konical using a long pipette, dispensing when the pipette is all the way at the bottom and gently to form two distinct layers of filtered cell supernatant and sucrose.
    a. **Alternative:** one can add sucrose first followed by gentle pipetting of filtered supernatant without disrupting the sucrose layer.
4. Centrifuge the solution in a Beckman Coulter Allegra X-15R Refrigerated Centrifuge at 4 °C for 2 h at 122,000 *x g* using Beckman Coulter SW28 swinging bucket rotor with adapters for Konical tubes.
5. After centrifugation, remove the tubes inside the BSC hood and confirm the presence of a small opaque pellet at the bottom of the tube. Slowly pour the media into 10-15 % sodium hypochlorite and use a clean Kimwipe® to wipe the inner walls of the conical tube to remove any traces of media and place the conical tube inverted on clean Kimwipes®.

6. Resuspend the eVLP pellet in 1/100–1/3000th of its original volume using 4 °C Opti-MEM™ buffer or cold PBS. Concentration can also be adjusted based on the desired application.
7. Spin the solution for five minutes at room temperature at 1000 *x g* to separate any remaining viral or protein debris and transfer the eVLP-containing supernatant to a sterile 1.5 mL polypropylene microcentrifuge tube.
8. Aliquot the eVLPs and store at −80 °C, or directly transduce cells and/or run analytical tests if prepared.

2.3 Cellular assays for eVLP validation

2.3.1 Materials and equipment

Plasmids		
Lentiviral nontargeting shRNA GFP	Addgene	155285
Cell Culture and eVLP Production Components (Media, Transfection, *etc.*)		
Phosphate-buffered saline (DPBS, 1X), Dulbecco's formula, without calcium, without magnesium	Thermo Fisher	J67802. K2
Corning® 100 mL Trypsin EDTA 1X	Corning	25-052-CI
DMEM, high glucose, GlutaMAX™ Supplement, pyruvate	Thermo Fisher	10569069
Gibco™ Fetal Bovine Serum, certified, heat inactivated, United States	Thermo Fisher	10082147
Trypan Blue Solution, 0.4%	Thermo Fisher	15250061
Chemicals		
Puromycin Dihydrochloride	GoldBio	P-600-100
UltraPure™ DNase/RNase-Free Distilled Water	Thermo Fisher	10977023
Critical Commercial Assays and Kits		
MiSeq Reagent Kit v2 (300-cycles)	Illumina	MS-102-2002
Consumables (equivalents can be used)		
Corning™ U-Shaped Cell Culture Flasks	Fisher Scientific	07-202-000
CELLTREAT Scientific Products 96 Well Tissue Culture Plate, Sterile	Fisher Scientific	50-202-142

Countess™ Cell Counting Chamber Slides	Thermo Fisher	C10228
Nunc™ Serological Pipettes	Thermo Fisher	
xTIP4 LTS Compatible Pipette Tips	Biotix	
Nunc™ 15 mL and 50 mL Conical Sterile Polypropylene Centrifuge Tubes	Thermo Fisher	339651
Snap Cap Low Retention Microcentrifuge Tubes, 1.5 mL graduated, sterile (polypropylene)	Thermo Fisher	3451
Instruments (equivalents can be used)		
Invitrogen™ Countess™ 3 Automated Cell Counter	Thermo Fisher	AMQAX2000
MiSeq™ System	Illumina	SY-410-1003

2.3.2 Qualitative validation of eVLP mediated genome editing activity

There are multiple approaches to qualitatively confirm genome editing activity after eVLP-mediated delivery. We will focus on BE-eVLPs. One common approach is to use fluorescent reporter systems in which the expression of a fluorescent protein is contingent upon precise gene editing (Katti et al., 2020; Tálas et al., 2021). Dow and colleagues developed a reporter system termed "GO" or "Gene On", where GFP expression is turned on only after precise editing with CBEs or ABEs (Katti et al., 2020). To create this system, they transduced HEK293T cells and mouse fibroblast cells with a silenced GFP gene, changing the AUG start sequence to ACG. GFP expression was observed only after editing with a CBE and sgRNA such that ACG within the editing window was converted to AUG (Katti et al., 2020). For ABE, they employed a similar approach, but instead corrected a premature TGA stop codon back to TGG to prevent early translation termination (Katti et al., 2020). In both cases, robust GFP activation was observed post-genome editing.

While the above system is powerful for testing genome editing activity, it might be limited by PAM availability for various Cas variants. In another example, Tálas and colleagues described a system they called "Base Editor Activity Reporter" or BEAR, which allows for monitoring of a fluorescent reporter in a sequence-unrestricted manner (Tálas et al., 2021). BEAR consists of a split GFP separated by an intron with a silenced 5′ splice donor, whose activity is restored by either ABEs or CBEs, resulting in full-length

GFP expression. The use of an intron enables the incorporation of PAM sequences compatible with the Cas variants of interest. Although sequencing methods provide a powerful readout of genome editing activity, fluorescent-based reporters offer a cost-effective, real-time approach for identifying base editing outcomes. These systems can be valuable tools when screening various parameters for efficient eVLP-mediated delivery, complementing more detailed sequencing-based analyses.

The production workflow for lentiviral particles and reporter cell line generation has been extensively documented (Elegheert et al., 2018; Katti et al., 2020; Pirona et al., 2020; Tálas et al., 2021; Zafra et al., 2018); therefore, we refer readers to those publications and resources for detailed protocols. Briefly, to create a base-editing reporter cell line, a lentiviral GFP plasmid can be modified using standard molecular biology techniques to introduce a premature stop codon or frameshift mutation, which can be corrected by a single base edit to restore GFP expression. Lentiviral particles carrying the mutated reporter plasmid are then produced and used to transduce the target cells, generating the reporter cell line. Once the cell line is established, the following general outline can be for the qualitative validation of BE-eVLP editing activity. The following protocol assumes use of HEK293T reporter cell line, however for other cells of interest, we recommend the users can play with cell density and transduction conditions. Importantly, for hard to transduce cells like primary cells, the users can use transduction enhancers (Delville et al., 2018) or spinoculation (Anastasov et al., 2016) to improve overall efficiencies.

2.3.3 BE-eVLP transduction of EGFP-HEK293T reporter cell line

1. Plate 2×10^4 cells in 100 μL of fresh cell culture media into 96-well plate.
2. After 24 h, add 1 μL and 5 μL of 100X BE-eVLPs concentrated with PEG mixture or sucrose ultracentrifugation dropwise to the cells.
3. Return the cells to the incubator at 37 °C, 5 % CO_2.
4. After 48-72 h post-addition of BE-eVLPs, visually inspect the flask for EGFP expression using a fluorescent microscope or use flow cytometry to quantify percentage of edited cells.

2.3.4 Quantitative validation of eVLP by sequencing

The use of next-generation high-throughput sequencing (HTS) analysis to quantitatively analyze the readout of genome editing around loci of interest has been extensively explained previously for both base editing and prime editing applications (Doman et al., 2022; Huang et al., 2021). In brief, the genomic DNA is extracted from mammalian cells and amplified by two rounds of PCRs. In PCR1, gene specific primers designed by NCBI's Primer-BLAST amplify the target site and add adapter sequences to introduce sample specific index primers (barcodes) in PCR2. In PCR2, the input from PCR1 is used to add Illumina barcodes. PCR1 primers are designed to anneal optimally at 61 °C with an amplicon size between 180–300 base pairs. Following PCR1 and PCR2, individual gels are run to verify presence of correct size bands. Similarly sized PCR2 products are pooled into a single library for subsequent gel extraction. The concentration of the gel extracted library is quantified by Qubit or a similar technique and diluted to 4 nM followed by sequencing using Illumina MiSeq™ instrument and MiSeq™ 300 v2 Kit. Results from sequencing can be analyzed with CRISPResso2 (Clement et al., 2019).

2.4 eVLP characterization

2.4.1 Materials and equipment

Reagents and chemicals		
Phosphate-buffered saline (DPBS, 1X), Dulbecco's formula, without calcium, without magnesium	Thermo Fisher	J67802. K2
Thermo Scientific™ Glycerol	Thermo Fisher	T17904K2
SDS, 10% Solution, RNase-free	Thermo Fisher	AM9822
Tris HCl (1 M), pH 7.5	Quality Biological	351 −006 −101
UltraPure™ DNase/RNase-Free Distilled Water	Thermo Fisher	10977023
Critical Commercial Assays and Kits		
QIAamp Viral RNA Mini Kit	Qiagen	52906
SuperScript™ III First-Strand Synthesis SuperMix for qRT-PCR	Thermo Fisher	11752050
FastScan™ Cas9 (*S. pyogenes*) ELISA Kit	Cell Signaling Technology	29666
QuickTiter™ MuLV Core Antigen ELISA Kit	Cell Biolabs, Inc.	VPK−156
Consumables (equivalents can be used)		
KingFisher™ Plastics for 96 deep-well format	Thermo Fisher	95040450
Nunc™ Serological Pipettes	Thermo Fisher	

Applied Biosystems™ MicroAmp™ Optical 8-Cap Strips	Fisher Scientific	43 −230 −32
Applied Biosystems™ MicroAmp™ 8-Tube Strip, 0.2 mL	Fisher Scientific	43 −230 −32
Nunc™ 15 mL and 50 mL Conical Sterile Polypropylene Centrifuge Tubes	Thermo Fisher	339651
Snap Cap Low Retention Microcentrifuge Tubes, 1.5 mL graduated, sterile (polypropylene)	Thermo Fisher	3451
96 Well Black/Clear Bottom Plate, Poly-D-Lysine Coated Surface, Pack of 5	Thermo Fisher	152037
Instruments (equivalents can be used)		
BioTek Synergy H1 Multimode Reader	Agilent	
DynaPro™ Dynamic Light Scattering (DLS) Plate Reader Instrument	DynaPro	
NanoSight Pro	Malvern Panalytical	
Applied Biosystems® StepOnePlus System	Thermo Fisher	

2.4.2 Biophysical characterization of eVLPs

The ability to evaluate particle uniformity at the expected size range after producing eVLPs is essential to check both sample purity and particle integrity. Two commonly used methods for assessing size and dispersity are nanoparticle tracking analysis (NTA) and dynamic light scattering (DLS) (González-Domínguez et al., 2020; Pankrac et al., 2018; Wolf et al., 2022). A major difference between NTA and DLS is that NTA is a direct particle measurement, recording particles moving across the frame of a microscope and digital display system, whereas DLS is an indirect bulk measurement, detecting light scattered by particles which is recorded by a sensor. Both methods rely on observing the Brownian motion of particles in liquid suspension, allowing for the calculation of a particle's hydrodynamic diameter using the Stokes-Einstein equation. NTA measurements can also give information about the particle number in solution.

For the analysis of eVLPs that have been 100-fold PEG concentrated, it is recommended to first test multiple dilutions of the sample in either DPBS or UltraPure™ Water to find an optimal particle concentration for analysis. Samples should be analyzed at room temperature. Since repetitive freeze-thawing can cause particle lysis and aggregation, samples that have been thawed more than once should not be analyzed as this can lead to inaccurate measurements. Measurement quality and proper sample preparation for DLS analysis can be validated by performing an additional 1:1 fold dilution in the same diluent and measuring the sample again. For an appropriately prepared and diluted sample, the measured particle size distributions should remain the same pre-and post-dilution, while the count rate should be halved.

When analyzing samples with NTA, it is similarly important to perform an appropriate dilution in either DPBS or UltraPure™ Water prior to analysis to ensure the microscope is not overloaded with samples flowing across the frame, or too diluted such that particle counting will lose significance. Approximately one hundred particles per frame is an appropriate number for analysis. Since eVLPs are expected to have a hydrodynamic diameter of approximately 100–120 nm, we recommend using nanosphere reference standards of approximately the same hydrodynamic diameter in the same analytical run, along with a blank matrix.

2.4.3 Ligand binding assays

Ligand binding assays are useful tools to characterize both the cargo (the RNP complex) and the structural components of eVLPs, which can give indications of eVLP concentration and potency (Banskota et al., 2022; Hamilton et al., 2021; Tarrés-Freixas et al., 2023). Some important considerations when performing a ligand binding assay on BE-eVLP structural and packaged components are (i) whether an appropriate dilution is performed to get the analyte within the linear range of the assay, (ii) whether the particle is efficiently lysed as to allow access to all of the analyte within the sample, (iii) whether there are multiple forms of the analyte present in solution, such as fully cleaved Cas9 and uncleaved gag-Cas9, and whether the assay can detect both forms completely, and (iv) whether the lysis buffer used to disrupt eVLP particles has interfering effects on the assay performance. Below are methods suitable for characterization of packaged SpCas9 concentration within eVLPs as well as capsid protein p30 quantification, if produced BE-eVLPs are 100-fold PEG concentrated.

anti-SpCas9 ELISA.

This protocol uses the FastScan™ Cas9 (*S. pyogenes*) ELISA kit from Cell Signaling Technology and follows the kit protocol with slight modifications in the sample preparation steps.

1. Bring all kit components to room temperature.

2. Create modified 2x Laemmli lysis buffer to efficiently lyse eVLPs and prepare samples for kit dilutions.
   a. To create 10 mL of 2x Laemmli lysis buffer, combine the following components in a sterile 15 mL conical tube and thoroughly vortex to mix:
      i. 2 mL of Glycerol
      ii. 4 mL of 10 % SDS, RNase-free
      iii. 1 mL of 1 M Tris HCl, pH 7.5
      iv. 3 mL of UltraPure™ water
3. Perform lysis of eVLP samples in 2x Laemmli lysis buffer in a sterile tissue culture hood.
   a. Use either freshly prepared eVLP samples or thaw samples kept at −80 °C and pipette briefly to evenly distribute eVLPs throughout the sample.
   b. In a sterile microcentrifuge tube, combine 2 μL of eVLP sample with 2 μL Laemmli lysis buffer.
   c. Vortex and spin down sample.
   d. Heat the diluted sample for five minutes at 60 °C to further promote particle lysis while avoiding Cas9 degradation due to excessive heating.
4. Follow the kit protocol, skipping the section titled "Preparing Cell Lysates".
   a. **Critical:** If analyzing 100-fold PEG purified eVLP samples, it is recommended to first try an overall sample dilution of 1000-fold, and to adjust the dilution in the future if the analyte is not within the range of the standard curve for quantification.
5. Analyze results in plate reader software according to the user manual and assay kit protocol.

anti-MLV (p30) ELISA.

The recommended protocol for performing the QuickTiter™ MuLV Core Antigen ELISA kit follows many of the same principles and considerations as the anti-Cas9 ELISA recommended protocol. Particles must again be efficiently lysed before proceeding with the specified kit instructions.

1. Bring all kit components to room temperature.
2. Create modified 2x Laemmli lysis buffer to efficiently lyse eVLPs and prepare samples for kit dilutions.
3. Perform lysis of eVLP samples in 2x Laemmli lysis buffer in a sterile tissue culture hood.
   a. **Note:** Use either freshly prepared eVLP samples or thaw samples kept at −80 °C and pipette briefly to evenly distribute eVLPs throughout the sample.

   b. In a sterile microcentrifuge tube, combine 2 μL of eVLP sample with 2 μL Laemmli lysis buffer.
   c. Vortex and spin down sample.
   d. Heat the diluted sample for five minutes at 60 °C.

4. **Critical:** Follow the kit protocol, skipping the section titled "MuLV Sample Dilution and Inactivation" and instead perform the following steps prior to the assay protocol:
   a. Make 10 mL of sample diluent by combining 9 mL of assay diluent (kit component) with 1 mL of Triton X-100 stock solution (kit component) in a sterile 15 mL conical polypropylene tube.
      i. Vortex the sample diluent.
      ii. **Note:** The amount of sample diluent prepared should be scaled for the needs of the assay run based on the amount of samples being analyzed.
   b. Dilute 2 μL of Laemmli-lysed eVLP sample into 998 μL of sample diluent and vortex the diluted sample.
   c. Combine 40 μL of diluted sample from the previous step with 360 μL of sample diluent in a clean polypropylene 1.5 mL tube and vortex.
      i. This amount allows for a sample to be plated in triplicate with 25 % overage to ensure all replicate wells can receive the same volume of diluted sample.
   d. Follow the kit protocol starting from "Assay Protocol".
      i. **Critical:** If analyzing 100-fold PEG purified eVLP samples, it is recommended to first try overall sample dilutions of 10,000-fold and 20,000-fold, and to adjust the dilution in the future if the analyte is not within the range of the standard curve for quantification.

5. Analyze results in plate reader software according to the user manual and assay kit protocol.
   a. **Note:** An estimate of eVLP particle count can be calculated by assuming 20 % of the measured p30 in samples are associated with eVLPs and that eVLPs have a copy number of 1800 molecules of p30 for every eVLP in solution (Banskota et al., 2022; Renner et al., 2020).

2.4.4 sgRNA quantification via RT-qPCR

Another useful tool to compare the packaging efficiency of RNPs into eVLPs is quantitative PCR, which can provide relative data for sgRNA present in an eVLP sample (An et al., 2024). This can be a useful tool as the levels of sgRNA not only correlate to Cas9/BE amounts present in the sample, but it could also be used to rapidly investigate whether different designs offer advantages in terms of packaging efficiency. The following steps and kits can be used to quantify sgRNA in eVLP samples:

1. Perform RNA extraction of BE-eVLPs with the QIAamp Viral RNA Mini Kit according to the kit protocol.
2. Perform reverse transcription of extracted sgRNA using the SuperScript™ III First-Strand Synthesis SuperMix for qRT-PCR according to the kit protocol with a sgRNA-specific primer.
3. Perform qPCR using SsoAdvanced Universal SYBR Green Supermix following recommendations from the user instruction manual on a compatible Applied Biosystems Real-Time PCR System™.
   a. A standard curve should be constructed using the same synthetic sgRNA or *in vitro* transcribed sgRNA at known concentrations.
      i. **Note:** Synthetic sgRNA can be amplified *via* PCR from target sgRNA plasmid using the same primers used for the RT-qPCR assay.
   b. cDNA input should be normalized to MuLV p30 content in the sample.
   c. **Critical:** It is important to design an efficient primer set to perform qPCR, there are multiple online tools to help in designing primer sets for qPCR based on a specified cDNA target such as the qPCR Primer Design tool from IDT.
4. Calculate the relative abundance of target sgRNA in samples using the standard curve and C_T values for standards and samples.
   a. **Note:** It is recommended that a stock standard curve is generated with primer efficiency of 90–110 % prior to testing with samples.

3. Optimization and troubleshooting

3.1 Transfection

It may be beneficial to include a positive transfection control sample to confirm that transfection of cells has occurred. Co-transfecting one culture of cells with BE-eVLP plasmids and pCMV-GFP (Addgene #11153) can provide visual confirmation of successful transfection *via* fluorescence microscopy. If transfection is still unsuccessful, it may be helpful to dilute the plasmids for transfection to increase the volume being pipetted during the transfection mix step. This can help to make the ratio of the plasmid mix more consistent, as attempting to pipette very small volumes introduces inaccuracies. While we have only tested jetPRIME® as a transfection reagent, it is likely that some transfection reagents may be more appropriate for large plasmid transfections. Therefore, we recommend testing different reagents to identify the most suitable one.

3.2 Characterization assays

3.2.1 Dynamic light scattering and nanoparticle tracking analysis

The expected yield of virus-like particles produced at 75 cm^2 cell culture flask scales and concentrated 100-fold is approximately 4.013E+ 08 particles/μL but this number may vary depending on the producer cell line, culture condition and transfection reagent. So, we recommend users to test various batches to establish a baseline with their method. Larger particles present in samples will scatter more light compared to smaller populations of particles, which will cause the signal generated by larger particles to be more intense, even when there are relatively fewer large particles compared to other size populations in the sample. This can result in smaller particles being masked during the analysis of signal intensities at different sizes, potentially leading to false conclusions. In order to avoid this risk, it may be beneficial to perform additional purification steps on eVLP samples, such as centrifugation or membrane filtration, to remove larger aggregates and unwanted material from the sample. This will make the final eVLP sample more monodisperse and easier to characterize using light scattering.

3.2.2 ELISA assays

Following this production method of eVLPs and the 100-fold PEG concentration, the expected yields of Cas9 and p30 proteins should be approximately 0.5 μM and 10–15 μM respectively when eVLP production is performed in a T75, but this number may vary depending on the producer cell line, culture condition and transfection reagent. So, we recommend users to test various batches to establish a baseline with their method. One of the most crucial considerations when conducting ELISA based methods on eVLP samples and their cargo is whether efficient particle lysis is occurring before the sample analysis. The SDS-containing modified Laemmli lysis buffer described earlier should be efficient in lysing eVLPs when combined with the suggested heating step. This solution is suitable for 100-fold concentrated eVLP samples, however, when dealing with less or more concentrated samples, the lysis method may need to be altered to avoid interference from SDS with the enzymes present in the ELISA assay. As a substitution, different surfactants and/or heating conditions should be tested for more diluted samples; and ELISA results should be compared against the lysis performance of the assay buffer provided in the kits, which contains its own mix of lysing agents.

In order to improve confidence in analytical results from this method, pipetting actions should not require the transfer of less than 10 μL of sample at a time if possible. Additionally, a validated standard that is separate from the standard curve should be included in each plate run as a quality control. For a more robust assay, including quality controls at high, medium, and low concentrations matching the standard curve ensures the assay's performance across the range of the curve.

3.2.3 Reverse transcription qPCR

There are many potential sources of errors when performing RT-qPCR that affect the assay's efficiency and therefore lead to false conclusions based on data generated in suboptimal conditions. One potential source of error could be caused by inefficient primer design. Secondary structures and primer-dimer formation can both reduce the efficiency of the PCR reaction as well as contribute to false signals if using an intercalating dye instead of custom probes. Assay validation should be performed prior to running actual samples using multiple different primer sets on the equivalent sequence to what is expected in the final reverse-transcribed product. A primer set with efficiency between 90–110 % should be chosen which is based on the slope of a primer set's C_T value and the log of each dilution the sample was run at, for which there are multiple online calculators.

In addition to primer set optimization, PCR conditions divergent from recommendations by the qPCR master mix manual should be tested. Changing the annealing temperature may help disrupt secondary structure formed by the sgRNA and improve the assay's efficiency. The extension time and number of cycles can also be adjusted and should be informed by the performance of validation runs including no-template controls for which signal should be close to background.

4. Summary

To summarize, we have illustrated concepts and recent advancements of gene editing strategies (nucleases, base editing and prime editing) and current viral and non-viral delivery platforms. More specifically, we have highlighted eVLPs as the novel mRNA and protein delivery system for gene editing agents. We also outlined detailed methods and protocols for eVLP production and subsequent validation and characterization. We hope that this detailed procedural workflow will serve as a guideline for any researchers using eVLPs for delivery of genome editors.

Acknowledgements

The authors acknowledge financial support from Boston University through startup funding support and Massachusetts Life Science Center Novel Therapeutics Delivery Grant. C.L. acknowledges financial support from the Translational Research in Biomaterials T32 training grant (T32EB006359).

Declaration of interests

S.B. has financial interests in Nvelop Therapeutics. The remaining authors declare no competing interests.

References

Advani, S. J., Weichselbaum, R. R., & Kufe, D. W. (2003). *Gene delivery systemsHolland-Frei Cancer Medicine* (6th ed.). BC Decker. https://www.ncbi.nlm.nih.gov/books/NBK12776/.

Alanis-Lobato, G., Zohren, J., McCarthy, A., Fogarty, N. M. E., Kubikova, N., Hardman, E., ... Niakan, K. K. (2021). Frequent loss of heterozygosity in CRISPR-Cas9-edited early human embryos. *Proceedings of the National Academy of Sciences of the United States of America, 118*(22), e2004832117. https://doi.org/10.1073/pnas.2004832117.

Alexander, I. E., Russell, D. W., & Miller, A. D. (1994). DNA-damaging agents greatly increase the transduction of nondividing cells by adeno-associated virus vectors. *Journal of Virology, 68*(12), 8282–8287. https://doi.org/10.1128/jvi.68.12.8282-8287.1994.

An, M., Raguram, A., Du, S. W., Banskota, S., Davis, J. R., Newby, G. A., ... Liu, D. R. (2024). Engineered virus-like particles for transient delivery of prime editor ribonucleoprotein complexes in vivo. *Nature Biotechnology,* 1–12. https://doi.org/10.1038/s41587-023-02078-y.

Anastasov, N., Höfig, I., Mall, S., Krackhardt, A. M., & Thirion, C. (2016). Optimized lentiviral transduction protocols by use of a poloxamer enhancer, spinoculation, and scFv-antibody fusions to VSV-G. In M. Federico (Vol. Ed.), *Lentiviral Vectors and Exosomes as Gene and Protein Delivery Tools. Methods in Molecular Biology: 1448.* New York, NY: Humana Press. https://doi.org/10.1007/978-1-4939-3753-0_4.

Anzalone, A. V., Randolph, P. B., Davis, J. R., Sousa, A. A., Koblan, L. W., Levy, J. M., ... Liu, D. R. (2019). Search-and-replace genome editing without double-strand breaks or donor DNA. *Nature, 576*(7785), 149–157. https://doi.org/10.1038/s41586-019-1711-4.

Arjomandnejad, M., Dasgupta, I., Flotte, T. R., & Keeler, A. M. (2023). Immunogenicity of recombinant adeno-associated virus (AAV) vectors for gene transfer. *BioDrugs: Clinical Immunotherapeutics, Biopharmaceuticals and Gene Therapy, 37*(3), 311–329. https://doi.org/10.1007/s40259-023-00585-7.

Banskota, S., Raguram, A., Suh, S., Du, S. W., Davis, J. R., Choi, E. H., ... Liu, D. R. (2022). Engineered virus-like particles for efficient in vivo delivery of therapeutic proteins. *Cell, 185*(2), 250–265.e16. https://doi.org/10.1016/j.cell.2021.12.021.

Bibikova, M., Golic, M., Golic, K. G., & Carroll, D. (2002). Targeted chromosomal cleavage and mutagenesis in drosophila using zinc-finger nucleases. *Genetics, 161*(3), 1169–1175. https://doi.org/10.1093/genetics/161.3.1169.

Böck, D., Rothgangl, T., Villiger, L., Schmidheini, L., Matsushita, M., Mathis, N., ... Schwank, G. (2022). In vivo prime editing of a metabolic liver disease in mice. *Science Translational Medicine, 14*(636), eabl9238. https://doi.org/10.1126/scitranslmed.abl9238.

Bollard, C. M., Russell, C., Cruz, Y., & Brenner, M. K. (2010). Chapter 19—gene therapy and allergy. In D. Y. M. Leung, H. A. Sampson, R. Geha, & S. J. Szefler (Eds.). *Pediatric allergy: Principles and practice* (pp. 211–222) (Second ed.). W.B. Saunders. https://doi.org/10.1016/B978-1-4377-0271-2.00019-5.

Bulcha, J. T., Wang, Y., Ma, H., Tai, P. W. L., & Gao, G. (2021). Viral vector platforms within the gene therapy landscape. *Signal Transduction and Targeted Therapy, 6*(1), 1–24. https://doi.org/10.1038/s41392-021-00487-6.

Bushman, F. D. (2020). Retroviral insertional mutagenesis in humans: Evidence for four genetic mechanisms promoting expansion of cell clones. *Molecular Therapy, 28*(2), 352–356. https://doi.org/10.1016/j.ymthe.2019.12.009.

Cai, Y., Bak, R. O., & Mikkelsen, J. G. (2014). Targeted genome editing by lentiviral protein transduction of zinc-finger and TAL-effector nucleases. *eLife, 3*, e01911. https://doi.org/10.7554/eLife.01911.

Cermak, T., Doyle, E. L., Christian, M., Wang, L., Zhang, Y., Schmidt, C., ... Voytas, D. F. (2011). Efficient design and assembly of custom TALEN and other TAL effector-based constructs for DNA targeting. *Nucleic Acids Research, 39*(12), e82. https://doi.org/10.1093/nar/gkr218.

Chen, P. J., Hussmann, J. A., Yan, J., Knipping, F., Ravisankar, P., Chen, P. F., ... Liu, D. R. (2021). Enhanced prime editing systems by manipulating cellular determinants of editing outcomes. *Cell, 184*(22), 5635–5652.e29. https://doi.org/10.1016/j.cell.2021.09.018.

Cieślar-Pobuda, A., Knoflach, V., Ringh, M. V., Stark, J., Likus, W., Siemianowicz, K., ... Łos, M. J. (2017). Transdifferentiation and reprogramming: Overview of the processes, their similarities and differences. *Biochimica et Biophysica Acta (BBA) - Molecular Cell Research, 1864*(7), 1359–1369. https://doi.org/10.1016/j.bbamcr.2017.04.017.

Clement, K., Rees, H., Canver, M. C., Gehrke, J. M., Farouni, R., Hsu, J. Y., ... Pinello, L. (2019). CRISPResso2 provides accurate and rapid genome editing sequence analysis. *Nature Biotechnology, 37*(3), 224–226. https://doi.org/10.1038/s41587-019-0032-3.

Crystal, R. G. (2014). Adenovirus: The first effective in vivo gene delivery vector. *Human Gene Therapy, 25*(1), 3–11. https://doi.org/10.1089/hum.2013.2527.

Davis, J. R., Banskota, S., Levy, J. M., Newby, G. A., Wang, X., Anzalone, A. V., ... Liu, D. R. (2024). Efficient prime editing in mouse brain, liver and heart with dual AAVs. *Nature Biotechnology, 42*(2), 253–264. https://doi.org/10.1038/s41587-023-01758-z.

Delville, M., Soheili, T., Bellier, F., Durand, A., Denis, A., Lagresle-Peyrou, C., ... Six, E. (2018). A nontoxic transduction enhancer enables highly efficient lentiviral transduction of primary murine T cells and hematopoietic stem cells. *Molecular Therapy. Methods & Clinical Development, 10*, 341–347. https://doi.org/10.1016/j.omtm.2018.08.002.

Di, J., Du, Z., Wu, K., Jin, S., Wang, X., Li, T., & Xu, Y. (2022). Biodistribution and non-linear gene expression of mRNA LNPs affected by delivery route and particle size. *Pharmaceutical Research, 39*(1), 105–114. https://doi.org/10.1007/s11095-022-03166-5.

Doman, J. L., Sousa, A. A., Randolph, P. B., Chen, P. J., & Liu, D. R. (2022). Designing and executing prime editing experiments in mammalian cells. *Nature Protocols, 17*(11), 2431–2468. https://doi.org/10.1038/s41596-022-00724-4.

Dong, W., & Kantor, B. (2021). Lentiviral vectors for delivery of gene-editing systems based on CRISPR/Cas: Current state and perspectives. *Viruses, 13*(7), 1288. https://doi.org/10.3390/v13071288.

Elegheert, J., Behiels, E., Bishop, B., et al. (2018). Lentiviral transduction of mammalian cells for fast, scalable and high-level production of soluble and membrane proteins. *Nature Protocols, 13*, 2991–3017. https://doi.org/10.1038/s41596-018-0075-9.

Fausther-Bovendo, H., & Kobinger, G. P. (2014). Pre-existing immunity against Ad vectors. *Human Vaccines & Immunotherapeutics, 10*(10), 2875–2884. https://doi.org/10.4161/hv.29594.

Fuenmayor, J., Gòdia, F., & Cervera, L. (2017). Production of virus-like particles for vaccines. *New Biotechnology, 39*, 174–180. https://doi.org/10.1016/j.nbt.2017.07.010.

Galla, M., Will, E., Kraunus, J., Chen, L., & Baum, C. (2004). Retroviral pseudotransduction for targeted cell manipulation. *Molecular Cell, 16*(2), 309–315. https://doi.org/10.1016/j.molcel.2004.09.023.

Gaudelli, N. M., Komor, A. C., Rees, H. A., Packer, M. S., Badran, A. H., Bryson, D. I., & Liu, D. R. (2017). Programmable base editing of A•T to G•C in genomic DNA without DNA cleavage. *Nature, 551*(7681), 464–471. https://doi.org/10.1038/nature24644.

Gee, P., Lung, M. S. Y., Okuzaki, Y., Sasakawa, N., Iguchi, T., Makita, Y., ... Yasuda, M., ... Hotta, A. (2020). Extracellular nanovesicles for packaging of CRISPR-Cas9 protein and sgRNA to induce therapeutic exon skipping. *Nature Communications, 11*, 1334. https://doi.org/10.1038/s41467-020-14957-y.

Godbout, K., & Tremblay, J. P. (2022). Delivery of RNAs to specific organs by lipid nanoparticles for gene therapy. *Pharmaceutics, 14*(10), 2129. https://doi.org/10.3390/pharmaceutics14102129.

González-Domínguez, I., Puente-Massaguer, E., Cervera, L., & Gòdia, F. (2020). Quality assessment of virus-like particles at single particle level: A comparative study. *Viruses, 12*(2), 223. https://doi.org/10.3390/v12020223.

Grieger, J. C., & Samulski, R. J. (2005). Packaging capacity of adeno-associated virus serotypes: Impact of larger genomes on infectivity and postentry steps. *Journal of Virology, 79*(15), 9933–9944. https://doi.org/10.1128/JVI.79.15.9933-9944.2005.

Hajiaghapour Asr, M., Dayani, F., Saedi Segherloo, F., Kamedi, A., Neill, A. O., MacLoughlin, R., & Doroudian, M. (2023). Lipid nanoparticles as promising carriers for mRNA vaccines for viral lung infections. *Pharmaceutics, 15*(4), https://doi.org/10.3390/pharmaceutics15041127.

Hamilton, J. R., Chen, E., Perez, B. S., Sandoval Espinoza, C. R., Kang, M. H., Trinidad, M., ... Doudna, J. A. (2024). In vivo human T cell engineering with enveloped delivery vehicles. Advance online publication *Nature Biotechnology*. https://doi.org/10.1038/s41587-023-02085-z.

Hamilton, J. R., Tsuchida, C. A., Nguyen, D. N., Shy, B. R., McGarrigle, E. R., Sandoval Espinoza, C. R., ... Doudna, J. A. (2021). Targeted delivery of CRISPR-Cas9 and transgenes enables complex immune cell engineering. *Cell Reports, 35*(9), 109207. https://doi.org/10.1016/j.celrep.2021.109207.

He, J., Yu, L., Lin, X., Liu, X., Zhang, Y., Yang, F., & Deng, W. (2022). Virus-like particles as nanocarriers for intracellular delivery of biomolecules and compounds. *Viruses, 14*(9), 1905. https://doi.org/10.3390/v14091905.

Herrera-Barrera, M., Ryals, R. C., Gautam, M., Jozic, A., Landry, M., Korzun, T., ... Sahay, G. (2023). Peptide-guided lipid nanoparticles deliver mRNA to the neural retina of rodents and nonhuman primates. *Science Advances, 9*(2), eadd4623. https://doi.org/10.1126/sciadv.add4623.

Herzog, R. W. (2019). Complexity of immune responses to AAV transgene products – example of factor IX. *Cellular Immunology, 342*, 103658. https://doi.org/10.1016/j.cellimm.2017.05.006.

Hirata, R., Chamberlain, J., Dong, R., & Russell, D. W. (2002). Targeted transgene insertion into human chromosomes by adeno-associated virus vectors. *Nature Biotechnology, 20*(7), 735–738. https://doi.org/10.1038/nbt0702-735.

Horie, T., & Ono, K. (2024). VERVE-101: A promising CRISPR-based gene editing therapy that reduces LDL-C and PCSK9 levels in HeFH patients. *European Heart Journal. Cardiovascular Pharmacotherapy, 10*(2), 89–90. https://doi.org/10.1093/ehjcvp/pvad103.

Huang, T. P., Newby, G. A., & Liu, D. R. (2021). Precision genome editing using cytosine and adenine base editors in mammalian cells. *Nature Protocols, 16*(2), 1089–1128. https://doi.org/10.1038/s41596-020-00450-9.

Joung, J. K., & Sander, J. D. (2013). TALENs: A widely applicable technology for targeted genome editing. *Nature Reviews. Molecular Cell Biology, 14*(1), 49–55. https://doi.org/10.1038/nrm3486.

Jouvenet, N., Lainé, S., Pessel-Vivares, L., & Mougel, M. (2011). Cell biology of retroviral RNA packaging. *RNA Biology, 8*(4), 572–580. https://doi.org/10.4161/rna.8.4.16030.

Kaczmarczyk, S. J., Sitaraman, K., Young, H. A., Hughes, S. H., & Chatterjee, D. K. (2011). Protein delivery using engineered virus-like particles. *Proceedings of the National Academy of Sciences of the United States of America, 108*(41), 16998–17003. https://doi.org/10.1073/pnas.1101874108.

Kalidasan, V., Ng, W. H., Ishola, O. A., Ravichantar, N., Tan, J. J., & Das, K. T. (2021). A guide in lentiviral vector production for hard-to-transfect cells, using cardiac-derived c-kit expressing cells as a model system. *Scientific Reports, 11*(1), 19265. https://doi.org/10.1038/s41598-021-98657-7.

Kanasty, R., Dorkin, J. R., Vegas, A., & Anderson, D. (2013). Delivery materials for siRNA therapeutics. *Nature Materials, 12*(11), 967–977. https://doi.org/10.1038/nmat3765.

Katti, A., Foronda, M., Zimmerman, J., Diaz, B., Zafra, M. P., Goswami, S., & Dow, L. E. (2020). GO: A functional reporter system to identify and enrich base editing activity. *Nucleic Acids Research, 48*(6), 2841–2852. https://doi.org/10.1093/nar/gkaa124.

Kedmi, R., Ben-Arie, N., & Peer, D. (2010). The systemic toxicity of positively charged lipid nanoparticles and the role of Toll-like receptor 4 in immune activation. *Biomaterials, 31*(26), 6867–6875. https://doi.org/10.1016/j.biomaterials.2010.05.027.

Kenjo, E., Hozumi, H., Makita, Y., Iwabuchi, K. A., Fujimoto, N., Matsumoto, S., ... Hotta, A. (2021). Low immunogenicity of LNP allows repeated administrations of CRISPR-Cas9 mRNA into skeletal muscle in mice. *Nature Communications, 12*, 7101. https://doi.org/10.1038/s41467-021-26714-w.

Knopp, Y., Geis, F. K., Heckl, D., Horn, S., Neumann, T., Kuehle, J., ... Galla, M. (2018). Transient retrovirus-based CRISPR/Cas9 all-in-one particles for efficient, targeted gene knockout. *Molecular Therapy. Nucleic Acids, 13*, 256–274. https://doi.org/10.1016/j.omtn.2018.09.006.

Koblan, L. W., Arbab, M., Shen, M. W., Hussmann, J. A., Anzalone, A. V., Doman, J. L., ... Liu, D. R. (2021). Efficient C•G-to-G•C base editors developed using CRISPRi screens, target-library analysis, and machine learning. *Nature Biotechnology, 39*(11), 1414–1425. https://doi.org/10.1038/s41587-021-00938-z.

Komor, A. C., Kim, Y. B., Packer, M. S., Zuris, J. A., & Liu, D. R. (2016). Programmable editing of a target base in genomic DNA without double-stranded DNA cleavage. *Nature, 533*(7603), 420–424. https://doi.org/10.1038/nature17946.

Kosicki, M., Tomberg, K., & Bradley, A. (2018). Repair of double-strand breaks induced by CRISPR-Cas9 leads to large deletions and complex rearrangements. *Nature Biotechnology, 36*(8), 765–771. https://doi.org/10.1038/nbt.4192.

Kymäläinen, H., Appelt, J. U., Giordano, F. A., Davies, A. F., Ogilvie, C. M., Ahmed, S. G., ... Dickson, G. (2014). Long-term episomal transgene expression from mitotically stable integration-deficient lentiviral vectors. *Human Gene Therapy, 25*(5), 428–442. https://doi.org/10.1089/hum.2013.172.

Lee, D., Liu, J., Junn, H. J., Lee, E.-J., Jeong, K.-S., & Seol, D.-W. (2019). No more helper adenovirus: Production of gutless adenovirus (GLAd) free of adenovirus and replication-competent adenovirus (RCA) contaminants. *Experimental & Molecular Medicine, 51*(10), 1–18. https://doi.org/10.1038/s12276-019-0334-z.

Leibowitz, M. L., Papathanasiou, S., Doerfler, P. A., Blaine, L. J., Sun, L., Yao, Y., ... Pellman, D. (2021). Chromothripsis as an on-target consequence of CRISPR-Cas9 genome editing. *Nature Genetics, 53*(6), 895–905. https://doi.org/10.1038/s41588-021-00838-7.

Li, Y., Bolinger, J., Yu, Y., Glass, Z., Shi, N., Yang, L., ... Xu, Q. (2019). Intracellular delivery and biodistribution study of CRISPR/Cas9 ribonucleoprotein loaded bioreducible lipidoid nanoparticles. *Biomaterials Science,* 7(2), 596–606. https://doi.org/10.1039/C8BM00637G.

Ling, S., Yang, S., Hu, X., Yin, D., Dai, Y., Qian, X., ... Cai, Y. (2021). Lentiviral delivery of co-packaged Cas9 mRNA and a vegfa-targeting guide RNA prevents wet age-related macular degeneration in mice. *Nature biomedical engineering, 5*(2), 144–156. https://doi.org/10.1038/s41551-020-00656-y.

Liu, B., Dong, X., Zheng, C., Keener, D., Chen, Z., Cheng, H., ... Sontheimer, E. J. (2024). Targeted genome editing with a DNA-dependent DNA polymerase and exogenous DNA-containing templates. *Nature Biotechnology, 42*(7), 1039–1045. https://doi.org/10.1038/s41587-023-01947-w.

Liu, J., Chang, J., Jiang, Y., Meng, X., Sun, T., Mao, L., ... Wang, M. (2019). Fast and efficient CRISPR/Cas9 genome editing in vivo enabled by bioreducible lipid and messenger RNA nanoparticles. *Advanced Materials (Deerfield Beach, Fla.), 31*(33), e1902575. https://doi.org/10.1002/adma.201902575.

Liu, P., Liang, S. Q., Zheng, C., Mintzer, E., Zhao, Y. G., Ponnienselvan, K., ... Xue, W. (2021). Improved prime editors enable pathogenic allele correction and cancer modelling in adult mice. *Nature Communications, 12*(1), 2121. https://doi.org/10.1038/s41467-021-22295-w.

Lu, B., Javidi-Parsijani, P., Makani, V., Mehraein-Ghomi, F., Sarhan, W. M., Sun, D., ... Atala, A. (2019). Delivering SaCas9 mRNA by lentivirus-like bionanoparticles for transient expression and efficient genome editing. *Nucleic Acids Research, 47*(8), e44. https://doi.org/10.1093/nar/gkz093.

Lukashev, A. N., & Zamyatnin, A. A. (2016). Viral vectors for gene therapy: Current state and clinical perspectives. *Biochemistry (Moscow), 81*(7), 700–708. https://doi.org/10.1134/S0006297916070063.

Lyu, P., & Lu, B. (2022). New advances in using virus-like particles and related technologies for eukaryotic genome editing delivery. *International Journal of Molecular Sciences, 23*(15), 8750. https://doi.org/10.3390/ijms23158750.

Lyu, P., Javidi-Parsijani, P., Atala, A., & Lu, B. (2019). Delivering Cas9/sgRNA ribonucleoprotein (RNP) by lentiviral capsid-based bionanoparticles for efficient 'hit-and-run' genome editing. *Nucleic Acids Research, 47*(17), e99. https://doi.org/10.1093/nar/gkz605.

Mangeot, P. E., Risson, V., Fusil, F., Marnef, A., Laurent, E., Blin, J., ... Ricci, E. P. (2019). Genome editing in primary cells and in vivo using viral-derived nanoblades loaded with Cas9-sgRNA ribonucleoproteins. *Nature Communications, 10*(1), 45. https://doi.org/10.1038/s41467-018-07845-z.

Michels, A., Ho, N., & Buchholz, C. J. (2022). Precision medicine: In vivo CAR therapy as a showcase for receptor-targeted vector platforms. *Molecular Therapy: the Journal of the American Society of Gene Therapy, 30*(7), 2401–2415. https://doi.org/10.1016/j.ymthe.2022.05.018.

Montini, E., Cesana, D., Schmidt, M., Sanvito, F., Ponzoni, M., Bartholomae, C., ... Naldini, L. (2006). Hematopoietic stem cell gene transfer in a tumor-prone mouse model uncovers low genotoxicity of lentiviral vector integration. *Nature Biotechnology, 24*(6), 687–696. https://doi.org/10.1038/nbt1216.

Nelson, J. W., Randolph, P. B., Shen, S. P., Everette, K. A., Chen, P. J., Anzalone, A. V., ... Liu, D. R. (2022). Engineered pegRNAs improve prime editing efficiency. *Nature Biotechnology, 40*(3), 402–410. https://doi.org/10.1038/s41587-021-01039-7.

Nishida, K., Arazoe, T., Yachie, N., Banno, S., Kakimoto, M., Tabata, M., ... Kondo, A. (2016). Targeted nucleotide editing using hybrid prokaryotic and vertebrate adaptive immune systems. aaf8729 *Science, 353*(6305), https://doi.org/10.1126/science.aaf8729.

Nooraei, S., Bahrulolum, H., Hoseini, Z. S., Katalani, C., Hajizade, A., Easton, A. J., & Ahmadian, G. (2021). Virus-like particles: Preparation, immunogenicity and their roles as nanovaccines and drug nanocarriers. *Journal of Nanobiotechnology, 19*(1), 59. https://doi.org/10.1186/s12951-021-00806-7.

Pankrac, J., Klein, K., McKay, P. F., King, D. F. L., Bain, K., Knapp, J., ... Mann, J. F. S. (2018). A heterogeneous human immunodeficiency virus-like particle (VLP) formulation produced by a novel vector system. *NPJ Vaccines, 3*(1), 1–10. https://doi.org/10.1038/s41541-017-0040-6.

Philippidis, A. (2024). CASGEVY makes history as FDA approves first CRISPR/Cas9 genome edited therapy. *Human Gene Therapy, 35*(1-2), 1–4. https://doi.org/10.1089/hum.2023.29263.bfs.

Pirona, A. C., Oktriani, R., Boettcher, M., & Hoheisel, J. D. (2020). Process for an efficient lentiviral cell transduction. *Biology Methods and Protocols, 5*(1), bpaa005. https://doi.org/10.1093/biomethods/bpaa005.

Prel, A., Caval, V., Gayon, R., Ravassard, P., Duthoit, C., Payen, E., ... Pagès, J. C. (2015). Highly efficient in vitro and in vivo delivery of functional RNAs using new versatile MS2-chimeric retrovirus-like particles. *Molecular Therapy. Methods & Clinical Development, 2*, 15039. https://doi.org/10.1038/mtm.2015.39.

Raguram, A., An, M., Chen, P. Z., & Liu, D. R. (2024). Directed evolution of engineered virus-like particles with improved production and transduction efficiencies. *Nature Biotechnology,* 1–13. https://doi.org/10.1038/s41587-024-02467-x.

Raguram, A., Banskota, S., & Liu, D. R. (2022). Therapeutic in vivo delivery of gene editing agents. *Cell, 185*(15), 2806–2827. https://doi.org/10.1016/j.cell.2022.03.045.

Renner, T. M., Tang, V. A., Burger, D., & Langlois, M. (2020). Intact viral particle counts measured by flow virometry provide insight into the infectivity and genome packaging efficiency of moloney murine leukemia virus. *Journal of Virology, 94*. https://doi.org/10.1128/jvi.01600-19.

Richter, M. F., Zhao, K. T., Eton, E., Lapinaite, A., Newby, G. A., Thuronyi, B. W., ... Liu, D. R. (2020). Phage-assisted evolution of an adenine base editor with improved cas domain compatibility and activity. *Nature Biotechnology, 38*(7), 883–891. https://doi.org/10.1038/s41587-020-0453-z.

Segel, M., Lash, B., Song, J., Ladha, A., Liu, C. C., Jin, X., ... Zhang, F. (2021). Mammalian retrovirus-like protein PEG10 packages its own mRNA and can be pseudotyped for mRNA delivery. *Science, 373*(6557), 882–889. https://doi.org/10.1126/science.abg6155.

Semple, S. C., Leone, R., Barbosa, C. J., Tam, Y. K., & Lin, P. J. C. (2022). Lipid Nanoparticle delivery systems to enable mRNA-based therapeutics. *Pharmaceutics, 14*(2), 398. https://doi.org/10.3390/pharmaceutics14020398.

Sioson, V. A., Kim, M., & Joo, J. (2021). Challenges in delivery systems for CRISPR-based genome editing and opportunities of nanomedicine. *Biomedical Engineering Letters, 11*(3), 217–233. https://doi.org/10.1007/s13534-021-00199-4.

Song, Y., Liu, Z., Zhang, Y., Chen, M., Sui, T., Lai, L., & Li, Z. (2020). Large-fragment deletions induced by Cas9 cleavage while not in the BEs system. *Molecular Therapy. Nucleic Acids, 21*, 523–526. https://doi.org/10.1016/j.omtn.2020.06.019.

Srivastava, A., Lusby, E. W., & Berns, K. I. (1983). Nucleotide sequence and organization of the adeno-associated virus 2 genome. *Journal of Virology, 45*(2), 555–564. https://doi.org/10.1128/jvi.45.2.555-564.1983.

Stadtmauer, E. A., Fraietta, J. A., Davis, M. M., Cohen, A. D., Weber, K. L., Lancaster, E., ... June, C. H. (2020). CRISPR-engineered T cells in patients with refractory cancer. *Science, 367*(6481), eaba7365. https://doi.org/10.1126/science.aba7365.

Strebinger, D., Frangieh, C. J., Friedrich, M. J., Faure, G., Macrae, R. K., & Zhang, F. (2023). Cell type-specific delivery by modular envelope design. *Nature Communications, 14*(1), 5141. https://doi.org/10.1038/s41467-023-40788-8.

Tálas, A., Simon, D. A., Kulcsár, P. I., Varga, É., Krausz, S. L., & Welker, E. (2021). BEAR reveals that increased fidelity variants can successfully reduce the mismatch tolerance of adenine but not cytosine base editors. *Nature Communications, 12*(1), 6353. https://doi.org/10.1038/s41467-021-26461-y.

Tarrés-Freixas, F., Aguilar-Gurrieri, C., Rodríguez de la Concepción, M. L., Urrea, V., Trinité, B., Ortiz, R., ... Blanco, J. (2023). An engineered HIV-1 gag-based VLP displaying high antigen density induces strong antibody-dependent functional immune responses. *NPJ Vaccines, 8*(1), 1–13. https://doi.org/10.1038/s41541-023-00648-4.

Turchiano, G., Andrieux, G., Klermund, J., Blattner, G., Pennucci, V., El Gaz, M., ... Cathomen, T. (2021). Quantitative evaluation of chromosomal rearrangements in gene-edited human stem cells by CAST-Seq. *Cell Stem Cell, 28*(6), 1136–1147.e5. https://doi.org/10.1016/j.stem.2021.02.002.

U.S. Food and Drug Administration. (2023, December 8). FDA approves first gene therapies to treat patients with sickle cell disease [Press release]. https://www.fda.gov/news-events/press-announcements/fda-approves-first-gene-therapies-treat-patients-sickle-cell-disease.

Voelkel, C., Galla, M., Maetzig, T., Warlich, E., Kuehle, J., Zychlinski, D., ... Baum, C. (2010). Protein transduction from retroviral gag precursors. *Proceedings of the National Academy of Sciences of the United States of America, 107*(17), 7805–7810. https://doi.org/10.1073/pnas.0914517107.

Wang, G. P., Berry, C. C., Malani, N., Leboulch, P., Fischer, A., Hacein-Bey-Abina, S., ... Bushman, F. D. (2010). Dynamics of gene-modified progenitor cells analyzed by tracking retroviral integration sites in a human SCID-X1 gene therapy trial. *Blood, 115*(22), 4356–4366. https://doi.org/10.1182/blood-2009-12-257352.

Wolf, T., Rosengarten, J., Härtel, I., Stitz, J., & Barbe, S. (2022). A Hydrodynamic approach to the study of HIV virus-like particle (VLP) tangential flow filtration. *Membranes, 12*(12), 1248. https://doi.org/10.3390/membranes12121248.

Yan, J., Oyler-Castrillo, P., Ravisankar, P., Ward, C. C., Levesque, S., Jing, Y., ... Adamson, B. (2024). Improving prime editing with an endogenous small RNA-binding protein. *Nature, 628*(8008), 639–647. https://doi.org/10.1038/s41586-024-07259-6.

Zafra, M., Schatoff, E., Katti, A., et al. (2018). Optimized base editors enable efficient editing in cells, organoids and mice. *Nature Biotechnology, 36*, 888–893. https://doi.org/10.1038/nbt.4194.

Zhang, S., Wang, Y., Mao, D., Wang, Y., Zhang, H., Pan, Y., ... Huang, P. (2023). Current trends of clinical trials involving CRISPR/Cas systems. *Frontiers in Medicine, 10*, 1292452. https://doi.org/10.3389/fmed.2023.1292452.

Zhang, W., Cao, S., Martin, J. L., Mueller, J. D., & Mansky, L. M. (2015). Morphology and ultrastructure of retrovirus particles. *AIMS Biophysics, 2*(3), 343–369. https://doi.org/10.3934/biophy.2015.3.343.

Zhi, S., Chen, Y., Wu, G., Wen, J., Wu, J., Liu, Q., ... Huang, J. (2022). Dual-AAV delivering split prime editor system for in vivo genome editing. *Molecular Therapy, 30*(1), 283–294. https://doi.org/10.1016/j.ymthe.2021.07.011.

Zu, H., & Gao, D. (2021). Non-viral vectors in gene therapy: Recent development, challenges, and prospects. *The AAPS Journal, 23*(4), 78. https://doi.org/10.1208/s12248-021-00608-7.

CHAPTER TWENTY

CRISPRoff epigenome editing for programmable gene silencing in human cell lines and primary T cells

Rithu K. Pattali[a], **Izaiah J. Ornelas**[a], **Carolyn D. Nguyen**[a], **Da Xu**[a], **Nikita S. Divekar**[a], **and James K. Nuñez**[a,b,*]

[a]Department of Molecular and Cell Biology, University of California, Berkeley, Berkeley, CA, United States
[b]Chan Zuckerberg Biohub San Francisco, San Francisco, CA, United States
*Corresponding author. e-mail address: jamesnunez@berkeley.edu

Contents

1. Introduction 518
2. Key considerations prior to use of crisproff 520
   2.1 Comparing epigenetic editing for gene silencing to gene knockout strategies 520
   2.2 Delivery method of CRISPRoff 521
   2.3 Gene targets and cell types 521
   2.4 Assessing efficiency and specificity of CRISPRoff 522
3. Selection of single guide RNA and cloning 522
   3.1 Design and selection of sgRNAs for CRISPRoff 522
   3.2 Cloning sgRNA plasmid 523
4. Targeted gene silencing in Hek 293T cells using crisproff plasmid 523
   4.1 Required materials 524
   4.2 Delivery of CRISPRoff by plasmid transfection 525
5. Crisproff mRNA *in vitro* synthesis 529
   5.1 Required materials 530
   5.2 *In vitro* synthesis of CRISPRoff mRNA 531
6. Considerations and troubleshooting 534
7. Targeted gene silencing in Hek 293T cells using Crisproff mRNA 535
   7.1 Required materials 535
   7.2 CRISPRoff mRNA nucleofection in HEK 293T cells 536
8. Troubleshooting and tips 538
9. Targeted gene silencing in Htert Rpe-1 cells using Crisproff mRNA 539
   9.1 Required materials 539
   9.2 CRISPRoff mRNA transfection in hTERT RPE-1 cells 540
10. Troubleshooting and tips 541
11. Targeted gene silencing in primary human donor derived T cells using Crisproff mRNA 542

Methods in Enzymology, Volume 712
ISSN 0076-6879, https://doi.org/10.1016/bs.mie.2025.01.010

11.1 Required materials 542
11.2 CRISPRoff mRNA and sgRNAs co-nucleofection in primary human T cells 544
12. Troubleshooting and tips 546
13. Conclusion 546
Acknowledgments 548
References 548

Abstract

The advent of CRISPR-based technologies has enabled the rapid advancement of programmable gene manipulation in cells, tissues, and whole organisms. An emerging platform for targeted gene perturbation is epigenetic editing, the direct editing of chemical modifications on DNA and histones that ultimately results in repression or activation of the targeted gene. In contrast to CRISPR nucleases, epigenetic editors modulate gene expression without inducing DNA breaks or altering the genomic sequence of host cells. Recently, we developed the CRISPRoff epigenetic editing technology that simultaneously establishes DNA methylation and repressive histone modifications at targeted gene promoters. Transient expression of CRISPRoff and the accompanying single guide RNAs in mammalian cells results in transcriptional repression of targeted genes that is memorized heritably by cells through cell division and differentiation. Here, we describe our protocol for the delivery of CRISPRoff through plasmid DNA transfection, as well as the delivery of CRISPRoff mRNA, into transformed human cell lines and primary immune cells. We also provide guidance on evaluating target gene silencing and highlight key considerations when utilizing CRISPRoff for gene perturbations. Our protocols are broadly applicable to other CRISPR-based epigenetic editing technologies, as programmable genome manipulation tools continue to evolve rapidly.

1. Introduction

CRISPR-based genome editing has revolutionized the ability to manipulate the genetic code of cells and organisms. While traditional CRISPR-Cas systems rely on inducing double-strand breaks (DSBs) or changing the sequence of the host genome, these genome editing technologies can be associated with unpredictable DNA repair outcomes, chromosomal aberrations, potential off-target editing, and cell toxicity (Doudna, 2020). An emerging technology to perturb gene function is to edit the epigenetic chemistry (i.e., DNA methylation and histone modifications) of the host genome that results in the modulation of gene expression without altering the underlying DNA sequence of the target gene(s). CRISPR-based epigenetic editors rely on catalytically dead Cas9 (dCas9) and the recruitment of chromatin modifiers to reprogram

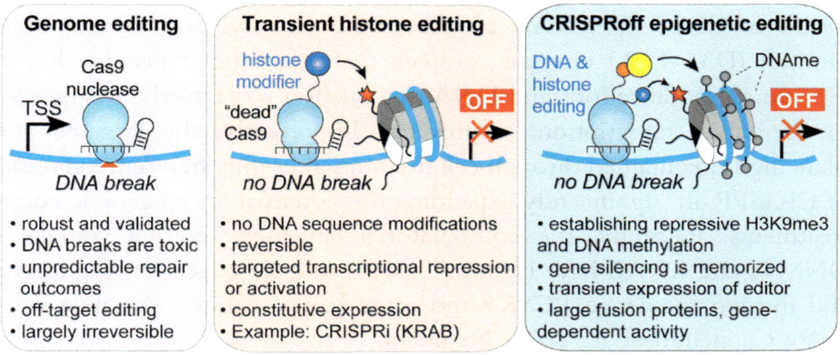

Fig. 1 A comparison of CRISPR-based technologies for gene inactivation. While Cas9 nuclease induces double-stranded DNA breaks, epigenetic editing relies on the recruitment of repressive epigenetic modifiers to catalytically dead dCas9 to transcriptionally repress the target gene. Transient histone editing requires constitutive expression of the dCas9-epigenetic repressor to sustain gene silencing, as exemplified by CRISPRi (KRAB). CRISPRoff combinatorially programs DNA methylation and H3K9me3 at the targeted gene promoter, resulting in gene repression that is memorized and heritable across cell division, without the need for constitutive expression of CRISPRoff to sustain gene silencing.

epigenetic marks at the targeted gene promoter that ultimately results in transcriptional repression or activation (Fig. 1).

Many CRISPR epigenetic editing technologies have been reported to date, each defined by a chromatin modifier fused to dCas9 (McCutcheon, Rohm, Iglesias, & Gersbach, 2024; Villiger et al., 2024). A commonly used epigenetic editor for transcriptional silencing is CRISPR interference (CRISPRi), which is composed of dCas9 fused to a Krüppel associated box domain (KRAB) repressor domain (Gilbert et al., 2013). KRAB serves as a scaffold for chromatin modifying proteins that establishes the histone modification H3K9me3 at the target gene promoter, leading to transcriptional repression of the target gene. For sustained target gene silencing, CRISPRi-mediated repression requires constitutive expression of dCas9-KRAB and the associated single-guide RNA (sgRNA), which is usually achieved by creating a stable cell line for in vitro experiments. This requirement limits its application in primary cells, complex tissues, and organisms that are challenging to engineer.

To overcome these limitations, we have developed CRISPRoff as an epigenetic editor that can heritably silence endogenous mammalian genes when targeted to their promoters without requiring constitutive expression of the CRISPR reagents (Nuñez et al., 2021). CRISPRoff is a protein fusion of dCas9 to KRAB and the DNA methyltransferase Dnmt3a-Dnmt3l (D3A-D3L) enzyme complex that establishes repressive histone H3K9me3 modifications and DNA methylation respectively at target loci for heritable transcriptional repression. Importantly, the targeted gene silencing is maintained through cell divisions after only transient expression of CRISPRoff, significantly expanding the potential for epigenetic editing treatments. Recently, the combinatorial epigenetic editing cocktail of DNMTs and KRAB have been utilized for durable gene silencing in vitro and in vivo to silence *PCSK9* and prion genes in mice (Amabile et al., 2016; Cappelluti et al., 2024; Neumann et al., 2024).

In this paper, we provide a comprehensive guide to utilizing CRISPRoff for targeted heritable gene silencing using epigenetic editing. We describe the steps involved in transiently delivering CRISPRoff into transformed cell lines, donor-isolated primary human T cells, and assessing its direct effects on transcriptional silencing. Additionally, we address potential challenges and troubleshooting strategies to ensure the successful application of this tool in various experimental contexts.

2. Key considerations prior to use of crisproff

Prior to utilizing CRISPRoff for gene silencing, we recommend considering several factors:

2.1 Comparing epigenetic editing for gene silencing to gene knockout strategies

When deciding between CRISPRoff-mediated gene silencing and CRISPR-Cas9-mediated gene knockout or other genomic manipulation technologies such base editing and prime editing (Cong et al., 2013; Mali et al., 2013; Komor, Kim, Packer, Zuris, & Liu, 2016; Anzalone et al., 2019), it is important to consider the goals of the experiment. CRISPR-Cas9 nuclease accompanied by a sgRNA introduces genomic alterations, such as indels, at the target site to disrupt the reading frame of the gene, rendering the gene dysfunctional. Cas9 can also introduce specific mutations when combined with a donor sequence, making it ideal for experiments that require precise genome edits. However, recent studies have elucidated that CRISPR nuclease mediated knockouts can lead to genomic instabilities such as chromosomal translocations and chromothripsis, highlighting the risks of using it in mammalian cells and organisms (Adikusuma, Williams, Grutzner, Hughes, & Thomas, 2017; Cullot et al., 2019; Fiumara et al., 2024; Kosicki, Tomberg, & Bradley, 2018; Tsuchida et al., 2023).

In contrast, CRISPRoff is a genetically safer alternative when the goal is to heritably silence a gene without introducing double-stranded DNA breaks, which is important when working with cells that are more susceptible to DNA damage, such as stem cells (Ihry et al., 2018) or primary cells (hematopoietic stem cells (HSCs)) (Dorset, Bak, 2023; Fiumara et al., 2024; Lee, Lozano, & Dunbar, 2021) or T cells (Nahmad et al., 2022; Tsuchida et al., 2023). Lastly, epigenetic editing allows the user to reverse the programmed epigenetic changes (Morita et al., 2016). For example, CRISPRoff mediated gene repression can be reversed by actively removing DNA methylation using the TET1 demethylase enzyme, resulting in transcriptional re-activation of the gene. However, we note that reactivation of synthetically repressed genes by DNA methylation remains an active area in the field, as we and others do not detect full reactivation of the gene with TET1. Altogether, the on/off flexibility makes CRISPRoff a potent, reversible, and programmable tool for long-term gene silencing without inducing genomic alterations.

2.2 Delivery method of CRISPRoff

Since transient expression of CRISPRoff leads to heritable gene silencing, it circumvents the need to generate a stable cell line for constitutively expressing the CRISPRoff fusion protein. In our experience, transfection of CRISPRoff plasmids into immortalized cell lines such as HEK 293T cells and hTERT RPE-1 cells leads to durable long-term gene silencing of target genes. However, as with any DNA transfection method, we observe cellular toxicity when delivering CRISPRoff plasmid DNA to primary cells and other cell types that are sensitive to plasmid DNA. To address this, we provide an alternative protocol for synthesizing and delivering CRISPRoff in vitro transcribed (IVT) mRNA into specific sensitive cell types such as primary human T cells, with potential applicability for in vivo experiments.

2.3 Gene targets and cell types

While CRISPRoff can silence the majority of genes within the human genome, not all genes are amenable to CRISPRoff mediated gene silencing, owing to significant variability in the local chromatin states across different genes and different cell types. For example, CRISPRoff requires

CpG dinucleotides for DNA methylation, often found as CpG islands of gene promoters (Deaton, Bird, 2011). Previously, we discovered that genes lacking annotated CpG islands can still be silenced with CRISPRoff, likely due to the presence of low levels of CpG dinucleotides that do not meet the CpG density cutoff of CpG islands. However, some genes are almost completely devoid of CpG dinucleotides and thus, are not amenable for CRISPRoff-mediated silencing (Nuñez et al., 2021). We recommend checking available genomic annotations of CpG content of the desired gene(s) to be repressed, such as by utilizing the CpG Island track in the UCSC Genome Browser, nested under the Expression and Regulation section. Additionally, the efficiency of CRISPRoff mediated gene silencing can vary between cell types. Immortalized or transformed cell lines, with their altered epigenetic landscapes, are more amenable to CRISPRoff treatment than primary cells, which may require additional optimization of editing and delivery conditions to achieve effective gene silencing.

2.4 Assessing efficiency and specificity of CRISPRoff

Finally, it is important to choose a quantitative method for assessing the efficiency of gene silencing after CRISPRoff treatment. We recommend quantitative PCR (qPCR), Western Blot, and/or flow cytometry to measure gene expression profiles. To assess the specificity of CRISPRoff and validate epigenetic changes, we recommend using various genomic techniques, such as RNA sequencing (RNA-seq), chromatin immunoprecipitation sequencing (ChIP-seq), and bisulfite sequencing to compare experimental and control samples.

3. Selection of single guide RNA and cloning

3.1 Design and selection of sgRNAs for CRISPRoff

The selection and design of a sgRNA is critical for ensuring the specificity and efficiency of CRISPRoff mediated gene silencing. Our previous experimental data show that active CRISPRoff sgRNAs are distributed broadly across the transcriptional start site (TSS), notably within a 1-kb window centered on the TSS of a target gene. Additionally, several bioinformatic tools (Doench et al., 2016; Horlbeck et al., 2016; Sanson et al., 2018) can be used to identify potential sgRNA sequences that minimize off-target effects while maximizing on-target activity. These tools predict sgRNA efficiency and specificity by evaluating factors such as GC content and accessibility of the target DNA as well as potential off-target sites. We recommend experimentally testing 3 to 5 different sgRNAs to determine the most optimal sgRNA sequence for any given gene. Additionally, simultaneously pooling the top 3 to 5 predicted sgRNAs targeting a gene promoter can improve the efficiency of target gene silencing.

3.2 Cloning sgRNA plasmid

sgRNA sequences are cloned into an expression plasmid from Addgene (ID#217306) that contains a U6 pol III promoter. The expression vector is a lentiviral plasmid that can be utilized either for constitutive sgRNA expression in the target cell line or for transient transfection alongside the CRISPRoff-expressing plasmid. In brief, the sgRNA plasmids are constructed by restriction cloning of protospacers downstream of a U6 promoter using *Bst*XI and BlpI cut sites, as described previously (Gilbert et al., 2014). The sgRNA expression plasmids also express a T2A-mCherry reporter to allow for the assessment of sgRNA plasmid transfection efficiency. Listed below are example positive control sgRNAs that we have validated for silencing non-essential endogenous genes, along with a non-targeting negative control sgRNA.

Name	Protospacer (5' to 3')
CD55 targeting sgRNA	GCTGGGCGTAGCTGCGACT
CD81 targeting sgRNA	GAGAGCGAGCGCGCAACGG
CLTA targeting sgRNA	CTCCCAGTCGGCACCACAG
Non-targeting control sgRNA	ACGACTAGTTAGGCGTGTA

4. Targeted gene silencing in Hek 293T cells using crisproff plasmid

Here, we outline the process of transfecting CRISPRoff-encoding plasmid DNA into HEK 293T cells to achieve gene silencing. In this example, CRISPRoff is directed to the promoter of a non-essential gene *CLTA*, which is tagged at its endogenous genomic locus with a GFP reporter in HEK 293T cells (Leonetti, Sekine, Kamiyama, Weissman, & Huang, 2016). First, we provide a protocol for co-transfecting two plasmid constructs: one expressing CRISPRoff (fused with BFP marker) and the

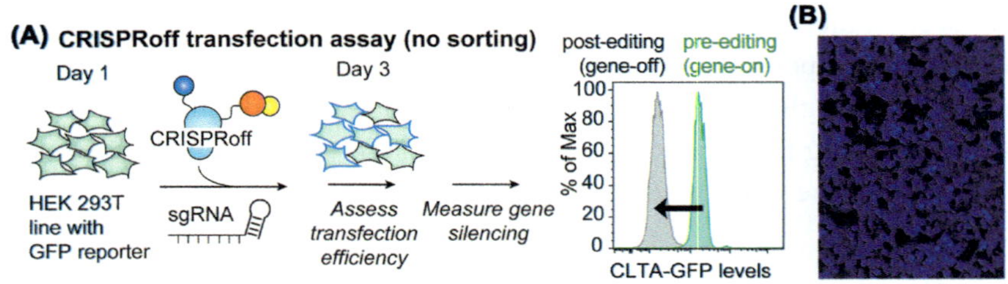

Fig. 2 CRISPRoff plasmid transfection assay in HEK 293T cells without sorting for transfected cells. (A) The workflow for targeting CRISPRoff to the endogenous gene CLTA tagged with GFP in HEK 293T cells. Cells are first transfected with plasmid DNA-expressing CRISPRoff, measured for transfection efficiency two days post-transfection, followed by flow cytometry analysis for CLTA-GFP expression at subsequent days. (B) An example of HEK 293T cells imaged two days after transfection of CRISPRoff plasmid DNA. CRISPRoff is fused directly to a BFP fluorescent marker that can be used for quantitative measurement of transfection efficiency.

other expressing an sgRNA (vector encoding an mCherry marker) targeting the *CLTA* promoter. Second, we describe an alternative approach where the HEK 293T cell line constitutively expresses an sgRNA targeting the *CLTA* promoter. In this approach, only the CRISPRoff-expressing plasmid is transfected, leading to robust *CLTA* silencing. For both experiments, the readout is the silencing of CLTA-GFP that can be quantitatively measured at the protein level in single cells using flow cytometry (Fig. 2A). Other detection methods such as qPCR or Western Blot can be used for verifying the extent of CLTA knockdown at the transcript and protein levels respectively.

4.1 Required materials

- CRISPRoff expression plasmid (Addgene #167981)
- HEK 293T cells (ATCC #CRL-3216)
- Dulbecco's modified eagle medium (DMEM) High Glucose (Thermo Fisher Scientific #11965118)
- 10% FBS (VWR #89510-186)
- Penicillin-Streptomycin-Glutamine (100X) (Gibco #1037801)
- *Trans*IT-LT1 Transfection Reagent (Mirus Bio #MIR 2225)
- Opti-MEM™ (Thermo Fisher Scientific #31985070)
- Hemocytometer or automated cell counter
- BD FACSymphony A1 Cell Analyzer (BD Biosciences)
- BD FACSAria™ Fusion Flow Cytometer (BD Biosciences)
- 24-well plates

- CO_2 incubator
- Centrifuge
- Sterile Eppendorf tubes
- 15 ml conical tubes
- Pipettes and tips

4.2 Delivery of CRISPRoff by plasmid transfection

This transfection protocol has been optimized in a 24-well format. Prior to transfection, maintain HEK 293T cells in DMEM with 10 % FBS and 1X Penicillin-Streptomycin-Glutamine. Passage the cells every 2–3 days, ensuring they remain at a confluency of 60–70 %.

4.2.1 Seeding HEK 293T cells for CRISPRoff plasmid transfection

Day 0: Seed HEK 293T cells to achieve ~60 – 70 % confluency the following day. We recommend seeding $\sim 0.9 \times 10^5 - 1.0 \times 10^5$ HEK 293T cells per well in a 24-well plate in a final volume of 600 µl per well. Adjusting the seeding density might be necessary depending on the timing of transfection and the method used for cell counting. For example, seed $\sim 1.0 \times 10^5 - 1.1 \times 10^5$ cells in the evening for transfection in the next morning; $\sim 0.8 \times 10^5 - 0.9 \times 10^5$ cells if seeding 20–24 h before transfection.

4.2.2 Transfection of CRISPRoff plasmids into HEK 293T cells

1. Day 1: Examine the cultured cells under a microscope to assess whether the cell density is approximately at a confluence of 60–70 %.
2. Pre-warm Opti-MEM™ and *Trans*IT-LT1 Transfection Reagent to room temperature for approximately 30–60 min before transfection.
3. If performing a co-transfection of CRISPRoff plasmid and an sgRNA vector, we recommend co-transfecting 300 ng of CRISPRoff plasmid and 150 ng of the target sgRNA vector into cells. CRISPRoff plasmids are diluted to be 300 ng/µl and sgRNA encoding plasmids are diluted to 150 ng/µl such that, we add 1 µl of CRISPRoff plasmid and 1 µl sgRNA vector (total of 2 µl) with the transfection reagents on Day 1.
4. Aliquot plasmid DNA into a PCR strip tube by adding:
   a. 1 µl (300 ng) of CRISPRoff encoding plasmid at 300 ng/µl.
   b. (if doing co-transfection) 1 µl (150 ng) of sgRNA vector at 150 ng/µl into the same PCR strip tube.
5. Mix 1.5 µl of the *Trans*IT-LT1 Transfection Reagent to 50 µl of pre-warmed Opti-MEM™. Incubate for 15 min at room temperature.

6. Add 51.5 μl of the *Trans*IT-LT1 Transfection Reagent-Opti-MEM™ mixture into the DNA. Incubate for 15 min at room temperature.
7. Add the mixture dropwise (~53 μl) to cells.
8. Gently tilt the plate to evenly distribute the transfection mixture without dislodging the cells.

4.2.3 Monitoring transfection using microscopy

1. Day 2: Use a fluorescent microscope to assess the transfection efficiency (Fig. 2B).
   a. BFP signal (CRISPRoff) will be low or undetectable on Day 2 and will increase substantially by Day 3.
   b. mCherry (sgRNA vector) signal will be high on Day 2.
2. Do not split cells or change media as this can dislodge the cells.

4.2.4 Assess transfection efficiency and gene silencing by flow cytometry

1. Day 3 and beyond: Dissociate the transfected HEK 293T cells following the manufacturer's protocol and seed cells into a fresh plate. For example, a 1:10 split into a new plate will be confluent in 2 days.
2. With dissociated cells, measure transfection efficiency (BFP and mCherry) as well as gene silencing (GFP) using the methods described in Section 2.4.
   a. We recommend assessing the transfection efficiency and gene silencing of CRISPRoff using flow cytometry. At Day 3, we typically detect >40 % BFP+ (Fig. 3A) and >60 % mCherry+ (Fig. 3B) cell populations with flow cytometry. Gene knockdown with CRISPRoff can be observed as early as Day 3, typically peaking between days 6 and 8, and in most cases, is sustained for an extended period. When collecting data over a time course (e.g., Days 3, 5, 7, 10, 14 post-transfection), we recommend normalizing the percentage of GFP-negative cells at each time point to the transfection efficiency measured on Day 3. For example, if only 50 % of cells are transfected with CRISPRoff, the relative maximum percentage of cells in the population that repress the target gene is 50 %. Furthermore, we recommend additional assessments of target gene silencing 2–3 weeks after transfection of CRISPRoff plasmid (Fig. 3C) to assess the durability of gene silencing.
3. If necessary, cells can be sorted for CRISPRoff and gRNA expressing cells if a more pure population is required (Fig. 4A).
   a. Sort ~0.5×10^5 – 1.0×10^5 cells per well for re-seeding into a 24-well plate.

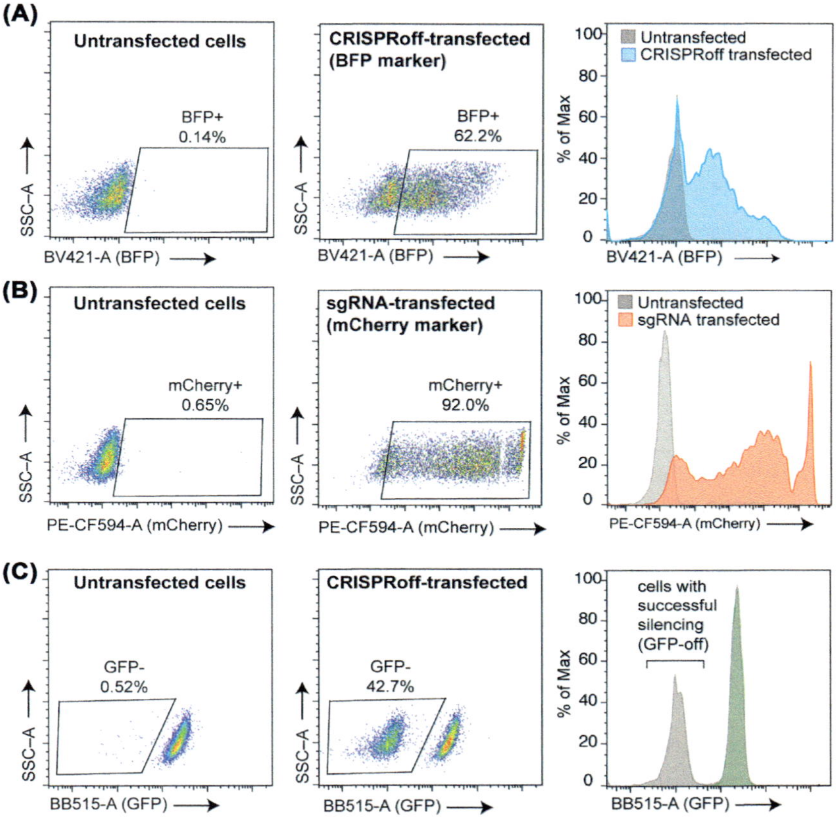

Fig. 3 Flow cytometry to measure transfection efficiency and loss of CLTA-GFP as a measure of successful gene silencing. (A) FlowJo software was used to analyze the flow cytometry data two days (Day 3 in protocol) post transfection. (Left) SSC versus BV421-A (BFP) was used to set the gate for CRISPRoff positive cells based on control samples. (Middle) BFP+ gate was used to determine the percent CRISPRoff positive (BFP+) cells from a control sample (cells above diagonal line are BFP+). (Right) Histogram of Left and Middle displaying transfection efficiency of CRISRPoff (BFP). (B) Similar as in (C) but for gating transfection efficiency of gRNA using PE-CF594-A (mCherry). (C) Example flow cytometry data measuring gene knockdown of reporter gene (CLTA) 21 days post transfection using BB515-A (GFP). (Left) SSC versus BB515-A (GFP) was used to set the gate for GFP negative cells based on control samples. (Middle) GFP- gate was used to determine the percent of CLTA negative (GFP-) for unsorted cells (cells below the diagonal line are GFP-). (Right) Histogram of Left and Middle displaying silencing efficiency of CRISPRoff (gray) compared to control sample (green).

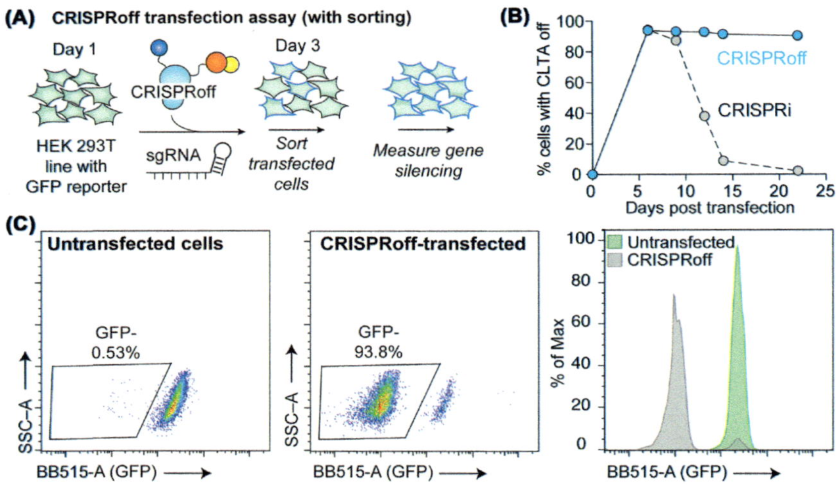

Fig. 4 CRISPRoff plasmid transfection assay in HEK 293T cells with sorting for transfected cells (A) Workflow for targeting CRISPRoff to endogenous gene tagged with GFP in HEK 293T cells with sorting transfected cells. (B) A time course analysis of CLTA silencing in HEK 293T cells following CRISPRoff transfection highlighting heritable gene silencing of CRISPRoff (blue) compared to CRISPRi (gray) (C) Similar as in (Fig. 3C) but cells were sorted at two days (Day 3 in protocol) post transfection for CRISPRoff and gRNA expressing cells.

b. After sorting for CRISPRoff and gRNA expressing cells, the entire population will consist of transfected cells, making it unnecessary to normalize to transfection efficiency at Day 3. This approach results in a cleaner experiment, with nearly all the cells having the gene silenced (Fig. 4B and C), leading to more consistent results compared to using unsorted cells.

4.2.5 Considerations and troubleshooting

1. If transfections are low from the CRISPRoff or sgRNA plasmid (<20 % BFP+/mCherry+), we recommend optimizing the seeding density and ensuring that the cells are evenly distributed throughout the well. Generally, we have experienced poor transfection efficiencies if the cells are too confluent (error in cell counting) or if cells are localized to the periphery of the well. After seeding the cells in a plate, tilt the plate to ensure that the cells do not attach to the outer edge of the well.
2. If performing plasmid transfection into a different cell type or plate size, we recommend following *Trans*IT-LT1 Transfection Reagent manufacturer's protocol.

3. As an alternative to transient expression of sgRNA plasmids, we recommend stable expression of the sgRNA vector via lentiviral transduction.
4. Optimization of sgRNA plasmid transfections may be required if using other sgRNA expression vectors.
5. While we provide an example of silencing endogenous genes with a reporter (GFP), antibody staining can be used in cases where no reporter line is available.

5. Crisproff mRNA *in vitro* synthesis

Messenger RNA (mRNA) delivery is an effective way to deliver CRISPR-based genome editing tools into mammalian cells (van Haasteren, Li, Scheideler, Murthy, & Schaffer, 2020; Madigan, Zhang, & Dahlman, 2023). Unlike DNA-based delivery systems, mRNA delivery can result in higher transfection efficiency and reduced cellular toxicity (Rohner, Yang, Foo, Goedel, & Chien, 2022). Furthermore, there is a reduced risk of insertional mutagenesis with mRNA-based delivery systems, making it an ideal candidate for mRNA therapeutics (Kowalski, Rudra, Miao, & Anderson, 2019). Direct delivery of mRNA encoding CRISPRoff into the cytoplasm utilizes the translational machinery of cells for short term, transient expression of CRISPRoff.

Previously, CRISPRoff has been synthesized as IVT mRNA and delivered into cells to successfully silence genes (Fig. 5) (Nuñez et al., 2021). Here, we provide a further optimized and detailed protocol for synthesizing CRISPRoff mRNA, followed by examples of transfecting and

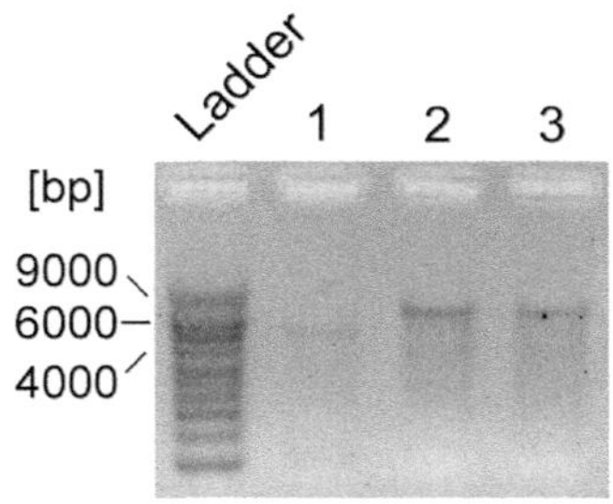

Fig. 5 Agarose gel electrophoresis post glyoxal denaturation of IVT mRNA to visualize mRNA size and quality. Lane 1 shows Cas9-BFP mRNA mRNA (~5.5 kb), Lane 2 shows CRISPRoff mRNA (~7.2 kb), and Lane 3 shows CRISPRoff DNMT3A catalytic mutant mRNA from Nuñez et al., 2021 (~7.2 kb).

nucleofecting CRISPRoff mRNA into mammalian immortalized cell lines and primary T cells. We observe that delivering CRISPRoff as mRNA achieves more robust epigenetic silencing, as mRNA delivery is significantly more efficient than plasmid DNA transfection. CRISPRoff mRNA is ~7000 nucleotides long, which exceeds the recommended length for many in vitro synthesis kits. To optimize both synthesis and delivery, we recommend using the IVT backbone from Aldevron (pALD-CV42 [T7]), the mMessage mMachine™ T7 ULTRA Transcription Kit (Thermo Fisher Scientific) for the IVT reaction and poly (A) tailing reactions.

5.1 Required materials

- CRISPRoff mRNA vector (generated by cloning CRISPRoff fusion gene) (Addgene #167981) into pALD-CV42 [T7] (Aldevron)
- mMessage mMACHINE™ T7 ULTRA Transcription Kit (Thermo Fisher Scientific #AM1345)
- RNaseZap™ RNase Decontamination Solution (Thermo Fisher Scientific #AM9780)
- RNase-free Microfuge Tubes (Thermo Fisher Scientific #AM12400)
- *Xho*I restriction enzyme (New England Biolabs #R0146S)
- 10X rCutSmart™ Buffer (New England Biolabs #B6004S)
- Thermocycler
- DNA Clean & Concentrator Kit (Zymo Research #D4033)
- Nuclease-free water
- 70 % ethanol
- PIPES, free acid
- bis-TRIS [bis-(2-hydroxyethyl) aminotris(hydroxymethyl) methane]
- EDTA
- Nanodrop spectrophotometer
- Qubit™ RNA High Sensitivity (HS) Assay Kit (Invitrogen #Q32852)
- Ambion NorthernMAX glyoxal loading dye Thermo Fisher Scientific (Invitrogen #AM8551)
- Millennial RNA Markers Thermo Fisher Scientific (Invitrogen #AM7150)
- 15 ml conical tubes
- Nuclease free eppendorf tubes
- Pipettes and tips
- Micro-centrifuge

5.2 *In vitro* synthesis of CRISPRoff mRNA

5.2.1 *Linearization of plasmid*

1. Day 1: Assemble the digestion reaction by adding the following to a PCR tube.

Component	Amount
CRISPRoff mRNA vector	8 ug
10X rCutSmart™ Buffer	5 μl
*Xho*I restriction enzyme	2 μl
Nuclease-free water	to 50 μl

2. Incubate the reaction in a thermocycler at 37 °C overnight.

5.2.2 *Concentration of linearized CRISPRoff DNA template*

1. Day 2: Use the DNA Clean and Concentrator Kit to concentrate the DNA. Perform all centrifugation steps at a speed of 10,000–16,000 g
   a. Divide the linearized CRISPRoff DNA template reaction mixture into two Eppendorf tubes, adding 25 μl of the mixture into each tube.
   b. Add 300 μl DNA Binding Buffer to each tube and mix by vortexing.
   c. Transfer the mixtures to spin columns placed in collection tubes and centrifuge for 30 s. Discard the flowthrough.
   d. Next, add 200 μl of DNA Wash Buffer to each column and centrifuge for 30 s. Repeat this wash step.
   e. For elution, use warm (50 °C) nuclease-free water. First, add 6 μl of nuclease-free water directly to the center of each column and incubate at room temperature for 1 min. Then, transfer the column to a 1.5 ml RNase-free Eppendorf tube and centrifuge for 30 s to collect the DNA. Repeat the elution by adding 4 μl of nuclease-free water to the center of the column, incubating for 1 min, and centrifuging for 30 s
   f. Combine the eluates from both columns to obtain a total of 20 μl of linearized DNA. Measure the DNA concentration using a Nanodrop spectrophotometer.

5.2.3 *In vitro transcription of CRISPRoff mRNA*

1. Spray RNaseZap™ RNase Decontamination Solution on bench, pipettes, tube racks, and gloves to remove RNase contamination.

2. We recommend assembling the transcription reaction with the mMessage mMACHINE™ T7 ULTRA Transcription Kit.
3. Thaw T7 2X NTP/ARCA, 10X T7 Reaction Buffer, and T7 Enzyme Mix on ice.
4. Add the following to a sterile PCR tube. The total reaction volume should equal 20 μl.

Component	Amount
Linearized CRISPRoff mRNA vector	1 μg
T7 2X NTP/ARCA	10 μl
10X T7 Reaction Buffer	2 μl
Nuclease-free water	to 20 μl

5. Gently flick the PCR tube or pipette up and down to mix, then spin the tube briefly to collect reaction mixture at the bottom of the tube.
6. Incubate the transcription reaction at 37 °C for 2 h.
7. Add 1 μl TURBO DNase to the reaction mixture to the transcription reaction to remove trace quantities of DNA contamination. Mix well and incubate the reaction at 37 °C for 15 min

5.2.4 Poly (A) tailing reaction

1. Thaw 5X *E*-PAP Buffer, 25 mM $MnCl_2$, and ATP Solution on ice.
2. For the Poly (A) tailing reaction, add the following.

Component	Amount
IVT Reaction mixture from above	20 μl
Nuclease-free water	36 μl
5X *E*-PAP Buffer	20 μl
25 mM $MnCl_2$	10 μl
ATP Solution	10 μl

3. Set aside 2.5 μl of the reaction mixture to use as a negative control for the RNA glyoxal gel.
4. Add 4 μl E-PAP to the rest of the above reaction mixture.

5. Gently flick the PCR tube or mix by pipetting up and down, then spin the tube briefly to collect the reaction mixture at the bottom of the tube.
6. Incubate the transcription reaction at 37 °C for 45 min, then place on ice.

5.2.5 Lithium chloride precipitation

1. Move the reaction mixture to a 1.5 ml RNase-free eppendorf tube.
2. Add 50 μl of LiCl Precipitation Solution to the reaction mixture and mix thoroughly.
3. Chill the reaction at −20 °C overnight.

5.2.6 CRISPRoff mRNA recovery and quantification

1. Day 3: Centrifuge the reaction mixture at maximum speed at 4 °C for 15 min
2. Discard the supernatant, taking care not to disturb the mRNA pellet.
3. Wash the mRNA pellet with 1 ml of 70 % ethanol and spin at 4 °C for 15 min
4. Remove ethanol, air dry the mRNA pellet briefly, and resuspend the pellet in 60 μl of nuclease-free water.
5. Measure mRNA concentration by nanodrop spectrophotometer or Qubit™ RNA High Sensitivity (HS) Assay Kit.

5.2.7 CRISPRoff mRNA visualization and quality check

In this protocol, we visualize the in vitro synthesized mRNA using denaturing gel electrophoresis (Fig. 5). To ensure the quality and length of the synthesized mRNA, we use glyoxal to denature and visualize CRISPRoff mRNA by size fractionation. About 1 μg of CRISPRoff mRNA is needed to efficiently visualize the mRNA on a gel. This step can be performed in conjunction with quantification of mRNA to ensure that the mRNA is of good quality prior to use. For best results, the synthesized mRNA must be always stored on ice before preparation for gel electrophoresis.

1. Prepare about 500 ml of BPTE buffer for casting and running the mRNA gel.
   10X BPTE buffer
   PIPES, free acid (100 mM)
   Bis-Tris, free base (300 mM)
   EDTA (10 mM)
   [fill up to 500 ml using ddH_2O]
   Adjust the pH of the final solution to 6.5 (See Note 1 and 2).
2. Clean the gel casting tray using distilled water.

3. Pour 1 % agarose gel in 1X BPTE buffer by adding agarose powder and dissolving it in the microwave (See Note 3, 4).
4. Let the gel set at room temperature to solidify.
5. Prepare the mRNA sample by mixing about 1 µg of CRISPRoff mRNA with glyoxal loading dye (Invitrogen #AM8551) at 1:1 vol.
6. To prepare the ladder, mix glyoxal loading dye with Invitrogen Millennial RNA Markers (Invitrogen #AM7150) at 1:1 ratio and bring to a desired volume.
7. Incubate the ladder and samples at 50 °C for 30 min, ensuring not to excessively heat the mRNA which can lead to its degradation.
8. Move the samples to ice quickly after incubation to maintain the denatured state of the mRNA (See Note 5).
9. Load the samples in the gel.
10. Run the gel in 1X BPTE buffer at 25 V for 50 min (See Note 6).
11. Image the gel in the ethidium bromide channel to compare RNA length and degradation.
12. Store the synthesized mRNA at −80 °C.

Notes.

1. It is essential to maintain the pH of the BPTE buffer at acidic pH to prevent reversing of glyoxal reactions.
2. Oakton pH700 pH meter was used to adjust the pH of the solution for accurate measurements.
3. No staining reagents are used in casting the gel.
4. Make sure to use a clean autoclaved flask for casting RNA gels.
5. Denatured mRNA can be stored on ice for 2–5 min prior to loading onto the gel.
6. Running the gel too fast can lead to smearing of mRNA, the gel can be run at lower voltages (e.g., 10 V) to prevent smearing.

6. Considerations and troubleshooting

1. Although we recommend the mMessage mMachine™ T7 ULTRA Transcription Kit from Thermo Fisher Scientific for the IVT reaction, other commercially available IVT may also be used, though they may require optimization.
2. We have observed that newly synthesized mRNAs exhibit superior performance compared to those subjected to long-term storage. To

mitigate the effects of degradation, it is advisable to avoid frequent freeze-thaw cycles by preparing smaller aliquots of CRISPRoff mRNA.

3. While IVT of CRISPRoff mRNA is generally highly reproducible, consider sourcing high-quality mRNAs from vendors like Aldevron and TriLink® for experiments requiring high consistency, such as in the case with in vivo studies.

7. Targeted gene silencing in Hek 293T cells using Crisproff mRNA

In this section, we describe the nucleofection of IVT synthesized CRISPRoff mRNA into HEK 293T cells that constitutively express an sgRNA targeting the promoter of the endogenous *CLTA* gene, which is tagged with a GFP reporter. The aim of this experiment is to evaluate the efficiency of CRISPRoff mRNA in transcriptionally silencing *CLTA*. By directly introducing CRISPRoff mRNA into the cells, we offer an alternative approach to plasmid DNA-based methods that achieves comparable gene silencing efficiency with potentially reduced cellular toxicity. The primary readout for this experiment is the reduction in CLTA-GFP expression, which can be quantitatively measured in single cells using flow cytometry (Fig. 6A). Additional detection techniques, such as qPCR or Western Blot, may also be employed to verify the extent of CLTA knockdown at the transcript and protein levels respectively.

7.1 Required materials

- HEK 293T cells (ATCC #CRL-3216)
- Dulbecco's modified eagle medium (DMEM) High Glucose (Thermo Fisher Scientific #11965118)
- 10 % FBS (VWR #89510-186)
- Penicillin-Streptomycin-Glutamine (100X) (Gibco #10378016)
- CRISPRoff mRNA (IVT, see Section 5.2)
- SF Cell Line 4D-Nucleofector™ X Kit S (Lonza #V4XC-2032)
- 96-well Shuttle™ Device (Lonza)
- Nucleocuvette™
- Fluorochrome-conjugated antibodies suitable for use in flow cytometry
- BD FACSymphony A1 Cell Analyzer (BD Biosciences)
- DPBS (1X) (Gibco #2027-04)
- Trypan Blue

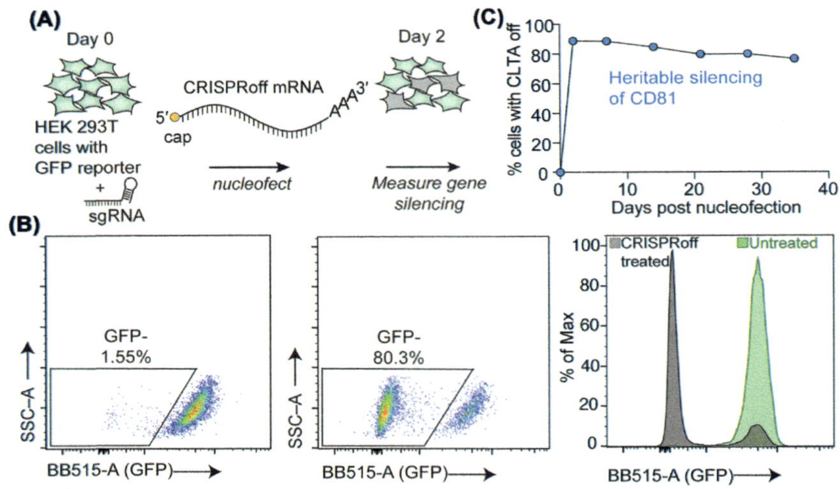

Fig. 6 CRISPRoff mRNA nucleofection assay in HEK 293T cells. (A) A schematic illustrating the workflow for nucleofecting HEK 293T cells with CRISPRoff mRNA, targeting the endogenous CLTA gene tagged with GFP. (B) A representative flow cytometry plot showing CLTA silencing (using BB515-A for GFP) in HEK 293T CLTA-GFP cells, 35 days post-nucleofection with CRISPRoff mRNA. (C) A time course analysis of CLTA silencing in HEK 293T cells following CRISPRoff mRNA nucleofection.

- Hemocytometer or automated cell counter
- 24-well plates
- CO_2 incubator
- Centrifuge
- Sterile Eppendorf tubes
- 15 ml conical tubes
- Pipettes and tips

7.2 CRISPRoff mRNA nucleofection in HEK 293T cells

Prior to nucleofection, maintain HEK 293T cells in DMEM with 10 % FBS and 1X Penicillin-Streptomycin-Glutamine. Passage the cells every 2–3 days, ensuring that they remain at a confluency of 60–70 %.

7.2.1 Nucleofection of CRISPRoff mRNA into HEK 293T cells

1. Day 0: Thaw and mix 2 μg of IVT synthesized CRISPRoff mRNA in a sterile RNase-free microcentrifuge tube. Store the mRNA on ice until nucleofection. Ensure that the volume of mRNA does not exceed 10 % of the total nucleofection volume (~22 μl).

2. Warm the SF Cell Line Nucleofector™ solution to room temperature for 15 min before nucleofection. For each reaction, supplement 16.4 μl of SF solution with 3.6 μl of its supplement.
3. Harvest HEK 293T cells and determine the cell count. For each nucleofection reaction, aliquot ~2.0×10^5 cells into a sterile micro-centrifuge tube or use a 15 ml conical tube for multiple reactions.
4. Centrifuge the cells at 500g for 5 min at room temperature and discard the supernatant. Wash the cells once with room temperature PBS at 500 g for 5 min, then discard the supernatant.
5. Resuspend the cells in the appropriate volume (18–20 μl) of SF Cell Line Nucleofector™ solution. Add the CRISPRoff mRNA to the resuspended cell solution. For example, mix 2 μl of the mRNA solution with cells resuspended in 18 μl of the SF Cell Line Nucleofector™ solution. Transfer the cells and mRNA mixture into a Lonza Nucleocuvette™, carefully avoiding the formation of bubbles. Gently tap the Nucleocuvette™ to ensure all cells are at the bottom of the Nucleocuvette™.
6. Nucleofect the cells using the Nucleofector™ program CM-130 on the 4D-Nucleofector™ System.
7. Following nucleofection, add 80 μl of the appropriate culture medium to each Nucleocuvette™ well. Transfer ~100 μl of the cell suspension into a well of a 24-well plate containing 400 μl of pre-warmed culture medium.
8. Incubate the plate at 37 °C with 5 % CO_2 for 24 h.

7.2.2 Assessing gene silencing by flow cytometry

Day 2 and beyond: Unlike plasmid transfections, BFP expression is not detected following nucleofection of CRISPRoff mRNA. Instead, we recommend assessing the efficiency of CRISPRoff gene silencing by directly measuring target protein knockdown using flow cytometry or Western Blot and/or assessing mRNA knockdown using qPCR.

1. Dissociate the nucleofected HEK 293T cells according to the manufacturer's protocol and measure fluorescence of CLTA-GFP on a flow cytometer.
2. For CRISPRoff mRNA delivery experiments, gene knockdown can be observed as early as day 2 post nucleofection, typically peaking between days 5 and 7 and in most cases, silencing remains heritable and stable over time (Fig. 6B).

3. Furthermore, we recommend additional assessments of target gene silencing 2-3 weeks after nucleofection of CRISPRoff mRNA (Fig. 6C) to assess the durability of long-term gene silencing.

8. Troubleshooting and tips

1. Always include a positive control such as an eGFP/mCherry reporter mRNA to assess nucleofection efficiency.
2. Include a non-targeting sgRNA control to assess specificity of gene silencing.
3. We recommend testing 3–5 guides per gene target and considering the pooling of multiple guides to enhance efficiency.
4. In this example, the sgRNA is stably expressed in the HEK 293T cell line. If a stable HEK 293T cell line expressing the target sgRNA is unavailable, consider co-nucleofecting the sgRNA along with CRISPRoff mRNA. Based on our experience, co-nucleofection of CRISPRoff mRNA (2 μg per nucleofection reaction) and sgRNA (2–3 μg per nucleofection reaction) per 2.0×10^5 cells also results in robust gene silencing.
5. While we provide an example of silencing endogenous genes with a reporter (GFP), antibody staining can be used in cases where no reporter line is available.
6. If nucleofection-related toxicity is observed, consider increasing the number of cells per nucleofection reaction.
7. In some cases, optimizing CRISPRoff mRNA (2–4 μg per nucleofection reaction) per 2.0×10^5 cells may be necessary to achieve the desired effective dose.
8. If gene silencing leads to cellular toxicity, consider seeding the cells in a 96-well plate instead of a 24-well plate post-nucleofection.
9. Although our protocol is optimized for the SF Cell Line 4D-Nucleofector™ X Kit S, further optimization of cell number and CRISPRoff mRNA amounts will be necessary when scaling up to more cells.
10. For cell types that are less suitable for nucleofection, we suggest encapsulating CRISPRoff mRNAs in commercial or in vitro-formulated lipid nanoparticles (LNPs) to enhance transfection efficiency and improve delivery.

9. Targeted gene silencing in Htert Rpe-1 cells using Crisproff mRNA

In this section, we outline the transfection of IVT synthesized CRISPRoff mRNA into hTERT RPE-1 cells, which stably express an sgRNA targeting the promoter of the endogenous *CD81* gene. hTERT RPE-1 cells, derived from human retinal pigment epithelial cells, maintain a near-diploid karyotype, making them a more physiologically relevant model for studying normal cellular processes (Hindul et al., 2022). The objective of this experiment is to assess the effectiveness of CRISPRoff mRNA in transcriptionally silencing *CD81*. The primary readout for this experiment is the reduction in CD81 expression, which can be quantitatively assessed in single cells through CD81 protein staining and flow cytometry (Fig. 7A). Additional methods, such as qPCR or Western Blot, may also be used to confirm the extent of CD81 knockdown at the transcript and protein levels respectively.

9.1 Required materials

- hTERT RPE-1 cells (ATCC #CRL-4000)
- DMEM/F-12, GlutaMAX™ supplement (Thermo Fisher Scientific #105650-018)

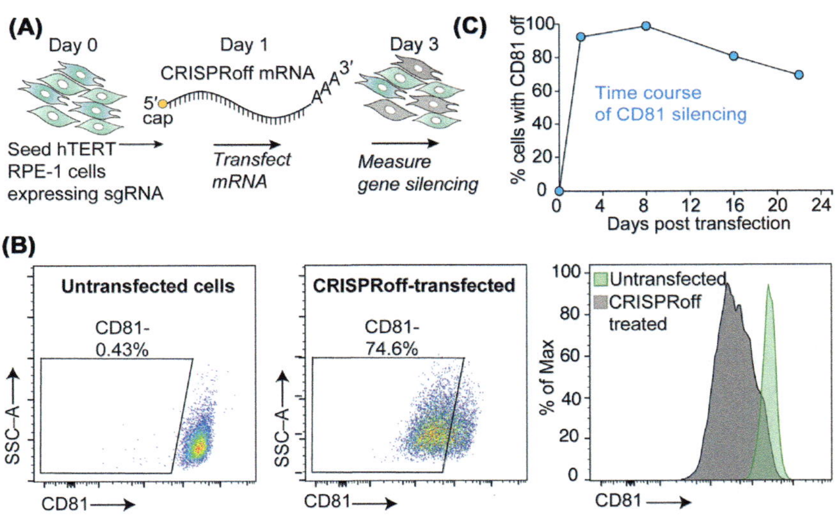

Fig. 7 CRISPRoff mRNA transfection assay in hTERT RPE-1 cells. (A) A schematic illustrating the workflow for transfecting hTERT RPE-1 cells with CRISPRoff mRNA. (B) A representative flow cytometry plot showing CD81 silencing in hTERT RPE-1 cells, 2 days post-transfection with CRISPRoff mRNA. (C) A time course analysis of CD81 silencing in hTERT RPE-1 cells following CRISPRoff mRNA transfection.

- 10 % FBS (VWR #89510-186)
- Penicillin-Streptomycin-Glutamine (100X) (Gibco #10378016)
- CRISPRoff mRNA (IVT, see Section 5.2)
- CRISPRoff *CD81* sgRNA as detailed in Section 3.2 (Ordered from Synthego)
- Lipofectamine™ MessengerMax™ Transfection Reagent (Invitrogen #LMRNA008)
- Fluorochrome-conjugated antibodies suitable for use in flow cytometry
- BD FACSymphony A1 Cell Analyzer (BD Biosciences)
- Trypan Blue
- Hemocytometer or automated cell counter
- 24-well plates
- CO_2 incubator
- Centrifuge
- Sterile Eppendorf tubes
- 15 ml conical tubes
- Pipettes and tips

9.2 CRISPRoff mRNA transfection in hTERT RPE-1 cells

Prior to the transfection, maintain human hTERT RPE-1 cells in DMEM/F-12 with 10 % FBS and 1X Penicillin-Streptomycin-Glutamine. Passage the cells every 2–3 days, ensuring that they remain at a confluency of 60–70 %.

9.2.1 Seeding hTERT RPE-1 cells for CRISPRoff mRNA transfection

1. Day 0: Seed 1×10^5 hTERT RPE-1 cells in a 24-well plate containing 0.5 ml of the appropriate cell culture medium. Adjusting the seeding density might be necessary depending on the timing of transfection and the method used for cell counting.

9.2.2 Transfection of CRISPRoff mRNA into hTERT RPE-1 cells

1. Day 1: Under a microscope, check seeded cell density to see if cells are ~60–70 % confluent.
2. Bring Opti-MEMTM and MessengerMAX™ Reagent to room temperature ~30 min before transfection.
3. In an RNase-free sterile microcentrifuge tube, for each transfection reaction, dilute 1.5 μl of MessengerMAX™ Reagent in 25 μl of pre-warmed Opti-MEM™ Medium. Mix thoroughly and incubate the transfection reaction at room temperature for 10 min

4. Mix 1 μg of CRISPRoff mRNA with 25 μl of pre-warmed Opti-MEM™ Medium in a sterile microcentrifuge tube.
5. Combine the diluted mRNA mixture with the diluted MessengerMAX™ Reagent, mix well, and incubate at room temperature for 5 min
6. Add ~50 μl of mRNA-lipid mixture dropwise to the cells.
7. Incubate the plates at 37 °C with 5 % CO2 for 24 h.

9.2.3 Assessing gene silencing by flow cytometry

Day 3 and beyond: Unlike plasmid transfections, BFP expression is not detected following transfection of CRISPRoff mRNA. Instead, we recommend assessing the efficiency of CRISPRoff gene silencing by directly measuring target protein knockdown using flow cytometry or Western Blot and/or assessing at the RNA level using qPCR.

1. Dissociate the transfected hTERT RPE-1 cells according to the manufacturer's protocol, perform antibody staining for CD81, and then measure CD81 knockdown using a flow cytometer.
2. For CRISPRoff mRNA delivery experiments, gene knockdown can be observed as early as day 3 (Fig. 7B), typically peaking between days 6 and 8 and in most cases, silencing remains heritable and stable over time.
3. Furthermore, we recommend additional assessments of target gene silencing 2-3 weeks after nucleofection of CRISPRoff mRNA (Fig. 7C) to assess the durability of long-term gene silencing.

10. Troubleshooting and tips

1. Always include a positive control such as eGFP or mCherry reporter mRNA to evaluate transfection efficiency.
2. Include a non-targeting sgRNA control to assess specificity of gene silencing.
3. We recommend testing 3–5 guides per gene and considering the pooling of multiple guides to enhance the efficiency of CRISPRoff-mediated silencing.
4. Consider increasing cell numbers per transfection reaction to account for nucleic acids mediated toxicity.
5. In some cases, optimization of CRISPRoff mRNA (1–3 μg per reaction) may be necessary to achieve the desired effective dose.

6. When scaling reactions to different volumes than those that we recommend, it is important to optimize the volumes of transfection reagents and the amounts of CRISPRoff mRNA.

11. Targeted gene silencing in primary human donor derived T cells using Crisproff mRNA

In this section, we describe the co-nucleofection of IVT synthesized CRISPRoff mRNA into primary human T cells, along with an sgRNA targeting the promoter of the endogenous *CD81* gene. Unlike immortalized cell lines, which often harbor genetic abnormalities, primary T cells offer a more accurate and relevant model for studying native immune functions. Derived directly from human peripheral blood mononuclear cells (PBMCs), primary T cells provide a physiologically relevant system, making them ideal for developing effective cancer immunotherapies (Stadtmauer et al., 2020).

The aim of this experiment is to evaluate the effectiveness of CRISPRoff mRNA in transcriptionally and heritably silencing *CD81* in activated primary T cells. We first provide an overview of the process for isolating primary human T cells from donor derived PBMCs and activating them to induce proliferation. In this example, we also outline the steps for co-delivering CRISPRoff mRNA and an sgRNA into primary T cells. This approach is particularly useful because creating stable sgRNA-expressing lentiviral lines in primary cells, such as T cells, can be challenging. The primary readout for this experiment is the reduction in CD81 protein, which can be quantitatively measured by antibody staining for CD81 followed by flow cytometry analysis (Fig. 8A). Additionally, qPCR or Western blot may be employed to verify CD81 knockdown at the transcript and protein levels respectively.

11.1 Required materials

- Human Peripheral Blood Mononuclear Cells, Frozen (STEMCELL Technologies #70025.3)
- X-VIVO™ 15 Serum-free Hematopoietic Cell Medium (Lonza #02-053Q)
- N-acetyl L-cysteine (Sigma-Aldrich #A7250)
- 2-mercaptoethanol (Gibco #21985023)
- 10 % FBS (VWR #89510-186)

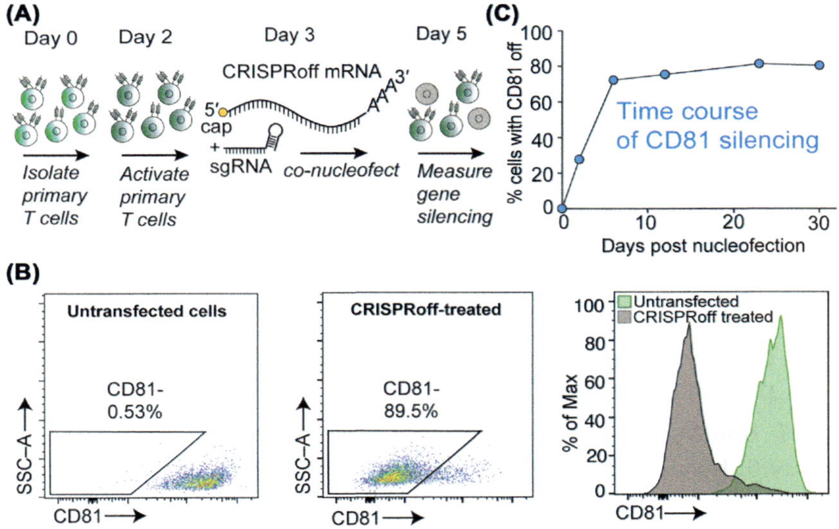

Fig. 8 CRISPRoff mRNA co-nucleofection assay in activated primary human T cells. (A) A schematic illustrating the workflow for isolating, activating, and co-nucleofecting primary human T cells with CRISPRoff mRNA and sgRNA. (B) A representative flow cytometry plot demonstrating CD81 silencing in primary human T cells, 16 days post-nucleofection with CRISPRoff mRNA. (C) A time course analysis of CD81 silencing in primary human T cells following CRISPRoff mRNA nucleofection.

- Penicillin-Streptomycin-Glutamine (100X) (Gibco #10378016)
- IL-2 (Peprotech #200-02)
- IL-7 (Peprotech #200-07)
- IL-15 (R&D Systems #BT-015-010)
- EasySep™ magnetic Cell Isolation kit (STEMCELL Technologies #17951)
- Dynabeads™ Human T-Activator CD3/CD28 for T Cell Expansion and Activation (Thermo Fisher Scientific #11131D)
- CRISPRoff mRNA (IVT, see Section 5.2)
- CRISPRoff *CD81* sgRNA as detailed in Section 3.2 (Ordered from Synthego)
- P3 Cell Line 4D-Nucleofector™ X Kit S (32 RCT) (Lonza #V4XP-3032)
- 96-well Shuttle™ Device (Lonza)
- Nucleocuvette™
- APC anti-human CD81 (TAPA-1) Antibody (Biolegend #349510)
- BD FACSymphony A1 Cell Analyzer (BD Biosciences)

- DPBS (1X) (Gibco #2027-04)
- Trypan Blue
- Hemocytometer or automated cell counter
- 24-well plates
- CO_2 incubator
- Centrifuge
- Sterile Eppendorf tubes
- 15 ml conical tubes
- Pipettes and tips

11.2 CRISPRoff mRNA and sgRNAs co-nucleofection in primary human T cells

11.2.1 Isolation of primary human T cells from PBMCs

1. Day 0: Isolate bulk CD3 + T lymphocytes from healthy donor PBMCs using the EasySep™ Magnetic Cell Isolation Kit, following the manufacturer's instructions.

11.2.2 Activation of primary human T cells

1. Day 2: Activate primary T cells using Dynabeads™ Human T-Activator CD3/CD28 for T Cell Expansion and Activation as per the manufacturer's protocol.
2. Culture activated Primary T cells in X-VIVO™ medium containing fetal bovine serum (5 %), 2-mercaptoethanol (50 μM), N-acetyl L-cysteine (10 mM), IL-2 (300 U/ml), IL-7 (5 ng/ml), and IL-15 (5 ng/ml).
3. We recommend using freshly isolated and activated primary T cells for CRISPRoff mRNA nucleofection. For cells already in culture, maintain activated primary human T cells by passaging them every 2–3 days and maintaining a cell density of 1×10^6 cells per ml.

11.2.3 Co-nucleofection of CRISPRoff mRNA and sgRNA into activated primary human T cells

1. Day 3: To prepare the CRISPRoff mRNAs-sgRNA mixture, mix 2.5 μg of the target gene sgRNA (*CD81* in this example) and 1 μg of CRISPRoff mRNA in a sterile RNase-free microcentrifuge tube and store on ice until ready for nucleofection. It is recommended that the mRNA-sgRNA mixture should not exceed 10 % of the total nucleofection reaction (~22 μl).
2. Warm up the P3 Primary Cell 4D-Nucleofector™ Solution 15 min prior to nucleofection. For each reaction, mix 16.4 μl P3 Primary Cell 4D-Nucleofector™ Solution with 3.6 μl of its supplement.

3. Harvest activated primary T cells and determine their cell density. For each nucleofection reaction, aliquot 1×10^6 cells into a sterile micro-centrifuge tube.
4. Centrifuge the cells at 500 g for 5 min at room temperature, then discard the supernatant. Wash the cells once with room temperature PBS at 500 g for 5 min and discard the supernatant.
5. Resuspend the cells in the appropriate volume (~18-20 µl) of P3 Primary Cell 4D-Nucleofector™ Solution. Add the sgRNA-mRNA mixture into the cells. For example, mix 2 µl of sgRNA-mRNA solution with cells resuspended in 18 µl of P3 Primary Cell 4D-Nucleofector™ Solution. Transfer the cells and mRNA-sgRNA mixture into one well of a Lonza Nucleocuvette™, being careful to avoid creating bubbles. Gently tap the Nucleocuvette™ to ensure all the cells settle at the bottom of the plate.
6. Nucleofect the cells using the Nucleofector™ program DS-137 on the 4D-Nucleofector™ System.
7. After nucleofection, add 80 µl of pre-warmed T cell culture medium into each Nucleocuvette™ well. Transfer ~100 µl of the cell suspension cells into a well of a 96-well plate containing 100 µl of pre-warmed T cell culture medium with growth factors.
8. Incubate the 96-well plate at 37 °C with 5 % CO_2 for 24 h.
9. Replace T cell media and growth factors as needed, approximately every 3-4 days.

11.2.4 Assessing gene silencing by flow cytometry

Day 5 and beyond: Unlike plasmid transfections, BFP expression is not detected following nucleofection of CRISPRoff mRNA. Instead, we recommend assessing the efficiency of CRISPRoff gene silencing by directly measuring target protein knockdown using flow cytometry or Western Blot and/or assessing mRNA knockdown using qPCR.

1. Harvest nucleofected primary human T cells, perform antibody staining for CD81, and then measure CD81 knockdown using a flow cytometer.
2. For CRISPRoff mRNA delivery experiments, gene knockdown can be observed as early as day 5, typically peaking between days 8 and 10 and in most cases, silencing remains heritable and stable over time (Fig. 8B).
3. Furthermore, we recommend additional assessments of target gene silencing 2-3 weeks after nucleofection of CRISPRoff mRNA (Fig. 8C) to assess the durability of long-term gene silencing.

12. Troubleshooting and tips

1. Always include a control eGFP/mCherry reporter mRNA control to assess nucleofection efficiency.
2. Include a non-targeting sgRNA control to determine the specificity of gene silencing.
3. We recommend testing 3-5 guides per gene and considering the pooling of multiple guides to enhance efficiency.
4. For primary cells such as T cells, we recommend using commercially available sgRNAs from vendors such as Synthego and IDT for optimal stability and performance. Synthego provides guides with default chemical modifications, including 2'-O-methyl (OMe) analogs on the first and last three bases of the guide RNA and 3' phosphorothioate (PS) internucleotide linkages between the first three and last two bases. IDT also offers these modifications, providing additional flexibility for customization based on experimental needs.
5. Consider increasing cell numbers per nucleofection reaction to account for nucleofection mediated toxicity.
6. Optimize sgRNA amounts (1–5 μg per reaction) and CRISPRoff mRNA (1–3 μg per reaction) per 1 million cells as needed to achieve the desired effective dose.
7. We recommend using freshly isolated T cells for silencing experiments; however, we have also observed stable silencing in primary T cells that were cryopreserved without prior stimulation, then thawed and used in CRISPRoff experiments.
8. Although our protocol is optimized for the P3 Cell Line 4D-Nucleofector™ X Kit S, further optimization of cell number and mRNA amounts will be necessary when scaling up.
9. For cell types that are less suitable for nucleofection, we suggest encapsulating CRISPRoff mRNAs in commercial or in vitro-formulated lipid nanoparticles (LNPs) to enhance transfection efficiency and improve delivery.

13. Conclusion

CRISPRoff is an emerging tool for heritable gene repression in the rapidly advancing field of epigenetic editing, offering a robust and reversible approach to gene silencing by altering the epigenome. By combining the precise DNA binding ability of CRISPR systems with the power to edit epigenetic marks, CRISPRoff serves as a highly programmable platform for targeted gene perturbation without altering the underlying DNA sequence or inducing DNA breaks. In this protocol, we present a detailed guide for utilizing CRISPRoff to program heritable gene silencing in human cells. Beyond CRISPRoff, other epigenetic editing tools have been reported and our protocol can be applied to the delivery of these technologies into mammalian cells (Alerasool, Leng, Lin, Gingras, & Taipale, 2022; Cano-Rodriguez et al., 2016; Chavez et al., 2015; DelRosso et al., 2023; Hilton et al., 2015; Kearns et al., 2015; Mahata et al., 2023; McCutcheon et al., 2024; O'Geen, Ren, Nicolet, Perez, & Halmai, 2017; O'Geen et al., 2019; Policarpi, Munafò, Tsagkris, Carlini, & Hackett, 2024; Tanenbaum, Gilbert, Qi, Weissman, & Vale, 2014). These emerging technologies extend the capabilities of CRISPR-based epigenetic editors, allowing researchers to activate or repress genes in a more controlled and distinct manner.

While CRISPRoff can silence a vast majority of genes in the human genome (Nuñez et al., 2021), not all genes are equally amenable to CRISPRoff mediated silencing. The effectiveness of this tool is affected by several factors, and it is important to recognize that CRISPRoff is still in the early stages of development. One key factor to consider is the local epigenetic landscape of the target gene (Yeo et al., 2018), as pre-existing epigenetic marks can potentially hinder or neutralize the effectiveness of CRISPRoff. Additionally, the specific cell type being targeted plays a major role, as genes in different cell types may have distinct epigenetic profiles. For example, primary cells and immortalized cell lines can respond variably to CRISPRoff induced epigenetic editing. Another key variable is the CpG density within the target gene promoter region, as promoter regions with low CpG content may be less susceptible to methylation-based silencing by CRISPRoff. All of these factors must be evaluated for optimal efficacy and success of CRISPRoff mediated gene silencing.

With these considerations in mind, we have shown in this protocol that CRISPRoff shows remarkable versatility across various immortalized cell types, such as HEK 293T and hTERT RPE-1 cells. Moreover, in this paper, we also highlight the efficacy of CRISPRoff in targeting an endogenous gene in primary donor-derived human T cells, thereby expanding its potential for developing potent immunotherapies. The ability of CRISPRoff to induce stable and heritable gene silencing highlights its promising application for in vivo epigenetic editing and offers exciting possibilities for its use as a therapeutic platform.

Acknowledgments

We thank members of the Nuñez lab for establishing and optimizing the protocols described in the manuscript. We thank Stephen Floor for advice on mRNA in vitro synthesis and gel analysis. This work was funded in part by research grants awarded to J.K.N.: a Hanna Gray Fellowship from the Howard Hughes Medical Institute (GT15341), NIH R35 MIRA (R35GM155044-01), CRISPR Cures for Cancer Initiative (Gladstone Institutes, UCB 20240097), a Spark Award from Bakar Labs (51474), and an Innovation Award from the Laboratory for Genomics Research (UCB 052611). J.K.N. is an Investigator of the Chan Zuckerberg Biohub San Francisco. R.K.P. is funded by graduate fellowships from the University of California Cancer Research Coordinating Committee and the Shurl and Kay Curci Foundation. I.J.O. is funded by graduate fellowships from the University of California, Berkeley Mentored Research Award, a Gilliam Fellowship from the Howard Hughes Medical Institute, and a National Science Foundation Graduate Research Fellowship. N.S.D. is funded by a postdoctoral fellowship from the California Institute for Regenerative Medicine (CIRM). C.D.N. acknowledges support from the SURF Rose Hills Fellowship from the University of California, Berkeley and the Amgen Scholars Program at the University of California, San Francisco.

References

Adikusuma, F., Williams, N., Grutzner, F., Hughes, J., & Thomas, P. (2017). Targeted deletion of an entire chromosome using CRISPR/Cas9. *Molecular Therapy, 25*(8), 1736–1738. https://doi.org/10.1016/j.ymthe.2017.05.021.

Alerasool, N., Leng, H., Lin, Z.-Y., Gingras, A.-C., & Taipale, M. (2022). Identification and functional characterization of transcriptional activators in human cells. *Molecular Cell, 82*(3), 677–695.e7. https://doi.org/10.1016/j.molcel.2021.12.008.

Amabile, A., Migliara, A., Capasso, P., Biffi, M., Cittaro, D., Naldini, L., et al. (2016). Inheritable silencing of endogenous genes by hit-and-run targeted epigenetic editing. *Cell, 167*(1), 219–232.e14. https://doi.org/10.1016/j.cell.2016.09.006.

Anzalone, A. V., Randolph, P. B., Davis, J. R., Sousa, A. A., Koblan, L. W., Levy, J. M., et al. (2019). Search-and-replace genome editing without double-strand breaks or donor DNA. *Nature, 576*(7785), 149–157. https://doi.org/10.1038/s41586-019-1711-4.

Cano-Rodriguez, D., Gjaltema, R. A. F., Jilderda, L. J., Jellema, P., Dokter-Fokkens, J., Ruiters, M. H. J., et al. (2016). Writing of H3K4Me3 overcomes epigenetic silencing in a sustained but context-dependent manner. *Nature Communications, 7*(1), 12284. https://doi.org/10.1038/ncomms12284.

Cappelluti, M. A., Mollica Poeta, V., Valsoni, S., Quarato, P., Merlin, S., Merelli, I., et al. (2024). Durable and efficient gene silencing in vivo by hit-and-run epigenome editing. *Nature, 627*(8003), 416–423. https://doi.org/10.1038/s41586-024-07087-8.

Chavez, Scheiman, A., Vora, J., Pruitt, S., Tuttle, B. W., P R Iyer, M., Lin, E., et al. (2015). Highly efficient Cas9-mediated transcriptional programming. *Nature Methods, 12*(4), 326–328. https://doi.org/10.1038/nmeth.3312.

Cong, L., Ran, F. A., Cox, D., Lin, S., Barretto, R., Habib, N., et al. (2013). Multiplex genome engineering using CRISPR/Cas systems. *Science (New York, N. Y.), 339*(6121), 819–823. https://doi.org/10.1126/science.1231143.

Cullot, Boutin, G., Toutain, J., Prat, J., Pennamen, F., Rooryck, P., et al. (2019). CRISPR-Cas9 genome editing induces megabase-scale chromosomal truncations. *Nature Communications, 10*(1), 1136. https://doi.org/10.1038/s41467-019-09006-2.

Deaton, A. M., & Bird, A. (2011). CpG islands and the regulation of transcription. *Genes & Development, 25*(10), 1010–1022. https://doi.org/10.1101/gad.2037511.

DelRosso, Tycko, N., Suzuki, J., Andrews, P., Aradhana, C., Mukund, A., et al. (2023). Large-scale mapping and mutagenesis of human transcriptional effector domains. *Nature, 616*(7956), 365–372. https://doi.org/10.1038/s41586-023-05906-y.

Doench, Fusi, J. G., Sullender, N., Hegde, M., Vaimberg, M., Donovan, E. W., et al. (2016). Optimized sgRNA design to maximize activity and minimize off-target effects of CRISPR-Cas9. *Nature Biotechnology, 34*(2), 184–191. https://doi.org/10.1038/nbt.3437.

Dorset, S. R., & Bak, R. O. (2023). The p53 challenge of hematopoietic stem cell gene editing. *Molecular Therapy Methods & Clinical Development, 30*, 83–89. https://doi.org/10.1016/j.omtm.2023.06.003.

Doudna, J. A. (2020). The promise and challenge of therapeutic genome editing. *Nature, 578*(7794), 229–236. https://doi.org/10.1038/s41586-020-1978-5.

Fiumara, Ferrari, M., Omer-Javed, S., Beretta, A., Albano, S., Canarutto, L., Varesi, D., et al. (2024). Genotoxic effects of base and prime editing in human hematopoietic stem cells. *Nature Biotechnology, 42*(6), 877–891. https://doi.org/10.1038/s41587-023-01915-4.

Gilbert, L. A., Horlbeck, M. A., Adamson, B., Villalta, J. E., Chen, Y., Whitehead, E. H., et al. (2014). Genome-scale CRISPR-mediated control of gene repression and activation. *Cell, 159*(3), 647–661. https://doi.org/10.1016/j.cell.2014.09.029.

Gilbert, Larson, L. A., Morsut, M. H., Liu, L., Brar, Z., Torres, G. A., et al. (2013). CRISPR-mediated modular RNA-guided regulation of transcription in eukaryotes. *Cell, 154*(2), 442–451. https://doi.org/10.1016/j.cell.2013.06.044.

Hilton, I. B., D'Ippolito, A. M., Vockley, C. M., Thakore, P. I., Crawford, G. E., Reddy, T. E., et al. (2015). Epigenome editing by a CRISPR-Cas9-based acetyltransferase activates genes from promoters and enhancers. *Nature Biotechnology, 33*(5), 510–517. https://doi.org/10.1038/nbt.3199.

Hindul, N. L., Jhita, A., Oprea, D. G., Hussain, T. A., Gonchar, O., Campillo, M. A. M., et al. (2022). Construction of a human hTERT RPE-1 cell line with inducible Cre for editing of endogenous genes. *Biology Open, 11*(2), bio059056. https://doi.org/10.1242/bio.059056.

Horlbeck, M. A., Gilbert, L. A., Villalta, J. E., Adamson, B., Pak, R. A., Chen, Y., et al. (2016). Compact and highly active next-generation libraries for CRISPR-mediated gene repression and activation. *eLife, 5*, e19760. https://doi.org/10.7554/eLife.19760.

Ihry, R. J., Worringer, K. A., Salick, M. R., Frias, E., Ho, D., Theriault, K., et al. (2018). P53 inhibits CRISPR–Cas9 engineering in human pluripotent stem cells. *Nature Medicine, 24*(7), 939–946. https://doi.org/10.1038/s41591-018-0050-6.

Kearns, N. A., Pham, H., Tabak, B., Genga, R. M., Silverstein, N. J., Garber, M., et al. (2015). Functional annotation of native enhancers with a Cas9–histone demethylase fusion. *Nature Methods, 12*(5), 401–403. https://doi.org/10.1038/nmeth.3325.

Komor, A. C., Kim, Y. B., Packer, M. S., Zuris, J. A., & Liu, D. R. (2016). Programmable editing of a target base in genomic DNA without double-stranded DNA cleavage. *Nature, 533*(7603), 420–424. https://doi.org/10.1038/nature17946.

Kosicki, M., Tomberg, K., & Bradley, A. (2018). Repair of double-strand breaks induced by CRISPR–Cas9 leads to large deletions and complex rearrangements. *Nature Biotechnology, 36*(8), 765–771. https://doi.org/10.1038/nbt.4192.

Kowalski, P. S., Rudra, A., Miao, L., & Anderson, D. G. (2019). Delivering the messenger: Advances in technologies for therapeutic mRNA delivery. *Molecular Therapy, 27*(4), 710–728. https://doi.org/10.1016/j.ymthe.2019.02.012.

Lee, B.-C., Lozano, R. J., & Dunbar, C. E. (2021). Understanding and overcoming adverse consequences of genome editing on hematopoietic stem and progenitor cells. *Molecular Therapy, 29*(11), 3205–3218. https://doi.org/10.1016/j.ymthe.2021.09.001.

Leonetti, M. D., Sekine, S., Kamiyama, D., Weissman, J. S., & Huang, B. (2016). A scalable strategy for high-throughput GFP tagging of endogenous human proteins. *Proceedings of the National Academy of Sciences, 113*(25), E3501–E3508. https://doi.org/10.1073/pnas.1606731113.

Madigan, V., Zhang, F., & Dahlman, J. E. (2023). Drug delivery systems for CRISPR-based genome editors. *Nature Reviews. Drug Discovery, 22*(11), 875–894. https://doi.org/10.1038/s41573-023-00762-x.

Mahata, B., Cabrera, A., Brenner, D. A., Guerra-Resendez, R. S., Li, J., Goell, J., et al. (2023). Compact engineered human mechanosensitive transactivation modules enable potent and versatile synthetic transcriptional control. *Nature Methods, 20*(11), 1716–1728. https://doi.org/10.1038/s41592-023-02036-1.

Mali, Yang, P., Esvelt, L., Aach, K. M., Guell, J., DiCarlo, M., et al. (2013). RNA-guided human genome engineering via Cas9. *Science (New York, N. Y.), 339*(6121), 823–826. https://doi.org/10.1126/science.1232033.

McCutcheon, S. R., Rohm, D., Iglesias, N., & Gersbach, C. A. (2024). Epigenome editing technologies for discovery and medicine. *Nature Biotechnology, 42*(8), 1199–1217. https://doi.org/10.1038/s41587-024-02320-1.

Morita, S., Noguchi, H., Horii, T., Nakabayashi, K., Kimura, M., Okamura, K., et al. (2016). Targeted DNA demethylation in vivo using dCas9–peptide repeat and scFv–TET1 catalytic domain fusions. *Nature Biotechnology, 34*(10), 1060–1065. https://doi.org/10.1038/nbt.3658.

Nahmad, Reuveni, A. D., Goldschmidt, E., Tenne, E., Liberman, T., Horovitz-Fried, M., et al. (2022). Frequent aneuploidy in primary human T cells after CRISPR–Cas9 cleavage. *Nature Biotechnology, 40*(12), 1807–1813. https://doi.org/10.1038/s41587-022-01377-0.

Neumann, E. N., Bertozzi, T. M., Wu, E., Serack, F., Harvey, J. W., Brauer, P. P., et al. (2024). Brainwide silencing of prion protein by AAV-mediated delivery of an engineered compact epigenetic editor. *Science (New York, N. Y.), 384*(6703), ado7082. https://doi.org/10.1126/science.ado7082.

Nuñez, J. K., Chen, J., Pommier, G. C., Cogan, J. Z., Replogle, J. M., Adriaens, C., et al. (2021). Genome-wide programmable transcriptional memory by CRISPR-based epigenome editing. *Cell, 184*(9), 2503–2519.e17. https://doi.org/10.1016/j.cell.2021.03.025.

O'Geen, H., Bates, S. L., Carter, S. S., Nisson, K. A., Halmai, J., Fink, K. D., et al. (2019). Ezh2-dCas9 and KRAB-dCas9 enable engineering of epigenetic memory in a context-dependent manner. *Epigenetics & Chromatin, 12*(1), 26. https://doi.org/10.1186/s13072-019-0275-8.

O'Geen, H., Ren, C., Nicolet, C. M., Perez, A. A., Halmai, J., Le, V. M., et al. (2017). dCas9-based epigenome editing suggests acquisition of histone methylation is not sufficient for target gene repression. *Nucleic Acids Research, 45*(17), 9901–9916. https://doi.org/10.1093/nar/gkx578.

Policarpi, C., Munafò, M., Tsagkris, S., Carlini, V., & Hackett, J. A. (2024). Systematic epigenome editing captures the context-dependent instructive function of chromatin modifications. *Nature Genetics, 56*(6), 1168–1180. https://doi.org/10.1038/s41588-024-01706-w.

Rohner, E., Yang, R., Foo, K. S., Goedel, A., & Chien, K. R. (2022). Unlocking the promise of mRNA therapeutics. *Nature Biotechnology, 40*(11), 1586–1600. https://doi.org/10.1038/s41587-022-01491-z.

Sanson, K. R., Hanna, R. E., Hegde, M., Donovan, K. F., Strand, C., Sullender, M. E., et al. (2018). Optimized libraries for CRISPR-Cas9 genetic screens with multiple modalities. *Nature Communications, 9*(1), 5416. https://doi.org/10.1038/s41467-018-07901-8.

Stadtmauer, E. A., Fraietta, J. A., Davis, M. M., Cohen, A. D., Weber, K. L., Lancaster, E., et al. (2020). CRISPR-engineered T cells in patients with refractory cancer. *Science (New York, N. Y.), 367*(6481), eaba7365. https://doi.org/10.1126/science.aba7365.

Tanenbaum, M. E., Gilbert, L. A., Qi, L. S., Weissman, J. S., & Vale, R. D. (2014). A protein-tagging system for signal amplification in gene expression and fluorescence imaging. *Cell, 159*(3), 635–5646. https://doi.org/10.1016/j.cell.2014.09.039.

Tsuchida, C. A., Brandes, N., Bueno, R., Trinidad, M., Mazumder, T., Yu, B., et al. (2023). Mitigation of chromosome loss in clinical CRISPR-Cas9-engineered T cells. 2023.03.22.533709 *bioRxiv*. https://doi.org/10.1101/2023.03.22.533709.

van Haasteren, J., Li, J., Scheideler, O. J., Murthy, N., & Schaffer, D. V. (2020). The delivery challenge: Fulfilling the promise of therapeutic genome editing. *Nature Biotechnology, 38*(7), 845–855. https://doi.org/10.1038/s41587-020-0565-5.

Villiger, L., Joung, J., Koblan, L., Weissman, J., Abudayyeh, O. O., & Gootenberg, J. S. (2024). CRISPR technologies for genome, epigenome and transcriptome editing. *Nature Reviews. Molecular Cell Biology, 25*(6), 464–487. https://doi.org/10.1038/s41580-023-00697-6.

Yeo, N. C., Chavez, A., Lance-Byrne, A., Chan, Y., Menn, D., Milanova, D., et al. (2018). An enhanced CRISPR repressor for targeted mammalian gene regulation. *Nature Methods, 15*(8), 611–616. https://doi.org/10.1038/s41592-018-0048-5.

9780443317521